FIFTH INTERNATIONAL SYMPOSIUM ON TRICHOPTERA

SERIES ENTOMOLOGICA

EDITOR

K.A. SPENCER

VOLUME 39

# PROCEEDINGS OF THE FIFTH INTERNATIONAL SYMPOSIUM ON TRICHOPTERA

LYON, FRANCE
21–26 JULY 1986

*Edited by*

MICHEL BOURNAUD and HENRI TACHET

1987 **DR W. JUNK PUBLISHERS**
a member of the KLUWER ACADEMIC PUBLISHERS GROUP
DORDRECHT / BOSTON / LANCASTER

**Distributors**

*for the United States and Canada*: Kluwer Academic Publishers, P.O. Box 358, Accord Station, Hingham, MA 02018-0358, USA
*for the UK and Ireland*: Kluwer Academic Publishers, MTP Press Limited, Falcon House, Queen Square, Lancaster LA1 1RN, UK
*for all other countries*: Kluwer Academic Publishers Group, Distribution Center, P.O. Box 322, 3300 AH Dordrecht, The Netherlands

**Library of Congress Cataloging in Publication Data**

International Symposium on Trichoptera (5th : 1986 :
Lyon, France)
Proceedings of the fifth International Symposium
on Trichoptera, Lyon (France) 21-26 July 1986.

(Series entomologica ; v. 39)
Includes indexes.
1. Caddis-flies--Congresses. I. Bournaud, Michel.
II. Tachet, Henri. III. Arens, Marie Françoise.
IV. Title. V. Series.
QL516.I57 1986 595.7'45 87-4218
ISBN 90-6193-620-9

ISBN 90-6193-620-9 (this volume)
ISBN 90-6193-190-8 (series)



PRINTED IN THE NETHERLANDS

# PROCEEDINGS
## OF
# THE FIFTH INTERNATIONAL SYMPOSIUM
## ON
# TRICHOPTERA
# LYON (FRANCE) 21 - 26 JULY 1986

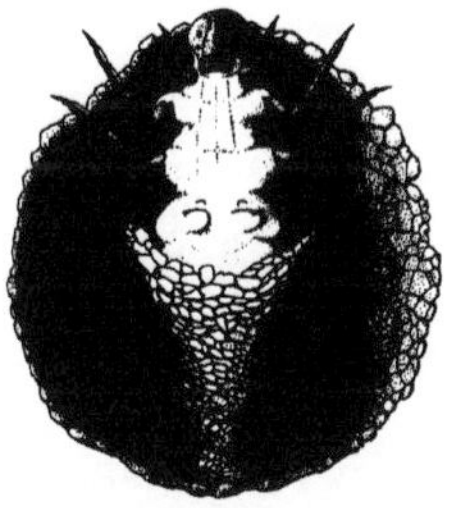

*Thremma gallicum* McLachlan

Laboratoire de Biologie animale et Ecologie
et Unité associée au C.N.R.S. n° 367 "Ecologie des eaux douces"
Institut d'Analyse des Systèmes Biologiques et Socio-Economiques
Université Claude Bernard, LYON I

1. B.S. SVENSSCN - 2. S.W. HAMILTON - 3. G. BECKER - 4. G. VAN DER VELDE - 5. R.H. WAGNER - 6. B. HIGLER - 7. K. KUMANSKI - 8. H. TACHET - 9. B. CELLOT - 10. P.W. SCHEFTER - 11. M. BOURNAUD - 12. M.A. GONZALEZ - 13. C. DENIS - 14. L. BOTOSANEANU - 15. S.T. DARLINGTON - 16. T. ITO - 17. B. GULLEFORS - 18. P. CHANTARAMONGKOL - 19. G. GISLASON - 20. M.G. GLAPSKA - 21. R.C. PETERSEN 22. C. TOMASZEWKI - 23. L. PETERSEN - 24. B. STATZNER - 25. P. ZWICK - 26. C. ROUX - 27. M.I. CRICHTON - 28. J. MURILLO I ROYO - 29. C. OTTO - 30. M.A. PUIG - 31. M. GONZALEZ del TANAGO - 32. A. NEBOISS - 33. D. GARCIA del JALON - 34. N.A. ERMAN - 35. M.C. MOLLES - 36. V.H. RESH - 37. M.C. BICCHIERAI - 38. J. MAJECKI - 39. F. CIANFICCONI - 40. G. SPINNELLI BATTA - 41. C. CORALLINI SORCETTI 42. K. TANIDA - 43. A. BOUZIDI - 44. N. WILLIAMS - 45. O. FOSSATI - 46. J.O. SOLEM - 47. F.M. GIBON - 48. L. RECASENS I ALBALADEJO - 49. M.J. WEVERS - 50. T. PITSCH - 51. P. STROOT - 52. P. GREEN - 53. J.S. RICHARDSON - 54. R.N. VINEYARD - 55. F. VAILLANT - 56. A. WELLS - 57. G.P. MORETTI - 58. P. BAGGE - 59. M. BALES - 60. R.M. BADCOCK - 61. S.W.L. da TERRA - 62. B.J. WILKINSON - 63. S. WILLIAMS 64. G. MARLIER - 65. O.S. FLINT - 66. J.A. GORE - 67. R.W. WISSEMAN - 68. M. OEMKE - 69. H. MALICKY - 70. E. PETERSSON - 71. D. DUDGEON 72. A.P. NIMMO - 73. J.C. MORSE - 74. B.J. ARMITAGE - 75. W. WICHARD - 76. T. ANDERSEN - 77. B. FAESSEL - 78. S. SMITH - 79. R. LEUVEN 80. J. WARD - 81. J. WEAVER III - 82. D. WILLIAMS - 83. O. WILLIAMS - 84. P. USSEGLIO-POLATERA - 85. L. COGERINO - 86. J. O'CONNOR.

Not in photograph : E. CASTELLA, B. FROCHOT, J. GIUDICELLI, R. HOUP, F. SCHMID, A. TAVARES, L.X. TIAN.
Registered but not present : N.H. ANDERSON, W.G. DOWNS, P. GATTAPONI, V.A. NELSON, C. SIEGENTHALER, G. WIGGINS.

# CONTENTS

# PREFACE

When we offered to host the 5th International Symposium on Trichoptera in Lyon in July 1986, we knew that for us the great Adventure was about to begin. On that July morning in 1983 in Clemson (South Carolina), we could only dimly imagine what the future held in store.

One of the worst moment came, in fact, when we were told by th official scientific authorities that no subsidies would be forthcoming. This refusal meant that we would be unable to give any financial help to those colleagues with only modest means at their disposal. Let us hope that the publication of the Proceedings will convince certain French scientists, and the powers that be, that the study of Trichoptera represents a valid and exciting field of research.

The American raid on Tripoli, which did not at the time seem to have any link with Trichoptera, threatened to act as a deterrent as far as the North American participants were concerned. However, the fears expressed by some of our colleagues (and not without reason, when we consider the series of terrorist attacks later carried out in France) were soon allayed, although the organisers of the Congress must confess to having kept a discreet eye open for any suspicious brown paper parcels or unattended luggage.

In spite of these difficulties and other minor problems, the 5th Symposium attracted 89 participants from 23 countries. This success was due in part to the impact of the previous Symposia (1974 Lunz am See, Austria ; 1977 Reading, England ; 1980 Perugia, Italy ; 1983 Clemson, U.S.A.), and also to an ever-increasing desire to participate.

20 posters and 57 lectures were presented. Two field trips took place, one to the Massif du Pilat (on the eastern edge of the Massif Central) and the other to the Massif de la Chartreuse (Prealps). These excursions gave participants the opportunity to get to know the landscapes and rivers of the Lyonnais region (as well as its culinary attractions). During the first excursion some participants were even lucky enough to collect a few specimens of *Thremma gallicum* McL., whose larva was the symbol of the Symposium.

The guest programme was followed by about forty courageous souls who discovered the charms (hidden, in the case of the *traboules*) of Lyon and its surroundings with a maximum of good humour and a minimum (we hope) of wear and tear.

This symposium brought together many participants from previous Symposia, but also welcomed for the first time several colleagues from the Far East : Dr T. Ito and K. Tanida from Japan, Dr Dudgeon from Hong Kong, Dr P. Chantaramongkol from Thailand, and Dr L.X. Tian from the People's Republic of China. Dr Tian's presence was the result of a process which had begun in Clemson ; at the time Dr Morse's efforts had had no immediate results, but the try, try and try again principle proved successful in the end. Dr Tian was able to come to Lyon thanks to a travel grant from the Agricultural University of Nanjing, and the dynamism of Dr Liu Dajun, the President of his University, whom we would like to thank here.

The scientific success (lectures, posters, and especially unofficial discussions) of a Symposium like ours depends essentially on the personalities of the participants. The quality of its organisation and cultural content (in the widest sense of the term) depends on the work carried out by the members of the Organising Committee, among whom we would like to thank in particular :

Mme Chantal Roux who organised the wine and cheese workshop on Monday July 21st, thereby giving back to the word Symposium its original meaning *(Symposium* in Greek means *banquet*).

Mr Bernard Cellot who was in charge of the poster session.

Professor François Vaillant from Grenoble who organised the field trip to the Massif de la Chartreuse.

Michèle Pariset and Glyn and Philippe Thoiron who worked on translations for the Symposium, and provided linguistic security for us and for the participants while it was in progress, thus ensuring that our beautiful French language did not pose any major problems for those who were not wholly familiar with it.

We would also like to thank all those who, in one way or another, helped us during the week the Symposium was held : Olivier Arens, Alain Berly, Emmanuel Castella, Laurent Cogerino, Gordon Copp, Odile Fossati, Sylvie Tachet and Philippe Usseglio-Polatera.

The preparation of the Proceedings has turned out to be a long but very exciting experience and here we would like to thank :

Mr André Arens (A2PC), who organised the typing of the manuscripts, Mr Bernard Seroul (CAP XXI) who saw to the first printing, and Mme Janine Tachet who dealt with the lay-out of figures and charts.

Mr W.R. Peters, who coordinated our dealings with the publisher.

Dr W. Junk, the publisher, who printed the definitive version of the Proceedings, and will organise its distribution.

This Symposium would not have been what it was without the valuable collaboration of Mme Marie-Françoise Arens who combined most efficiently the roles of secretary, hostess and, lastly, proof-reader of the Proceedings.

We received two proposals for the 6 th Symposium, one from Dr C. Otto (Umeå, Sweden), and the other from Dr C. Tomaszewski (Łodz, Poland). Both sites were equally attractive, but the participants finally decided in favour of Łodz because of the difficulties experienced by our colleagues from Eastern Europe in attending Symposia.

Thus the Adventure which began in July 1983 is about to come to an end with the publication of the Proceedings. To quote an expression borrowed from the Olympic Games by the Head of our Laboratory, Professor A.L. Roux, and used in his opening speech, we will now *hand over the torch* to Dr C. Tomasewski and his colleagues, and we wish them well in their task of organising the **6th** Symposium in **1989** in **Poland.**

Spotkamy się więc, żeby mówić jeszcze o chruścikach.

Lyon le 22 Décembre 1986

Michel BOURNAUD Henri TACHET

# LIST OF PARTICIPANTS

ANDERSEN, Mr. Trond, Universitetsbibliotek i Bergen, avd Realfagbygget, Allégt.41, 5000 Bergen, Norway
* ANDERSON, Mr. Norman H., Oregon State University, Entomology Department, Corvallis, Oregon 97331, U.S.A.
ARMITAGE, Mr. Brian J., P.O. Box 880, Athens, Al 35611, U.S.A.
BADCOCK, Miss Ruth M., University of Keele, Department of Biological Sciences, Keele, Staffordshire ST5 5BG, England
BAGGE, Mr. Pauli, Department of Biology, University of Jyväskylä, Vliopistonk 9, 40100 Jyväskylä 10, Finland
BALES, Mr. Michael, University of Keele, Department of Biological Sciences, Keele,Staffordshire, ST5 5BG, England
BECKER, Mr. Georg, Limnologische Flusstation des Max Planck Institute für Limnologie, Postfach 260, 6407 Schlitz, Germany (F.R.).
BICCHIERAI, Mrs. Maria Clara, Istituto di Zoologia (Fac. Scienze MM.FF.NN.) Università, Via Elce di Sotto, O6100 Perugia, Italy.
BOTOSANEANU, Mr. Lazare, University of Amsterdam, Plantage Middenlaan 64, 1018 DH Amsterdam, The Netherlands.
BOURNAUD, Mr. Michel, Laboratoire de Biologie animale et Ecologie, Université Claude Bernard, 43 Boulevard du 11 novembre 1918, 69622 Villeurbanne Cedex, France.
CASTELLA, Mr. Emmanuel, Laboratoire de Biologie animale et Ecologie, Université Claude Bernard, 43 Boulevard du 11 novembre 1918, 69622 Villeurbanne Cedex, France.
CELLOT, Mr. Bernard, Laboratoire de Biologie animale et Ecologie, Université Claude Bernard, 43 boulevard du 11 novembre 1918, 69622 Villeurbanne Cedex, France.
CHANTARAMONGKOL, Mrs. Porntip, Department of Biology, University of Chiengmai, Chiengmai, Thailand.
CIANFICCONI, Miss Fernanda, Istituto di Zoologia (Fac . Scienze MM.FF.NN.) Università, Via Elce di Sotto, 06100 Perugia, Italy.
CORALLINI SORCETTI, Mrs. Carla, Istituto di Zoologia, (Fac. Scienze MM.FF. NN.), Università, Via Elce di Sotto, 06100 Perugia, Italy.
CRICHTON, Mr. Michael Ian, University of Reading, Department of Pure and Applied Zoology, Whiteknights, P.O. Box 228, Reading RG6 2AJ, England.
DARLINGTON, Mr. Steven Terence, Plymouth Polytechnic, Drake Circus, Plymouth, Devon PL4 8AA, England.
DENIS, Mr. Christian, Laboratoire de Biologie animale, Avenue Général Leclerc, Université de Rennes I, B.P. 25 A, 35042 Rennes, France.
* DOWNS, Mr. Wilbur G., Yale University, Medical School, 60 College Street, New Haven, Connecticut 06510, U.S.A.
DUDGEON, Mr. David, Department of Zoology, University of Hong Kong, Hong Kong
ERMAN, Mrs. Nancy A., University of California, Department of Forestry and Resource Management, 145 Mulford Hall, Berkeley, Ca 94720, U.S.A.
FAESSEL, Mr. Bernard, C.E.M.A.G.R.E.F., 3 quai Chauveau, 69009 Lyon, France.
FLINT, Mr. Oliver S., Smithsonian Institution, Washington, D.C. 20560, U.S.A.
FOSSATI, Mrs. Odile, Laboratoire de Biologie animale et Ecologie, Université Claude Bernard, 43 boulevard du 11 novembre 1918, 69622 Villeurbanne Cédex, France.

FROCHOT, Mr. Bernard, Laboratoire d' Ecologie, Université de Dijon, 21000 Dijon, France.

GARCIA de JALON, Mr. Diego, Departamento de Zoologia y Entomologia, ETSI Montes, Universidad Politécnica de Madrid, 28040 Madrid, Spain.

*GATTAPONI, Mr. Probo, Istituto di Zoologia (Fac. Scienze MM.FF.NN.), Università, Via Elce di Sotto, 06100 Perugia, Italy.

GIBON, Mr. François-Marie, Orstom, B.P. 2528, Bamako, Mali.

GÍSLASON, Mr. Gisli Mar, Liffraedistofnun Haskolans, Institute of Biology, Grensasvegur 12, 108 Reykjavik, Iceland.

GIUDICELLI, Mr. Jean, Laboratoire de Biologie animale et Ecologie, Faculté des Sciences et Techniques de Saint-Jérome, rue H. Poincaré, 13397 Marseille Cedex, France.

GLAPSKA, Miss Mag. Grazyna, Department of Evolutionary Biology, University of Lodz, ul Banacha 12/16, 90-237 Lodz, Poland.

GONZALEZ, Mr. Marcos Andres, Departamento de Zoologia, Facultad de Biologia, Universidad de Santiago, Santiago de Compostela (La Coruna), Spain.

GORE, Mr. James A., University of Tulsa, Faculty of Biological Sciences, Tulsa, Ok 74101, U.S.A.

GREEN, Mr. Philip, University College, Department of Zoology, P.O. Box 78, Cardiff CF1 1XL, United Kingdom.

GULLEFORS, Mr. Bo, Department of Entomology, University of Uppsala, Box 561, 75122 Uppsala, Sweden.

HAMILTON, Mrs. Marjorie Rothschild, Clemson University, Department of Biological Sciences, Clemson, S.C. 29634-1903, U.S.A.

HAMILTON, Mr. Steven W., Clemson University, Department of Entomology, S.C. 29634-0365, U.S.A.

HIGLER, Mr. Bert, R.I.N., P.O. Box 46, 3956 ZR Leersum, The Netherlands.

HOUP, Mr. Ronald, Division of Water, Kentucky State Government, 519 N Lexington Avenue, Wilmore, Kentucky 40390, U.S.A

ITO, Miss Tomiko, Hokkaido Fish Hatchery, 3-373, Kita-Kashiwagi, Eniwa, Hokkaido 061-14, Japan.

KUMANSKI, Mr. Krassimir, National Natural History Museum, Boulevard Russki 1, 1000 Sofia, Bulgaria.

LEUVEN, Mr. Rob S.E.W., Laboratory of aquatic Ecology, Catholic University of Nymegen, Toernooiveld, 6525 ED Nymegen, The Netherlands.

MAJECKI, Mr. Janusz, University of Lodz, Department of evolutonary Biology, ul. banacha 12/16, 90-237 Lodz, Poland.

MALICKY, Mr. Hans, Biologische Station Lunz, 3293 Lunz am See, Austria.

MARLIER, Mr. Georges, Chemin des Selves, La Combe du Cade, 83300 Draguignan, France.

MOLLES, Mr. Manuel Carl, University of New Mexico, Department of Biology, Albuquerque, New Mexico 87131, U.S.A.

MORETTI, Mr. Giampaolo, Istituto di Zoologia (Fac. Scienze MM.FF.NN.), Università, Via Elce di Sotto, 06100 Perugia, Italy.

MORSE, Mr. John C., Clemson University, Department of Entomology, Clemson S.C. 29634-0365, U.S.A.

MURILLO I ROYO, Mr. Joaquim, Departament d'Ecologia, Fac. Biologia, Univ. Barcelona, Avenida Diagonal 645, 08028 Barcelona, Spain.

NEBOISS, Mr. Arturs, Museum of Victoria, Department of Entomology, 71 Victoria Crescent, Abbotsford, 3067 Victoria, Australia.

*NELSON, Mr. Vernon A., Southern Connecticut State University, 501 Crescent Street, New Haven, Connecticut 06515, U.S.A.

NIMMO, Mr. Andrew P., University of Alberta, Entomology Department, Edmonton, Alberta T6G 2E3,Canada.

O'CONNOR, Mr. James, National Museum of Ireland, Kildare Street, Dublin 2, Ireland.
OEMKE, Mr. Mark, Michigan State University, Department of Zoology, East Lansing, Michigan 48824, U.S.A.
OTTO, Mr. Christian, Department of Ecological Zoology, Umea University 901 87 Umea, Sweden.
PETERSEN, Mrs. Lena, Limnology Institute, Box 65, 221 00 Lund, Sweden
PETERSEN, Mr. Robert C., Limnology Institute, Box 65, 221 00 Lund, Sweden.
PETERSSON, Mr. Erik, Department of Zoology, Section of Entomology, Box561, 751 22 Uppsala, Sweden.
PITSCH, Mr. Thomas, Technische Universität Berlin, Institut für Biologie, FB 14, Sekr. FR1-1, Franklinstrasse 28-29, 1000 Berlin, Germany (F.R.).
PUIG I GARCIA, Miss Ma Angels, Universidad de Murcia, Fac. Biologia, Departamento de Ecologia, Murcia, Spain.
RECASENS I ALBALADEJO, Miss Laura, Departament d'Ecologia, Fac. Biologia Univ. Barcelona, Avenida Diagonal 645, 08028 Barcelona, Spain.
RESH, Mr. Vincent H., University of California, Department of Entomology, Berkeley, California 94720, U.S.A.
RICHARDSON, Mr. John S., Institute of Animal Resource Ecology and Department of Zoology, University of British Columbia, Vancouver, British Columbia V6T 1W5, Canada.
ROUX, Mrs. Chantal, Université Claude Bernard, Laboratoire de Biologie animale et Ecologie, 43 boulevard du 11 novembre 1918, 69622 Villeurbanne Cedex, France.
SCHEFTER, Mrs. Patricia Wright, Royal Ontario Museum, Department of Entomology, 100 Queen's Park, Toronto, Ontario M5S 2C6, Canada.
SCHMID, Mr. Fernand, Institut de la Recherche Biosystématique, Ministère de l'Agriculture, Neathy Building, Ottawa, Ontario K1A OGC, Canada.
* SIEGENTHALER, Mrs. Claudine, Chemin des Vignes 8, 1009 Pully, Switzerland.
SMITH, Mr. Stamford, Central Washington University, Department of Biology, Ellensburg Wa 98926, U.S.A.
SOLEM, Mr. John O., University of Trondheim, the Museum, Erl. Skakkesgt. 47, 7000 Trondheim, Norway.
SPINELLI BATTA, Mrs. Giulana, Istituto di Zoologia (Fac. Scienze MM.FF.NN.), Università, Via Elce di Sotto, 06100 Perugia, Italy.
STATZNER, Mr. Bernhard, Zoologisches Institut I, Universität, Postfach 6380, 7500 Karlsruhe, Germany (F.R.).
STROOT, Mr. Philippe, Facultés Universitaires Notre-dame de la Paix, Rue de Bruxelles 61, 5000 Namur, Belgium.
SVENSSON, Mr. Björn S., Lundhagsvägen 37, 178 00 Ekerö, Sweden.
TACHET, Mr. Henri, Université Claude Bernard, Laboratoire de Biologie animale et Ecologie,43 boulevard du 11 novembre 1918, 69622 Villeurbanne Cedex, France.
TANIDA, Mr. Kazumi, Department of Life Sciences, University of Osaka Prefecture, College of Integrated Arts and Sciences, Mozu-ume-machi, Sakai, Osaka 591, Japan.
TAVARES, Miss Annette, University of Toronto, Division of life Sciences, Scarborough Campus, West Hill, Ontario MIC 1A4, Canada.
TERRA, Mr. Luiz S.W., Estaçao Aquicola, 4481 Vila do Conde Codex, Portugal.
TIAN, Mr. Li Xin, Nanjing Agricultural University, Department of Plant Protection, Nanjing, Jiangsu, China (P.R.).
TOMASZEWSKI, Mr. Cesary, Department of Evolutionary Biology, University of Lodz, ul Banacha 12/16, 90 237 Lodz, Poland.

USSEGLIO-POLATERA, Mr. Philippe, Université Claude Bernard, Laboratoire de Biologie animale et Ecologie, 43 boulevard du 11 novembre 1918, 69622 Villeurbanne Cedex, France.

VAILLANT, Mr. François, U.S.M.G. de Grenoble, 118 allée de Pont Croissant, Montbonnot, 38330 Saint-Ismier, France.

VAN DER VELDE, Mr. Gerard, Laboratory of Aquatic Ecology, Catholic University of Nymegen, Toernooiveld, 6525 ED Nymegen, The Netherlands.

VINEYARD, Mr. Richard Noël, University of Toronto, Royal Ontario Museum, Department of Entomology, 100 Queen's Park, Toronto, Ontario M5S 2C6 Ontario, Canada.

WAGNER, Mr. Rüdiger Hagen, Limnologische Flussstation des MPI für Limnologie, Postfach 260, 6407 Schlitz, Germany (F.R.).

WARD, Mr. James, Colorado State University, Department of Zoology, Fort Collins, Colorado 80523, U.S.A.

WEAVER, Mr. John, University of New Hampshire, Department of Entomology, Durham, NH 03824, U.S.A.

WELLS, Mrs. Alice, University of Adelaide, Department of Zoology, G.P.O Box 498, Adelaide, South Australia 5001, Australia.

WEVERS, Mrs. Mary Jo, Oregon State University, Department of Fisheries and Wildlife, Corvallis, OR 97331, U.S.A.

WICHARD, Mr. Wilfried, Siebengebirgsstrasse 221, 5300 Bonn 3, Germany (F.R.)

* WIGGINS, Mr. Glenn, Department of Entomology, Royal Ontario Museum, 100 Queen's Park, Toronto Ontario M5S 2C6, Canada.

WILKINSON, Miss Bridget June, 15 Manor Drive, Southgate, London N14 5JH, England.

WILLIAMS, Mr. Dudley, Division of Life Sciences, Scarborough Campus, University of Toronto, West Hill, Ontario MIC 1A4, Canada.

WILLIAMS, Mrs. Nancy, Division of Life Sciences, Scarborough Campus, University of Toronto, West Hill, Ontario MIC 1A4, Canada.

WISSEMAN, Mr. Robert William, Oregon State University, Department of Entomology, Corvallis, OR 97331, U.S.A.

ZWICK, Mr. Peter, Limnologische Flussstation des Max Planck Instituts für Limnologie, Postfach 260, 6407 Schlitz, Germany (F.R.).

* not present in Lyons

## LIST OF POSTER DISPLAYS (□) AND ORAL COMMUNICATIONS

N.H. ANDERSON and R.W. WISSEMAN
Recovery of the Trichoptera fauna near Mt St Helens five years after the 1980 eruption.

B.J. ARMITAGE
A new north American species and comments on intraspecific variation in the genus *Goera* (Trichoptera : Limnephilidae).

R.M. BADCOCK, M.T. BALES and J.D. HARRISON.
Observations on gill number and respiratory adaptation in caddis larvae.

P. BAGGE
Emergence and distribution of Hydroptilidae in the littoral and outlet biocoenose of lake Konnevesi (Central Finland).

M.T. BALES, R.M. BADCOCK and J.D. HARRISON
Respiratory rates and distribution in caddis larvae in relation to acclimation to temperature.

M.C. BICCHIERAI and G.P. MORETTI
The prosternal horn of Trichoptera larvae.

L. BOTOSANEANU and I.M. LEVANIDOVA
The remarkable genus *Palaeagapetus* Ulmer (Hydroptilidae : Trichoptera).

□P. CHANTARAMONGKOL
The caddisflies of Sri Lanka

F. CIANFICCONI and G.P. MORETTI
Trichoptera colonization in the irrigated meadows of Norcia (Umbria), Italy.

C. CORALLINI SORCETTI and G.P. MORETTI
Further observations on the parasites and epibionts found in Italian Trichoptera larvae.

□M.I. CRICHTON
A study of egg masses of *Glyphotaelius pellucidus* (Retz.) (Trichoptera: Limnephilidae).

□M. DAKKI and H. TACHET
Les larves d'*Hydropsyche* (Hydropsychidae) du Maroc.

S.T. DARLINGTON, A.M. GOWER, L. EBDON
Studies on *Plectrocnemia conspersa* (Curtis) in copper contamined streams in South West England.

□K. KUMANSKI
Is the fauna of Trichoptera in Cuba sufficiently investigates ?

R.S.E.W. LEUVEN, J.A.M. VANHEMELRIJK and G. VAN DER VELDE
The distribution of the Trichoptera in Dutch soft waters differing in acidity

J. MAJECKI
Remarks on the life history of *Anabolia furcata* Brauer (Trichoptera)

□H. MALICKY
Ecological and eidonomic trends in Mediterranean Stenophylacini.

□B. MOHAMMAD, J. MAJECKI and C. TOMASZEWSKI
The caddisflies (Trichoptera) of Primeval Forest in Białowieża.

M.C. MOLLES and L.S.W. da TERRA
Temporal versus spatial variation in trichopteran faunas of northern Portugal.

J. C. MORSE and R.W. HOLZENTHAL
Higher classification of Triplectidinae (Trichoptera : Leptoceridae).

A. NEBOISS
Preliminary comparison of New Guinea Trichoptera with North Sulawesi and North Queensland Fauna.

V.A. NELSON
Light trap collections of Trichoptera near the source of Hopp Brook, New Haven, Connecticut.

□A. P. NIMMO
Division of *Psychoglypha* (Trichoptera : Limnephilidae) to two sub-genera.

J.P. O'CONNOR
A review of the Irish Trichoptera

M.P. OEMKE
The effects of temperature and diet on the larval growth of *Glossosoma nigrior*

C. OTTO
Adaptive egg laying behaviour in two species of caddisflies.

L.B.M. PETERSEN
Direct observations of *Hydropsyche* prey selection

R.C. PETERSEN, Jr
Seston quality as a factor influencing trichopteran populations.

E. PETERSSON and J.O. SOLEM
Male mate recognition in Leptoceridae.

□T. PITSCH
Contribution to larval taxonomy, ecology and distribution of the Central European species of the genus *Philopotamus* (Trichoptera : Philopotamidae).

## EDITORIAL COMMENT

These Proceedings group 65 contributions (lectures and posters have been considered as two different, but equally valid, means of expression). The abstracts of unpublished contributions are not presented here.

We should like to take the opportunity to thank all the authors for the way they prepared their manuscripts, thus making our task much easier. However, because of the heterogeneity of the manuscripts submitted, and in order to improve the presentation, we decided to re-typed all the papers. This decision obviously increased the risk of new errors (i.e. ones not in the original document), but proof-reading by the authors meant that the risk was reduced to a minimum.

We were also obliged to re-type some of the tables if they were likely to be unreadable when printed, or differed too much from the others. Since re-reading of these tables by the authors would have meant longer delays, we took it upon ourselves to proof-read them.

As our knowledge of English is still rather basic, we asked Dr M.I. Crichton to read over some of the manuscripts submitted by non-English speakers. We would like to thank him for doing this and thus eliminating any mistranslations or stylistic errors, while respecting the author's meaning.

The contributions have been divided up into 8 main themes : 1 Morphology and Anatomy, 2 Taxonomy, 3 Fossil forms, 4 Biogeography, 5 Phylogeny, 6 Biology and Physiology, 7 Light traps and flight patterns, 8 Ecology. There is of course bound to be some overlapping of themes as any division is to some extent arbitrary. The different indexes should allow the readers to find the themes or authors they are interested in as rapidly as possible. Contributions have been grouped under each theme heading in such a way as to bring together all those that deal with similar subjects.

The discussions that followed some of the contributions have not been published in this document, either because during the oral papers there was little time for discussion, or because the authors tend to modify their texts according to remarks made in public or private, and thus publication of the discussion itself is unnecessary.

The index of authors has been limited to the authors of contributions, since the index of authors quoted (without their dates) would be relatively worthless. We have separated the subject index and the geographical index. As the Symposium was concerned with Trichoptera, they and they only are to be found in the taxonomic index.

We believe that these Proceedings present a representative panorama of currrent developments in research on Trichoptera, and will form an indispensable data base not only for Trichoptera specialists, Entomologists and Hydrobiologists, but also for all Biologists, thanks to the wide variety and originality of the themes contained in it.

# LIST OF CONTRIBUTIONS

**4. BIOGEOGRAPHY**



**5. PHYLOGENY**

## 6. BIOLOGY AND PHYSIOLOGY

### 6.1. REPRODUCTION.



### 6.2. METABOLISM



### 6.3.INTERSPECIFIC RELATIONS



## 7. LIGHT TRAPS AND FLIGHT PATTERNS

### 7.1 COLLECTIONS

8.2. ZONATION AND DISTRIBUTION



8.3. COMMUNITIES



8.4. ENVIRONMENTAL PERTURBATIONS

# 1 - MORPHOLOGY AND ANATOMY

# PROSTERNAL HORN AND GILSON'S GLAND IN CERTAIN LIMNEPHILID LARVAE

M.C. BICCHIERAI and G.P. MORETTI

## SUMMARY

Morphology, histology and ultrastructure of the prosternal horn and Gilson's gland in certain limnephilid larvae were examined.

## INTRODUCTION

The prosternal horn and Gilson's gland that opens at its tip have been studied by various authors (Gilson 1896, Henseval 1896, Branch 1922, Brocher 1923, Hubault 1924, Quennedey 1968, 1969). It is known that these structures are present in the case-bearing larvae of four families of Trichoptera : Limnephilidae, Goeridae, Lepidostomatidae and Phryganeidae.

In Phryganeidae Gilson's gland bifurcates into two bundles of glandular tubes in each thoracic segment ; only the prothoracic gland opens into the prosternal horn which is stumpy, transparent and poorly sclerotized. A histochemical and ultrastructural study of the gland in *Agrypnia varia* Fab. was made by Quennedey (1969).

In Limnephilidae, Goeridae and Lepidostomatidae the prosternal horn is sclerotized, more or less brownish, and Gilson's gland is a simple, median tube limited to the prothorax.

In this study the scanning electron microscope (SEM) was used to examine the prosternal horn of *Micropterna sequax* McL., *Allogamus antennatus* McL., *Potamophylax cingulatus* Steph., *Halesus radiatus* Curt., *Grammotaulius nigropunctatus* Retz., *Glyphotaelius pellucidus* Retz., *Limnephilus flavicornis* Fabr., *Limnephilus rhombicus* L., *Limnephilus lunatus* Curt. and *Limnephilus coenosus* Curt. Gilson's gland was examined with a light microscope and a transmission electron microscope (TEM) mainly on specimens of *M. sequax*.

Although there is proof that Gilson's gland produces a lipidic secretion, its role remains unknown. Our laboratory observations confirm what was previously reported (Quennedey 1968) : larvae from which the prosternal horn and in some cases Gilson's gland were removed, were able to rebuild their cases in the time and in a way that are not unlike those of the control larvae. Thus, our present knowledge does not allow us to make any hypotheses concerning the specific role of these structures in case building activity even though other elements, such as the position of the horn and the form of its tip, could lead us to such considerations.

## MATERIALS AND METHODS

The limnephilid larvae from different localities were reared in the laboratory. For light microscopy, larvae were fixed in Bouin, Duboscq-Brasil and Carnoy fluids. The prosternal horn and Gilson's gland were removed and stained "in toto" with Mayer's haemalum-eosin.

For scanning electron microscopy, larvae were fixed with ethanol 70% or with 4% glutaraldehyde in 0.1 M phosphate buffer, pH 7.2. The prosternal horn was subsequently dissected, dried, mounted on a specimen holder, gold-coated and examined in a Philips SEM 501.

For transmission electron microscopy, the larval prothorax was removed and fixed for 6-8 h at 4°C either with the above mentioned glutaraldehyde solution or with 4% paraformaldehyde-2.5% glutaraldehyde mixture in 0.1 M cacodylate buffer, pH 7.2. The latter fixative gave the better results. Specimens were then washed in the buffer, post-fixed for 1 h at 4°C with 1% osmium tetroxide in the same buffer as the fixative, dehydrated in a graded ethanol series and embedded in Epon-Araldite. Thin sections were stained in saturated alcoholic uranyl acetate and in lead citrate solutions before being observed and photographed under a Philips EM 400.

Semi-thin sections, stained with methylene blue in 1% borax, were examined using a light microscope.

RESULTS AND DISCUSSION

The prosternal horn is a cuticular appendage, more or less sclerotized, that originates in the median line of the prosternum, anterior to the pigmented areole. It is curved forward, towards the head of the larva, gradually tapering from the base to the tip, and may extend as far as the gula, depending on the inclination of the head.

In *Allogamus antennatus, Potamophylax cingulatus* and *Halesus radiatus* (Pl. I, 1,5) the prosternal horn has a smooth cuticle with few basal spinules that point towards the tip. The prosternal horn of *Micropterna sequax* has spinules on the lower third and that of *Grammotaulius nigropunctatus, Glyphotaelius pellucidus, Limnephilus flavicornis, L. rhombicus, L. coenosus* and *L. lunatus* (Pl. I, 2) is covered with spinules over most of its length, except for the apical portion which is always smooth (Pl. I, 4). The insertion point of the horn and its surrouding area are extremely spinulous (Pl. I, 6). An excentric, narrow duct runs inside the length of the prosternal horn, close to its internal surface (Pl. I, 8). The apical opening of the duct is subterminal, at least in Limnephilidae, because a less sclerotized, pointed process rises above it externally (Pl. I, 3). At higher magnification the tip appears longitudinally grooved forming a series of finger-like projections which give a crested aspect to the terminal edge. The cuticle of the apical process is often pleated (Pl. I, 3, 7).

At the base of the prosternal horn is Gilson's gland (Pl. II, 1), which runs sometimes sinuously along the middle of the prothorax between the nerve cord and the alimentary canal. Before connecting with the prosternal horn, the apical portion of the gland passes ventrally to the prothoracic ganglia (Pl. II, 7, 9). The tubular gland has a cul-de-sac ending (Pl. II, 2) and consists of a single layer of large secretory cells arranged around a central lumen which is continuous with the duct of the prosternal horn (Pl. II, 3, 4, 8).

Transverse sections of the gland normally show two secretory cells, each with a large, highly chromatophilic, basal nucleus and an apical microvillar border covered by a cuticular intima which limits the glandular lumen (Pl. II, 6). Where the cuticle presents a uniform thickness the lumen has a circular cross section (Pl. II, 7) but where its thickness varies the lumen has an irregular cross section, sometimes even a four-leaf clover shape.

In transverse section the horn is nearly circular, bordered by a thick cuticle beneath which is an epidermal layer. Internally the excentric duct shows an irregular lumen and is surrounded by the duct cells and a few small tracheae (Pl. II, 5).

The last instar larva of *Micropterna sequax* was used for the ultrastructural study of Gilson's gland. The prosternal horn of this larva is brownish, about 500 µm long, and the gland is a tube of secretory cells up to 1.5 mm long with a lumen of 10-15 µm in diameter.

Under TEM examination the large secretory cells (Pl. III, 1) show the apical plasma membrane forming numerous, long (3 µm) microvilli which are often sinuous. The microvillar tips are darkened by the presence of hemidesmosomes. Each microvillus contains a central canal that appears continuous with a tubule of smooth endoplasmic reticulum (Pl. III, 4 ; Pl. IV, 1).

The basal region of the secretory cells is separated from the haemolymph by a basement lamella (40 nm thick) and the basal plasma membrane is invaginated away from it by deep, tortuous infoldings of 0.1 µm in diameter (Pl.III, 2, Pl. IV. 2). This amplification of the basal surface is believed to facilitate fluid exchange between the cell and the haemolymph.

The intercellular space is free in the basal part while septate junctions are seen in the apical part (Pl. IV, 3, 5).

The gland cells have large (up to 40 x 20 µm) basal nuclei, rounded or elliptical in shape, often with a lobed surface. Dense chromatin clumps are evenly distributed in the karioplasm where the nucleoli are clearly visible (Pl. III, 1, 2 ; Pl. IV, 8). The cytoplasm consists mainly of smooth endoplasmic reticulum (SER) both in the cisternal and in the tubular form (Pl. III, 3 ; Pl. IV, 8). The cisternae are frequently distended with a scarcely electron dense material. In the apical region of many cells, beneath the microvilli, numerous, concentric cisternae of SER form membranous whorls enclosing fine granular material (Pl. III, 4 ; Pl. IV, 3). The rough endoplasmic reticulum is scarcely developed and adjoins cisternae of SER (Pl. IV, 8). Mitochondria are abundant and polymorphous (Pl. III, 2). Golgi complexes and free ribosomes are observed in the cytoplasm as well as some lysosomes and bundled microtubules. There are also a few membrane-bound vacuoles with a scanty matrix.

The cuticle overlying the secretory cells consists of a lamellated endocuticle, 1.5-4 µm thick, and a perforated epicuticle, 30 nm thick (Pl. IV, 3). Under the epicuticle there is a layer, about 0.2 µm thick, formed by small tubules, 12 nm in diameter (Pl. IV, 4). The fine granular or amorphous lipidic secretion is visible just below the epicuticle, among the small tubules, and in the glandular lumen (Pl. IV. 6, 7).

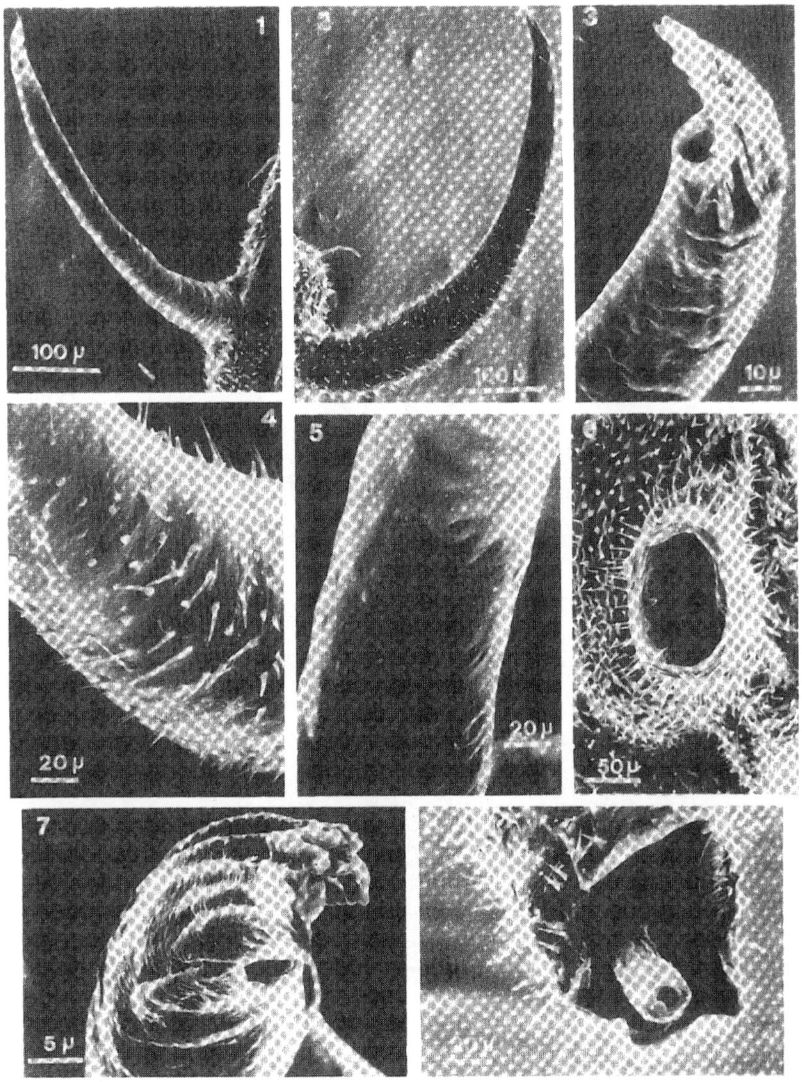


Plate I. SEM micrographs. PROSTERNAL HORN. 1. prosternal horn in *Halesus radiatus*. 2. Prosternal horn in *Limnephilus lunatus*. 3. Apical portion of the horn in *Micropterna sequax*. 4. Spinulous cuticle of the horn in *Glyphotaelius pellucidus*. 5. Smooth cuticle of the horn in *Allogamus antennatus*. 6. Insertion point of the horn in *Limnephilus flavicornis*. 7. Apex of the horn in *Potamophylax cingulatus*. 8. Internal excentric duct of the horn in *Grammotaulius nigropunctatus*.

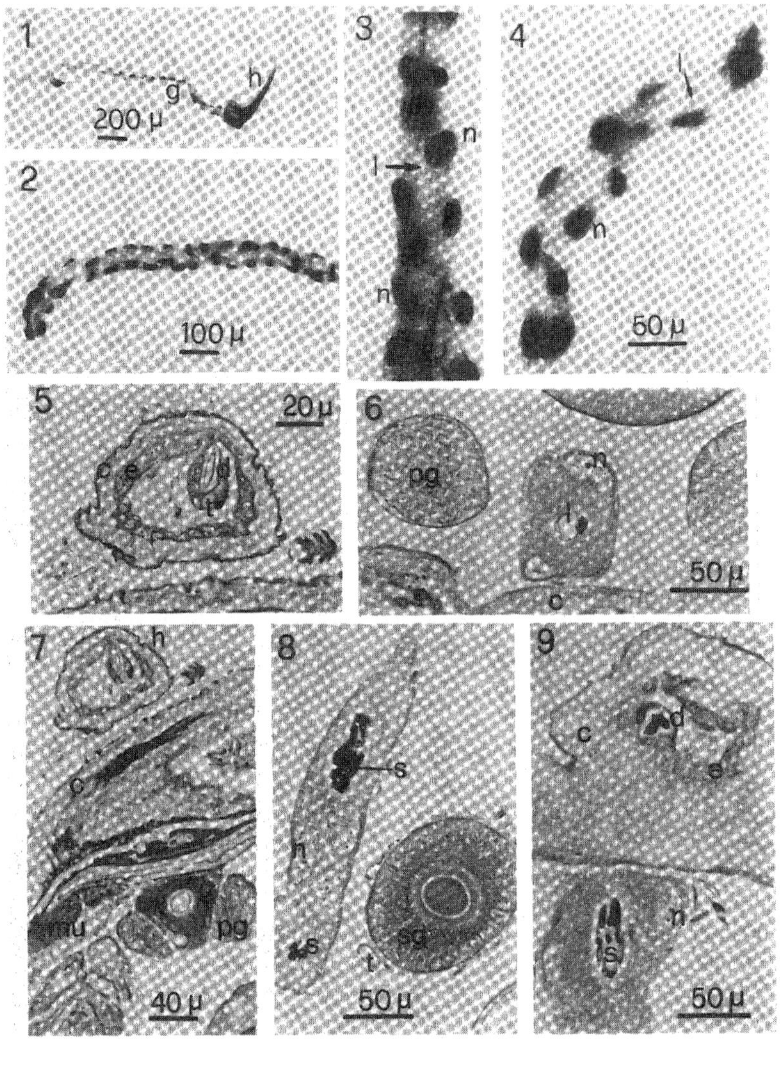


Plate II. Light microscope examination. 1. prosternal horn and Gilson's gland in *Micropterna sequax*. 2. Tubular Gilson's gland. 3 and 4. Secretory cells disposed in a single layer around the lumen. 5. Cross section of the prosternal horn. 6. Cross section of Gilson's gland. 7. General view of horn and gland in cross section. 8. Oblique section of gland ; the irregular lumen is filled with secretion. 9. Apical portion of Gilson's gland connecting with prosternal horn.

c = cuticle, d = duct of the prosternal horn, e = epidermal layer, g = Gilson's gland, h = prosternal horn, l = glandular lumen, mu = muscle, n = nucleus, pg = prothoracic ganglion, s = secretion, sg = silk gland, t = trachea.

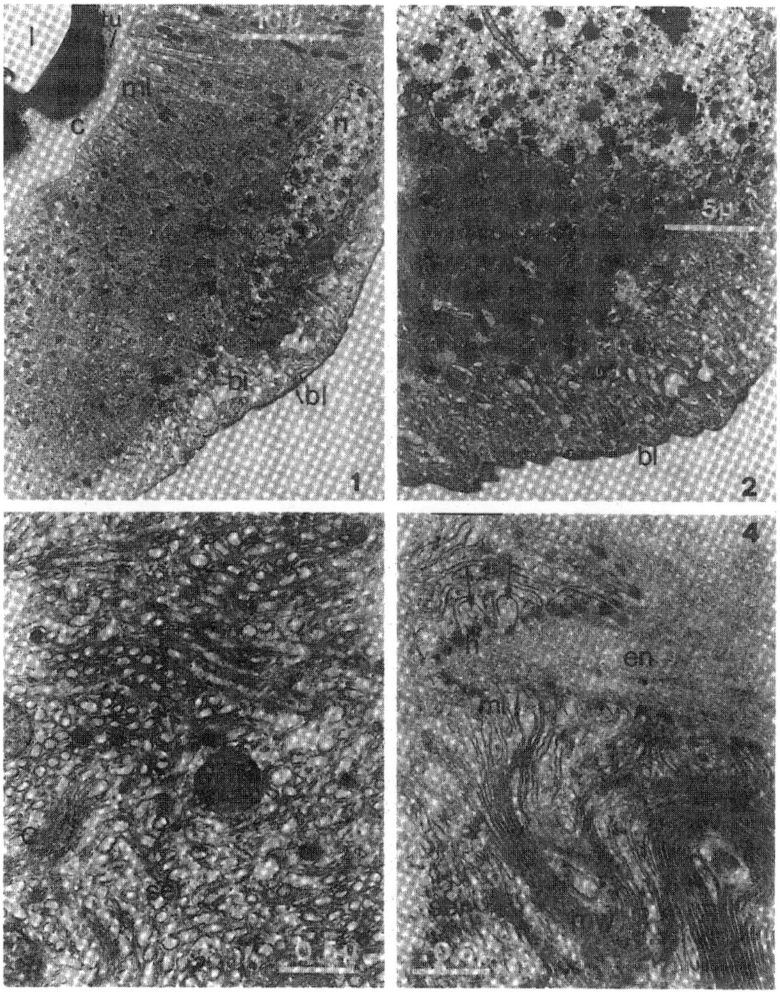


Plate III. TEM micrographs. 1. Transverse section of Gilson's gland. 2. Gland cell : lobed nucleus and cytoplasm rich in organelles. 3. Cell cytoplasm. 4. Concentric membranous whorls in the apical region of the gland cell. Arrows show the central canals of the microvilli in cross section. c = cuticle, bl = basement lamella, bi = basal infoldings, en = endocuticle, G = Golgi complex, h = hemidesmosomes, l = lumen, li = lysosomes, m = mitochondria, mi = microvilli, mw = membranous whorls, n = nucleus, ser = smooth endoplasmic reticulum, tu = tubules.

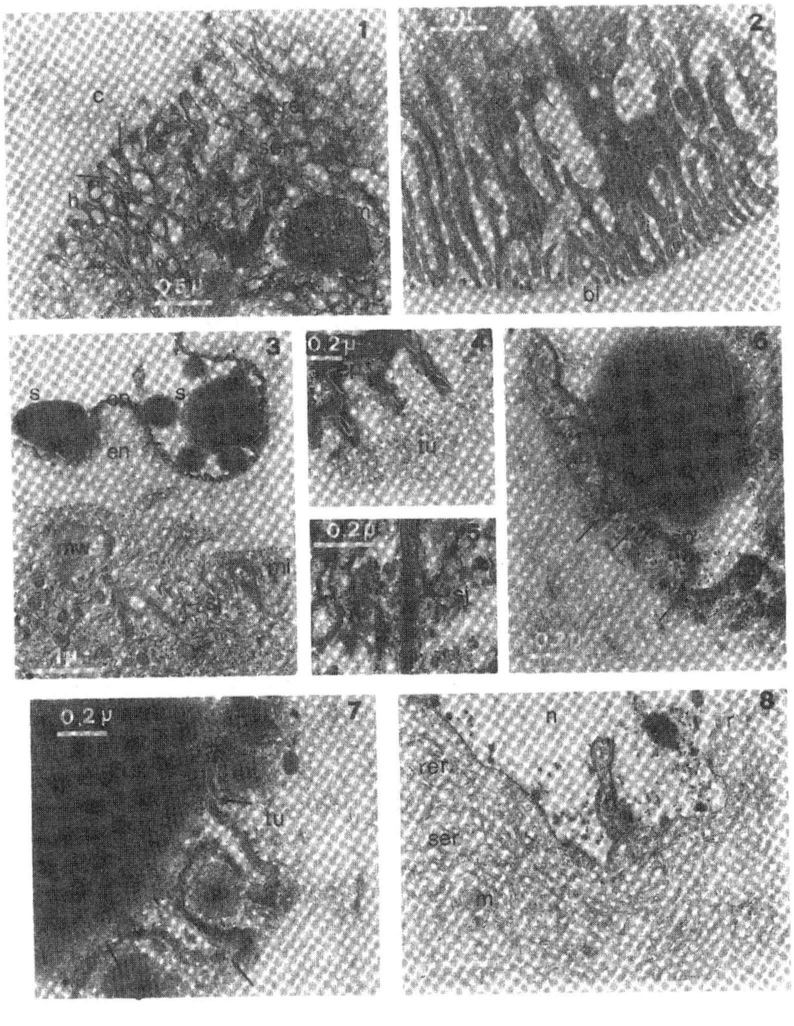


Plate IV : TEM micrographs. GLAND CELL. 1. Apical border of microvilli with their central canals (arrows). 2. Basal region of cell. 3. Apical region of cell. 4. Small tubules under epicuticle. 5. Septate junction in upper intercellular space. 6.Perforated (arrows) epicuticle. 7. Secretion passing into the glandular lumen (arrows). 8. Extensive development of cisternal and tubular SER and short profiles of RER.

c = cuticle, bl = basement lamella, en = endocuticle, ep = epicuticle, h = hemidesmosomes, i = infoldings, m = mitochondrion, mi = microvilli, mw = membranous whorls, n = nucleus, rer = rough endoplasmic reticulum, s = secretion, ser = smooth endoplasmic reticulum, sj = septate junctions, tu = tubules.

It is highly probable that the secretion is produced by the SER - the tubules of which penetrate the microvilli - and diffuses across the plasma membrane of the microvilli and through the cuticle to the glandular lumen.

REFERENCES

Branch, H. 1922. A contribution to the knowledge of the internal anatomy of Trichoptera. **Ann.Ent. Soc. Amer.,** 15 : 256-275.

Brocher, F. 1923. La corne prosternale des larves de Trichoptères. **Ann. Biol. Lac.,** 12 : 83-91.

Gilson, G. 1896. On segmentally disposed thoracic glands in the larvae of the Trichoptera. **J. Linn. Soc. London,** 25 : 407-412.

Henseval, M. 1896. Etude comparée des glandes de Gilson. **La Cellule,** 11 : 329-354.

Hubault, E. 1924. Recherches sur la structure intime des étuis des larves de certains Trichoptères. **Ann. Biol. Lac.,** 13 (1-2).

Quennedey, A. 1968. Anatomie comparée des glandes de Gilson des larves de Trichoptères. **Bull. Soc. Zool. France,** 93 : 467-477.

Quennedey, A. 1969. Les glandes de Gilson des larves de *Phryganea varia* Fabr. (Insecta, Trichoptera). Etude histochimique et ultrastructurale. **J. Microscopie,** 8 : 479-496.

# THE CLINGING APPARATUS OF TRICHOPTERA PUPAE : NUMBER, STRUCTURE, AND SEM OBSERVATIONS OF THE DENTICULATIONS ON THE UROTERGAL PLATES

G. SPINELLI BATTA and G.P. MORETTI

## SUMMARY

A morphological comparison by light microscopy and SEM is proposed of the anchoring apparatus of pupae of someTrichoptera species present in Italy. The study was extended to the cuticular structures surrounding the tergal sclerite plates.

## INTRODUCTION

Studies were carried out on the anchoring apparatus of the urotergites in segments of pupae of certain families of Trichoptera present in Italy. The function of this section of the clinging apparatus is completely separate from that of the dorsal process of the first abdominal segment, although it is clear that the presegmental and postsegmental plates of the tergal sclerites also play a part in the initial phase of emergence from the case.

## METHODOLOGY

The study was in part carried out on permanent prepared specimens with observation by light microscopy and subsequent drawings in a light chamber, in part on exuviae prepared on stubs and observed, after gold-coating, under SEM Philips 501. This last method of study allowed for the identification, in greater detail, of the cuticular structures and the shape of the teeth on the sclerified plates.

## DISCUSSION

From a survey of the single species examined, starting with the family Rhyacophilidae, a certain resemblance was noted between the two species *Rhyacophila italica* Moret. and *R. dorsalis* Curt. so that it was impossible to give any diagnostic value at a specific level to the pupal exuviae of these two species. In both, plates are present from the third to the seventh tergite, while the fourth and fifth also present postsegmental plates. The teeth of the anterior plates point backwards, while those of the posterior plates point forwards, (Pl.. I, 1). Teeth vary in number from ten to thirty, being hooked, rather pointed, closely grouped together and arranged in five or six rows. All the teeth originate from a sclerified plate at the base of which, small cusps may be noted, (Pl. I, 2). The surrounding cuticular area is characterized by a netlike alveolar structure.

From scanning electron microscope observations of *Ptilocolepus granulatus* Pict. (Pl. I, 3) (Glossosomatidae) it may be noted that the sculptured shapes of the cuticle resemble more closely those of the Rhyacophilidae than Glossosomatidae, as in the case of the denticulated sclerites.

An excellent example of the phenomenon may be found in *Agapetus laniger* Pict. (Pl. IV, 1) where the posterior plates of the fourth tergite are elongated rather than rounded, with numerous short, stumpy teeth thus differentiating them from those of Rhyacophilidae. Thus the diagnostic characteristic for these two similar families is valid.

In the case of Hydropsychidae, all the three species were considered : *H. pellucidula* Curt. (Pl. I, 4, 5, 6), *H. modesta* Nav. (Pl. I, 7) and *Cheumatopsyche lepida* Pict. (Pl. I, 8). Differences may be noted between the three species in the number, shape and disposition of the teeth. In the first two species plates may be found from the third to the seventh tergites, in the third species, from the third to the eighth. The postsegmental plates of the third and fourth tergite differ from those of the preceding families, the third tergite plates having in fact a characteristically elliptical alignment which is not evident in the fourth. Furthermore, in Hydropsychidae the denticulation varies greatly both in number and shape as well as in structure, and one may distinguish stumpy, long, robust and grooved teeth, at times with a particular rayed arrangement. The cuticular structure is smooth near the plate but spine-like on the outskirts. An abundant series of macrochaetae is clearly visible, a phenomenon lacking in *H. modesta* the cuticle of which has few spines, the plates presenting evident differences in the teeth. The two species are therefore clearly distinguishable, particularly from *Cheumatopsyche lepida*, the smaller number of teeth of which are visibly grooved and inserted at the top of a pyramid shaped cusp. The cuticle is almost altogether free of spines and setae.

In *Plectrocnemia conspersa* Curt. (Pl. II, 1,2), the pattern of the plates is entirely different from that of the preceding species : the teeth are in fact arranged in a single series, being extremely curved and pointed in the fifth postsegmental and the sixth presegmental ; stubby with a blunt apex and with clearly visible grooves in others. The cuticle has a typical tiled or overlapping structure.

Of the Psychomyiidae the exuviae of *Tinodes antonioi* Bots. Vig. Tat. (Pl. IV, 2) have plates bearing scattered, short, stout teeth with the exception of the fifth postsegmental plates where they are shorter and far more numerous.

In Limnephilidae the plates have a relatively small number of teeth if presegmental, while if postsegmental, they are rich in cusps and arranged in a triple series. This pattern is to be found in both *Limnephilus lunatus* McL. (Pl.II, 4,5), and in *Leptodrusus budtzi* Ulm. (Pl. II, 3) as well as in *Chaetopteryx gessneri* McL. (Pl. II, 6) where only small differences may be noted such as the more or less angular shape of the teeth in the latter species, or the overlapping cuticle which is similar to the pattern found in Rhyacophilidae.

Among the Goeridae, *Silo nigricornis* Pict. (Pl. II, 7, 8) provides evidence of a change in the cuticular structure which is spine-like and in the teeth of the plates which are limited to two or three, and are extremely curved except for those of the postsegmental plates arranged in triple series or at times forked or bifid.

In *Ceraclea fulva* Ramb. (Pl. IV, 3) (Leptoceridae) the system of the plates is present from the third to the seventh segment, the teeth following the pattern of the preceding species : few on the presegmental plates though slightly more numerous and arranged in a series on the postsegmentals.

In the Sericostomatidae we have an example which is diagnostically distinguishable in *Sericostoma italicum* Moret. (Pl. III, 1, 2) (Pl. IV. 4) and *S. siculum* McL (Pl. IV, 5) but not in *S. pedemontanum* McL., which is similar to *S. italicum*. The plates in this last species are in fact longer and narrower than in *S. siculum* where the denticulation is more numerous, while on the contrary, there is a close resemblance between *S. italicum* and *S. pedemontanum*. Furthermore, the three species have in common a cuticle with rows of spinules and toothlike plates which are set in visible crater-like depressions.

In the Beraeidae, two genera were examined : *Beraeodes* and *Beraea* which are very much alike. In *Beraeodes minutus* L.(Pl. III, 3,4) the plates present on the third and sixth tergites are distinguishable by oval areas on which there is a single tooth in the presegmental plates and two in the postsegmentals which are conicle and stumpy with a slightly narrowed base ; the cuticle is irregularly pleated and lacking in spines but with a slight overlap. In *Beraea maurus* Curt. (Pl. IV, 6) the oval plates are also armed with one to three teeth accompanied by small proximal teeth. On the eighth tergite two spiny areas are visible on which three trichoid sensilla stand out.

In *Helicopsyche sperata* McL. (Pl. III, 5,6) we may note the same arrangement of the plates from the third to the sixth tergite. Denticulation is scarce ; slightly pointed or blunt teeth displaying however a clearly visible longitudinal crest. The horn-shaped teeth on the plates of the fifth tergite are highly characteristic.

The Odontoceridae close the survey of the families examined, and are characterized as having evident generic distinctions. In *Odontocerum albicorne* Scop. (Pl. III, 7,8) there are oval, sclerified oval-shaped plates armed with a single tooth except in the case of the postsegmentals of the fifth segment which have two, well apart from one another ; the teeth on the presegmental plates are characteristically spatula-shaped and slightly grooved.

## CONCLUSIONS

In this initial survey of the sclerites of abdominal tergites we note the wide variety of their structural patterns in Trichoptera pupae and a certain convergence in structural forms among different families.

It should, however, be taken into account that these families belong to the suborder Annulipalpia, rather than to that of the Integripalpia.

Furthermore, the shape and arrangement of the abdominal anchoring system is conditioned by coleobiosis on the basis of the pupal case structure.

Correct phyletic analysis may only be proposed following the conclusion of this research.

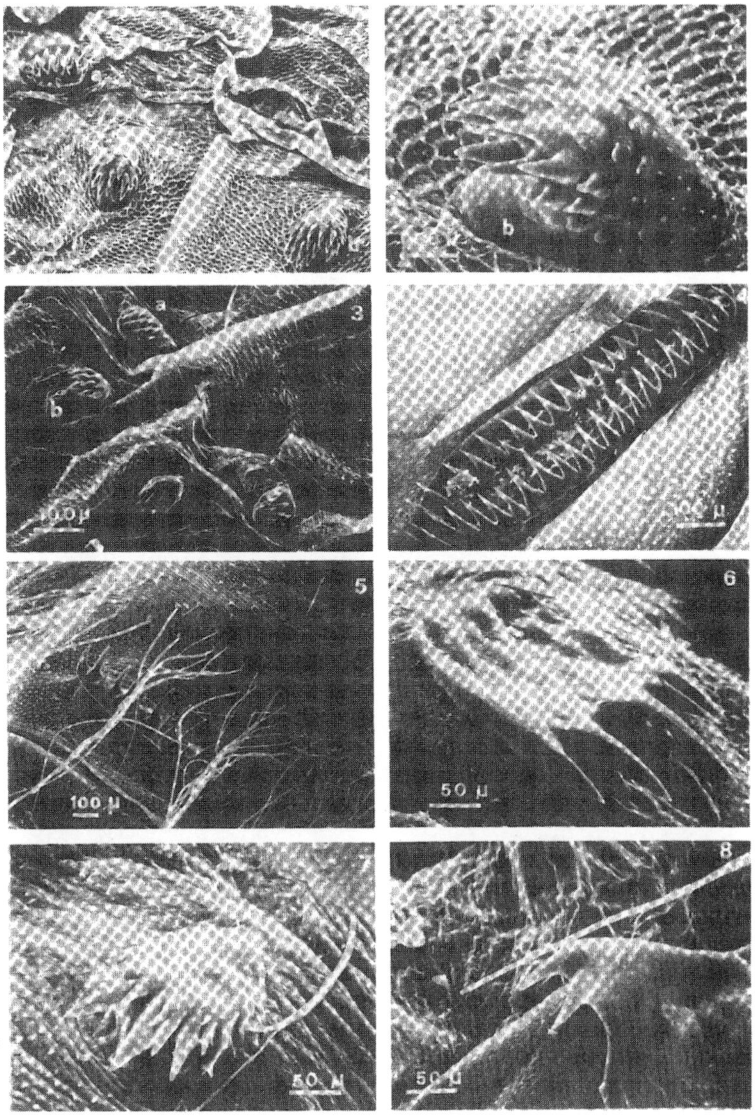


Plate I. SEM micrographs. 1-2, *Rhyacophila italica* Moret. 3, *Ptilocolepus granulatus* Pict. ; 4, *Hydropsyche pellucidula* Curt. : III posterior plate ; 5, *H. pellucidula* : IV posterior plate ; 6, *H. pellucidula* : V anterior plate ; 7, *H. modesta* Nav. : III anterior plate ; 8, *Cheumatopsyche lepida* Pict. : III anterior plate.
(a : posterior plate, b : anterior plate, c : cuticle).

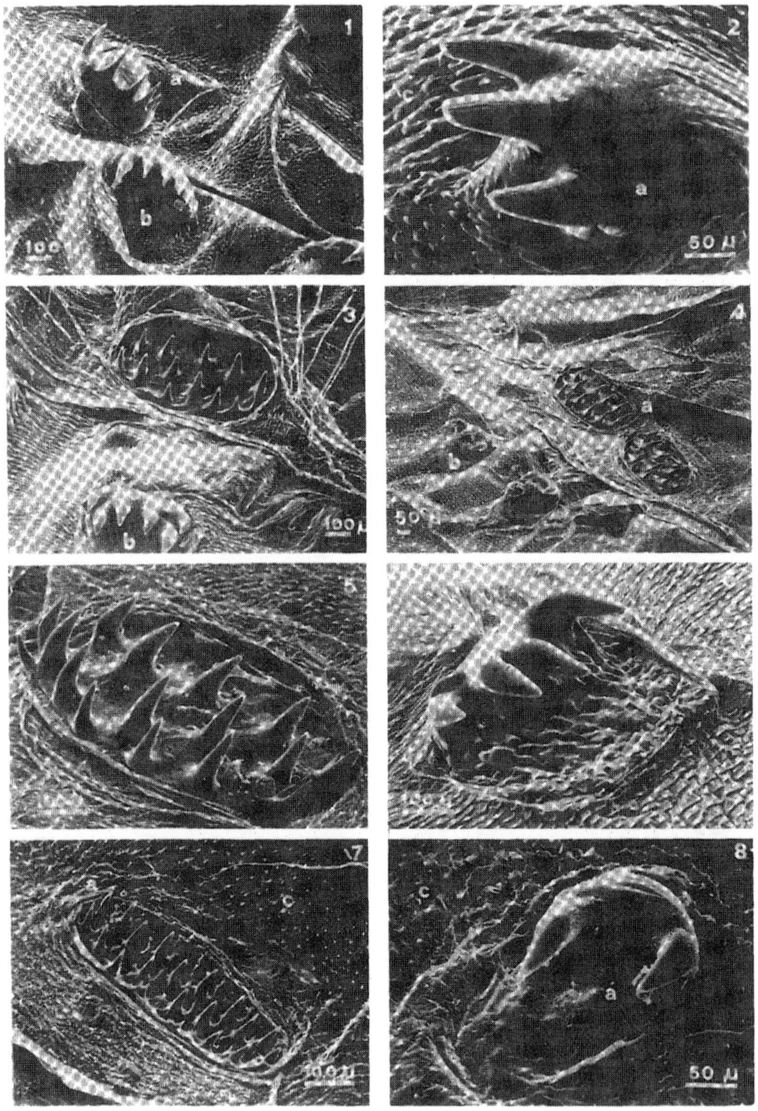

Plate II. SEM micrographs. 1-2, *Plectrocnemia conspersa* Curt. ; 3, *Leptodrusus budtzi* Ulm. ; 4, *Limnephilus lunatus* McL. ; 5, *L. lunatus* : V posterior plate ; 6, *Chaetopteryx gessneri* McL; : III anterior plate ; 7-8, *Silo nigricornis* Pict. (a : posterior plate, b : anterior plate, c : cuticle).

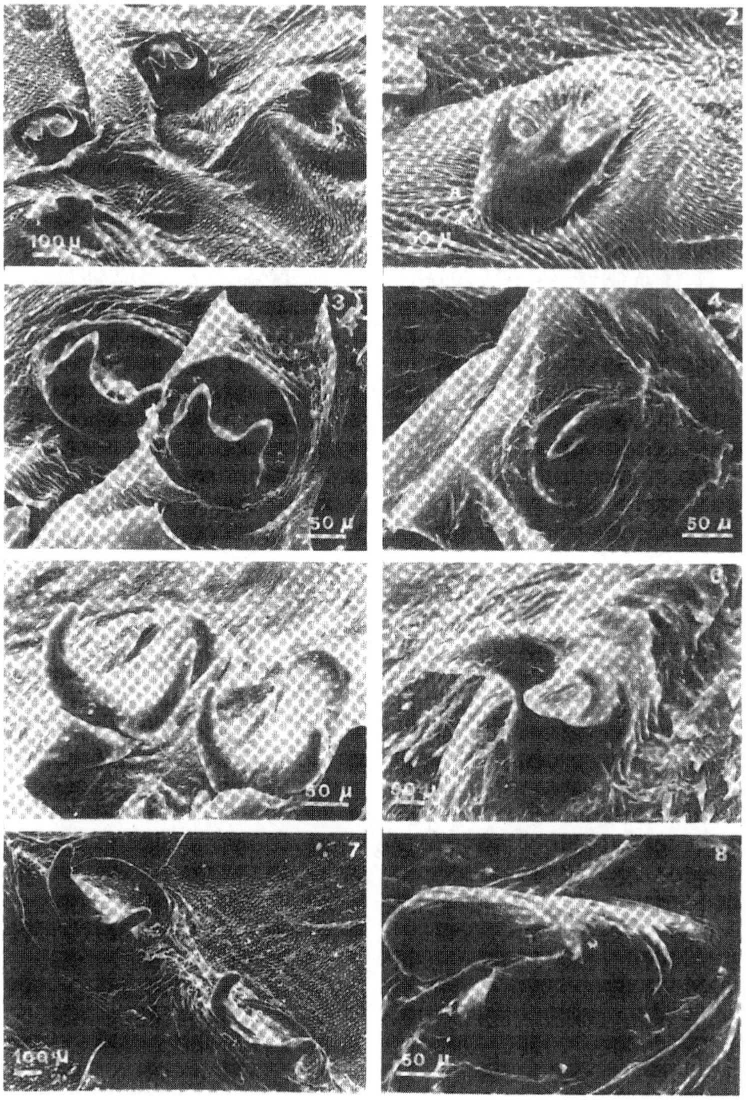


Plate III. SEM micrographs. 1-2, *Sericostoma italicum* Moret. ; 3, *Beraeodes minutus* L. : V posterior plates ; 4, *B. minutus* : III anterior plate ; 5, *Helicopsyche separata* McL. V posterior plates ; 6, *H. sperata* : III anterior plate ; 7, *Odontocerum albicorne* Scop. : V posterior plates ; 8, *O. albicorne* : III anterior plate. (a : posterior plate ; b : anterior plate).

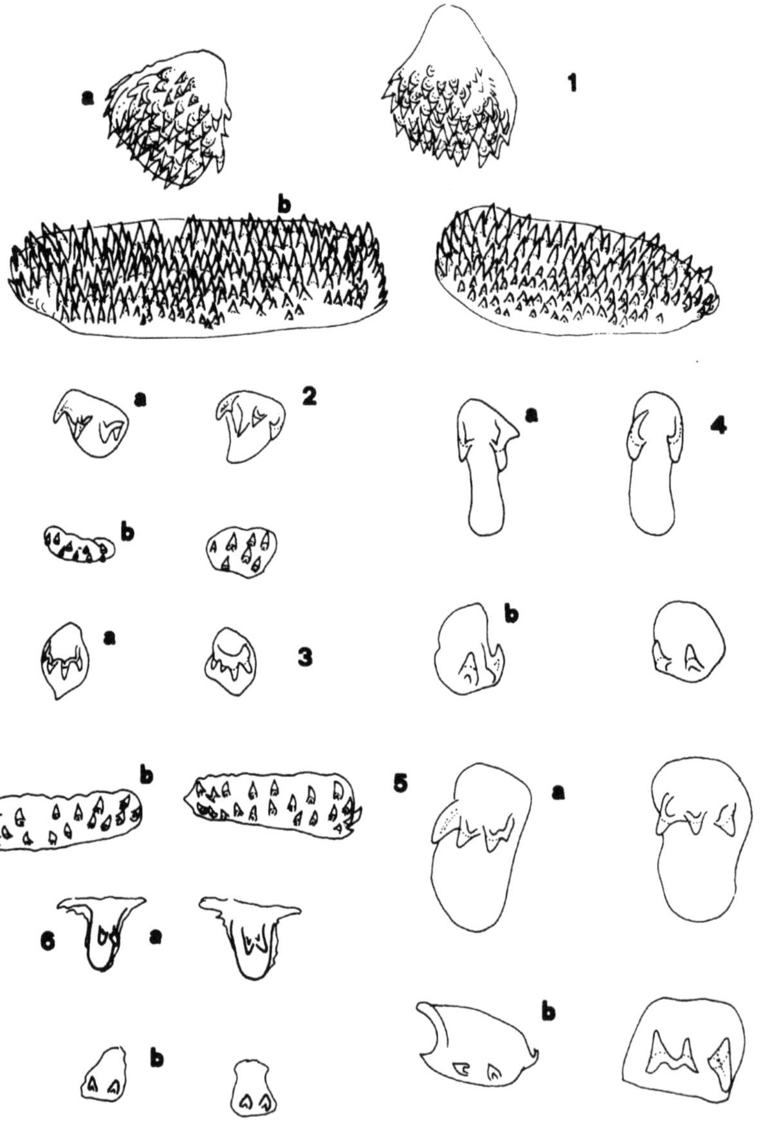


Plate IV. Light microscope examination. 1, *Agapetus laniger* Pict. ; 2, *Tinodes antonioi* Bots. Vig. Tat. ; 3, *Ceraclea fulva* Ramb. ; 4, *Sericostoma italicum* Moret. ; 5, *Sericostoma siculum* McL. ; 6, *Beraea maurus* Curt. (a : anterior plates ; b : posterior plates)

## REFERENCES


Hickin, N.E. 1967. Caddis larvae-Larvae of the British Trichoptera. Hutchinson of London.

Lepneva, S.G. 1971. Larvae and pupae of Integripalpia, Trichoptera. Fauna U.S.S.R. Acad. Sc. U.S.S.R.

Lestage, J.A. 1921. Les larves et nympes aquatiques des Insectes d'Europe. *In* Rousseau, J. Lebegue, Bruxelles.

Rutherford, J.E. 1985. An illustrated key to the pupae of six species of *Hydropsyche* (Trichoptera : Hydropsychydae) common in southern Ontario streams. **Great Lakes Entom.,** 18 : 123-132.

# RESEARCH ON THE MUSCULATURE AND SENSE ORGANS OF THE PROTUBERANCES OF THE FIRST ABDOMINAL SEGMENT IN CADDIS LARVAE (TRICHOPTERA)

C. TOMASZEWSKI, H. FÜLLER and G. GLAPSKA

## SUMMARY

Tomaszewski (1981) hypothesized that the protuberances of the 1st abdominal segment of caddis larvae are of special importance in triggering case-building activity. This hypothesis was called in question by Rowlands and Hansell (1984), and so the authors carried out these anatomical and histological investigations of the protuberances in order to understand their structure and function. The musculature and mechanoreceptors of the protuberances in the larvae of *Anabolia furcata* Brau. and *Halesus* sp. are described.

## INTRODUCTION

In spite of numerous investigations on the structure of caddis larvae, the question as to what releases case-building activity still remains open. On the basis of experiments on building behaviour, Tomaszewski (1981) concluded that the protuberances of the 1st abdominal segment control building activity. A larva removed from its case and with the first abdominal segment covered with a plastic band did not begin to build a new case. Thus, Tomaszewski put forward the hypothesis that pressure on the protuberances signals the presence of the case, whereas the lack of this contact releases building activity.

This hypothesis has been rejected by Rowlands and Hansell (1984), because their experiments with slightly modified methods brought about different results. More detailed studies on the function of the protuberances based on their structure appeared to be necessary, so the present authors began these studies on their functional morphology.

More detailed information on musculature and innervation of the first abdominal segment is found in Hasenfuss (1973) but only for the free-living larva of *Rhyacophila nubila* which does not possess the protuberances. Thus, the aim of the present paper is to explain the topography of the musculature and sensory innervation of the protuberances in a few species of case-building larvae.

## MATERIAL AND METHODS

Adult larvae of *Anabolia furcata* Brau., *Halesus* sp., *Oligotricha striata* (L.) and *Chaetopteryx villosa* (Fabr.) were investigated. For an understanding of the topography of the musculature several methods were used. In order to follow the course of muscles, preserved larvae were studied under a stereomicroscope so that the muscles of the protuberances were gradually uncovered in several methods of preparation from different sides. The muscle topography was checked again in thick anisotropic sections of the first abdominal segment as well as in series of microtome sections. The muscles were given the letters A, B, C, K, L, M and N so that they could be homologized with muscles of the fundamental structural scheme of other segments.

The method of staining with rongalit-white according to Witting (1955) and later practised by Hasenfuss (1973) was used for showing up the innervation. After introduction of vital stains, larvae ready were fixed and complete biological or thick sections were made from them.

For histological studies, series of transverse microtome sections of 7 and 10 µm of this abdominal segment were made. Bouin's fluid was used as a preservative and the sections were stained in haemalum eosine.

## RESULTS

Histological sections showed that the protuberances contain a large space filled with haemolymph and very few organ structures (Fig. 1). In *Anabolia* the completely sclerotized cuticle of 22-28 µm thickness is provided with thin spinelike bristles of 4 -9 µm length on the distal area of the protuberances.

There is a flat epidermis of about 5 µm thickness whose cells have only a few cuticular sensilla entering the haemolymph space. There are few muscles, thin nerve fibres and single tracheae and tracheoles, and large lobes of fat body in the interior of the protuberances.

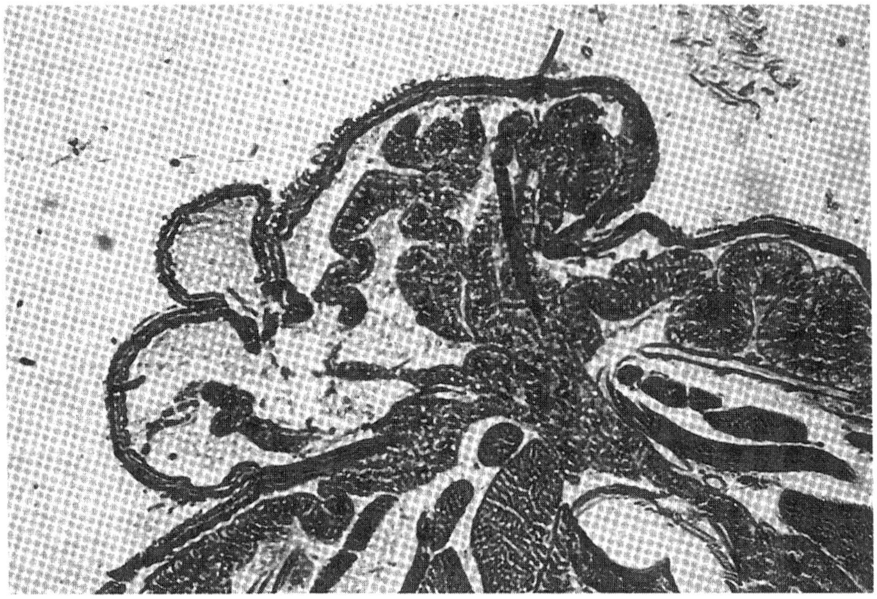

Figure 1. Transverse section of the lateral protuberance of the larvae of *Anabolia furcata*. Enlargement x 56.

## Musculature

The musculature of the protuberances of the first abdominal segment of the species studied shows the common, basic scheme of the structure which is, to some extent, specific for the species. In some individuals of the same species, there are differences in the width and localization of the points of muscle insertion. The description below is based on the structure in *Anabolia furcata* (Fig. 2). Within the dorsal protuberance there are three bilaterally symmetrical pairs of muscles which serve as retractors.

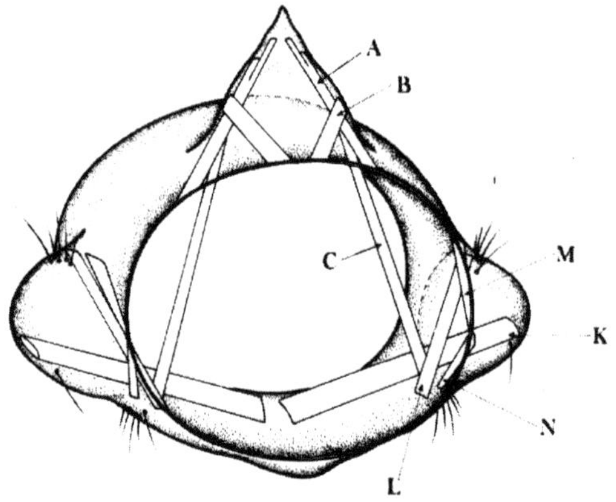


Figure 2. Schematic diagram of musculature of the protuberances of the 1st abdominal segment of the larva of *Anabolia furcata*. A, B, C, K, L, M, N - muscles.

Muscle A with two or three branches is attached directly below the dorsal apex of the protuberance and runs in a ventrocaudal direction towards the posterior edge of the segment. It is inserted in the dorsolateral area of the inter-segmental fold.

Muscle B is set in the lower half of the wall of the protuberance and runs as a wide band-like muscle beside muscle A in a mediocaudal direction to end near the dorsomedial line of the inter-segmental fold.

Muscle C is fixed at the base of the dorsal protuberance, below and somewhat in front of muscle B. It is a long thin muscle which, depending on the longitudinal musculature of the segment, runs in a curve through the base of the lateral protuberances and ends on their ventral edges. In the figure, its course is presented in a schematic way.

Each of the lateral protuberances has four specific muscles which certainly take part in their pulsation. Muscle K serves as a retractor of the lateral protuberances. It is a broad, band-like muscle with five or six branches inserted below the protuberance tip, which runs in front of muscle C in a ventromedial direction and ends in the anterior part of the sternum beside the mid-line.

Muscle L is fixed to the dorsal edge of the protuberance above a group of dorsal bristles. It runs through the base of the protuberance and ends on its ventral edge behind muscle C.

Muscle M is also fixed to the dorsal edge of the protuberance directly behind muscle L. This muscle expands ventrocaudally and finishes on the posterior edge of the protuberance above directly muscle M.

Muscle N is inserted on the posterior edge of the protuberance under muscle M and is directed ventrocaudally to finish on the ventral edge of the protuberance at some distance from muscle L.

**Sense organs**

For a better understanding of the functioning of the protuberances of the first abdominal segment, a knowledge of their sense organs is essential. Because their surfaces make contact between the body wall and the case, search should be made for mechanoreceptors which provide information on presence or absence of contact. In our studies, particular attention was given to stretch receptors, scolopidial organs or specific cuticular sensilla.

A systematic survey of microtome sections of the four species did not reveal any scolopidial organs or stretch receptors in the protuberances. Information on contact of the first abdominal segment with the case is likely to be delivered by cuticular sensilla. The specific role is played by tactile hairs which, independently of the convexity of the protuberances, are variously stimulated. Each of the four species has on the first segment of the abdomen a specific pattern of tactile hairs which in *Anabolia furcata* was closely examined.

The lateral protuberances of *Anabolia* in the proximal area are provided with dorsal and ventral groups of tactile hairs. In different individuals the number of these cuticular sensilla in both groups may vary. However, their localization and arrangement are more consistent. In addition to the two groups of tactile hairs there is a single seta on the ventral side below the apex of each lateral protuberance. The course of the axons of the sense cells can be studied in thick sections stained with rongalit-white. Axons of both hair groups join in two thin sensory nerves which, in the dorsal and ventral areas of the protuberances run in the body in a proximal direction. The axon of the single seta joins with the nerve which runs down from the dorsal hair group. The dorsal protuberance of *Anabolia* does not have tactile hairs, but there are two bilaterally symmetrical groups of tactile hairs at its base.

Special search was made for campaniform sensilla. In *Halesus*, two big sensilla of this type were found, in the posterior of each lateral protuberance, close to each other. They are clearly seen under the stereo-microscope, and one is shown in histological section in figure 3.

DISCUSSION

These observations on the topography of the musculature and the sensory innervation of the first abdominal segment protuberances should be considered in terms of homology and function. The paper of Hasenfuss (1973) is of special significance for the comparison of the musculature since it provides a detailed description of the muscles in *Rhyacophila* which has no protuberances. The homologies of these muscles are given belows.

| | | | | | | | |
|---|---|---|---|---|---|---|---|
| *Anabolia furcata* | A | B | C | K | L | M | N |
| *Rhyacophila nubila* (Hasenfuss 1973) | M7 | M8 | M29 | M28 | M30 | ? | ? |

Muscles A and B of the dorsal protuberance correspond to medial, short, oblique muscles of the outer dorsal longitudinal musculature, muscle C to outer, front lateral M29 muscle in *Rhyacophila* . Four muscles of the lateral protuberances belong to the group of lateral dorsoventral muscles. Muscle K is homologous to the outer anterior lateral M28, muscle L to the outer posterior lateral M30. Homologization of muscles M and N has not been possible. The function of the protuberances must be related to the activity of other muscles of the body and to changing haemolymph pressure.

Information on the position of the lateral protuberances and their contact with the wall of the case may be transmitted to the larva by the tactile hairs and campaniform sensilla. The large campaniform sensilla of *Halesus* would be suitable for a study of their biological function and electrophysiological investigation. Campaniform sensilla on the dorsal protuberance in *Halesus* and other species have still to be discovered.

The hypothesis put forward by Tomaszewki (1981) for the control of building activity could be confirmed by a physiological explanation of the function of the receptors.

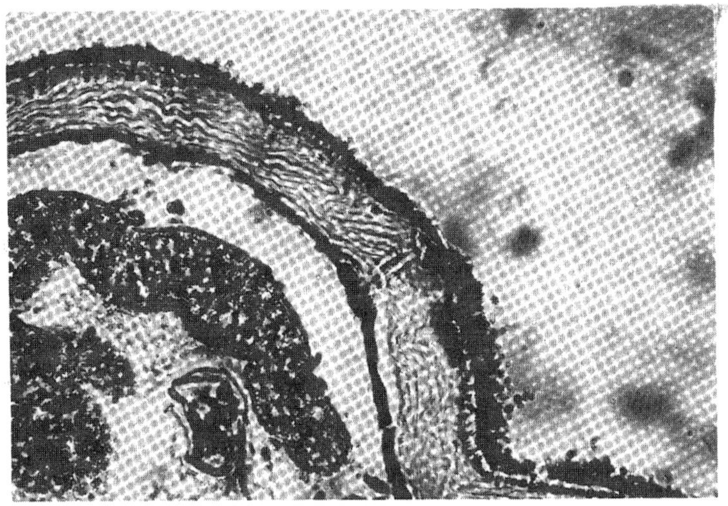

Figure 3. Section of the lateral protuberance of *Halesus* sp. with a campaniform sensillum. Enlargement x 208

AKNOWLEDGEMENT

The authors thank Ms Gerta Klare, Dipl. Biol., Jena, for her valuable help in preparating the material.

REFERENCES

Hasenfuss, I. 1973. Vergleichend-morphologische Untersuchung der sensorischen Innervierung der Rumpfwand der Larven von *Rhyacophila nubila* Zett. (Trichoptera). **Zool. Jb. Anat.**,90 : 1-54.

Rowlands, M.L.J. & Hansel, M.H. 1984. Abdominal Contact as a Case Building Control in a Limnephilid Larva. **Proc. 4th Intern. Symp. Trichoptera,** J.C. Morse (ed.), Junk, The Hague : 321-323.

Tomaszewski, C. 1981. The Principles of Case Building Behaviour in Trichoptera. **Proc. 3rd Intern. Symp. Trichoptera,** G.P. Moretti (ed.), Junk, The Hague : 365-373.

Witting, G. 1955. Untersuchungen am Thorax von *Perla abdominalis* Burm. (Larve und Imago) unter besonderer Berücksichtigung des peripheren Nervensystems und des Sinnesorgane. **Zool. Jb. Anat.**, 74 : 491-570.

# 2. TAXONOMY

# LES LARVES D'*HYDROPSYCHE* DU MAROC (TRICHOPTERES, HYDROPSYCHIDAE)

M. DAKKI and H. TACHET

SUMMARY

In Morocco, six species of *Hydropsyche* have been identified (*H. fezana, H. lobata, H. maroccana, H. pellucidula, H. obscura* and *H. resmineda* ). The larvae of these six species are described and compared with similar European species.

## INTRODUCTION

Six espèces d'*Hydropsyche* ont été récoltées au Maroc jusqu'à présent (Dakki 1978, 1986a) (*H. fezana, H. lobata, H. maroccana, H. pellucidula, H. obscura* and *H. resmineda* ) [1]. Ces six espèces sont connues à l'état larvaire. Le présent travail donne la description des larves de ces six espèces. Les données concernant leur biogéographie et leur écologie sont présentées dans Dakki 1982, 1986 a et b.

## METHODES

Pour ces descriptions nous avons utilisé les caractères qui ont été retenus dans un précédent travail (Bournaud *et al.* 1982). Compte tenu des affinités existantes entre la faune marocaine et la faune européenne (Dakki 1986b) nous avons comparé les espèces marocaines avec les larves d'espèces européennes voisines.

## DESCRIPTION DES ESPECES

Nous diviserons les espèces en deux groupes en considérant la structure du bord antérieur de l'apotome frontoclypéal.

**Espèces à bord antérieur de l'apotome frontoclypéal droit.**

- Espèces dépourvues de branchies sur le 7° segment abdominal.

*H. obscura* Navas 1935 (= *H. atlanta* ). (Fig. 1)

Cette larve a été identifiée par J. Giudicelli (com. pers.). L'apotome frontoclypéal est aussi large dans sa partie médiane qu'au bord antérieur. La coloration de la face dorsale de la capsule céphalique est sombre et rappelle celle que l'on observe chez l'espèce européenne *H. instabilis* Curtis. Le submentum présente des lobes latéraux courts et larges. Les prosternites postérieurs sont sub-carrés et pigmentés.

*H. fezana* Navas 1935.

L'apotome frontoclypéal est plus large au milieu qu'au bord antérieur. La coloration de la capsule céphalique est du même type que celle de l'espèce européenne *H. siltalai* Döhler. Le submentum présente des lobes latéraux courts et larges. Les prosternites postérieurs sont sub-carrés et pigmentés.

- Espèces avec des branchies sur le 7° segment abdominal.

*H. lobata* McLachlan 1884. (Fig. 2)

La larve a été décrite par Garcia de Jalón (1981). L'apotome frontoclypéal est aussi large au milieu qu'au bord antérieur. La zone située en arrière des prétentorinas est de teinte sombre formant un "v". Le submentum présente des lobes latéraux relativement allongés. Les prosternites postérieurs sont subcarrés et pigmentés. De larges bandes sombres médio-dorsales sur le pro- et le mésonotum.

(1) - *H. punica* Malicky 1981 n'a pas encore été trouvée au Maroc.

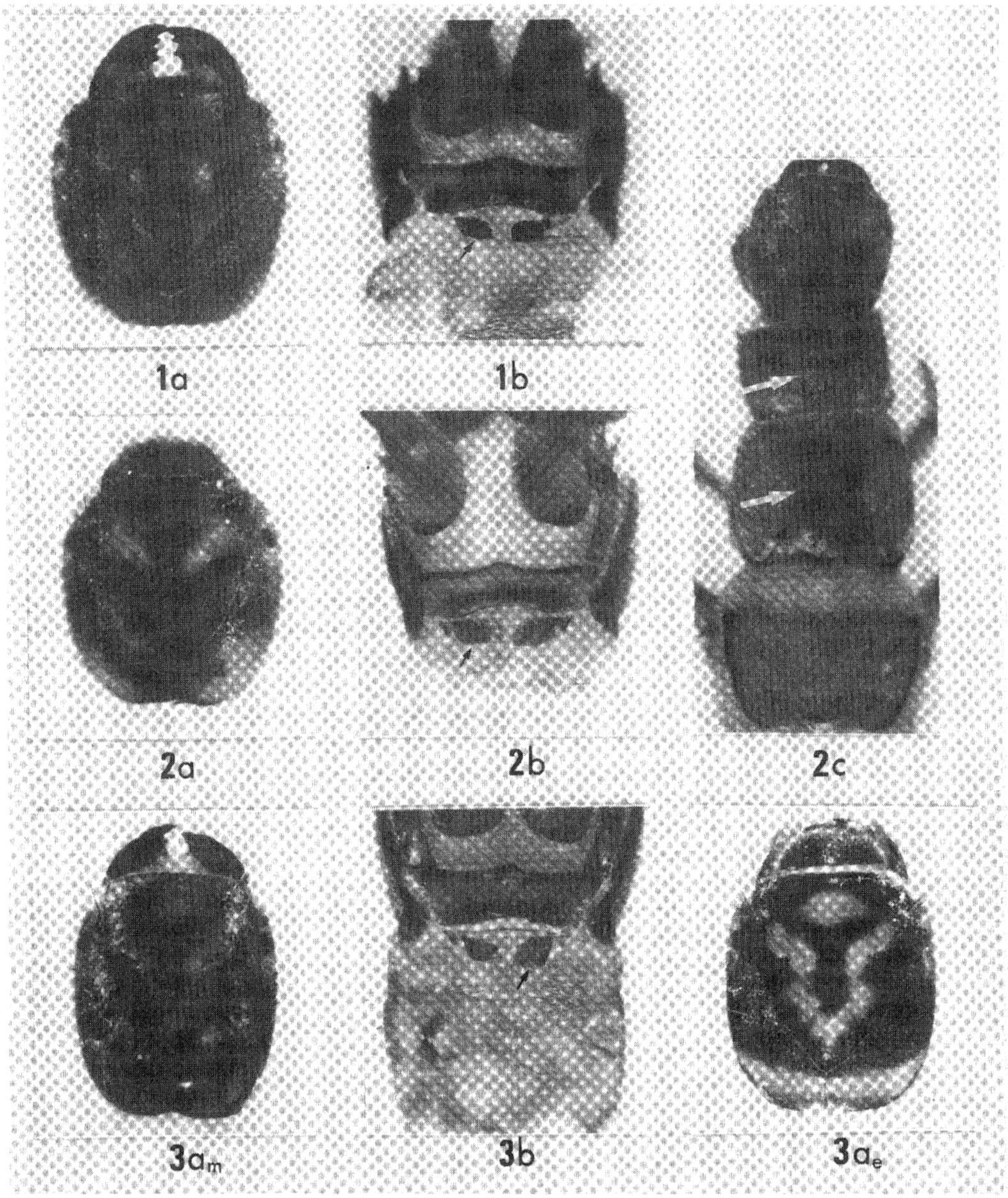


Figure 1 : *H. obscura*. a : capsule céphalique en vue dorsale, b : prosternites postérieurs.

Figure 2 : *H. lobata*. a : capsule céphalique en vue dorsale, b : prosternites postérieurs, c : capsule céphalique et thorax en vue dorsale.

Figure 3. *H. pellucidula*. $a_m$ : capsule céphalique en vue dorsale, espèce marocaine, $a_e$ : capsule céphalique en vue dorsale, espèce européenne, b : prosternites postérieurs, espèce marocaine.

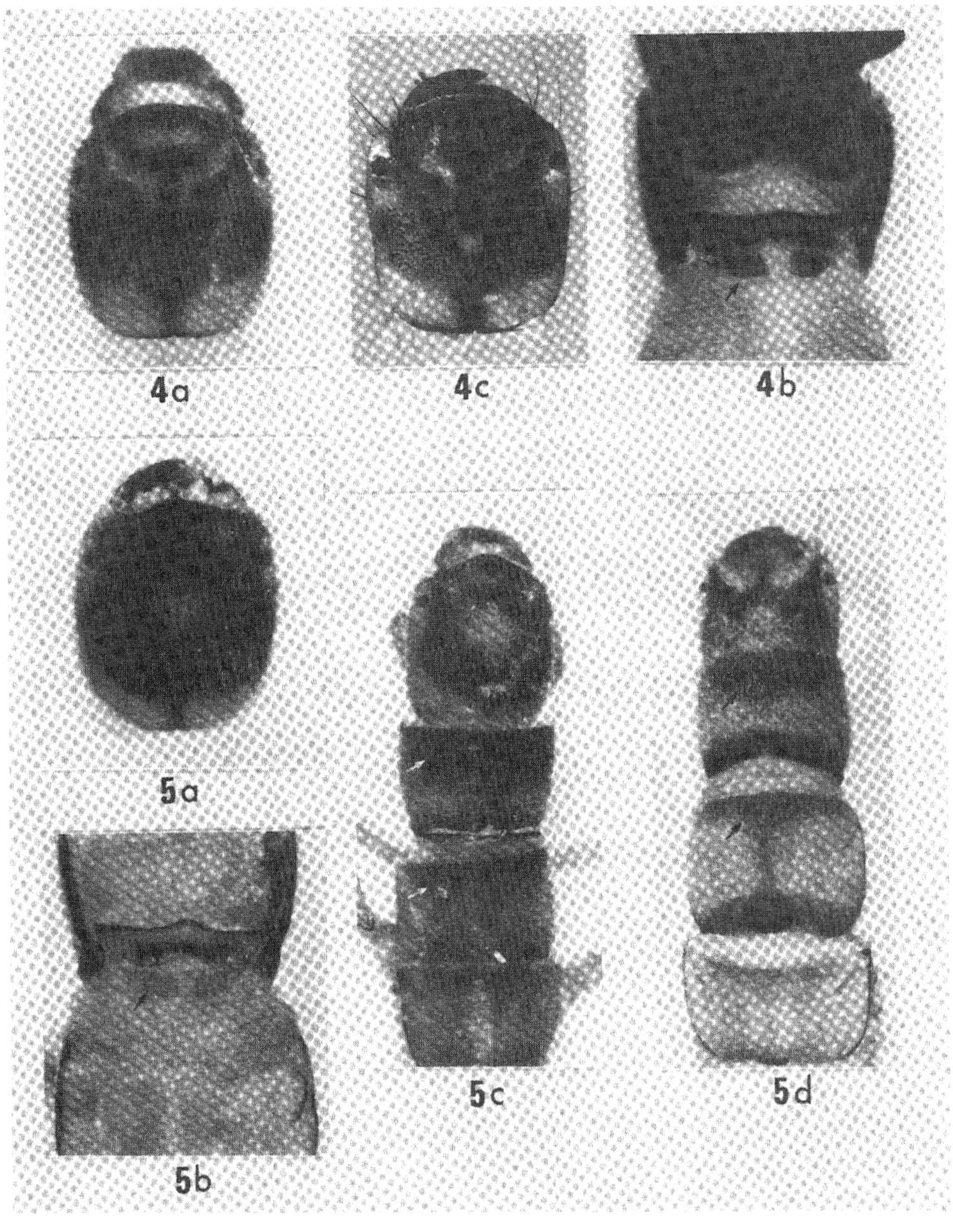


Figure 4. : *H. maroccana.* a : capsule céphalique en vue dorsale, b : prosternites postérieurs, c : *H. exocellata* : capsule céphalique en vue dorsale.

Figure 5. : *H. resmineda.* a : capsule céphalique en vue dorsale, b : prosternites posterieurs, c : capsule céphalique et thorax en vue dorsale, d : *H. contubernalis* : capsule céphalique et thorax en vue dorsale.

*H. pellucidula* Curtis 1834. (Fig. 3)
Les larves marocaines sont tout à fait semblables aux larves européennes dont elles différent cependant par une pigmentation générale de la capsule céphalique plus sombre. Le submentum comme chez les larves européennes a des lobes latéraux relativement étroits et allongés. Les prosternites postérieurs sont subcarrés et pigmentés. Le problème soulevé par la découverte dans de nombreuses régions d'Europe de l'espèce voisine *H. dinarica* Marinkovic 1979 est trop récent pour qu'il soit possible d'affirmer que cette espèce n'existe pas au Maroc, mais tous les exemplaires que nous avons pu observer étaient bien des *H. pellucidula*.

**Espèces à bord antérieur de l'apotome frontoclypéal convexe**

Il existe deux espèces de ce type au Maroc.

*H. maroccana* Malicky 1977. (Fig. 4)
La coloration de la capsule céphalique est proche de celle de l'espèce européenne *H. exocellata* Dufour mais la distinction entre parties claires et parties sombres est moins nette. Les prosternites postérieurs sont subcarrés et pigmentés. Les lobes latéraux du submentum sont relativement allongés.

*H. resmineda* Malicky 1977. (Fig. 5)
Le bord antérieur de l'apotome frontoclypeal est fortement convexe. Il y a toujours une zone claire plus ou moins développée de forme losangique dans la région centrale de l'apotome. Les prosternites postérieurs sont allongés transversalement et translucides comme chez les espèces européennes *H. ornatula* et *H. contubernalis*. Les lobes latéraux du submentum sont relativement allongés. Le bord antérieur du pro- et du mésonotum présente une bande sombre comme chez l'espèce européenne *H. contubernalis*.

## CONCLUSION

Les similitudes qui ont pu être relevées pour les larves entre espèces européennes et marocaines n'ont pas toujours leur équivalent au niveau des adultes. C'est ainsi que si *H. maroccana* et *H. exocellata* présentent d'incontestables ressemblances à l'état de larve et à l'état d'adulte de même que *H. fezana* et *H. siltalai* impliquant ainsi une étroite parenté, il n'en est pas toujours de même. En effet, *H. resmineda* dont les larves par certains caractères se rapprochent d'*H. contubernalis* est, à l'état adulte, beaucoup plus proche d'*H. exocellata* et d'*H. maroccana*. Il est donc essentiel, si d'éventuelles relations de filiation doivent être recherchées, de prendre en compte à la fois larves et adultes.

## REFERENCES


Bournaud, M. Tachet, H & Perrin, J.F. 1982. Les hydropsychidae (Trichoptera) du Haut-Rhône entre Genève et Lyon. **Annls. Limnol.**, 18 : 61-80.

Dakki, M. 1972. Trichoptères du Maroc. **Bull. Int. Sci. Rabat.** 6 : 139-155.

Dakki, M. 1978. Le genre *Hydropsyche* au Maroc (Trichoptera, Hydropsychidae). **Bull. Inst. Sci. Rabat.**, 3 : 111-120.

Dakki, M. 1986a. Biotypologie et gradient thermique spatio-temporels, étude sur un cours d'eau du Moyen Atlas (Maroc). **Bull. Ecol.**, 17, (sous presse).

Dakki, M. 1986b. Recherches hydrobiologiques sur le Haut Sebou (Moyen Atlas). Thèse Doctorat, Rabat, Maroc 214 p.

Garcia de Jalón, D. 1981. Description of *Hydropsyche* larvae found in the Iberian peninsula. **Proc. 3rd. Internatn. Symp. Trichoptera,** G.P. Moretti (ed.), Junk, The Hague : 87-92.

# STUDIES OF NEOTROPICAL CADDISFLIES, XXXVI : THE GENUS *CALOSOPSYCHE* IN CENTRAL AMERICA, WITH DESCRIPTIONS OF ITS IMMATURE STAGES (TRICHOPTERA : HYDROPSYCHIDAE)

O. S. FLINT, Jr. and J. BUENO SORIA


## SUMMARY

The genus *Calosopsyche* was erected by Ross & Unzicker for 5 Antillean species on the basis of the structure of their male and female genitalia. We describe the adults, larvae and pupae of *C. continentalis* (Panama, Costa Rica), and adults only of *C. ardisia* (Costa Rica), *C. bicuspis* (Costa Rica) and *C. elachista* (Panama, Costa Rica). In addition *Hydropsyche dearmasi* Boto. and *Plectropsyche sandrae* Flint are transfered to *Calosopsyche* (new combinations), the male genitalia of *C. carinifera* are illustrated for the first time, and the genus is divided into two species groups.

The genus is easily distinguished by a combination of venational and genitalic characters in the adult stage. The larvae are recognized by the undivided submentum, the single process of the trochantin with a small dorsal knob, the gastric mill with only small spines, and the pelage of the abdomen. The pupae are also recognizable by the structure of the hook plates, especially the narrow, high plates on segments 5-8. We recognize the genus as valid, and probably occupying a rather primitive position in the Hydropsychinae.


## INTRODUCTION

The recognition and definition of the genera of New World Hydropsychinae has become more and more contentious in recent years. An early attack on the problem, and still a controversial one, was that of Ross and Unzicher (1977). In this work, the (mostly) New World units were studied with the result that three new genera, *Mexipsyche*, *Streptopsyche*, and *Calosopsyche* were proposed, and *Hydropsyche* was divided into two parts with *Symphitopsyche* (and its new subgenus, *Ceratopsyche*) recognized for what had been the *bifida* group. Much of the reasoning and results of this paper were roundly criticized by Schmid (1979). Trichopterists are still arguing over whether to recognize this last division of *Hydropsyche*, with Schuster (1984) presenting strong arguments for its acceptance at the generic level, but Schefter *et al.* (1986) relegate it again to the subgeneric level. The fate of the three new Latin American genera has not generated much thought, however, probably because they are not frequently reported in the literature nor are many people working with them. At our last meeting, Bueno (1984) described the immature stages of *Mexipsyche*, but beyond the fact that they show a very close relationship to *Hydropsyche*, was unable to come to any firm conclusion as to possible synonymy.

We here present data on another of the Neotropical genera, *Calosopsyche*. In 1984 Bueno discovered larvae and pupae, including metamorphotypes, and collected adults of an unknown hydropsychid genus and species in Costa Rica. In 1967 Flint had collected adults of several species unplaceable to genus in Costa Rica and Panama. Upon joint study it was apparent that we both had the same genus, and even the same species. Our first impression was that these represented an undescribed genus, but study of the Ross and Unzicher paper revealed close similarities to the Antillean genus *Calosopsyche*. Further study has convinced us that this is the correct generic assignment, and that the genus is valid.

### Genus *Calosopsyche* Ross and Unzicker

*Calosopsyche* Ross and Unzicker, 1977 : 309.

Type species : *Hydropsyche calosa* Banks, original designation.

**Adult.** Medium sized to large (forewing 6 to 14 mm), brown caddisflies with rather stout bodies. Antennae generally as long as wings, rarely (*carinifera*) distinctly longer. Tibial spurs 2,4, 4 ; male with apical segment of anterior tarsus asymetrical, with a lateral tuft of dark hair situated on a small protuberance. Maxillary palpus with 5 segments : basal short, second long, third slightly shorter than second, fourth slightly shorter than third, fifth not as long as basal four combined. Wings essentially as in *Hydropsyche* (figure 1), with following notable points : forewing with apex of Cu2 and 1A united before wing margin, hindwing with apex of Sc and R1 united and continuing as a single vein to near margin, lacking crossvein m, with basal portions of M and Cu1 well separated. Male with a small raised lobe on fifth abdominal sternum into which a small, saclike gland empties. Male genitalia with tenth tergum short and rather high, with three setiferous warts arranged from top to bottom near midlength, beyond which extends a thin, lightly sclerotized plate. Clasper with apical segment as long as, or longer than, basal segment, often with apex divided. Phallus with base large, bulbous, and constricted above base ; apex with a pair of lateral, platelike lobes between whose bases are phallotremal sclerites. Female genitalia with clasper receptacle on ninth segment generally very large and cavernous, placed anteroventrally on the segment.

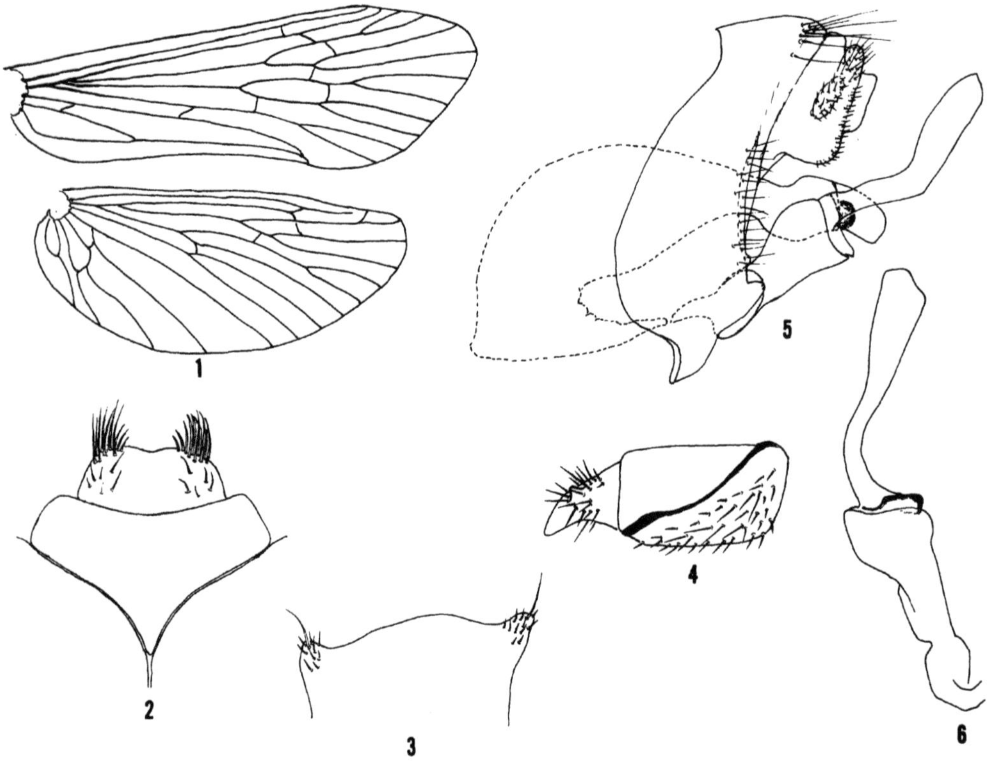


Figures 1-6. 1, *C. calosa*, wings. 2-4, *C. continentalis*, larva : 2, submentum and ventral apotome ; 3, anterior margin of frontoclypeus ; 4, foretrochantin. 5-6, *C. carinifera*, male genitalia : 5, lateral ; 6, clasper, posteroventral.

**Larva,** Labrum simple, with anterolateral brushes, dorsal surface densely hairy with most setae much enlarged. Mandibles (Fig.9) rather broadly trianguloid, exterior face with narrow dorsal and ventral ridges, between which are numerous short, broad setae ; left mandible with a setal brush on dorsomesal face. Labium with submentum (Fig.2) not deeply cleft. Frontoclypeus (Fig. 3) with anterior margin produced mesally and asymmetric laterally, lateral margins not noticeably constricted at midlength. Trochantin (Fig. 4) a single conical projection with a small dorsal knob. Prosternum (Fig. 8) transverse, with well-developed posterolateral sclerites. Pronotum divided on midline, meso- and metanota entire. Gastric mill with many posteriorly directed small spines ; lacking tooth-plates. Abdomen lacking hairs along lateral line : surface lacking minute spines and scale hairs, covered with decumbent, black pointed setae and scattered, erect, longer black setae. Eighth and ninth sterna (Fig. 11) each with paired sclerites bearing many small, pointed pale setae and an apical fringe, of long, black enlarged setae, ninth segment in addition with small dorso- and ventrolateral sclerites with some specialized setae. Abdominal segments 3-7 each with 1-3 small, membranous knobs along lateral line. Gills present on meso- and metasterna, and abdominal sterna 1-6, all gills densely branched ; 1 pair of gills on mesosternum and first abdominal sternum, 2 pairs on metasternum and abdominal sterna 2-7, lateralmost pair of gills often with an indication of a subdivision in half. Anal gills present. Anal prolegs short, with many short, dark, enlarged setae ; brush well-developed ; claw without ventral teeth.

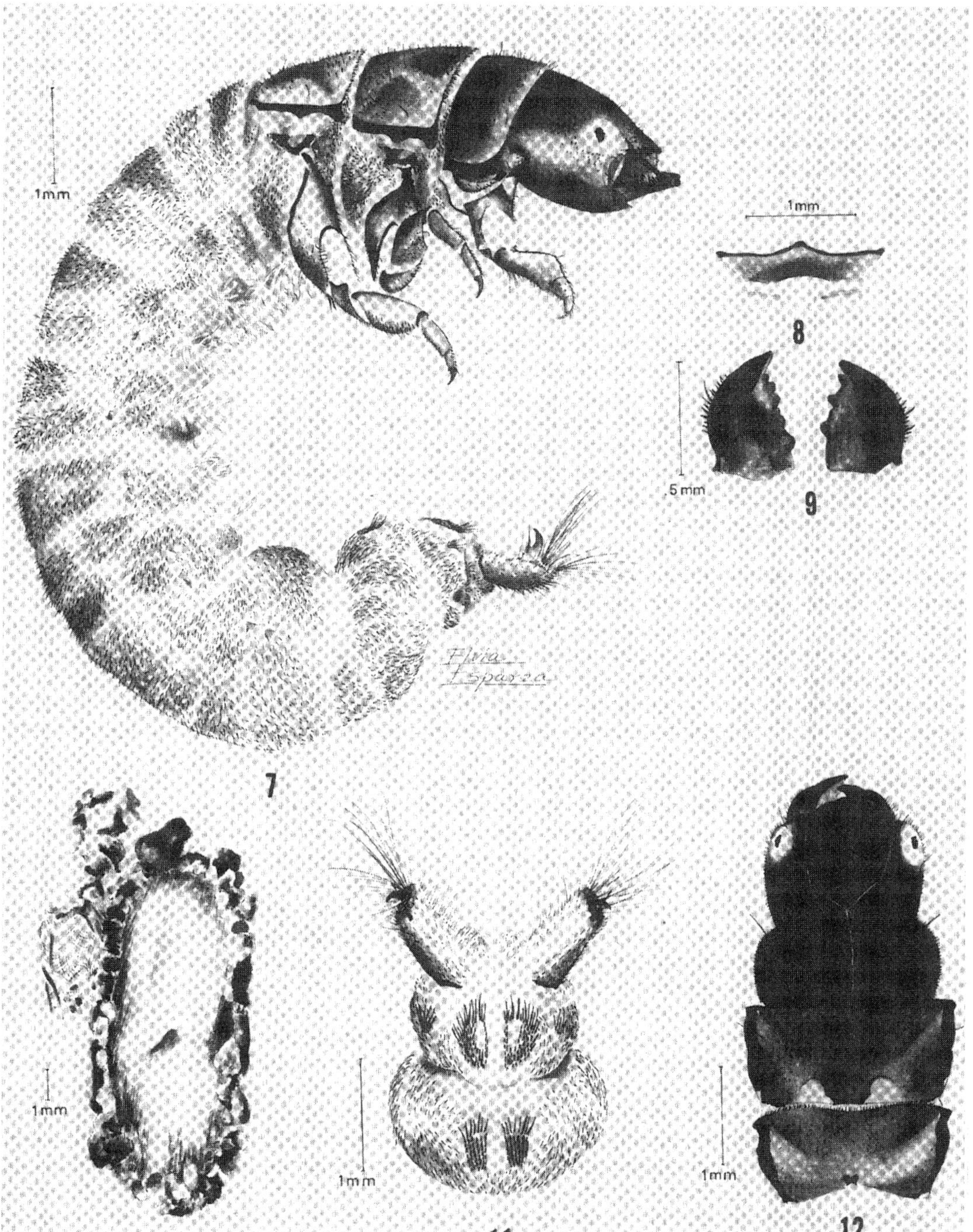


**Figures 7-12.** *C. continentalis*, larva : 7, lateral view ; 8, prosternal sclerites ; 9, mandibles; 1O, pupal shelter, ventral ; 11, apex of abdomen, ventral ; 12, head and thorax, dorsal.

**Pupa.** Labrum with rounded basolateral lobes, each bearing a dense cluster of setae, mesal region with many dark setae. Mandibles rather straight, with 4 or 5 strong teeth over apical half of inner margin. Head with many setae, especially anteriad. Antennae surpassing tip of abdomen, with apices curved up over abdomen. Tibia and tarsus of midleg broad, and tarsus bearing a fringe on outer side basally. Meso- and metanota mesally bearing many dark setae as do coxae and trochanters of all legs. Abdominal terga with few, short setae, except for strong fringes of long setae on segments 4 and 5 and progressively sparser fringes on 6-8, venter with many dark setae. Abdomen with broad single or double lateral processes on middle segments, ventrally with many gills. Hook plates anteriorly on segments 2-8 and posteriorly on 3 and 4 ; anterior plates on 2-4 very broad, with a single row of hooks, plates on 5-8 very narrow, with a small cluster of hooks on a raised mesal portion of plate, posterior plate of 3 broadly transverse with 2 rows of hooks, that of 4 much narrower with single row of hooks. Apical processes hairy, upturned with a small apicolateral point and a larger apicomesal point.

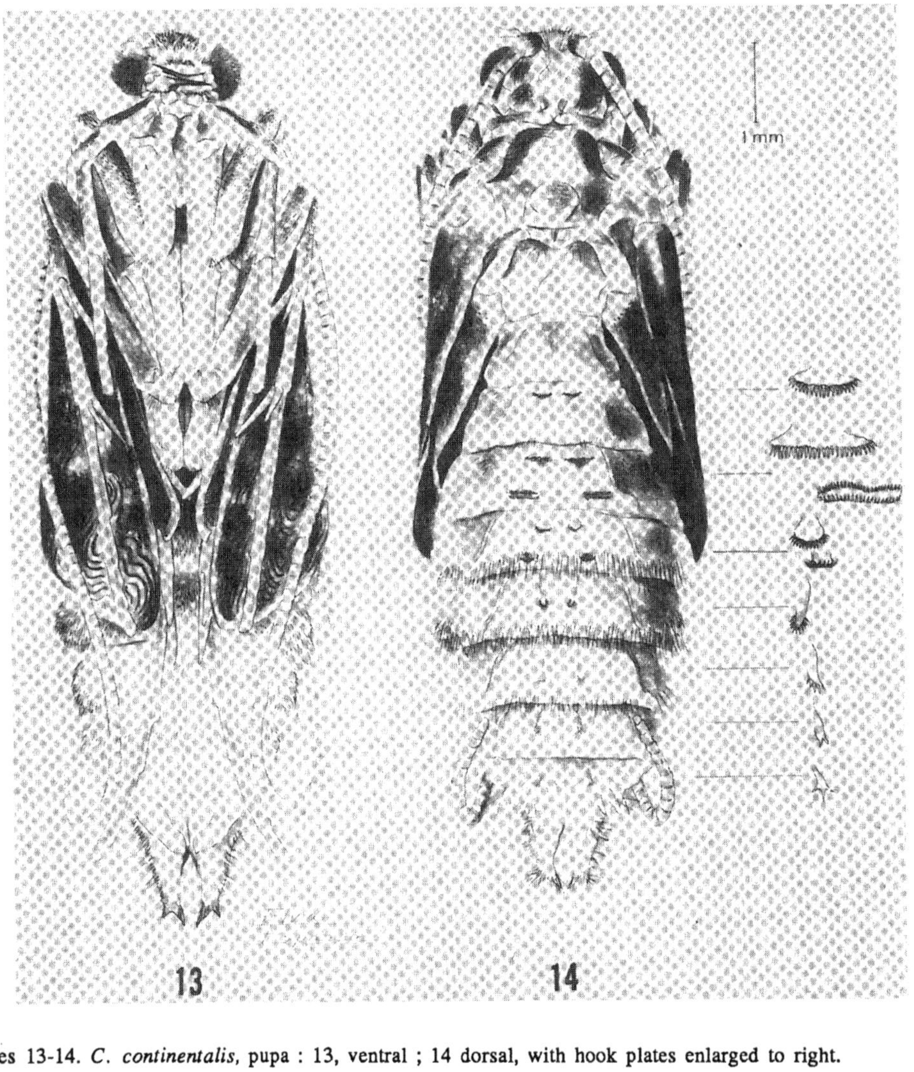


Figures 13-14. *C. continentalis*, pupa : 13, ventral ; 14 dorsal, with hook plates enlarged to right.

**Discussion.** The above characterizations are based on adults of both the Antillean and continental species. However, we have only the immature stages of a single continental species, the insular ones being totally unknown. We hope that discovery of these stages in the insular species will not necessitate a reassesment of the conclusions here presented.

The adults do offer a distinctive combination of venational characteristics. Basically the venation agrees with *Hydropsyche* (when we use this name we are referring to the restricted usage which excludes the *Symphitopsyche-Ceratopsyche* lineage) in that the apices of Cu2 and 1A are united before the wing margin in the forewing. It differs from both in lacking crossvein m and having the basal sections of M and Cu1 well separated in the hindwing. The glandular apparatus of the fifth sternum appears to be like that of *Hydropsyche*. The structure of the male genitalia is also very distinct, especially that of the tenth tergum and phallus. In respect to the phallus the genus would seem more closely allied to *Cheumatopsyche, Potamyia* and *Plectropsyche* than the *Hydropsyche* genera. The tenth tergal structure is quite unique although it suggests *Plectropsyche* as does the venation.

The larvae offer a distinctive collection of characters, although they readily key to the Hydropsychinae. The undivided submentum is otherwise only found in *Smicridea* in this subfamily as is the shape of the anterior margin of the frontoclypeus. The structure of the trochantin is like that of some examples of *Potamyia,* and quite unlike that of any other genera in the subfamily. The structure of the prosternum is typical of *Hydropsyche* or *Ceratopsyche.* The gastric mill is quite unlike most Hydropsychinae which bear tooth-plates, but more like the Arctopsychinae, Diplectroninae, or *Leptonema albovirens* (Boon 1985). The pelage of the abdomen is distinctive, although the sclerites of the terminal sterna are typical of most genera but different from that of *Smicridea.* As a consequence of this assortment of characters it does not run easily in published keys.

The pupae appear to be rather characteristic of the *Hydropsyche-Ceratopsyche* lineage in most characters, especially the mandibles and apical appendages. However, the positioning of the hook plates agrees with *Ceratopsyche,* but their form appears to be unique, especially the narrow, high plates of the posterior segments.

Ecologically the genus appears to be rather restricted. McElravy *et al.* (1981,1982) report the environmental parameters in considerable detail at the Fortuna dam site in northern Panama. Based on appearances at other sites, we believe the ecological characteristics at this site are typical, to a greater or lesser degree, of all known sites where examples have been taken. Briefly these are a mountainous region of high rainfall, with little seasonal variation in rainfall or temperature. These sites were once heavily forested, though now these forests are much reduced.

Based on this unique assemblage of characters which render adults, larvae and pupae easily recognized, we accept *Calosopsyche* as a distinctive monophyletic unit meriting generic status. We further recognize two groups of species in the genus, record it for the first time from Central Àmerica from where we describe 4 new species, one with its larva and pupa.

The *calosa* group

This group contains all the Antillean species. They are distinguished from the continentalis group by the shape of the apical clasper segment. In all the species for which the male is known this segment is unmodifed apically, it is either rounded or truncate (Fig. 6), not bifid or bearing a lobe on one side.

The following species are included in this group :

*Calosopsyche calosa* (Banks) 1938 : 300. -Flint, 1962 : 23 (♂ genitalia). -Ross & Unzicker 1977 : 309 (♀ genitalia, to *Calosopsyche*). Cuba.

*Calosopsyche cubana* (Flint) 1962 : 24 (♀ genitalia). -Botosaneanu 1977 : 280 (♂ genitalia). -Ross & Unzicker 1977 : 309 (to *Calosopsyche)*. Cuba.

*Calosopsyche darlingtoni* (Flint) 1962 : 23-24 (♀ genitalia). -Ross & Unzicker 1977 : 309 (to *Calosopsyche* ). Cuba.

*Calosopsyche dearmasi* (Botosaneanu) 1980 : 105-106 ( ♂ genitalia). -NEW COMBINATION. Cuba.

*Calosopsyche batesi* (Flint) 1962 : 25-26 ( ♀ genitalia) . -Ross & Unzicker 1977 : 309 (to *Calosopsyche).* Haiti.

*Calosopsyche carinifera* (Flint) 1962 : 27 (♀ genitalia). -Ross & Unzicker 1977 : 309 (to *Calosopsyche).* -MALE GENITALIA HEREIN. Dominican Republic.

The *continentalis* group

This group is limited to the countries of Costa Rica and Panama in Central America. In all species the apical segment of the clasper is modified, usually by being bifid, or bearing a small lobe to one side.

We include the following species :

*Calosopsyche sandrae* (Flint) 1967 : 15 (♂genitalia). - NEW COMBINATION. Costa Rica.

*Calosopsyche continentalis* Flint & Bueno, herein (♂♀genitalia, larva, pupa). Costa Rica, Panama.

*Calosopsyche ardisia* Flint & Bueno, herein (♂ , ♀ genitalia). Costa Rica.

*Calosopsyche bicuspis* Flint & Bueno, herein (♂ , ♀ genitalia). Costa Rica.
*Calosopsyche elachista* Flint & Bueno, herein (♂ , ♀ genitalia). Costa Rica, Panama.

DESCRIPTIONS OF THE SPECIES

*Calosopsyche carinifera* (Flint)
Figures 5-6

A series of 17 males and 14 females from Dominican Republic, Convento, 12 km south of Constanza, 6-13 June 1969, Flint & Gomez, permits us to figure the male of this species for the first time.

*Calosopsyche continentalis* Flint & Bueno, new species
Figures 2-4, 7-16, 25.

Genus B, undescribed species A. -McElravy *et al* ., 1981 : 153.
Genus A, undescribed species A. -McElravy *et al* ., 1982 : 307.

This species is related to *bicuspis* and *sandrae* on the basis of the deeply divided apex of the clasper. From *bicuspis* it may be distinguished the apical segment of clasper which is asymmetrically forked and the less inflated base of the phallus. In *sandrae* the outer fork at the apex of the phallus is very small and the inner fork is strongly produced. The females of *sandrae* are unknown, but the females of *bicuspis* are easily distinguished by the shape of the clasper receptacle although both have a well developed, second pocket dorsally. In *continentalis* the receptacle is very widely open to the posterior with hardly any dorsal flange.

**Adult.** Length of forewing, ♂ 9-10 mm, ♀ 10.5-13 mm. Color brown, appendages and venter of body, yellow brown ; forewings brown, densely irrorate with flecks of yellow-brown hair. Male genitalia : Tenth tergum with wart *a* erect and mostly free, wart *b* elongate and distinct, wart *c* elongate and distinct, lateral plate slightly bilobed with ventral lobe largest. Clasper with apical segment in posteroventral aspect distinctly angled, apex deeply divided with lateral ramus much narrower than mesal. Phallus with base enlarged and constricted ; apical lobes longer than high. Female genitalia : Clasper receptacle deep, rounded, dorsally with a second cuplike receptacle.

**Larva.** Up to 15 mm long, by 2.5.mm wide. Head dark reddish-brown , pale around eyes and posteriad along lateral margin, surface with indistinctly paler muscle scars. Surface of genae laterally and anteriorly with many small, pale, bladelike setae which are longest around eyes and near anterior margin ; ventral stridulatory grooves very numerous and closely spaced, not deeply impressed. Anterior leg reddish brown, mid- and hindlegs yellow brown. Pronotum reddish-brown, meso- and metanota yellowish-brown with some darker areas; all with black lateral margins, pronotum with posterior margin black and meso- and metanota with posteromesal black marks. All nota densely covered with shorter, pale and longer, dark, bladelike setae. Sclerites of eighth and ninth abdominal sterna yellowish-brown, with many short, pale, bladelike setae and posterior margins with longer, dark setae. Anal prolegs yellowish-brown, black basally, with many enlarged, black setae.

**Pupa.** Hook plate 2A narrow, with a raised, transverse row of 15 hooks across center ; 3A very broad and produced posteriad, with a raised, transverse row of 25 hooks across it at anterior third ; 4A narrower, produced anteriad and posteriad, with a raised, transverse row of 15 hooks across center ; 5A very narrow, elongate, with a raised central portion bearing 6 hooks ; 6A very narrow, elongate, with a very high central portion bearing 5 hooks ; 7A very narrow, elongate with a very high central portion bearing 5 hooks ; 8A very narrow, elongate with a raised central portion bearing 3 hooks ; 3P broadly ovoid with a row of 40 hooks along posterior margin and row of 35 hooks posteriad of anterior margin ; 4P a third as wide as 3P, produced anteriad, with a row of 8 hooks along posterior margin. Apical processes in posterior view with lateral point half length and third width of mesal point, surface minutely spiculate.

**Material.** Holotype, male : Panama, Pcia. Cocle, El Valle, 15 July 1967, O.S. Flint, Jr. Holotype in NMNH. Paratypes : Same data, 9♂ , 5♀ . Pcia. Panama, Cerro Campana, 11-14 July 1967, O.S. Flint, Jr., 2♂, 1♀. Pcia. Chiriqui, Fortuna dam site (8° 44'N, 82° 16'W), 1050m, November 1976-December 1977, H.Wolda, 37♂♂ and♀♀ (O.T.U. # 46). Costa Rica, Pcia. Cartago, km10 Turrialba-Siquirres Road, 24 April 1984, J.Bueno *et al.* 5♂, 10♀. Pcia. Alajuela, Rio Pizote, ca. 5km (air) S Brasilia (10.972 N, 85.345 W), 390m, 12 Mar 1986, Holzenthal & Fasth, 2♂ , 26♀ ; Reserva Bosque Nubosa Monteverde (10.30N, 84.74 W), 950m, 1 Mar 1986, Holzenthal & Fasth, 1♂ . Pcia Guanacaste, Rio Negro, P.N. Rincon de la Vieja (10.765 N, 85.313 W), 810m, 3 Mar 1986, Holzenthal & Fasth, 1♂ ; 1♀ . Pcia Puntarenas, Rio Guineal, 1km(air) E Finca Helechales (9.076 N, 83.092 W), 840m, Holzenthal, Morse & Fasth, 8♂, 4♀ . Pcia. San

Jose, P.N. Braulio Carrillo, 6.2 km NE admin. build. (10.09 N, 83.97 W), 1100m, 6 Feb 1986, Holzenthal & Morse, 1 ♀. Other : Pcia. Cartago, km10 Turrialba-Siquirres Road, 24 April 1984, J.Bueno *et al.* 9 larvae, 3 pupae, 1♂, 3 ♀ metamorphotypes. Material in CUC, IBUNAM, U CB and NMNH.

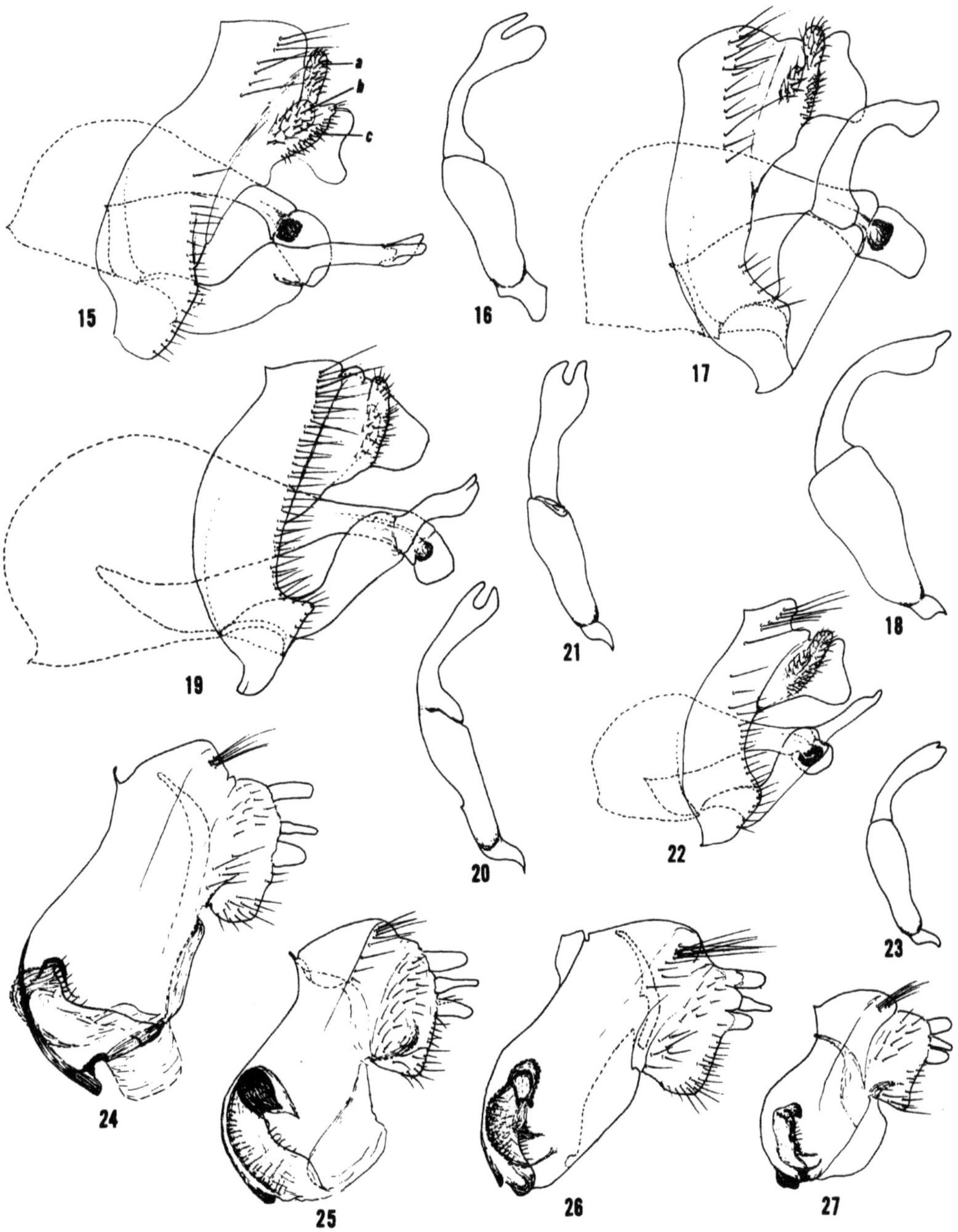


Figures 15-27. Male genitalia, lateral, and clasper, posteroventral. 15-16, *C. continentalis* ; 17-18, *C. ardisia* ; 19-21, *C. bicuspis* (21, clasper of paratype from Monteverde) ; 22-23, *C. elachista*. Female genitalia, lateral. 24 *C. ardisia* ; 25, *C. continentalis* ; 26, *C. bicuspis* ; 27, *C. elachista*.

**Biology.** The immature stages were collected at an elevation of 550 meters in a steeply descending stream about 6 meters wide with a rather deep and fast flow. They were in a fast flowing, tumbling section and found on a large boulder wholly under water. They built their cases on the boulder with the opening facing into the current and the net to one side of the opening. A colony of 25-30 individuals was located on one big stone, with 30 % of it in the pupal stage including a few metamorphotypes. The water appeared to be of good physical and chemical quality.

*Calosopsyche ardisia* Flint & Bueno, new species
Figures 17-18, 24

This species is the most distinctive continental species, in that the apical clasper segment is strongly curved, and undivided, but bears a process from one side of the apex. The female clasper receptacle is very distinctive, being oblique, very large and complex in structure.

**Adult.** Length of forewing, ♂ 11mm, ♀ 13-14.5mm. Color light brown ; forewing almost completely covered with pale, yellow brown hair. Male genitalia : Tenth tergum with wart *a* free for upper third, wart *b* elongate, distinct, wart *c* elongate, very narrow : lateral plate very slightly bilobed, with ventral lobe produced. Clasper with apical segment in posteroventral aspect distinctly angled, apex produced into a small fingerlike lobe on external margin. Phallus with base enlarged and constricted ; apical lobes longer than high. Female genitalia : Clasper receptacle set obliquely, deep, anterior margin oblique with another oblique internal shelf, with anterodorsal angle cuplike.

**Material.** Holotype, male : Costa Rica, (Pcia. Alajuela), Turricares (sic., Turrucares). Holotype in MCZ. Paratypes : same data, 1 ♀ ; Rio La Paz, route 9, 7.6km N Vara Blanca (10.208 N, 84.166 W), 1340m, 13 Feb 1986, J.C. Morse, 1 ♀ . [Pcia. Cartago] Orosi, Garlepp Coll., 1 ♀ ; Turrialba, 5000 ft, 31 Aug, Wm. Schaus Coll., 1 ♀ . Pcia. Puntarenas, Rio. Bellavista, ca. 1.5km NW Las Alturas (8.951 N, 82.846 W), 1400m, 18 Feb 1986, Holzenthal, Morse & Fasth, 2 ♀ . [Pcia. unknown], La Trinidad, 1 ♀ .Paratypes in CUC, MCZ and NMNH.

*Calosopsyche bicuspis* Flint & Bueno, new species
Figures 19-21, 26

The male is easily recognized by the apical segment of the clasper whose tip is divided into two nearly equal rami, and the extremely inflated base of the phallus. The female is similar to that of *continentalis*, but distinguished by the narrower entrance to the clasper receptacle which bears a strong dorsal flange.

**Adult.** Length of forewing, ♂ 9-11mm, ♀10.5-13mm. Color brown, appendages and venter of body, yellow brown ; forewings brown, densely irrorate with yellow-brown hair, with several whiter flecks at anal angle. Male genitalia : Tenth tergum with wart *a* free for upper third, wart *b* elongate, indistinct, wart *c* elongate, very narrow ; lateral plate with dorsal lobe lacking, ventral lobe produced. Clasper with apical segment in posteroventral aspect constricted and angled near base (in paratype from Monteverde, this segment is nearly straight and of uniform width throughout), apex produced into a pair of small, almost equal-sized lobes. Phallus with base tremendously enlarged and deeply constricted ; apical lobes about as long as high. Female genitalia : Clasper receptacle deep, rounded, with a dorsal cuplike structure ; with a dorsal flange serving to narrow entrance to receptacle.

**Material.** Holotype, male : Costa Rica, Pcia. Alajuela, Rio La Paz, route 9, 7.6km N Vara Blanca (10.208 N, 84.166 W), 1340m, 13 Feb 1986, J.C.Morse. Holotype in NMNH. Paratypes : Same data, 1♂ ; 1 ♀ ; Rio La Paz Pequena, route 9, 7.8km N Vara Blanca (10.211 N, 84.166 W), 1320 m, 13 Feb 1986, Holzenthal, Morse & Fasth, 2 ♀ .Pcia. San Jose, P.N. Braulio Carrillo, 6.2km NE Admin.build. (10.09 N, 83.97 W), 1100m, 6 Feb 1986, Holzenthal & Morse, 2♀. Pcia. Puntarenas, Monteverde area (10°18'N, 84°49'W), 1400-1700m, 6-14 June 1973, Erwin & Hevel, 1♂. Paratypes in CUC and NMNH.

*Calosopsyche elachista* Flint & Bueno, new species
Figures 22-23, 27

Genus A, undescribed species A. -McElravy *et al.*, 1981 : 153.

Genus B, undescribed species A. -McElravy *et al.*, 1982 : 307, 310.

This easily recognized species has only a slightly divided tip to a barely curved apical clasper segment, the tenth tergal warts mostly fused, and small apical lobes to the phallus. The clasper receptacle of the female is comparatively narrow, vertical, and bears a dorsal caplike structure.

**Adult.** Length of forewing, ♂ 5.5-7mm, ♀ 8-9mm. Color brown, appendages and venter of body, yellow brown ; forewings brown, sparsely flecked with yellow-brown hair. Male genitalia : Tenth tergum with wart *a* free for upper third, warts *b* and *c* elongate, completely fused and joined to venter of *a* ; lateral plate very slightly bilobed, with ventral lobe produced. Clasper with apical segment in posteroventral aspect slightly curved, apex slightly divided into a pair of small lobes. Phallus with base enlarged and constricted ; apical lobe as long as high. Female genitalia : Clasper receptacle vertical, moderately deep, posterior margin vertical, with a distinct dorsal cap.

**Material.** Holotype, male : Panama, [Pcia. Panama], Cerro Campana, 11-14 July 1967, O.S. Flint, Jr . Holotype in NMNH. Paratypes : Same data, 1 ♂, 1 ♀. [Pcia. Cocle], El Valle, 15 July 1967, O.S. Flint, Jr., 3♂. Pcia. Chiriqui, Fortuna dam site, (8°44'N, 82°16'W), 1050 m, November 1976-December 1977, H. Wolda, 963 ♂♂ & ♀♀ (O.T.U. #61). Costa Rica, [Pcia. Cartago], Turrialba, 2-5 November 1967, E.L. Todd, 1♀ . [Pcia. Cartago], 3 mi. west Turrialba, 18-21 June 1967, Flint & Ortiz, 1 ♂ . [Pcia. Alajuela], 14.1 mi. east Esparta, 23 July 1967, O.S. Flint, Jr., 1 ♂ Paratypes IBUNAM, UCB and NMNH.

**Biology.** McElravy *et al.* (1981, 1982) have reported this species in great abundance at the Fortuna dam site. It was, in fact, the most frequently collected species at that site, comprising some 13.3% of the total trichopteran catch. They have provided a good climatic and ecological description of the site, which must be at near optimum for this species.

## ACKNOWLEDGEMENTS

The authors thank Mrs. Elvira Esparza for the drawings of the larva and pupa. We also thank R.W. Holzenthal and J.C.Morse of Clemson University (CUC), E. Barrera C. and M. Cordoba of the Universidad Nacional Autonoma de Mexico (IBUNAM), A.F. Newton, of the Museum of Comparative Zoology, Harvard (MCZ), E.P. McElravy and V.H. Resh of the University of California, Berkeley (UCB), and T.L. Erwin and G.F. Hevel of the National Museum of Natural History (USNM) for collecting or loaning material used in this study. Aid for field work was provided by NSF Grant GB-2616 to the senior author and CONACYT grant No. 14Ø1Ø5 G2Ø3 to the junior author.

## REFERENCES

Banks, N. 1938. New West Indian neuroteroid insects. **Rev. de Entomol.**, 9 :285-3Ø4.

Boon, P.J. 1985. Gastric mill structure and its phylogenetic significance in larval Hydropsychidae (Trichoptera). **Zool. J. Linn. Soc.**, 84 : 301-324.

Botosaneanu, L. 1977. Trichoptères (imagos) de Cuba, capturés par moi-même en 1973 (Insectes, Trichoptera). **Fragmenta Entomolog.**, 13 : 231-284.

Botosaneanu, L. 1980. Trichoptères adultes de Cuba collectés par les zoologistes cubains (Trichoptera). **Mitt. Munch. Entomol. Ges.**, 69: 91-116.

Bueno-Soria, J. 1984. Description of the larva and pupa of Proc. **4th Intern. Symp. Trichoptera** J.C. Morse (ed.), Junk, The Hague : 49-55.

Flint, O.S., Jr. 1962. Antillean species of *Hydropsyche* (Trichoptera). **Bull. Brooklyn Entomol. Soc.**, 57 : 22-27.

Flint, O.S. Jr. 1967. Studies of Neotropical caddis flies, IV : new species from Mexico and Central America. **Proc. U. S. Nat. Mus.**, 123(3608) : 1-24.

McElravy, E.P., Resh, V.H., Wolda, H. & Flint, O.S., Jr. 1981. Diversity of adult Trichoptera in a "non-seasonal" tropical environment. **Proc. 3rd Intern. Symp. Trichoptera**, G.P. Moretti (ed.), Junk, The Hague : 149-156.

McElravy, E.P., Wolda, H., & Resh, V.H. 1982. Seasonality and annual variability of caddis adults (Trichoptera) in a "non-seasonal" tropical environment. **Arch. Hydrobiol.**, 94: 3Ø2-317.

Ross, H.H. & Unzicker, J.D. 1977. The relationships of the genera of American Hydropsychinae as indicated by phallic structures (Trichoptera, Hydropsychidae). **J. Georgia Entomol. Soc.**, 12 : 298-312.

Schefter, P.W., Wiggins, G.B. & Unzicker, J.D. 1986. A proposal for Assignment of *Ceratopsyche* as a subgenus of *Hydropsyche*, with new synonyms and a new species (Trichoptera : Hydropsychidae). **J. N. Am. Benthol. Soc.**, 5 : 67-84.

Schmid, F. 1979. On some new trends in Trichopterology. **Bull. Entomol. Soc. Canada**, 11 : 48-57.

Schuster, G.A. 1984. *Hydropsyche? - Symphitopsyche? - Ceratopsyche? :* a taxonomic enigma. **Proc. 4th Intern. Symp. Trichoptera**, J.C. Morse (ed.), Junk, The Hague : 339-345.

# SETAL CHARACTERS IN LARVAL DIAGNOSES FOR SOME NEARCTIC SPECIES OF *CHEUMATOPSYCHE* (TRICHOPTERA : HYDROPSYCHIDAE)

P.W. SCHEFTER and G. B. WIGGINS

## SUMMARY

Identification of *Cheumatopsyche* larvae to the species level has remained an intractable taxonomic problem in spite of the importance of these species in hydrobiological work. New sets of setal and cuticular characters, similar to those applied to larval diagnoses in *Hydropsyche* (Schefter and Wiggins 1986) offer promise of some advance in this problem. Character suites derived from setae of the head and notal sclerites, and minute spines on the notal sclerites and abdominal cuticle have diagnostic value in species investigated. Of fourteen species in which larvae were associated and analyzed, ten were found to have diagnostic combinations of characters.

## INTRODUCTION

A preliminary investigation of Nearctic *Cheumatopsyche* larvae suggests new character suites which might be useful for taxonomy in this difficult group. A study of *Hydropsyche* (*Ceratopsyche*) larvae by Schefter and Wiggins (1986) resulted in diagnoses and descriptions for most North American species based on character suites using primary and secondary setae as well as colour patterns. *Cheumatopsyche* larvae generally lack colour variations, and chaetotaxy of legs and anal prolegs is, at the present level of resolution, not useful in species identification. There are, however, subtle differences in morphology and distribution of secondary setae, many of which were illustrated by Statzner (1984). We have found, in addition, that size and distribution of minute cuticular spines located on the notal sclerites and abdominal segments can provide characters useful in diagnoses of *Cheumatopsyche* species.

## METHODS

Larvae were cleaned, cleared in KOH, mounted in glycerine on depression slides, and observed at 100x with a compound microscope. Excessive handling may cause setae to be dislodged, but the spines, outgrowths of the cuticle, remain intact.

Setal types used in this study are as in Schefter and Wiggins (1986) :

*hl* = hairlike setae, flexible, single or divided, short or long ;

*bl* = brushlike setae, short and divided apically ;

*ms* = minute spines, cuticular outgrowths on notal sclerites and dorsum of abdomen.

## DISCUSSION

In Nearctic *Cheumatopsyche* two species groups, or complexes as defined by Gordon (1974) in her study of adults, are recognizable as larvae : the *sordida* group lacking a median indentation or notch in the anterior margin of the frontoclypeus, and the *gracilis* group, with a notch which is barely discernible in some species and pronounced in others. Within each group, species may be distinguished by various combinations of the following characters :

- the pronotum bears short (Figure 1) or long (Figure 2) hair-like (*hl*) setae on the anterior margin ;
- the mesonotum lacks (Figure 3) or bears (Figure 4) minute spines, visible as tiny dots between setae ;
- the metanotum lacks (Figure 5) or bears (Figure 6) minute spines ;
- the dorsum of the larval abdomen usually bears numerous small spines (Figure 8), which appear as tiny dots on the cleared cuticle viewed with transmitted light. In some species the spines are enlarged (Figure 7), especially seen dorsally on the first three abdominal segments, while in others the spines are either absent or so small as to be inconspicuous on the cuticle (Figure 9) ;
- the frontoclypeus bears secondary setae usually near the posterior angle, and those illustrated in this study are short undivided *hl* setae (Figure 10), long single and bifid *hl* setae (Figure 11), or short *bl* setae (Figure

12). These setal types may also occur on the pronotum.

In North America, *Cheumatopsyche* occur in many lotic environments, and yet only one fourth of the known species are associated as larvae. Discrimination of species among larvae is necessary, and we hope that character suites such as those illustrated in this study will prove useful to those working with these abundant and diverse organisms.

REFERENCES

Gordon,A.E.1974. A synopsis and phylogenetic outline of the Nearctic members of *Cheumatopsyche*. **Proc. Acad. Nat. Sci. Phila.,** 126(9) : 117-160.

Schefter, P.W. & Wiggins,G.B. 1986. A systematic study of the Nearctic larvae of the *Hydropsyche morosa* group (Trichoptera :Hydropsychidae). **Roy. Ont. Mus. Life Sci. Misc. Publ.,** Toronto, Canada : 1-100.

Statzner,B. 1984. Keys to adult and immature Hydropsychinae in the Ivory Coast (West-Africa) with notes on their taxonomy and distribution (Insecta : Trichoptera). **Spixiana,** 7 : 23-50.

***Overleaf***

Figures 1-12. *Cheumatopsyche* larvae, various species, showing setal and cuticular characters.

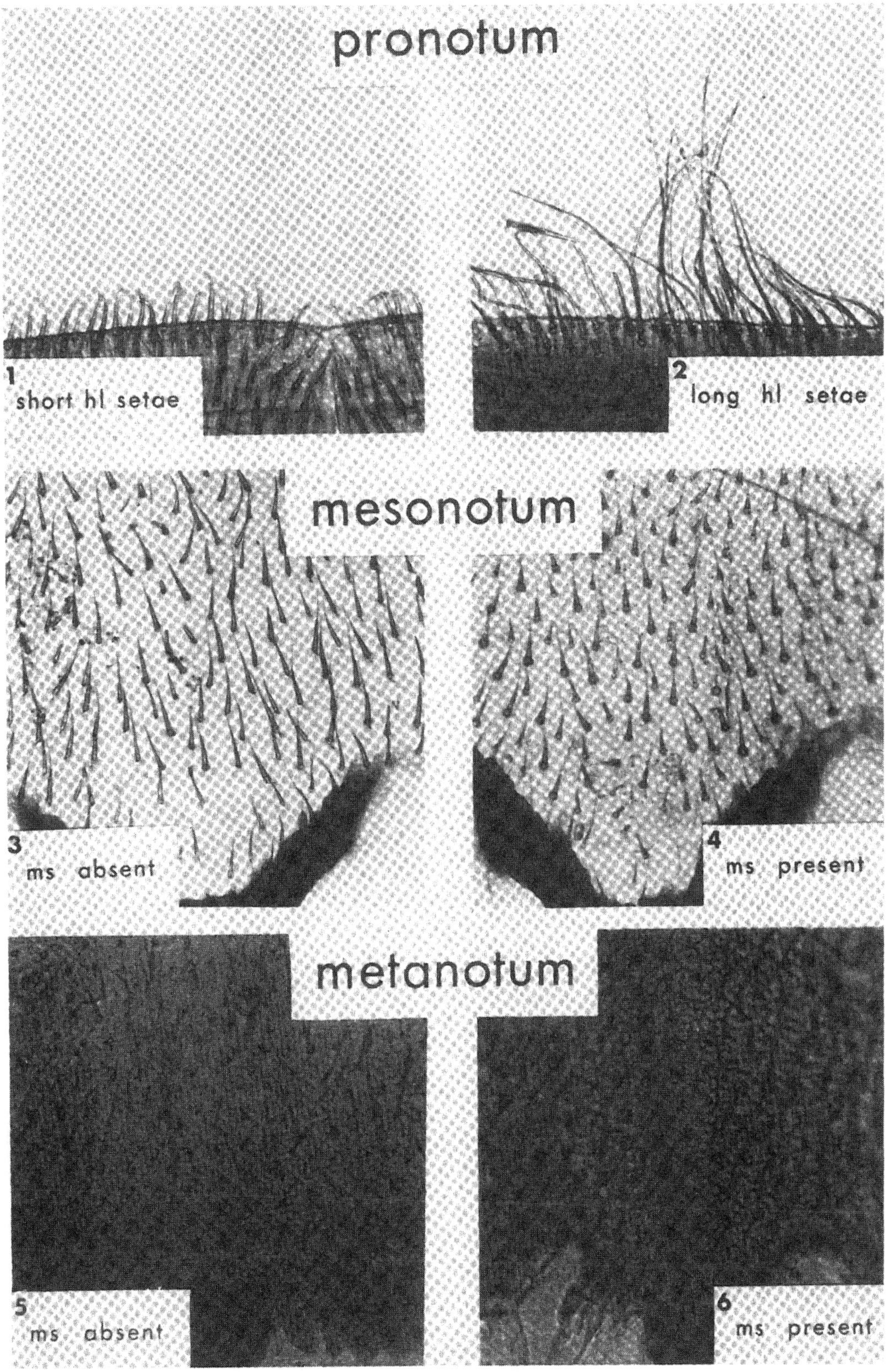
pronotum
1
short hl setae
2
long hl setae
mesonotum
3
ms absent
4
ms present
metanotum
5
ms absent
6
ms present

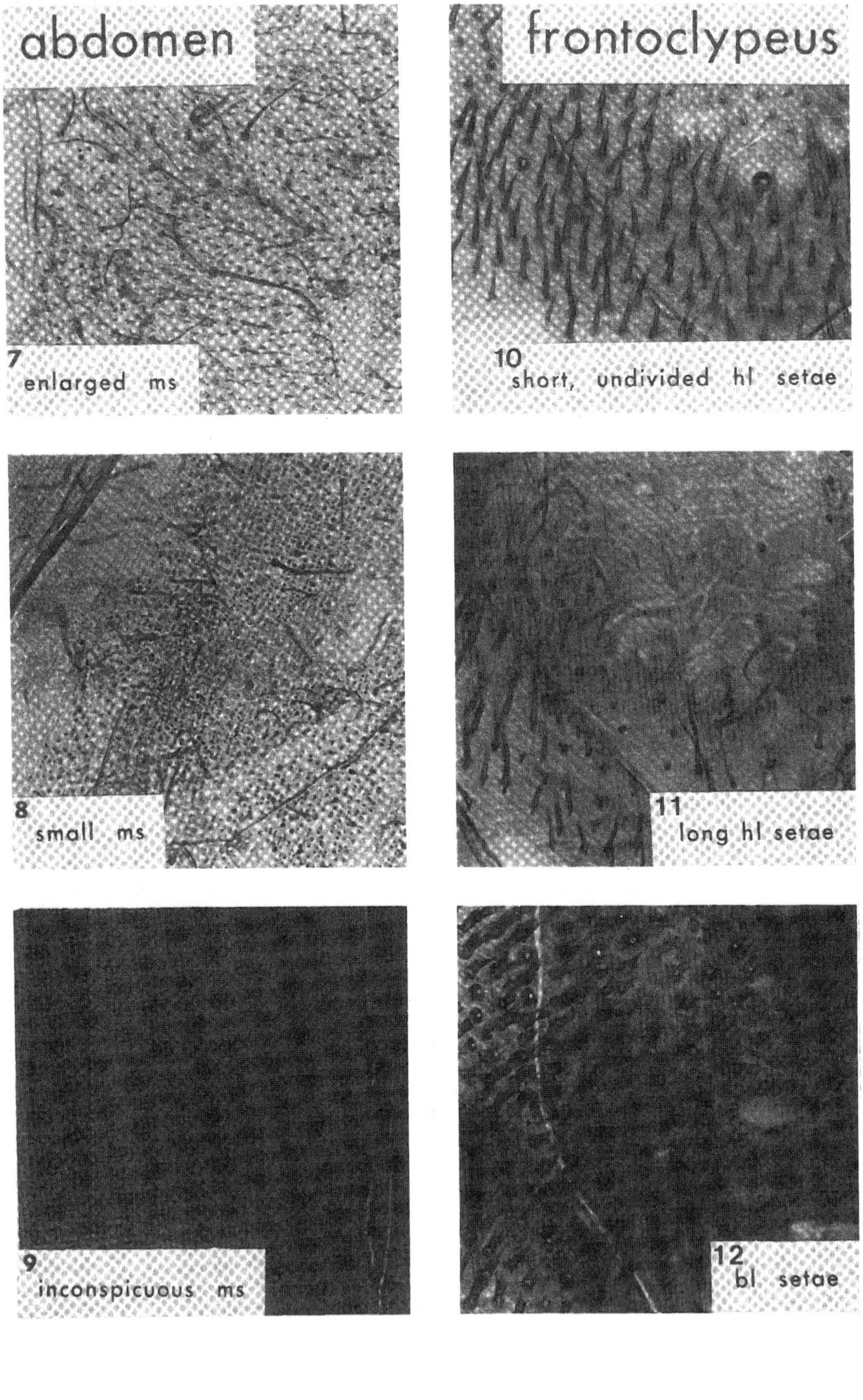
abdomen
frontoclypeus
7
enlarged ms
10
short, undivided hl setae
8
small ms
11
long hl setae
9
inconspicuous ms
12
bl setae

# THE REMARKABLE GENUS *PALAEAGAPETUS* ULMER, 1912 (HYDROPTILIDAE)

L. BOTOSANEANU & I.M. LEVANIDOVA


## SUMMARY

*Palaeagapetus* Ulmer, 1912 is a small genus of very primitive Hydroptilids which was considered until 1956 as belonging to the Glossosomatidae. First described for a fossil species in Oligocene Baltic amber. The next 3 described species were Neartic, two of them western, a third one eastern (but there is only one western species, *P. guppyi* Schmid being a synonym of *P. nearcticus* Banks). In 1986 the surprising discovery was published of the first Palearctic species, from Hokkaido, northern Japan. The species -all montane crenobionts- are very similar inasmuch as body characters (neuration included) are concerned, the same being apparently true for larval morphology and for the interesting larval habits (the known larvae are strongly associated, for food and case construction, with certain aquatic liverworts). Nevertheless, species are readily distinguished by male genitalia.

Enclosed is the description of a 2nd living Palearctic species, caught from a spring in the Far Eastern part of Soviet Union (it is unclear if a damaged specimen caught on Sakhalin Island is conspecific with it). Study of male genitalia suggests that the new species belongs to a lineage including the fossil species from Baltic amber and the species from eastern N. America ; the western Nearctic species certainly belongs to a distinct lineage ; and the Japanese species seems to be isolated. The genus probably originated in Eurasia in Cretaceous (or Eocene) times. Suppositions are made on the possible time and way of spreading of *Palaeagapetus* lineages from Eurasia to North America.


## INTRODUCTION

*Palaeagapetus* was described by Ulmer (1912 : 35-36) for a fossil species from Oligocene Baltic amber, *P. rotundatus* ; the species was placed in Rhyacophilidae Glossosomatinae, and *Palaeagapetus* was considered as Glossosomatine, or as Rhyacophiline, until 1956 when Ross (p. 14) placed it in the Hydroptilidae, a view presently generally accepted and strongly supported by study of larvae (f.i. : Marshall, 1979 : 157-159). Together with another very small genus, the purely Palearctic *Ptilocolepus* Kolenati, 1848, it constitutes the primitive subfamily Ptilocolepinae Martynov, 1913, in several respects a surviving connecting link between Glossosomatidae and Hydroptilidae. Three living Nearctic species were described after the type species : *nearcticus* Banks (1936 : 265, figs. 1-3), *celsus* Ross (1938 : 111-112 ; 1944 : figs. 105, 135), and *guppyi* Schmid (1951 : 1-2) ; *nearcticus* occurs in western North America from California to Vancouver Island ; *guppyi* is, without any doubt, a synonym of *nearcticus*, as we could ascertain by examination of the male holotype of *nearcticus* (kept in the M.C.Z., Harvard University ; kindly sent to L.B. by Dr. S.R.Shaw) and comparison with description and illustration of *guppyi* ; *celsus* inhabits the Appalachian Mountains of North Carolina and Tennessee and the Laurentians of Quebec, and it is likely that it will be found in montane areas between these two extremes (Wiggins 1977 : 148). The discovery of the first living Palearctic species, *P. ovatus* Ito & Hattori, 1986, from Hokkaido, Japan, came as a surprise. Larvae and pupae of only two species are presently known : *P. celsus* (Flint 1962 ; Wiggins 1977 : 148-149) and *P. ovatus* (Ito & Hattori 1986) ; they share a peculiar larval morphology and the delightful, oval, flat cases made from pieces of liverworth leaves ; the tight association of the larvae of these montane crenobiont insects with liverworts for substrate, food, and building materials, is a very interesting peculiarity of *Palaeagapetus.*

We shall describe here the 2nd Paleartic species of the genus.

## *PALAEAGAPETUS FINISORIENTIS* NEW SPECIES

Locality, material. - male holotype (in alcohol, in the Zoological Institute of Leningrad), collected on 5th July 1973 by Levanidova ; Vodopadny spring, bassin of Kedrovaya River, "Kedrovaya Pad" Nature Reserve, Vladyvostok district, U.S.S.R.

A second specimen of *Palaeagapetus* was, from a spring on the Chekhov Mt., Sakhalin Island (29th June

1973, leg. Zhiltzova) ; its genitalia lacking, it was not labelled as type.

Description of male - Wing expanse ca. 8.5 mm. Head and thorax dark brown, forewings brownish. Spurs 2, 4, 4. Short but strong median "tooth" distally on abdominal sternites VI and VII. Wings with shape and venation typical for *Palaeagapetus* (discoidal cell closed in both wings, forewing with forks 1-5, hindwings lacking f4).

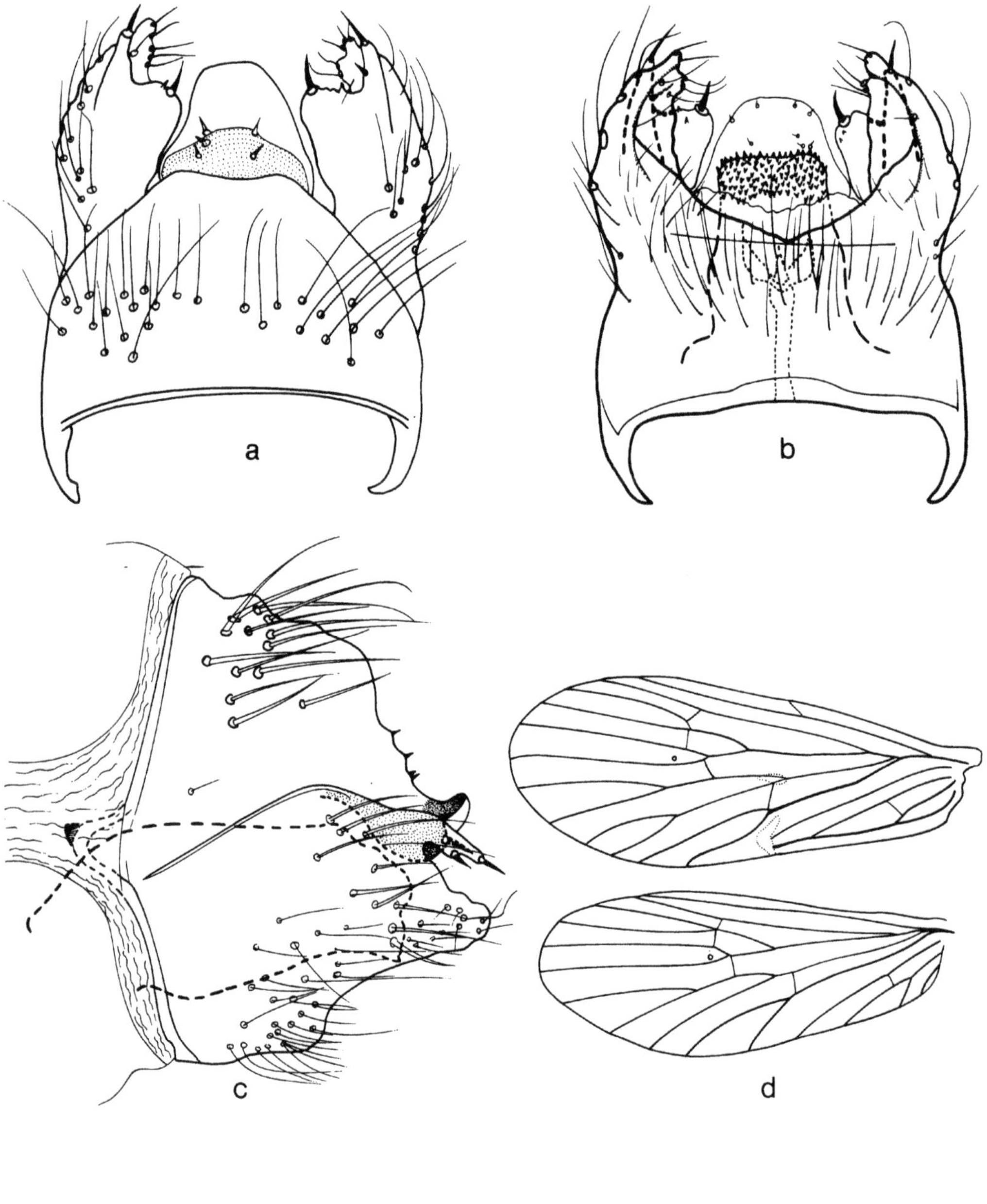


Figure 1. *Palaeagapetus finisorientis* n. sp. a, b, c) male genitalia, dorsal, ventral, lateral view. d) wings.

Male genitalia. IXth segment forming a compact genital capsula, without strongly protruding appendages ; medio-proximally with relatively short but strong projections ; in lateral view, IXth tergite and sternite of almost the same size, very distinctly separated by oblique limit. IXth tergite dome-shaped, terminally shallowly excised, dorsal border in lateral view very irregular, long hairs in proximal half. IXth sternite with very irregular lower border in lateral view, with deep triangular distal excision. No appendage rooted on IXth tergite, which is almost directly continued by the relatively small Xth segment situated at a lower level (in lateral view slightly upturned towards obtuse end ; in dorso-ventral view with truncate apex ; proximal part less sclerotized than distal part, and with some short spines). Distally, IXth sternite continued by a complex of not strongly protruding lobes representing the inferior appendages ; 3 lobes can be distinguished on each side ; two of them are forming a dorsal pair : a lateral lobe, digitiform, with one apical and sometimes also one preapical spine, and a median lobe, much broader and shorter, with one apical spine ; the dorso-lateral, digitiform lobe is separated by a deep and narrow sinus from a large and broad, hairy ventral lobe with irregular margins. Phallic apparatus short and broad, apically truncate, distal part apparently bilobed, probably membranous and covered with minute spines, terminal part of ductus ejaculatorius with sclerotized trident.

Derivatio nominis. - The specific name alludes to the species' location in the Extreme-Oriental part of Eurasia.

## DISCUSSION

*Palaeagapetus* species are very uniform inasmuch as body and wing characters are concerned ; nevertheless, comparison of male genitalia enables us to speculate about character polarity, about primitive and derived conditions ; it is plausible that compact male genitalia, without strongly protruding appendages, are representing the primitive condition, whereas strongly protruding, complex appendages are the derived condition. We can summarize the situation of the various species in the following manner :

I. *rotundatus* : no appendages of IXth tergite ; inferior appendages simply bilobed.

II. *celsus* : no appendages of IXth tergite ; inferior appendages with only slightly protruding lobes, dorsal lobe with superior and inferior branch, and obtuse ventral lobe.

III. *finisorientis* : no appendages of IXth tergite ; inferior appendages with only slightly protruding lobes, dorsal lobe with median and lateral branch, and obtuse ventral lobe.

IV. *ovatus* : no appendages of IXth tergite ; inferior appendages very strongly protruding and slender (with small lobes in median position).

V. *nearcticus* : IXth tergite with very strong, lateral, tribranched appendages ; inferior appendages slightly protruding, simply bilobed.

It is obvious : that the most primitive condition is that of the fossil *rotundatus* ; that the conditions in *celsus* and in *finisorientis*can be easily derived from that in *rotundatus* ; that *ovatus*, on one side, and *nearcticus*, on the other side, are the most apomorphous species, representing two quite distinct lineages.

We shall let to others the pleasure of constructing a perfect cladogram for *Palaeagapetus* : we agree with Berry (1982 : 59) that "... cladism ... neither aids nor hinders evolutionary biology" ; instead, we shall try to understand something of the evolutionary pathways of this genus.

*Palaeagapetus* originated possibly in Cretaceous times in northen Palearctic ; unfortunately, there is no fossil evidence from this epoch ; but Ross (1956 : 22) emphasizes the striking ressemblance between wings of living species of this genus and the fossil Necrotauliidae from the lower Jurassic.

During late Cretaceous, Europe and Asia were probably attached by land to North America, and it is reasonable to suppose that some primeval stock of *Palaeagapetus* migrated to North America at this time, or , even more probably, during early Eocene ; there is extensive evidence for mammals (Webb, 1985) that there was much immigration during the Wasatchian mammal age (53-49 Ma) between the Holarctic continents through a North Atlantic corridor, the so-called "De Geer Route" across Greenland and the Faroes, an area that must have been richly forested and continuously productive. Such a lineage could have produced the recent *celsus* from mountains of eastern North America. Another lineage progressing along Siberia to the east could have led to the extreme-oriental *finisorientis* which, phylogenetically, is clearly on the same level with *celsus*.

More recent events have finally led to *nearcticus* from western North America ; again excellent evidence from land mammals (Webb, 1985) shows that with the onset of Oligocene (Chadronian mammal age, ca. 36 Ma) there was much immigration from Eurasia to North America by way of the Bering Strait (this could have started somewhat earlier, during late Eocene) ; but possibly the best opportunities for a *Palaeagapetus* lineage to cross the Bering Strait were reached during mid-Miocene (f.i. : Hemingfordian, ca. 18 Ma) when the greatest waves of immigrants from Eurasia established themselves in North America, and when the Bering corridor

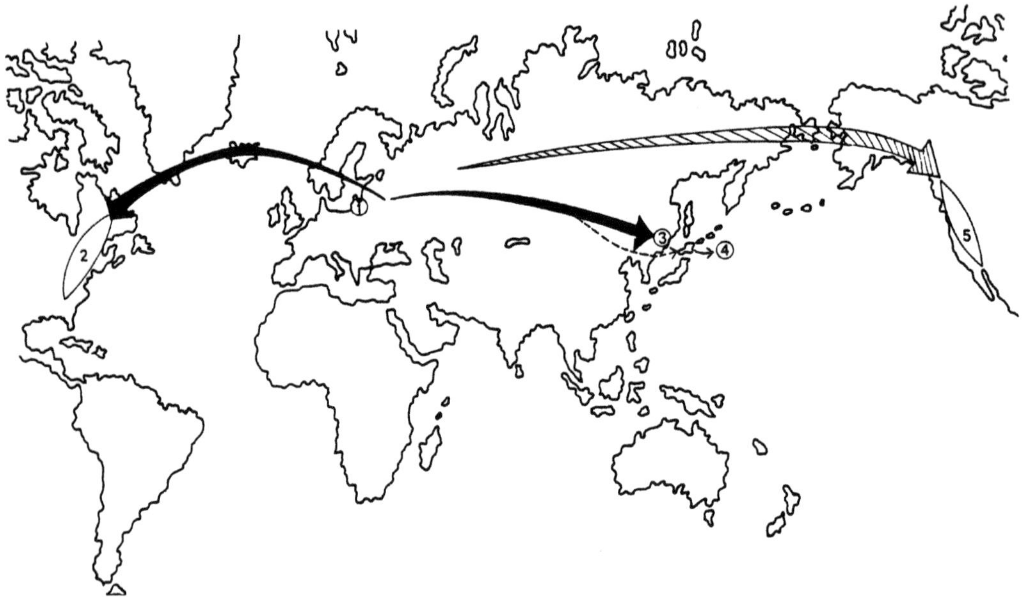


Figure 2. Possible ways and times of spreading of *Palaeagapetus* lineages from northern Eurasia (full arrows : late Cretaceous or early Eocene events; hatched arrow : Oligocene-Miocene events; dashed arrow : possibly more recent events). 1 : + *rotundatus* ; 2 : *celsus* ; 3 : *finisorientis* ; 4 : *ovatus* ; 5 : *nearcticus*.

must have been broad and richly varied.

Difficult is the interpretation of the presence on Hokkaido, Japan, of a very apomorphous species, *ovatus*, certainly not belonging to the same lineage as *nearcticus*, and whose affinities with species like *rotundatus*, *celsus*, and *finisorientis* are not easy to demonstrate. It is not impossible that this is a recent offshot of the lineage having produced *finisorientis*. Anyway, the presence on Hokkaido and in the district of Vladyvostok of two phylogenetically quite different species, desserves special attention.

## REFERENCES


Banks, N. 1936. Four new Trichoptera from the United States. **Arch. morph. taxon. Ent. Berlin-Dahlem,** 3(4) : 265-268, .

Berry, R.J. 1982. Neo-Darwinism. Studies in Biology no. 144. Edward Arnold. 68 p.

Flint, O.S. 1962. The immature stages of *Palaeagapetus celsus* Ross (Trichoptera, Hydroptilidae). **Bull. Brooklyn Ent. Soc.,** 57 (2) : 40-44.

Ito, T. & Hattori, T. 1986. Description of a new species of *Palaeagapetus* (Trichoptera, Hydroptilidae) from northern Japan, with notes on bionomics. **Kontyû, Tokyo,** 54 (1) : 143-151.

Marshall, J.E. 1979. A review of the genera of the Hydroptilidae (Trichoptera). **Bull. B.M. (N.H.), Ent. Ser.,** 39 (3) : 135-239.

Ross, H.H. 1938. Description of Nearctic Caddis Flies (Trichoptera) with special reference to the Illinois species. **Bull. Illinois Nat. Hist. Surv.,** 21 (4) : 101-183.

Ross, H.H. 1944. The Caddis Flies, or Trichoptera, of Illinois. **Bull. Illinois Nat. Hist. Surv.,** 23 (1) : 1-326.

Ross, H.H. 1956. Evolution and Classification of the Mountain Caddisflies. **Univ. Illinois Press,** Urbana. 213 p.

Schmid, F. 1951. Quelques nouveaux Trichoptères néarctiques. **Bull. Inst. Sci. nat. Belg.,** 27 (7) : 1-16.

Ulmer G. 1912. Die Trichopteren des Baltischen Bernsteins. **Beitr. Naturk. Preussens, Schr. phys. - ök. Ges. Königsbg.,** 10 : 1-380.

Webb, S.D. 1985. Main pathways of mammalian diversification in North America : 201-217. *In* : The great interamerican biotic interchange, F.G. Stehli & S.D. Webb (eds), Plenum Press, N. York and London. 532 p.

Wiggins, G.B. 1977. Larvae of the North American Caddisfly Genera (Trichoptera). **Univ. Toronto Press,** Toronto and Buffalo. 401 p.

# A NEW GENUS AND SPECIES OF LIMNEPHILINE TRICHOPTERA FROM THE FAR EAST OF THE U.S.S.R. (TRICHOPTERA : LIMNEPHILIDAE)

G.B. WIGGINS

SUMMARY

Single male and female specimens of a limnephilid caddisfly found in the collections of the Zoological Institute, Leningrad, had been identified by A.V. Martynov as of a new genus and species. The specimens bore no name, nor evidently were they mentioned in the literature. The name *Lepnevaina* gen. nov. *signata* sp. nov. is proposed. The new genus belongs to the Limnephilinae, but the characters are highly aberrant. In the male, extraordinary development of the superior appendages, and loss of all but a vestige of the intermediate appendages, result in genitalic morphology that shows affinity with no other genus in the Limnephilinae. In the female, segments IX and X are unusually short, and the vulval lobes circular and enlarged. A prominent ovoid setal wart on the frons is distinctive in both sexes. Evidence for relationships is reviewed, and the genus assigned provisionally to the tribe Limnephilini.

## INTRODUCTION

While studying Trichoptera in the collections of the Zoological Institute of the USSR Academy of Sciences, Leningrad, in 1984, I came upon a series of two distinctive specimens, a male and a female, belonging to the Limnephilidae. These bore labels dating from their study by A.V. Martynov, designating them as representing a new species in a new genus. Since a search of the literature has failed to show that these were ever described, I do so here.

### *Lepnevaina* gen. nov.

Type-species *Lepnevaina signata* sp. nov.

Because objective distinction between generic and specific characters in a monotypic genus is not possible, the entire description follows under the species name.

### *Lepnevaina signata* gen. et sp. nov.

**Adult.** This is a distinctive caddisfly with rather short wings bearing stout hairs ; overall colour yellowish brown. Length of fore wing 7mm in both sexes.

Head (Fig. 1, 2) with all dorsal setal warts present (*sensu* Wiggins 1984) ; dorsomedian epicranial suture present but not well developed ; distinctive in having a single, large, ovoid median setal wart on the frons extending from the antennal sockets to the edge of the anterior tentorial pits, prominently raised and bearing long stout setae. Compound eyes moderately large, occupying most of the length of the head in dorsal aspect. Antennae broken in both specimens, scape much shorter than head.

Thorax (Fig. 2) with one pair of large pronotal setal warts dorsally, a small setal wart at the extreme ventrolateral edge of the pronotum ; mesoscutal setae confined to a pair of slender and elongate setal warts, but with a second pair of very small setal warts ; mesoscutellum rather short, one-third the length of the mesonotum, setae confined to a pair of ovoid setal warts. Wings uniform yellowish brown, covered with short yellowish hairs and longer darker hairs on the veins, membrane lacking colour pattern ; fore wings somewhat rounded in outline ; venation (Fig. 3) typical for the Limnephilinae and without distinguishing characters. Legs with tibial spurs 1, 3, 4 ; stout black setae on tarsi and tibiae ; fore femur of male with dense band of short black setae along ventral edge.

Male genitalia (Fig. 4). Segment VIII with tergum more heavily sclerotized than other abdominal segments, considerably longer and more robust than sternum, but lacking densely packed short setae dorsally. Segment IX in lateral aspect narrowed dorsally and ventrally, widest laterally ; inferior appendages fused with IX, reduced to a small lobe. Segment X with superior appendages elongate, very large, tapering but truncate apically, apicodorsal edge dentate ; intermediate appendages lacking but probably represented by a pair of low ridges on the body of segment X, a flattened plate on each side of the anal opening. Phallus with parameres arising from the base of a slender aedeagus, each paramere terminating in an inflated scoop-shaped lobe bearing an elongate apical point and a shorter dorsal point.

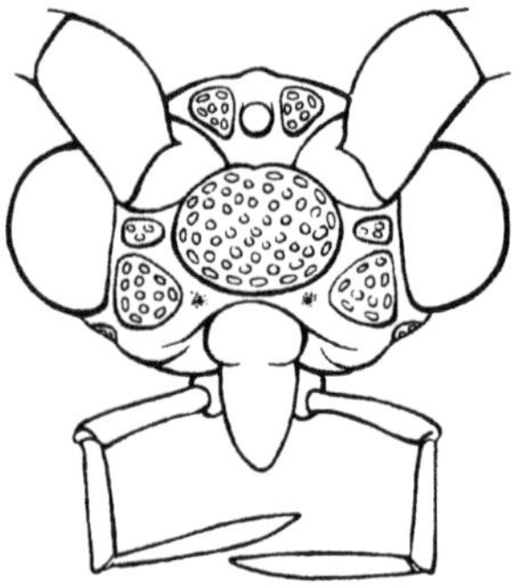

1

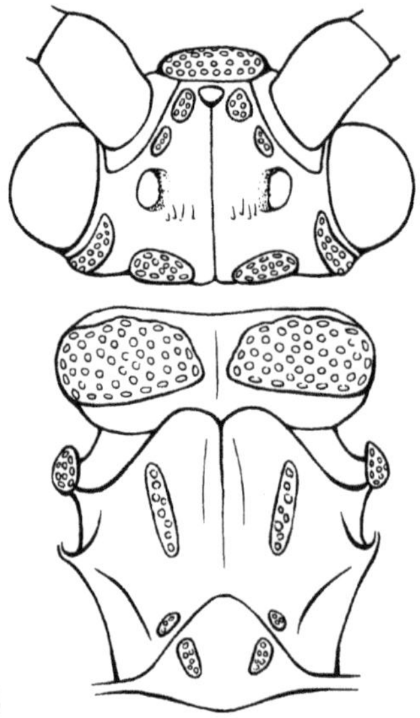

2

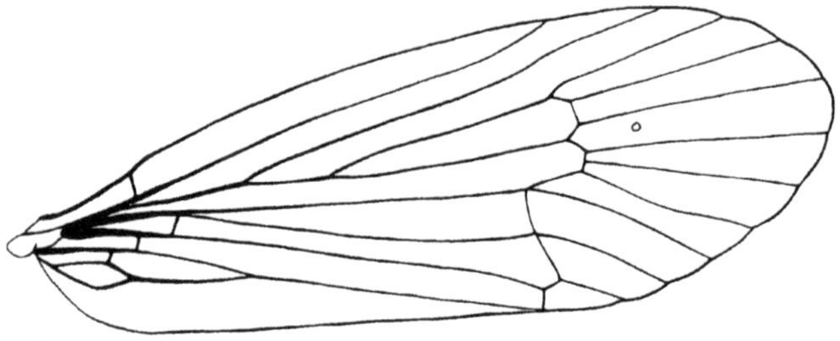

3

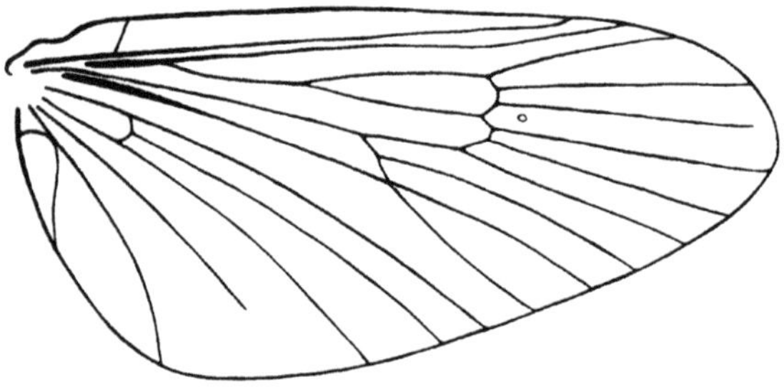

Figures 1-3 : *Lepnevaina signata* 1, Head of holotype male, frontal. 2, Head, pronotum, and mesonotum of holotype male, dorsal. 3, Wings of allotype female.

Female genitalia (Fig. 5). Segment VIII excised ventromesally at the origin of prominent vulval lobes ; median lobe broken at base in allotype female, lateral lobes circular in outline. Segment IX short, fused with X which is also short, both without prominent apical lobes ; ventrolateral lobes of sternum IX densely setate, meeting on the median line between the anal opening and the vulval lobes ; segment X sclerotized entirely around the anal opening, the rim slightly extended.

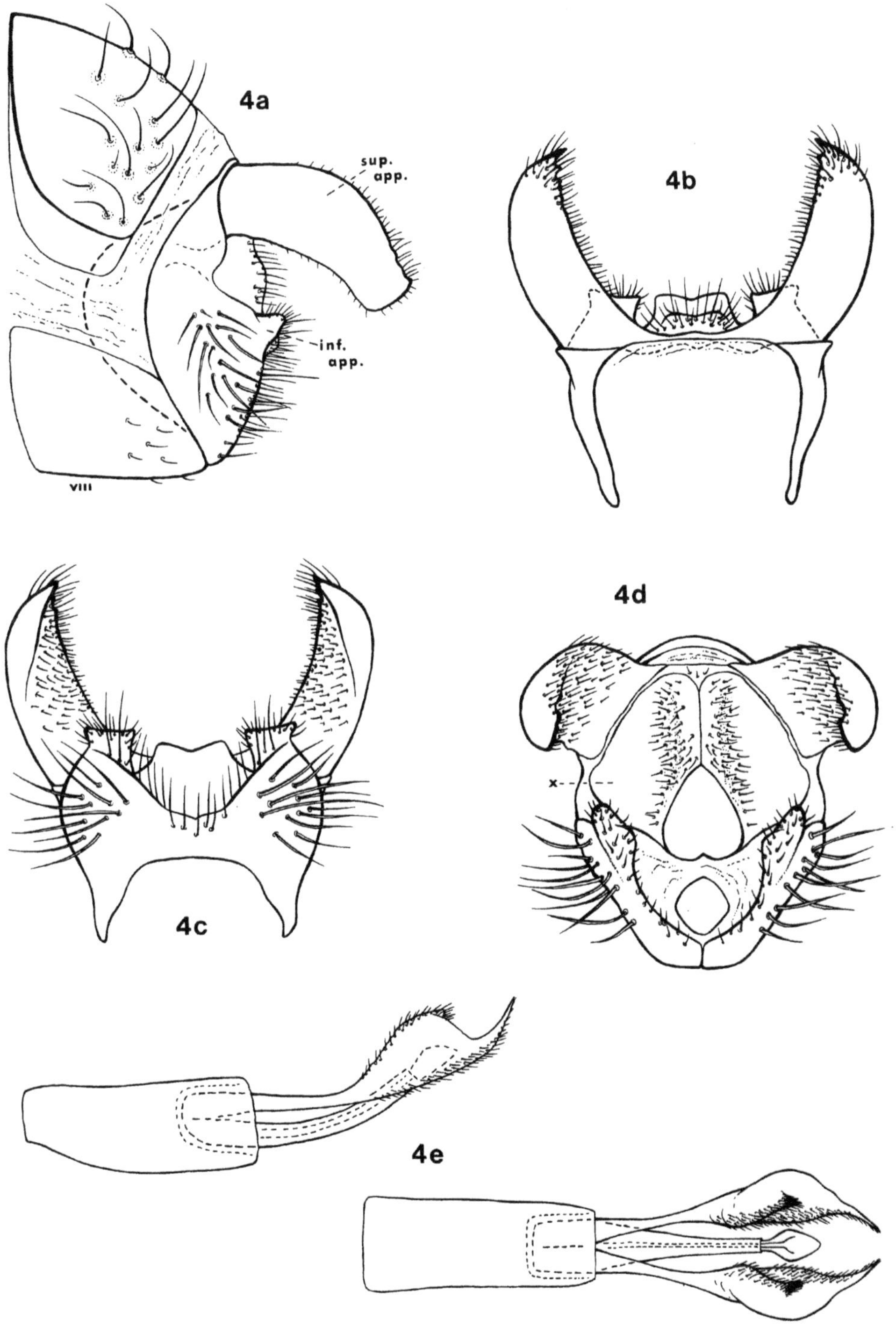


Figure 4 : Male genitalia of *Lepnevaina signata,* holotype. a, lateral ; b, dorsal ; c, ventral ; d, caudal ; e, phallus in lateral and dorsal aspects ; inferior (inf. app.) and superior appendages (sup. app.), segments indicated by Roman numerals.

**Immature stages and biology.** No information is available on the immature stages, and inference to habitat can be based only on the label data "bank of the River Ulas".

**Types.** HOLOTYPE, Male : USSR, Far East, Khanchai, bank of the River Ulas, Ct. of Buluhuna, 22.VII.1898, Klements. ALLOTYPE, Female : same data as male.

Specimens (pinned) in the collection of the Zoological Institute, USSR Academy of Sciences, Leningrad.

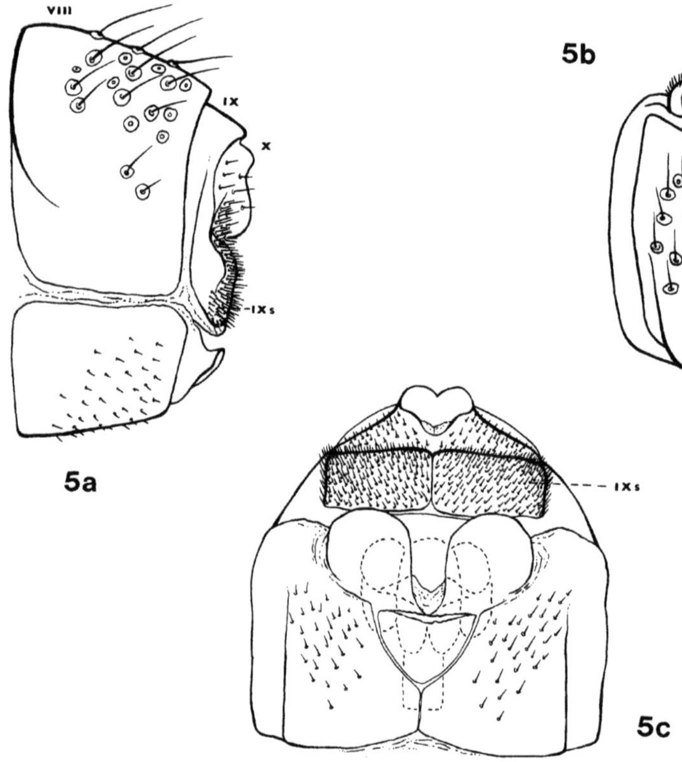


Figure 5 : Female genitalia of *Lepnevaina signata*, allotype. a, lateral ; b, dorsal ; c, ventral (median vulval lobe broken) ; sternum of segment IX (IXs).

**Etymology.** Although my first inclination in naming this genus was to denote its original recognition by A.V. Martynov, review of occupied generic names revealed that Martynov's very substantial contributions to systematic entomology were already recognized in several generic names. Hence, I dedicate the name of this genus to S.G. Lepneva for contributions to knowledge of the Trichoptera of the USSR. She was for many years curator of the collection of Trichoptera in the Zoological Institute, following A.V. Martynov in that post. The specific name derives from the Latin *signata* - - marked, signifying the prominent setal wart on the frons of the insect.

SYSTEMATIC CONSIDERATIONS

*Lepnevaina* can be readily assigned to the subfamily Limnephilinae but affinity with other genera is problematical. The prominent median setal wart on the frons in both sexes is unusual if not unique in the Limnephilinae. In the male, extraordinary development of the superior appendages, coupled with the virtual disappearance of the intermediate appendages, is unique in the Limnephilinae. In the female, the marked foreshortening of segments IX and X is unusual in the Limnephilinae, as is the relatively large size and circular shape of the vulval lobes.

Within the subfamily Limnephilinae, two of the four tribes, Limnephilini and Stenophylacini, are possible assignments for *Lepnevaina*. Because the tribes are rather loosely constituted and not defined by apomorphic characters, objective placement in any of them is not possible.

The somewhat rounded outline of the fore wings in *Lepnevaina* is characteristic of the Stenophylacini, but is a plesiomorphic feature in the Limnephilidae. Although diversification of female genitalic structure in most genera of the Stenophylacini has not led to enlargment of the three vulval lobes, which are greatly enlarged and diagnostic in *Lepnevaina*, somewhat similar structure does occur in females of the European stenophylacine genera *Acrophylax*,*Chionophylax*, and *Parachiona*; and in these same genera segments IX and X are foreshortened and the lateral lobes of the ninth sternum are fused. These features suggest the structure in *Lepnevaina*, and probably represent apomorphic but homoplastic states. Female genitalic structure in *Parachiona picicornis* Pictet is noteworthy in this similarity (Schmid 1951, Fig. 32-35). However, male genitalic structure in none of these genera reveals characters shared with *Lepnevaina*, nor are any evident in the Stenophylacini generally. The slender parameres in this tribe are unlike those in *Lepnevaina*.

In the Limnephilini, females in most genera have three prominent vulval lobes and have the lateral lobes of sternum IX fused as in *Lepnevaina*. Although females in most species have segments IX and X extended as a tube around the anal opening, these segments are foreshortened in some *Limnephilus*, as they are in *Lepnevaina*. Male genitalic structure in the limnephiline genera *Anabolia*, *Asynarchus*, and *Limnephilus* shows a strong apomorphic tendency for enlargement of the superior appendages and reduction of the inferior appendages, as does *Lepnevaina*, although not for reduction of the intermediate appendages. Parameres in most Limnephilini and in *Lepnevaina* are expanded apically in a spoon-shaped structure, probably an apomorphic structure.

More exhaustive assessment of the affinity of this genus based on derived characters must await phylogenetic analysis of the Limnephilidae as a whole. Until that can be done, evidence now available is consistent with provisional treatment of *Lepnevaina* as an aberrant genus in the Limnephilini.

ACKNOWLEDGEMENTS

This study is supported by an operating grant from the Natural Sciences and Engineering Research Council of Canada. I thank L.A. Zhiltzova for kindly facilitating my study of the collections of the Zoological Institute in Leningrad, and F. Schmid for comments on the manuscript, noting that he does not share the view taken here. Illustrations were prepared by Zile Zichmanis.

REFERENCES

Schmid, F. 1951. le groupe de *Enoicyla*. **Tijdskr. Entomol.,** 94:207-226

Wiggins, G.B. 1984. Trichoptera, Chap. 16, *In* Merritt, R.W. and Cummins, K.W. (eds). An introduction to the aquatic insects of North America, Second Ed., Kendall/Hunt Publ. Co., Dubuque: 271-311.

# *PSYCHOGLYPHA (MONOPHYLAX)* NEW SUBGENUS (TRICHOPTERA, LIMNEPHILIDAE, LIMNEPHILINI)

A.P. NIMMO

## SUMMARY

*Psychoglypha* (*Monophylax*) new subgenus is erected to contain *Psychoglypha mono* Denning on the basis of male genitalic characters. The remaining known species of *Psychoglypha* are assigned to *Psychoglypha* (*Psychoglypha*).

Denning (1970) described *Psychoglypha mono,* noting its distinctness from other known species of the genus, and enumerating distinctive characters. At that time he went no further. However, he agrees (pers.com.) that the distinctive characters of *P. mono* support a division of the genus as here proposed. Recently, on examining a specimen of *P. mono* I failed at first to recognise it even to genus until I chanced to note the wing pattern typical of the genus (very faint in the specimen), whereupon identification was straightforward using Denning (1970). Having the genitalic distinctiveness of this species thus forcibly brought to my attention, I concluded that it represented a distinctive division, or subgenus of *Psychoglypha,* which is here characterised and named *Psychoglypha* (*Monophylax*) new subgenus : the *Psychoglypha* 'guardian' of Mono County, California, for which the species is named. The remaining known species are thus assigned to *Psychoglypha* (*Psychoglypha*).

Genitalic characters which distinguish *Psychoglypha* (*Monophylax*) *mono* Denning from *Psychoglypha* (*Psychoglypha*) species are : sternum IX short, not curved or projected posterad (Figure 1) ; inferior appendages (claspers) widely separated basally, long, slender, blade-like (Figure 1, 2) ; internal lateral concavity between inferior and superior appendage bases absent (Figure 1) ; intermediate appendages (segment X) short, squat, broadly rounded in lateral aspect (Figure 1) ; and space between inferior appendages occupied by large, dorsally concave sclerite, with membranous articulation to sternum IX (Figure 1, 2).

To the above, noted by Denning (1970), I add : clasper blades directed mesad in ventral aspect (Figure 2), and tergum VIII with no trace of short, spine-like, or finer setae in brush-like formation along postero-dorsal edge. This last may not be unique to *Psychoglypha* (*Monophylax*) as Denning (1970) does not show this part in the illustrations of his several new species. Also, though certainly not unique to *Psychoglypha* (*Monophylax*), the superior appendages (cerci) show no sign whatsoever of the bifid structure exhibited by most other species of the genus.

Other than the genitalic characters listed above, *P. mono* adults appear in no other way to differ from the other known species of *Psychoglypha.* The overall habitus of the species is typical of the genus.

## REFERENCE

Denning, D.G. 1970. The genus *Psychoglypha* (Trichoptera : Limnephilidae). **Can. Ent.**, 102 : 15-30.

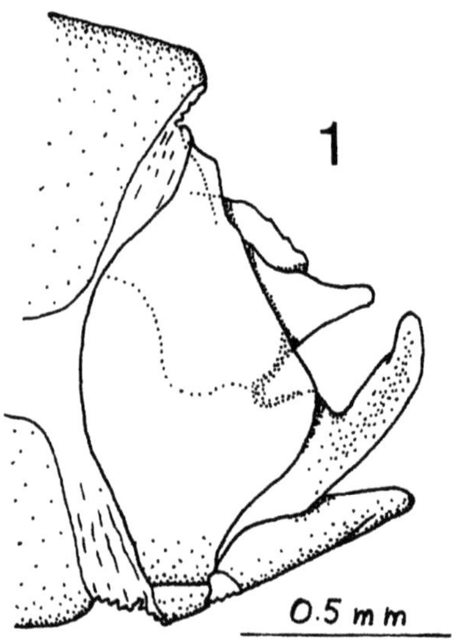


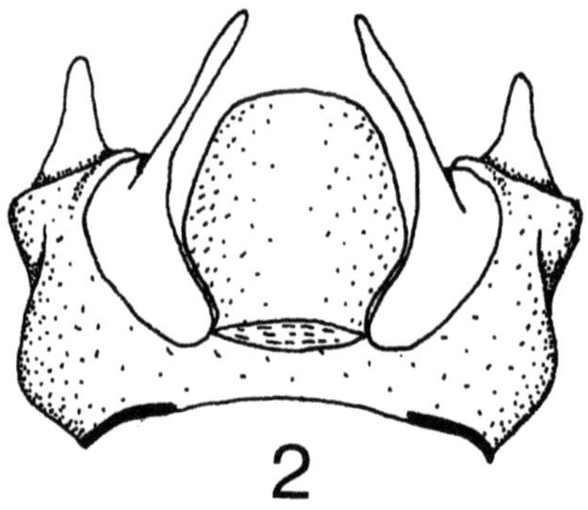


Figures 1 - 2. *Psychoglypha* (*Monophylax*) *mono* male genitalia.
Figure 1, lateral aspect.
Figure 2, ventral aspect.

# 3. FOSSIL FORMS

# CADDISFLIES AND QUATERNARY PALAEOECOLOGY - WHAT HAVE WE LEARNED SO FAR ?

N.E. WILLIAMS

## SUMMARY

The past ten years have seen the emergence of the Trichoptera as a viable, palaeoecological tool. Caddisfly fossils from sites in North America, Great Britain and Europe have been studied. The results of these studies are relevant not only to the field of palaeoecology but to ecology in general, taxonomy and zoogeography.

Quaternary caddisfly fossils are extremely abundant and usually well preserved in waterlaid sediments. Individual sclerites can be identified by reference to shapes, textures, colour and muscle scar patterns and setal distributions. Some species pairs difficult to separate as whole larvae e.g. *Halesus digitatus* (Schrank) and *H. radiatus* (Curtis), are more easily separated as individual sclerites.

Study of the biological literature and distributional data relevant to a caddisfly assemblage yields information at two levels. The probable local habitat and the climate can be described. Interglacial caddisfly assemblages show considerable similarity to modern communities in terms of species composition and associations, and relative abundance of taxa. However, comparisons with assemblages of the other plant and animal groups suggest that there are several important factors to be considered in interpreting caddisfly assemblages, particularly those from sediments deposited during cold periods : for example, caddisflies may be slower to migrate than certain terrestrial insects, in particular the beetles, and some glacial assemblages may therefore be dominated by "hangers on". Lotic and lentic species may also migrate at different rates.

Sufficient Quaternary sites have now been studied to give some concrete evidence for the history of movements of various caddisfly species. For exemple, we now know that *Hydropsyche bulgaromanorum* Malicky and other hydropsychid species presently found only on continental Europe, occurred in Britain 2,000 years B.P. and during previous interglacials.

Caddisfly fossils also have the potential to offer information on intraspecific variation over time. Questions concerning the *Hydropsyche bifida/morosa* complex in North America for example, might benefit from study of the glacial and integlacial material available.

It is hoped that these examples may encourage those interested in modern organisms, communities and populations to take advantage of the information offered by Quarternary fossils.

## INTRODUCTION

Over the past ten years or so several people have begun looking at caddisfly sclerites from sediments deposited during various glacial and interglacial periods. Palaeoecological interpretations of fossils from sites in North America (Williams and Morgan 1977, Williams *et al.* 1981), Great Britain (Wilkinson 1981, 1984) and Europe (Elias and Wilkinson 1983) have been published, while studies of many additional sites are in progress. I would like here to mention briefly what has been done with Quaternary fossils in terms of palaeoecology but also to point out the kinds of information which they are beginning to offer to all Trichopterologists ; and to suggest some of the uses to which I think they should be put.

## DISCUSSION

It is now quite evident that Quaternary caddisfly fossils are not at all rare. In fact they are extremely abundant and usually well preserved in most waterlaid sediments.

We have learned over the past decade or so, how to identify individual sclerites. The level of identification of sclerites is presently limited by lack of larval/adult associations and availability of reference material, not by lack of characters on sclerites. There are many characters available. Fossils are clean and flat and the characters of chitin are not obscured by the presence of other tissues. Some of these characters can be seen in the frontoclypeus of various hydropsychid species (Figure 1). The shape of the sclerite is important and may include measurements of length/width ratios, proportions of anterior and posterior parts, symmetry, and details such as degree of scalloping of the anterior edge. Sclerite textures vary between species from very smooth to reticulated, bumpy or spiny, protuberances being of various shapes, size and densities. Colour

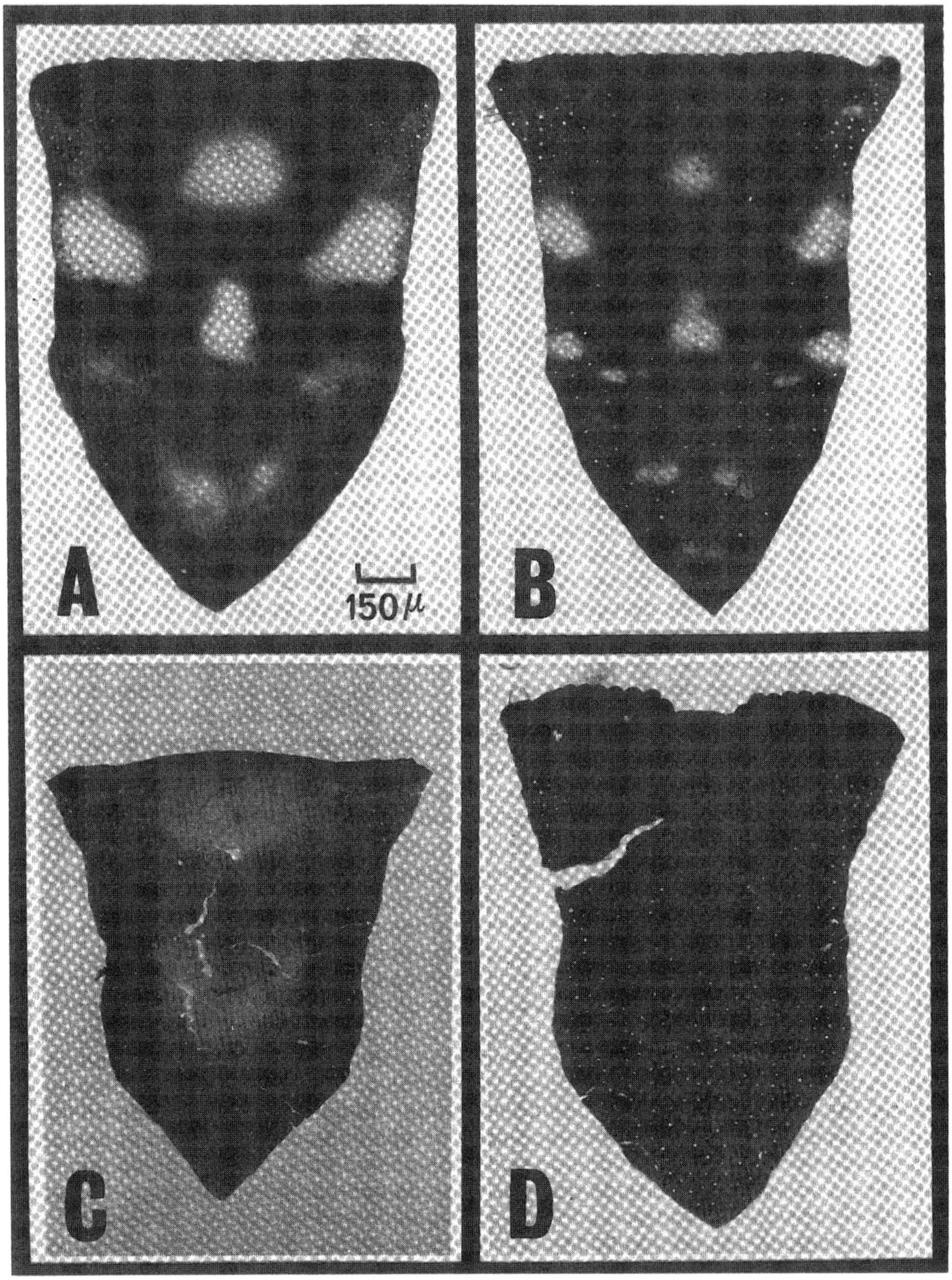


Figure 1. Frontoclypeal sclerites. (A). modern specimen of *Hydropsyche alternans* (Walker) from Lake Winnipeg (B). fossil *Hydropsyche morosa* Hagen from Au Sable River, Michigan, (approximate age 4,000 B.P.) (C). fossil *Hydropsyche confusa* (Walker) from the Don Valley, Toronto, Ontario (approximate age 100,000+ B.P.)(D). fossil *Cheumatopsyche* sp. from the Don Valley.

patterns, positions and arrangements of muscle scars, positions of primary setae and sensory pits, and positions and density of secondary setae are also used. The frontoclypeus is particularly rich in characters but other sclerites are used.

Modern larval taxonomy can benefit from fossil study. Some species pairs difficult to separate as whole larvae are in fact easier to distinguish as individual sclerites. *Halesus radiatus* (Curtis) and *H. digitatus* (Schrank) for example, can be separated on the basis of the shape of the posterior past of the frontoclypeus (distinctly squared in the case of *H. digitatus*). Good photographs are the best means of comparing fossil and modern sclerites and for this reason I am assembling an atlas of modern sclerites, beginning with photos of the frontoclypeus, pronotum and mesonotum of limnephilid and hydropsychid species.

In terms of palaeoecological interpretation, sites in North America, Britain and Europe have shown that by referring to the ecological literature relevant to assemblages of caddisfies and to the known distributions of individual taxa it is possible to acquire information at two levels (Williams in press). First a probable local habitat can be described and second, deductions can be made about the past climatic conditions. It is not necessary to have species names on all taxa. Many generic and even family identifications contribute information. The quality of interpretation depends on careful analysis and synthesis of a great variety of available information on such topics as distribution, feeding, behaviour, water chemistry, hydrology and biogeography, on comparison of assemblage with modern communities and on co-operation with sedimentologists and biologists working with other groups of organisms.

Caddisflies are joining ranks with beetles (used for about 40 years) as good indicators of past habitats and climates. There are good indications of ecological stability in caddisfly communities over many thousands of years. Fossil assemblages tend to have species composition and associations and relative abundance of taxa similar to modern communities. For example, Ross (1944) described the association of *Brachycentrus lateralis* (Say) and *B. numerosus* (Say) in the Kankakee River of Illinois and the Ohio River in Kentucky in recent times. Fossils of these two species occurred together in a Michigan river 4,000 years ago and in an Ontario river during the last interglacial (Williams in prep.). In fact interglacial assemblages in general seem to be very similar to modern communities. At this point however, caddisfly palaeoecology becomes less straightforward and far more interesting. There is now evidence from several sites to suggest that caddisflies do not always respond to climatic change by rapid immigration and emigration. A comparison of beetle and caddisfly fossils from a British riverine deposit at Queensford believed to be about 39,000 years old, will illustrate. The mainly terrestrial beetles are almost all arctic and alpine species and include no tree beetles (Briggs *et al.* 1985). However, the caddisflies identified to species (Williams unpublished data) still occur in northern and central Europe and Britain. This was however an unusually small caddisfly assemblage consisting of numerous *Hydropsyche contubernalis* McLachlan specimens and lesser numbers of *Ecclisopteryx guttulata* (Pictet), *Melampophylax mucoreus* (Hagen), a probable *Potamophylax* species and an unidentified limnephiline. This low diversity and the presence of algal feeders do suggest a possible interpretation. The beetles (known to disperse rapidly) are probably true indicators of a cold (perhaps rapidly cooling) macroclimate. However, a shallow river in a treeless environment, could provide relatively warm water in summer and a rich supply of algae. Since there was previous soil development (during the preceding warmer period) nutrients for algal growth would be available. In addition, groundwater might aid in maintaining a relatively stable water temperature in winter. Also relevant is the biogeographical evidence which suggests that lotic limnephilids desperse more slowly than lentic species (Otto 1982, Williams unpublished data).

This simplified example, I think serves to illustrate the complexity of factors interacting to produce a fossil assemblage but also hints at the store of information Quaternary fossils have to offer to those interested in ecology and biogeography. Palaeoecology can contribute to our understanding of insect responses and indeed there are questions discussed in the literature and at previous Trichoptera symposia on which fossils can now cast some light. For example, the presence of *Hydropsyche bulgaromanorum* Malicky in Britain in the past (Malicky 1984) now has further confirmation. My studies have yielded fossils of this species from three sites ranging in age from less than 2,000 years B.P. to more than 200,000 years B.P. It should be quite possible to collect in addition very recent sediments to check out the time of disappearance of this and other species from rivers such as the Thames. Why not study distributional shifts of species using the concrete (or rather chitinous) evidence available ?

Studies of intraspecific variation can also make use of fossils to add the element of time. I have for example, interglacial, glacial and post-glacial specimens of the North American species *Hydropsyche morosa* Hagen. Some of these are fairly typical eastern forms while others show degrees of variation in shape and colour patterns of the sclerites. Schefter and Unzicker (1984) discussed variability in the *morosa/bifida* complex and suggested that a single species living in boreal or analogous habitats occurred in North America prior to the last glaciation. Later glaciation divided and separated populations in western and eastern mountain chains leading to differentiation. I suggest that this and similar theories could be profitably investigated by looking at appropriate interglacial, glacial and post-glacial fossils.

In conclusion, I hope that these few briefly outlined examples have illutrated the potential contributions which I feel fossil caddisfly sclerites have to offer not only to palaeoecologists but to those studying modern entire animals, communities and populations.

REFERENCES

Briggs, D.J., Coope, G.R. & Gilbertson, D.D.1985. The chrônology and environmental framework of early man in the Upper Thames Valley. **Brit. Archaeol. Rep.,** 15 : 1-176.

Malicky, H. 1984. The distribution of *Hydropsyche guttata* Pictet and *H. bulgaromanorum* Malicky (Trichoptera : Hydropsychidae), with notes on their bionomics. **Entomologists's Gaz.,** 35 : 257-264.

Otto, C. 1982. Habitat, size and distribution of Scandinavian limnephilid caddisflies. **Oikos,** 38 : 355-360.

Ross, H.H. 1944. The caddisflies or Trichoptera of Illinois, Illinois, **Nat. Hist. Surv. Bull.,** 23 : 1-326.

Schefter, P.W. & Unzicker, I.D. 1984. A review of the *Hydropsyche morosa-bifida* complex in North America (Trichoptera : Hydropsychidae). **Proc. 4th Int. Symp. Trichoptera,** J.C. Morse (ed.), Junk, The Hague : 331-335.

Wilkinson, B. 1981. Quaternary sub-fossil Trichoptera larvae from a site in the English Lake District. **Proc. 3rd Int. Symp. Trichoptera,** G.P.Moretti (ed.), Junk, The Hague : 409-419.

Wilkinson, B. 1984. Interpretation of past environments from sub-fossil caddis larvae. **Proc. 4th Int. Symp. Trichoptera,** J.C.Morse (ed.), Junk, The Hague : 447-452.

Williams, N.E. In Press. The use of caddisflies (Trichoptera) in Palaeoecology. **Palaeogeog. Palaeoclimatol. Palaeoecol.**

Williams, N.E. & Morgan, A.V. 1977. Fossil caddisflies (Insecta : Trichoptera) from the Don Formation, Toronto, Ontario, and their use in paleoecology. **Can. J. Zool.,** 55 : 519-527.

Williams, N.E., Westgate, J.A., Williams, D.D., Morgan, A. & Morgan, A.V. 1981. Invertebrate fossils (Insecta : Trichoptera, Diptera, Coleoptera) from the Pleistocene Scarborough Formation at Toronto, Ontario, and their palaeoenvironmental significance. **Quat. Res.,** 16 : 146-166.

# TRICHOPTERA SUB-FOSSILS FROM TEMPERATE RUNNING WATER SEDIMENTS

B.J. WILKINSON

## SUMMARY

Sub-fossil caddis larval remains were extracted from sediments collected from two archaeological sites, one by a brook in Birmingham, the other from the banks of the river Severn at Stoutport. Both watercourses are now polluted, the brook is very badly polluted.

The species identified can be found today in the English Midlands, e. g. *Cheumatopsyche lepida* and *Hydropsyche contubernalis* from the river, *Silo pallipes* and *Drusus annulatus* from the brook. However no exact parallel can be drawn with the present day as the now rare species *Lasiocephala basalis* was present in both samples.

The majority of sub-fossil Trichoptera bearing deposits contain a lake or pond fauna, as lakes and ponds form the most frequently preserved sediments, (Elias and Wilkinson 1983). However a considerable percentage are from the alluvial plains of rivers, or, more rarely, of streams or brooks. Two such sites are described in this paper.

Both of these sites are associated with English archaeological remains (Table 1) : the Stourport Marina excavation took place on mediaeval deposits on the banks of the River Avon, while the Cob Lane excavation took place on Bronze Age brookside deposits radiocarbon dated at 3120 ± 70 years BP. in Bournville Birmingham. The study was done in association with an as yet unpublished analysis of the Coleoptera from these two sites by Peter Osborne of Birmingham University. The Cob Lane material was collected in two samples, one a large bulk sample, and another smaller one a short distance away. The two samples were biologically identical and so were merged for entomological purposes. The Stourport material came as a series of samples (Table 2). These samples show few signs of much change in conditions, with the exception of the frequency of *Lasiocephala basalis* in sample 9. The Stourport samples were not all of the same size.

A flowing water assemblage does not usually simply represent those insects that lived over the place of deposition. Most fossils have been washed down a short distance, in general only a few metres. A small percentage will either have been carried a great distance, from the head of the river, or have been carried into the river by a smaller tributary. Thus, stream and brook species may be found in rivers, but not vice versa.

An additional possible problem is the reworking of deposits. If the river or stream later cuts into its own deposits, any sub-fossil remains may then combine with the fresh material giving a mixed deposit. However this is quite rare ; but if it occurs differential weathering will give the two groups of Trichoptera sub-fossils a different texture and appearance. Moreover, sub-fossil remains are brittle and will break up if they are dried out or transported any substantial distance.

The pattern within the river deposit may be further complicated by braiding, or even by oxbow lake formation. Canalization and dredging of rivers in recent centuries has also changed the natural flow patterns. Of the two water bodies considered in this paper the Cob Lane brook probably appears superficially much the same as it did in the Bronze Age, while the River Avon at Stourport which is now in one steep banked channel was probably braided to some extent and will certainly have been much shallower in places. Both water courses are now heavily polluted. The Avon no longer carries the fauna it did in mediaeval times, and a two hour collecting visit by the author some years back in the Bournbrook about 2 kilometres below Cob Lane yielded a lone *Gammarus*.

As would be expected from the source of the material the species are predominantly those of flowing water, Hydropsychids and Goerids are common. Rhyacophilids, Glossosomatids, Lepidostomatids and Odontocerids are also found.

Hydropsychid remains are prominent in both sites. The specimens were identified using modern material from the author's collection, and that of the British Museum (Natural History), and using the papers by

Hildrew and Morgan, Szczesny and other more recent papers. The sclerite used for identification was usually the frontoclypeus but the parietals and ventral sclerites also proved useful on occasion. On the frontoclypeus features used included the type of anterior margin, the position and number of setigerous pores and the overall shape. These features were used by Szczesny in his key to the Central European Hydropsychid larvae, and by subsequent authors, and by Moseley in his key to sub-fossil Hydropsychids. Incidentally, the *H. fulvipes* in Moseley's paper is actually *H. exocellata* (Moseley pers. comm.). The European literature on Hydropsychid larvae is well enough developed to allow specific identifications even in this large and complex group.

Table 1. Sub-fossil Trichoptera from Bournville and Stourport compared with modern Trichoptera from Hay on Wye.

| | Bournville | Stourport | Hay on Wye |
|---|---|---|---|
| *Rhyacophila (dorsalis)* group | 3 | - | - |
| *Glossosoma* sp. | - | 2 | - |
| *Agapetus* sp. | 1 | - | - |
| *Philopotamus* cf. *montanus* Don. | - | 1 | - |
| *Hydropsyche angustipennis* Curtis | 2 | - | - |
| *Hydropsyche contubernalis* McL. | - | 23 | * |
| *Hydropsyche instabilis* Curtis | - | 2 | - |
| *Hydropsyche pellucidula* Curtis | 2 | 42 | * |
| *Hydropsyche siltalai* Dohler | 1 | 3 | - |
| *Cheumatopsyche lepida* Pictet | - | 7 | * |
| *Polycentropus flavomaculatus* Pictet | - | 1 | * |
| *Polycentropus irroratus* Morton | 2 | - | - |
| *Cyrnus trimaculatus* Curtis | 1 | - | - |
| *Psychomyia pusilla* Fabr. | - | 1 | * |
| *Lype reducta* Hagen | 8 | 1 | - |
| *Tinodes* sp. | 1 | 2 | - |
| *Brachycentrus subnubilus* Curtis | 4 | 6 | - |
| *Drusus annulatus* Stephens | 14 | 2 | - |
| *Limnephilus* sp. | - | 1 | * |
| *Anabolia* cf. *nervosa* | - | 1 | * |
| *Potamophylax* cf. *latipennis* | 10 | 6 | - |
| *Halesus* sp. | 4 | 3 | - |
| *Micropterna* cf. *sequax* | - | 1 | - |
| *Chaetopteryx* cf. *villosa* | 13 | 2 | * |
| *Goera pilosa* Fabr. | 1 | 1 | * |
| *Silo nigricornis* Pictet | 2 | - | - |
| *Silo pallipes* Fabr. | 14 | 7 | - |
| *Lepidostoma hirtum* Fabr. | - | 14 | * |
| *Lasiocephala basalis* Kol. | 1 | 14 | - |
| *Sericostoma* cf. *personatum* | 21 | 17 | * |
| *Beraea pullata* Curtis | 1 | - | - |
| *Odontocerum albicorne* Scop. | 14 | - | - |

N.B. seven Leptocerids were also found at Hay.

Other groups are not as well covered. Many of the species marked 'genus' cf. 'species' are so named because the author could not obtain either specimens or adequate illustrations of specimens of various mainland European species, eg. *Chaetopteryx* spp. Others are both distinctive, and in northern Europe unique, eg. *Odontocerum albicorne*.

Determination of the Glossosomatids from these two, and other flowing water sites to species proved to be impossible. Despite apparently clear distinguishing features the keys gave little help with fragmentary fossil remains and were difficult enough to apply to complete modern specimens.

Both of the assemblages represent a fauna typical of their sites if they were unpolluted. I have placed alongside the Stourport records the results of a small collection of Trichoptera larvae I made in early May

1983, to find only two major discrepancies between the two faunae. The first is the total lack of Leptocerids among the Stourport sub-fossils. The second is the absence of *Lasiocephala basalis* from the Wye at Hay.

Table 2. Sub-fossil Trichoptera remains from Stourport marina mediaeval deposits. (Minimum number of individuals).

| | 12 | 11 | 9 | 8 | 3 | 2 | tree frags. |
|---|---|---|---|---|---|---|---|
| *Glossosoma* sp. | - | - | 1 | - | 1 | - | - |
| *Philopotamus montanus* | - | - | 1 | - | - | - | - |
| *Hydropsyche contubernalis* | 1 | 2 | 8 | 11 | - | 1 | - |
| *Hydropsyche instabilis* | 1 | - | - | - | 1 | - | - |
| *Hydropsyche pellucidula* | 5 | 7 | 13 | 11 | 4 | 2 | - |
| *Hydropsyche siltalai* | 1 | 1 | 1 | - | - | - | - |
| *Cheumatopsyche lepida* | - | - | 4 | 2 | 1 | - | - |
| *Polycentropus flavomaculatus* | - | - | 1 | - | - | - | - |
| *Psychomyia pusilla* | - | ? | 1 | - | - | - | - |
| *Lype reducta* | - | - | 1 | - | - | - | - |
| *Tinodes* sp. | - | - | 2 | - | - | - | - |
| *Brachycentrus subnubilus* | - | - | 2 | - | - | - | - |
| *Drusus annulatus* | - | - | 2 | - | - | - | - |
| *Limnephilus* sp. | - | ? | 1 | - | - | ? | - |
| *Anabolia* cf. *nervosa* | 1 | - | 1 | - | - | - | - |
| *Potamophylax* cf. *latipennis* | - | 1 | 1 | 2 | 1 | ? | - |
| *Halesus* sp. | - | - | 2 | 1 | - | - | - |
| *Micropterna* cf. *sequax* | - | - | - | 1 | - | - | - |
| *Chaetopteryx* cf. *villosa* | - | - | - | - | 1 | 1 | - |
| *Goera pilosa* | - | 1 | - | - | - | - | - |
| *Silo pallipes* | - | 2 | 1 | 3 | 1 | - | - |
| *Lepidostoma hirtum* | - | 1 | 10 | 1 | - | - | 2 |
| *Lasiocephala basalis* | - | - | 14 | - | - | - | - |
| *Sericostoma* cf. *personatum* | 1 | 1 | 8 | 3 | 1 | 3 | - |

Figure 1. Pronotum of *Lepidostoma hirtum* from Stourport.

The lack of Leptocerids is worrying since it is very possibly due to imperfect collecting techniques. The sieve mesh size used in the processing of sub-fossil materials is 300 microns. This permits some of the smaller sclerites to be washed through the mesh. Some larger transparent sclerites, for example those of some species of *Ceraclea*, could be missed in sorting. Certainly from the number of modern specimens collected in the field it would appear that in most sub-fossil sites Leptocerids are consistently under-represented.

*Lasiocephala basalis* Kol. is present in both of these sites and is also well represented in other flowing water sub-fossil assemblages. The first specimen found was that from Bournville, represented by a single pronotal half. Initial scanning of the literature proved fruitless. Further study of the specimen revealed a structure reminiscent of that of the pronotum of *Lepidostoma hirtum*. A specimen of *Lasiocephala basalis* at the British Museum (Natural History) was compared with the sub-fossil. The two proved to be almost identical. When further specimens of this pronotum were encountered in the Stourport deposits with an associated frontoclypeus, the frontoclypeus too proved to be nearly identical to that of the British Museum specimen.

*Lasiocephala basalis* is not common in Britain today, in contrast to the apparent situation only a few hundred years ago. While the Bronze Age was warmer than at present in the Midlands, mediaeval times were, if anything, cooler, so temperature is unlikely to have been a decisive factor. Destruction of the local habitat by man is more likely to have been important, perhaps linked with the slow destruction of the forests in post mediaeval times, either by destroying shade, or by an increase in the number of suspended particles through soil erosion, which would damage the clarity of the water.

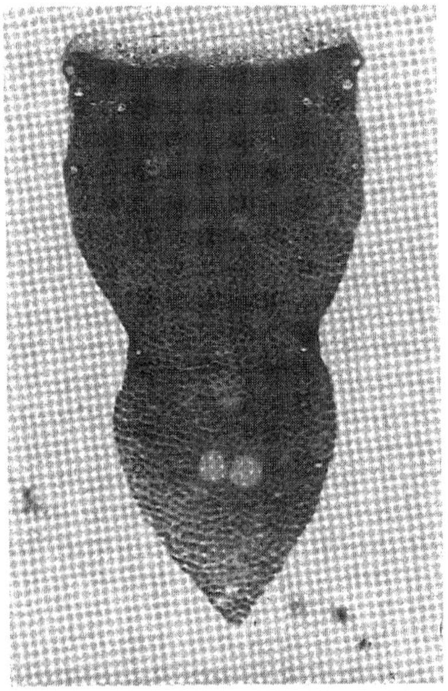

Figure 2. *Lasiocephala basalis* frontoclypeus

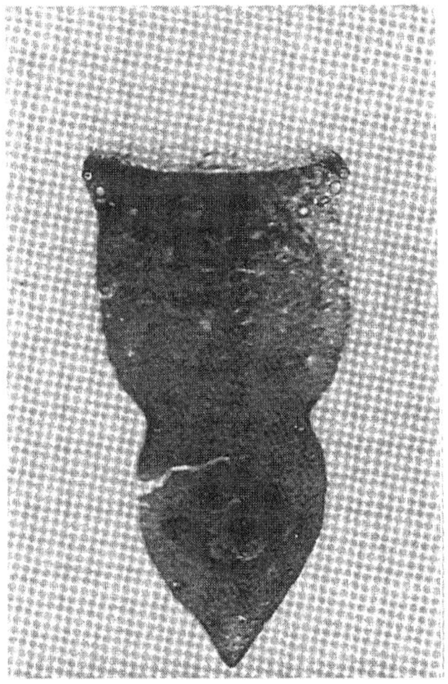

Figure 3. *Lepidostoma hirtum* frontoclypeus.

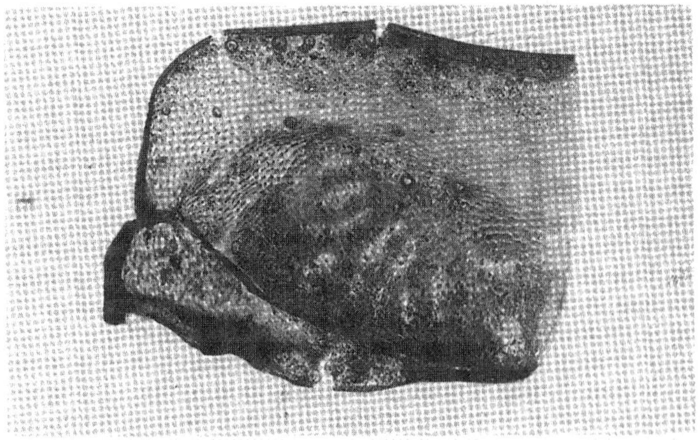

Figure 4. *Lasiocephala basalis* pronotum from Stourport.

## REFERENCES

Botosaneanu, L. & Malicky, H. 1978. Trichoptera. *In* Limnofauna Europaea, J. Illies (ed.), 2nd ed.Gustav Fischer Verlag, Stuttgart.

Edington, J.M. & Hildrew, A.G. 1981. A key to the Caseless Caddis larvae of the British Isles : with notes in their ecology. **Fresh. Biol. Ass. Sc. Pub.**, 43, 92 p.

Elias, S.A. & Wilkinson, B.J. 1983. Lateglacial insect fossil assemblages from Lobsigensee (Swiss plateau). Studies in the Late Quaternary of Lobsigensee 3. **Rev. Paleobiol.**, 2(2) : 189-204.

Hildrew, A.C. & Morgan, J.C. 1974. The taxonomy of the British Hydropsychidae (Trichoptera). **J. Ent. (B).** 43 (2) : 217-229.

Moseley, K.A. 1978. A preliminary report on Quaternary fossil Caddis larvae (Trichoptera). **Quat. News.** 26 : 2-12.

Szczesny, B. 1974. Larvae of the genus *Hydropsyche* (Insecta : Trichoptera) from Poland. **Polsk. Archiv. Hydrobiol.**, 21 (3/4) : 387-390.

# CARIBBEAN AMBER CADDISFLIES - BIOGEOGRAPHICAL ASPECTS

W. WICHARD

## SUMMARY

The fossil fauna of Trichoptera of the Dominican amber and the present-day fauna of the Greater Antilles are composed of nearctic, neotropical and Ethiopian elements. All three elements were already present in the early Miocene. This indicates that the colonization by caddisflies from these three biogeographical regions progressed simultaneously with the development of the Greater Antilles and that the stormy development favoured endemic forms.

## INTRODUCTION

Besides the vicariance model of Caribbean biogeography (Rosen 1975) and the facts of Caribbean tectonics (Perfit and Heezen 1978), Dominican amber represents a third factor which will be of great importance to the future clarification of the origin of the Caribbean and especially the colonization of the Greater Antilles by plants and animals. Its numerous and well-preserved inclusions provide an excellent insight into the living world of the Caribbean in the early Miocene, i.e. 20-23 million years ago (Baroni Urbani and Saunders 1980). Dominican amber documents a surprising richness of fauna, especially of insects, from that early Miocene period, in which the tectonic movements gradually ceased and the Greater Antilles slowly reached their present state, but when Jamaica was still submerged until the middle of the Miocene. The Trichoptera play a subordinate role in the insects found in Dominican amber.

## BIOGEOGRAPHICAL ASPECTS

The caddisflies account for less than 0.1% of all inclusions and so are rare, like all aquatic insects in the Dominican amber (Wichard 1984). Up to now (Table 1) I know of 112 caddisflies from various collections. Though this is a very small number, it is enough to permit an initial comparison with the present-day fauna in the Greater Antilles. 8 of the recent 12 families and 12 of the recent 26 genera are represented in the inclusions. One genus is described new from Dominican amber. This indicates, firstly, that the early Miocene fauna was similar to that of the present and, secondly, that, as the number of Trichoptera found in Dominican amber increases, further genera occurring in the Greater Antilles today will probably be identified, unless certain still missing genera only immigrated after the early Miocene, particularly as the Greater Antilles were still "*in statu nascendi* " at that time.

This raises the question of the origin of the trichopteran fauna in the Caribbean and in the Greater Antilles. Flint (1978) used the present fauna to answer this biogeographic question and presented three relationships. The fauna of the Greater Antilles is composed of neotropical, nearctic and Ethiopian elements, whereby the neotropical element predominates. According to the fossil caddisflies in the Dominican amber all three elements were already present in the early Miocene. This indicates that the colonization from these three biogeographical regions already occurring at an early date, during or at the beginning of the emergence of the Greater Antilles. Thus one must assume that the colonization by caddisflies progressed simultaneously with the development of the Greater Antilles and that this stormy development favoured endemic forms.

Particularly striking is the preponderance of the genus *Antillopsyche* (Polycentropodidae), accounting for almost 50% of all the amber caddisflies. Almost all the specimens belong to one species : *Antillopsyche oliveri* Wichard 1985. The high number of individuals could be due to : 1. the particular affinity of the imagoes for resin, which presumably flowed on the Hymenaea trees and 2. a genuine dominance of distribution of the genus *Antillopsyche*. However, the evidence of *Antillopsyche* in the early Miocene from Dominican amber is especially interesting because this dominant genus is endemic today and belongs to the subfamily Pseudoneuroclipsinae, whose distribution is otherwise restricted to the Ethiopian and Oriental region. The genus *Antillopsyche* is possibly a very early relict of the Mesozoic when Africa was in close contact with

South America. Either the Miocene amber forest had already been colonized by this genus in the Mesozoic, in which case the Greater Antilles are partly older than currently assumed, or - less likely - the genus immigrated by the roundabout South America route in the early Cenozoic, in which case, hope exists that closely related members of the subfamily Pseudoneuroclipsinae will still be found on the neotropical continent.

Table 1. The Trichoptera of the Greater Antilles (Flint 1968, Botosaneanu 1979, Kumanski 1987) and of the Dominican amber (Wichard unpublished).

| FAMILY | GENUS | GREATER ANTILLES | DOMINICAN AMBER | - Specimens - |
|---|---|---|---|---|
| Atopsychidae | *Atopsyche* | x | x | - 1 - |
| Glossosomatidae | *Campsiophora* | x | x | - 5 - |
| | *Cariboptila* | x | | |
| | *Cubanoptila* | x | x | - 5 - |
| Philopotamidae | *Chimarra* | x | x | - 17 - |
| Xiphocentronidae | *Xiphocentron* | x | | |
| Polycentropodidae | *Antillopsyche* | x | x | - 53 - |
| | *Cernotina* | x | x | - 5 - |
| | *Polycentropus* | x | | |
| Psychomyiidae | *Austrotinodes* | x | | |
| Hydropsychidae | *Leptonema* | x | x | - 3 - |
| | *Macronema* | x | | |
| | *Smicridea* | x | | |
| | *Calosopsyche* | x | | |
| | *Palaehydropsyche* | - | x | - 1 - |
| Hydroptilidae | *Leucotrichia* | x | | |
| | *Alisotrichia* | x | x | - 2 - |
| | *Neotrichia* | x | | |
| | *Hydroptila* | x | | |
| | *Oxyethira* | x | x | - 2 - |
| | *Orthotrichia* | x | | |
| | *Ochrotrichia* | x | x | - 9 - |
| Leptoceridae | *Oecetis* | x | | |
| | *Nectopsyche* | x | x | - 4 - |
| Odontoceridae | *Marilia* | x | | |
| Calamoceratidae | *Phylloicus* | x | | |
| Helicopsychidae | *Helicopsyche* | x | x | - 5 - |
| | | 26 | 13 | - 112 - |

The neotropical element forms the second largest group of caddisflies in Dominican amber with almost 50% of all specimens. In contrast to the*Antillopsyche* the amber caddisflies in this group are distributed among almost all the remaining genera. Thus all eight families found in the amber are predominately or exclusively of neotropical origin and have immigrated from South America. An example of the richness of species in the neotropical element is provided by the genus *Chimarra* which occurs as the subgenus *Curgia* in the Greater Antilles and in the rest of the Neotropis. The 17 caddisflies of the genus *Chimarra* embedded in amber belong to 7 fossil species (Wichard 1983 a, b). They indicate the evident trend in the neotropical element, i.e., the proliferation of the species, which was possibly stimulated by the simultaneous development of the Caribbean.

Barely 1% of all fossil Trichoptera in Dominican amber of the early Miocene belong to the nearctic element, which is represented by the subfamily Hydropsychinae. Revision of this taxon by Ross und Unzicker (1977) has placed the genus *Calosopsyche* systematically nearer the genus *Plectropsyche*. *Calosopsyche* is chiefly found in the Greater Antilles but also in Panama (Flint and Bueno Soria 1987). *Plectropsyche* is distributed in Central America, but does not occur in the Greater Antilles. The only caddisfly of nearctic origin found in Dominican amber is *Palaehydropsyche fossilis* Wichard 1986 and belongs to the *Hydropsyche* branch and is closely related to both genera, *Calosopsyche* and *Plectropsyche*. Comparison of the frequency of occurrence of Ethiopian and neotropical elements in Dominican amber with that of the nearctic element suggests that the latter with its related *Hydropsyche* branch was only able to colonize the Greater Antilles relatively late, perhaps only since the Oligocene/Miocene or later. This assumption will be supported if Dominican amber reveals no further specimens of *Calosopsyche* in the future.

## REFERENCES


Baroni Urbani, C. & Saunders, J.B. 1980. The fauna of the Dominican Republic amber : the present status of knowledge. - **Mem. 9a Conf. geol. Caribe, Santo Domingo**, 1 : 213-223.

Botosaneanu, L. 1979. The Caddisflies (Trichoptera) of Cuba and of Isla De Pinosa : a synthesis. - **Stud. Fauna Curaçao other Caribbean Islands**, 59 : 33-62.

Flint, O.S. 1968. The Caddisflies of Jamaica (Trichoptera). - **Bull. Inst. Jamaica, Sc. Ser.**, 19 : 5-67.

Flint, O.S. 1978. Probable origins of the West Indian Trichoptera and Odonata fauna. - **Proc. 2nd Int. Symp. Trichoptera**, M.I. Crichton (ed.), Junk, The Hague : 215-223.

Flint, O.S. & Bueno Soria J. (in press). Studies on Neotropical Caddisflies, XXXVII : The Genus *Calosopsyche* in Central America, with Descriptions of its Immature Stages (Trichoptera : Hydropsychidae). - **Proc. 5th Int. Symp. Trichoptera**.

Kumanski, K. 1987 (in press) On Caddisflies (Trichoptera, Insecta) of Cuba. - **Acta Zool. Bulg.**, 35.

Perfit, M.R. & Heezen, B.C. 1978. The geology and evolution of the Cayman Trench. - **Geol. Soc. Amer. Bull.**, 89 : 1155-1174.

Rosen, D. 1975. A vicariance model of Caribbean biogeography. - **Syst. Zool.**, 24 : 431-464.

Ross, H.H. & Unzicker, J.D. 1977. The Relationships of the Genera of American Hydropsychinae as indicated by Phallic Structures (Trichoptera, Hydropsychidae). - **J. Georgia Ent. Soc.**, 12 : 298-312.

Wichard, W. 1983a. Köcherfliegen des Dominikanischen Bernsteins - II. Fossile Arten der Gattung *Chimarra* (Trichoptera, Philopotamidae). - **Mitt. Münch. Ent. Ges.**, 72 : 137-145.

Wichard, W. 1983b. Köcherfliegen des Dominikanischen Bersteins - III. *Chimarra succini* n.sp. (Stuttgarter Bernsteinsammlung Trichoptera, Philopotamidae). - **Stuttgarter Beitr. Naturk., B**, 95 : 1-8.

Wichard, W. 1984. Fossil Caddisflies in Fossil Resins. - **Proc. 4th Int. Symp. Trichoptera**, J.C. Morse (ed.), Junk, The Hague : 441-444.

Wichard, W. 1985. Köcherfliegen des Dominikanischen Bersteins - IV. *Antillopsyche oliveri* spec. nov. (Trichoptera, Polycentropodidae). **Stud. Neotrop. Fauna Environment**, 20 : 117-124.

Wichard, W. 1986. Köcherfliegen des Dominikanischen Bernsteins - V. *Palaehydropsyche fossilis* gen. n. sp. n. (Trichoptera, Hydropsychidae). - **Stud. Neotrop. Fauna Environment**, 21.

# 4. BIOGEOGRAPHY

# A REVIEW OF THE IRISH TRICHOPTERA

J.P. O'CONNOR

## SUMMARY

A provisional list of the Irish Trichoptera is presented along with a synopsis of the known range of each species. The more interesting distributions are discussed. Other aspects of the Irish trichopterous fauna are also reviewed.

## INTRODUCTION

Ireland is the largest island on the European coasts with the exception of Great Britain, the area of which is nearly three times as great. Britain intervenes between Ireland and continental Europe in such a way as to render it difficult for continental influences to reach Ireland except by passage across the larger island, a circumstance which has affected profoundly the history and present condition of Ireland, its fauna and flora included (Praeger 1950).

The island is a low-lying country whose central lowlands, which commonly lie between 60 m and 120 m above sea-level, are diversified by isolated hills and plateaux and surrounded by a discontinuous upland rim which rarely exceeds 910 m. Since Ireland is an outpost in the Atlantic Ocean, maritime influences predominate. Harsh frosts are rare ; only the very centre of the country will know more than one day in the year when temperatures will not rise above freezing point throughout the twenty-four hours. High summer temperatures are rare : 33°C is the highest ever recorded. The annual range is only 9°C in the south-west and 10°C in the east. The amount of rainfall is not excessive, 1400 mm in the south-west and 700 mm in the east, but the number of days on which it falls is very high. Numerous rain-days and infrequent droughts mean that humidity is high and evaporation low (Mitchell 1976). Due to its saucer-like shape and the high humidity, Ireland is an island extremely rich in aquatic habitats suitable for Trichoptera. It abounds in lakes, rivers, streams, bogs and other wetlands. Because of its unique geographical position, topography, climate etc., the composition of its fauna and flora is of great interest to biogeographers and it has been assigned a separate zone (17) in Limnofauna Europaea (Illies 1978).

Since my progress report on studies of Irish Trichoptera (O'Connor 1978a), research has continued on the group in Ireland and it is now possible to present a provisional list of the species occurring on the island. Much Irish material preserved in museums in Great Britain and Ireland has been reidentified and this has yielded new information including deletions due to misdeterminations. Since a detailed account of the Irish trichopterous fauna is now in preparation, I will only deal here with those species known with certainly from the island. A total of 143 species has been authentically recorded and these are listed below. A brief summary of their distribution is included. This is based on a county basis (Fig. 1). More detailed distributional data will be published later. The nomenclature follows Barnard (1985).

## A PROVISIONAL LIST OF IRISH TRICHOPTERA

### Rhyacophilidae

1. *Rhyacophila dorsalis* (Curtis) : widespread ; 2. *R. munda* McLachlan : widespread ; 3. *Glossosoma boltoni* Curtis : widespread ; 4. *G. conformis* Neboiss : at present only known from Wicklow in eastern Ireland ; 5. *Agapetus delicatulus* McLachlan : only known from the southern half of the island ; 6. A. *fuscipes* Curtis : widespread ; 7. *A. ochripes* Curtis : widespread.

### Hydroptilidae

This group is still underrecorded in Ireland.

8. *Agraylea multipunctata* Curtis : widespread ; 9. *Allotrichia pallicornis* (Eaton) : Dublin and Wicklow in eastern Ireland ; 10. *Hydroptila angulata* Mosely : Galway and Kerry ; 11. *H. cornuta* Mosely : Cavan, Clare, Kerry and Wicklow ; 12. *H. forcipata* (Eaton) : widespread ; 13. *H. martini* Marshall : Cavan, Kerry, Kildare,

Westmeath and Wexford. 14. *H. occulta* (Eaton) : Cavan and Tipperary ; 15. *H. pulchicornis* Pictet : Cavan, Clare and Kerry ; 16. *H. simulans* Mosely : western and south-western Ireland ; 17. *H. sparsa* Curtis : widespread ; 18. *H. tigurina* Ris : pupae collected in the River Caragh, Kerry (O'Connor 1978b) ; 19. *H. tineoides* Dalman : widespread ; 20. *Oxyethira falcata* Morton : Cavan, Clare, Galway and Mayo ; 21. *O. flavicornis* (Pictet) : widespread ; 22. *O. frici* Klapalek : Cavan, Clare, Galway, Kerry and Mayo ; 23. *O. sagittifera* Ris : Galway and Kerry ; 24. *O. simplex* Ris : Galway and Kildare ; 25. *O. tristella* Klapalek collected by Mosely (1930) in Galway and by the author in Kerry in July 1986 ; 26. *Ithytrichia clavata* Morton : adults taken in a light-trap on the River Curraheen, Kerry (O'Connor 1982) ; 27. *I. lamellaris* Eaton : widespread ; 28. *Orthotrichia angustella* (McLachlan) : Clare and Kerry ; 29. *O. costalis* (Curtis) : Clare and Kerry.

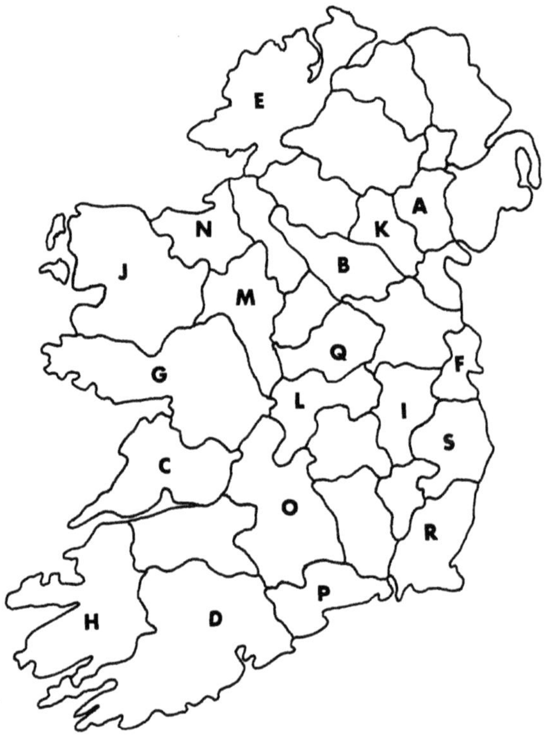


Figure 1. The Irish counties. The following are cited in the text : Armagh-A ; Cavan-B ; Clare-C ; Cork-D ; Donegal-E ; Dublin-F ; Galway-G ; Kerry-H ; Kildare-I ; Mayo-J ; Monaghan-K ; Offaly-L ; Roscommon-M ; Sligo-N ; Tipperary-O ; Waterford-P ; Westmeath-Q ; Wexford-R ; Wicklow-S.

**Philopotamidae**

30. *Philopotamus montanus* (Donovan) : widespread ; 31. *Wormaldia mediana* McLachlan : Cork, Kerry and Wicklow ; 32. *W. occipitalis* (Pictet) : widespread ; 33. *W. subnigra* McLachlan : Cavan, Kerry and Westmeath ; 34. *Chimarra marginata* (L.) : widespread.

**Psychomyiidae**

35. *Lype phaeopa* (Stephens) : widespread ; 36. *L. reducta* (Hagen) : Cork, Kerry, Monaghan and Tipperary ; 37. *Metalype fragilis* (Pictet) : Cavan, Clare, Galway, Roscommon and Westmeath ; 38. *Psychomyia pusilla* (Fabricius) : widespread ; 39. *Tinodes dives* (Pictet) : male taken beside a swift flowing stream at Ben Bulben, Sligo (O'Connor and Good 1984) ; 40. *T. maclachlani* Kimmins ; widespread ; 41. *T. maculicornis* (Pictet) : Cavan, Clare, Donegal, Monaghan and Westmeath where it is associated with calcareous waters (O'Connor and Wise 1980) ; 42. *T. unicolor* (Pictet) : Clare, Donegal and Wicklow where it is associated with calcareous streams ; 43. *T. waeneri* (L) : widespread.

**Ecnomidae**

44. *Ecnomus tenellus* (Rambur) : widespread.

**Polycentropodidae**

45. *Cyrnus flavidus* McLachlan : records from central, south-western and western Ireland ; 46. *C. insolutus* McLachlan : occurs in Lough Derrygeeha, Clare (O'Connor 1977) ; 47. *C. trimaculatus* (Curtis) : widespread; 48. *Holocentropus dubius* (Retzius) : Clare, Cork, Kerry and Monaghan ; 49. *H. picicornis* (Stephens) : widespread ; 50. *Neureclipsis bimaculata* (L.) : widespread ; 51. *Plectrocnemia conspersa* (Curtis) : widespread; 52. *P. geniculata* McLachlan : widespread ; 53. *Polycentropus flavomaculatus* (Pictet) : widespread ; 54. *P. irroratus* (Curtis) : widespread, 55. *P. kingi* McLachlan : widespread.

**Hydropsychidae**

56. *Cheumatopsyche lepida* (Pictet) : widespread ; 57. *Hydropsyche angustipennis* (Curtis) : widespread ; 58. *H. contubernalis* McLachlan : widespread but local ; 59. *H. instabilis* (Curtis) : Dublin and Wicklow ; 60. *H. pellucidula* (Curtis) : widespread ; 61. *H. siltalai* Dohler : widespread ; 62. *Diplectrona felix* McLachlan : widespread.

**Phryganeidae**

63. *Agrypnia obsoleta* (Hagen) : widespread ; 64. *A. pagetana* Curtis : widespread ; 65. *A. varia* (Fabricius) : widespread ; 66. *Oligotricha striata* (L.) : only known from two specimens taken respectively in Roscommon and Westmeath near the turn of the century. The identity of both specimens has been confirmed ; 67. *Phryganea bipunctata* Retzius : widespread ; 68. *P. grandis* L. : widespread.

**Lepidostomatidae**

69. *Crunoecia irrorata* (Curtis) : widespread ; 70. *Lasiocephala basalis* (Kolenati) : Kerry ; 71. *Lepidostoma hirtum* (Fabricius) : widespread.

**Limnephilidae**

72. *Apatania auricula* (Forsslund) : widespread in Kerry with a single record from Clare ; 73. *A. muliebris* McLachlan : two adults collected in Kerry have been determined respectively as *A. cimbrica* Nielsen and *A. nielseni* Schmid but until the status of these taxa are clarified, they are included here as forms of *A. muliebris*; 74. *A. wallengreni* McLachlan : Donegal, Galway, Mayo and Westmeath ; 75. *Drusus annulatus* (Stephens) : widespread ; 76. *Ecclisopteryx guttulata* (Pictet) : Wicklow : 77. *Halesus digitatus* (Schrank) : Cavan, Dublin, Kerry and Kildare ; 78. *H. radiatus* (Curtis) : widespread ; 79. *Hydatophylax infumatus* (McLachlan) : an adult light-trapped on the River Shannon, Cavan ; 80. *Mesophylax impunctatus* McLachlan : Clare and Kerry ; 81. *Micropterna lateralis* (Stephens) : widespread ; 82. *M. sequax* McLachlan : widespread ; 83. *Potamophylax cingulatus* (Stephens) : widespread ; 84. *P. latipennis* (Curtis) : widespread ; 85. *Stenophylax permistus* McLachlan : widespread ; 86. *Chaetopteryx villosa* (Fabricius) : widespread ; 87. *Anabolia nervosa* (Curtis) : widespread ; 88. *Glyphotaelius pellucidus* (Retzius) : widespread ; 89. *Grammotaulius nigropunctatus* (Retzius) : widespread ; 90. *Limnephilus affinis* Curtis : widespread ; 91. *L. auricula* Curtis : widespread ; 92. *L. binotatus* Curtis : widespread but with few records ; 93. *L. centralis* Curtis : widespread ; 94. *L. coenosus* Curtis : Galway and Kerry ; 95. *L. decipiens* (Kolenati) : Cavan, Galway and Monaghan ; 96. *L. elegans* Curtis : Clare, Kerry, Mayo and Westmeath ; 97. *L. flavicornis* (Fabricius) : widespread ; 98. *L. fuscinervis* (Zetterstedt) : Cavan, Clare, Kildare, Mayo, Monaghan, Offaly and Westmeath. These counties lie in a central band across the country. Larvae were found among leaf and stem debris of reed which had accumulated in erosion holes on the bare peat substratum at the edge of a reed swamp of a small lake in Cavan (Wallace *et al.* 1985) ; 99. *L. griseus* (L.) : widespread ; 100. *L. hirsutus* (Pictet) : widespread ; 101. *L. ignavus* McLachlan : widespread but local ; 102. *L. incisus* Curtis : widespread ; 103. *L. lunatus* Curtis : widespread ; 104. *L. luridus* Curtis : widespread ; 105. *L. marmoratus* Curtis : widespread ; 106. *L. nigriceps* (Zetterstedt) : Armagh, Cavan and Westmeath ; 107. *L. pati* O'Connor : Donegal, Mayo and Westmeath ; 108. *L. rhombicus* (L.) : widespread ; 109. *L. sparsus* Curtis : widespread ; 110. *L. stigma* Curtis : widespread ; 111. *L. vittatus* (Fabricius) : widespread ; 112. *Phacopteryx brevipennis* (Curtis) : taken in an alder marsh in Cavan. This species may have been underrecorded because the adults are retiring and difficult to find (Wallace *et al.* 1983).

**Goeridae**

113. *Goera pilosa* (Fabricius) : widespread ; 114. *Silo nigricornis* (Pictet) : widespread ; 115. *S. pallipes* (Fabricius) : widespread.

**Beraeidae**

116. *Beraea maurus* (Curtis) : widespread ; 117. *B. pullata* (Curtis) : widespread ; 118. *Beraeodes minutus* (L.) : Cavan, Kerry and Tipperary.

**Sericostomatidae**
119. *Sericostoma personatum* (Spence in Kirby and Spence) : widespread.
**Odontoceridae**
120. *Odontocerum albicorne* (Scopoli) : widespread.
**Molannidae**
121. *Molanna albicans* (Zetterstedt) : Armagh, Cavan, Monaghan, Roscommon and Westmeath.
**Leptoceridae**
122. *Atripsodes albifrons* (L.) : widespread ; 123. *A. aterrimus* (Stephens) : widespread ; 124. *A. bilineatus* (L.) : the identity of specimens taken in Westmeath before 1900 has been confirmed but there are no recent records ; 125. *A. cinereus* (Curtis) : widespread ; 126. *A. commutatus* (Rostock) : widespread ; 127. *Ceraclea albimacula* (Rambur) : widespread ; 128. *C. annulicornis* (Stephens) : authentic old records from Donegal, Roscommon and Westmeath. Recently recorded in Kerry ; 129. *C. dissimilis* (Stephens) : widespread ; 130. *C. fulva* (Rambur) : widespread ; 131. *C. nigronervosa* (Retzius) : widespread ; 132. *C. senilis* (Burmeister) : widespread ; 133. *Mystacides azurea* (L.) : widespread ; 134. *M. longicornis* (L.) : widespread ; 135. *Adicella reducta* (McLachlan) : Clare, Cork, Kerry, Tipperary, Waterford and Wicklow ; 136. *Triaenodes bicolor* (Curtis) : widespread ; 137. *Ylodes reuteri* (McLachlan) : only known from three adults collected at Rosslare, Wexford, in 1902 (O'Connor 1979) ; 138. *Oecetis furva* (Rambur) : widespread ; 139. *O. lacustris* (Pictet) : widespread ; 140. *O. notata* (Rambur) : old confirmed records from Dublin and Waterford with a recent one from Tipperary ; 141. *O. ochracea* (Curtis) : widespread ; 142. *O. testacea* (Curtis) : widespread ; 143. *Setodes argentipunctellus* McLachlan : still only known from two adjoining lakes in Kerry where it was first discovered over one hundred years ago (O'Connor and Wise 1984).

DISCUSSION

The origin of the Irish biota has been the subject of much discussion and speculation (see Sleeman *et al.*, in press). The fauna and flora, leaving aside species introduced by man, is the result of survival, evolutionary change and migration during the two million years approximately which make up the Pleistocene epoch (Watts, in press). From the distributional data available (Botosaneanu and Malicky 1978, Barnard 1985), in terms of the number of species present, the Irish fauna represents 73.3% of those of Great Britain and the Western European Lowlands while it is 66.2% of that of the Central Lowlands. It is most similar to the British fauna but three Irish species, *T. maculicornis, A. auricula* and *L. fuscinervis,* have not been taken there. By contrast, 110 species are common to both the Western and Central Lowlands. A total of seven Irish species occur only in the Western Lowlands while 16 are found only in the Central Lowlands. Altogether, nine species (*H. tigurina, A. delicatulus, W. mediana, M. fragilis, T. dives, T. maclachlani, P. kingi, E. guttulata* and *M. lateralis*) are not recorded from either area. *L. pati* has not been included in the above analysis since reliable data are not yet available on its European distribution. It is evident therefore that the Irish trichopterous fauna is formed from different components of the European fauna.

Lack of space prevents a detailed analysis of the Irish Trichoptera here but certain conclusions may be outlined. For a number of coleopterous and dipterous groups, Speight (in press) has demonstrated that the Irish fauna is essentially an extension of the fauna occurring in the western seaboard countries of Europe and that it has arrived *via* Great Britain. This author has also argued that the lack of extant endemic taxa among those surveyed and the great similarity between Irish and British lists provide no support for any notion that the species present today in Ireland have been isolated for any significant length of time. The composition of the Irish Trichoptera supports these assertions. Since the ice departed some 10,000 years ago natural immigration rates among the Trichoptera should be sufficient to explain the presence of all but a handful of species. The latter are now of course the interesting ones. However, it is surprising that some British and mainland European species [e.g. *Brachycentrus subnubilus* Curtis, *Limnephilus extricatus* McLachlan and *Mystacides nigra* (L.)] have not successfully colonised Ireland.

Today, there is increasing destruction of the Irish wetlands caused by *inter alia* arterial and field drainage, turf production, afforestation and the intensification of agriculture (e.g. Flanagan and Toner 1975, Lennox and Toner 1980, Bruton and Convery 1982, Jeffrey 1984, Anon. 1985). It is timely therefore that protection should be considered for the rarer Irish Trichoptera. In this context, it is satisfying that the Irish population of *S. argentipunctellus* now lies within a national park. *H. tigurina* and *C. insolutus* would be also suitable candidates for protection.

REFERENCES

Anon. 1985. Review group on forestry. Dublin : Stationery Office.

Barnard, P.C. 1985. An annotated check-list of the Trichoptera of Britain and Ireland. **Entomologist's Gaz.**, 36 : 31-45.

Botosaneanu, L. & Malicky, H. 1978. Trichoptera. *In* , J. Illies (ed.) Limnofauna Europeae. 2nd edition. Stuttgart : Gustav Fischer Verlag.

Bruton, R. & Convery, F. J. 1982. Land drainage policy in Ireland. Dublin : The Economic and Social Res. Inst.

Flanagan, P.J. and Toner, P.F. 1975. A preliminary survey of Irish lakes. Dublin : An ForasForbartha.

Illies, J. ed. 1978. Limnofauna Europaea. 2nd ed. Stuttgart, Gustav Fischer Verlag.

Jeffrey, D.W. ed. 1984. Nature conservation in Ireland. Progress and problems. Dublin : Royal Irish Acad.

Lennox, L.J. & Toner, P.F. 1980. The national survey of Irish rivers. A third report on water quality. Dublin : An Foras Forbartha.

Mitchell, F. 1976. The Irish landscape. London, Collins.

Mosely, M.E. 1930. Irish Trichoptera, Neuroptera etc. **Entomologist,** 63 : 30-34.

O'Connor, J.P. 1977. Lough Derrygeeha, Co. Clare, a new locality for *Cyrnus insolutus* McLachlan (Trichoptera : Polycentropodidae). **Entomologist's Rec. J. Var.,** 89 : 309-310.

O'Connor, J.P. 1978a. A progress report on studies of Irish Trichoptera. **Proc. 2nd Int. Symp. Trichoptera,** M.I. Crichton (ed.), Junk, The Hague : 303-308.

O'Connor, J.P. 1978b. *Hydroptila tigurina* Ris new to Ireland with notes on *Apatania wallengreni* McLachlan and *Limnephilus binotatus* Curtis (Insecta : Trichoptera). **Ir. Nat. J.,** 19 :191-192.

O'Connor, J.P. 1979. *Triaenodes reuteri* McLachlan new to Ireland with a second Irish record of *Apatania wallengreni* McLachlan (Insecta : Trichoptera). **Ir. Nat. J.,** 19 : 403-404.

O'Connor, J.P. 1982. *Ithytrichia clavata* (Trichoptera : Hydroptilidae), a caddisfly new to Ireland. **Ir. Nat. J.,** 20 : 548-549.

O'Connor, J.P. & Good, J.A. 1984. *Tinodes dives* (Pictet) : a caddisfly new to Ireland from Ben Bulben, Co. Sligo. **Entomologist's Rec. J. Var.,** 96 : 108-109.

O'Connor, J.P. & Wise, E.J. 1980. The larva of *Tinodes maculicornis* (Trichoptera :Psychomyiidae), with notes on the species' distribution and habitat in Ireland. **Freshwat. Biol.,** 10 : 367-370.

O'Connor, J.P. & Wise, E.J. 1984. Observations on the Trichoptera of the Killarney Lakes, Co. Kerry. **Ir. Fish. Invest., Ser. A.,** 24, 15 p.

Praeger, R. L. 1950. Natural History of Ireland. London, Collins.

Sleeman, D.P., Devoy, D.J. & Woodman, P.C. eds. (in press). Proceedings of the postglacial colonisation conference. **Occ. Pub. Ir. Biogeog. Soc.,**1.

Speight, M.C.D. (in press). Use of invertebrates, as exemplified by certain insect groups, in considering hypotheses about the history of the Irish postglacial fauna. *In* Sleeman, D.P. *et al.* (eds). Proc. postglacial colonisation conference. **Occ. Pub. Ir. Biogeog. Soc.,** 1 : 60-66.

Wallace, I.D., Wallace, B., O'Connor, J.P. & 0'Connor M.A. 1985. The larva of *Limnephilus fuscinervis* (Trichoptera : Limnephilidae), with notes on the species' distribution and habitat in Ireland. **Ir. Nat. J.,** 21 : 397-400.

Wallace, I.D. , Wallace, B.& O'Oconnor, M.A. 1983. *Phacopteryx brevipennis* new to Ireland with notes on *Oxyethira simplex* and *Beraeodes minutus* (Insecta : Trichoptera). **Ir. Nat. J.,** 21 : 168-169.

Watts, W. A. (in press) Origin of the Irish flora and fauna. *In* Sleeman, D.P. *et al.* (eds.). Proceedings of the postglacial colonisation conference. **Occ. Pub. Ir. Biogeog. Soc.,** 1 : 9-14.

# AN ATTEMPT TO EVALUATE THE STATE OF THE CADDIS FLY FAUNA OF BELGIUM

Ph. STROOT


## SUMMARY

An evaluation of the conditions of the caddis fly fauna of Belgium is undertaken by comparing its present-day state to the former situation. Rare species appear to be particularly threatened, a few even seem extinct now. Among the more common species, many also show a marked decrease, either throughout the country or only in the more altered lower regions. The causes of their decrease or vanishing are also presented and discussed.


## INTRODUCTION

During the last International Symposium, Botosaneanu (1981) presented a comprehensive report summarizing most of the available information on endangered and vanishing Trichoptera. His particularly worrying picture especially pointed out huge lacunas in our knowledge of the state of many caddis fly faunas, and emphasized the need for more concern in protection oriented studies. In this connection, the general revision and survey carried out by the present author in Belgium provides a choice material for a global evaluation of the state of a caddis fly fauna in a densely populated north-western European country.

## INVESTIGATION AREA - MATERIAL AND METHODS

Located between France, Germany and the Netherlands, Belgium is a small country (30,507 $km^2$) traditionally divided into three main parts, Low, Middle and High Belgium (Fig. 1). These divisions are principally based on altitude, which increases from sea-level in the north-west to more than 500 m in the south-east (the Hautes-Fagnes, culminating at 692 m). Very densely populated, Belgium is heavily marked by both agricultural and industrial practices. Only the Ardennes (Fig. 1), mainly covered with forests, remain relatively spared by human activities.

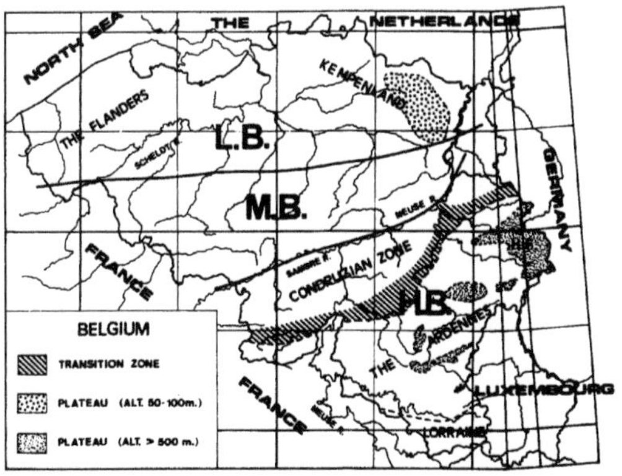


Figure 1. Main regions of Belgium represented in the 50 km U.T.M. grid. L.B., M.B. and H.B. stand for Low, Middle and High Belgium.

The recent general revision and faunal survey of the caddis flies in Belgium resulted in an atlas of 225 distribution maps (Stroot 1984b) and an annotated catalogue (Stroot 1985a). All the gathered information, more than 8000 data, serve for the present evaluation.

Because of the hazardous character of such pluriannual chorological information, a distinction has to be made between rare species, whose capture seems accidental and whose absence of record during a given period may not be significant, and more common species, for which a further treatment is possible (Stroot 1985b).

The arbitrary criterion of rareness is the occurrence in only three or fewer 10 km U.T.M. squares. The eventual decrease of the most common species is evaluated through an index based on a $X^2$ metric. This index measures for each species the differences between the proportion of its data "prior to" and "since" and the proportion of the sum of all data during each of these two periods. Its functioning is explained in Stroot & Depiereux (in prep.).

In accordance with the common usage in the former E.I.S. provisional atlases (see e.g. Heath & Leclercq 1981) and their analyses (Leclercq *et al.* 1980), 1950 was chosen as a hinge to compare the previous to the recent state of the fauna. As concerns the Belgiaàn caddis flies, this date also agrees with the previous general survey (Marlier 1949) and with a gap in the numbers of records (from 1946 to 1958).

RESULTS : PRESENT STATE OF THE FAUNA

Among the 200 species recorded in Belgium (Stroot 1985a), 178 were found before 1950 and 170 afterwards. It means that 22 species have only been collected since 1950, but also that 30 species were exclusively found before. The relative proportion of data prior to and since 1950 (respectively 37,7 and 62,3%) and the recent discovery of 22 species new to the national fauna indicate a greater intensity of the exploration during the last decades. But the fact that 30 species have not been recorded in the recent period suggests the hypothesis of a general impoverishment of the caddis fly fauna in Belgium. As stressed before though, this hypothesis should be envisaged separately for rare and more common species.

**Rare species**

According to our criterion, 48 species, i.e. 24% of the fauna, are considered as rare. Whereas 15 of these species were only discovered recently, mostly during the last ten years, 27 of them were exclusively recorded before 1950. The general impoverishment pointed out above could therefore principally be imputed to the apparent vanishing of rare species. However, on account of their trifling probability of encounter, it seems excessive and hazardous to regard these 27 species as really extinct or nearing extinction.

It may be the case for some of them, though. So, for instance, *Rhyacophila hirticornis* McL. was only collected in 1918 in a well localized torrent-like acidic stream of the Ardennes (Stroot 1985a) but could not be found there recently, despite intense personal searching. Traces of organic enrichment increasing towards the springs suggest the influence of a domestic pollution by the hamlet and the farm surrounding the spring-head. But the clear-cut deforestation done around the streamlet could also be responsible for the elimination of the species.

Another verified case is the recent extinction of a maybe unique population of *Apatania muliebris* McL. s.l. This population was discovered in 1979 in a small rheocrene spring (Stroot 1984a), and could belong to an endemic form of the *muliebris* complex (Botosaneanu, pers. comm.). But its habitat was completely destroyed in 1981 through the digging of a fishing-pond right into the brooklet. Of course no more *Apatania* can be found there any more. Even if the discovery of other populations in the surroundings should not be totally excluded, this seems to be a dramatic example of a recent extinction of a scientifically interesting taxon. In the circumstance, it is a form of a true glacial relict (Nielsen 1950, 1976b), probably in process of speciation. Let us note that Nielsen (1976b) also reports the extinction of two endemic subspecies of *A. muliebris* from Danish springs.

*Hydropsyche bulgaromanorum* Mal. could also have disappeared from Belgium. This typical potamobiontic species (Malicky 1977 ; Lecureuil *et al.* 1983) was only recorded in Belgium in 1917, near the Scheldt River (Stroot 1985a), in which no *Hydropsyche* seems to occur any more. According to Malicky (1984), *H. bulgaromanorum* is widespread throughout Europe. It could therefore also be expected in the Meuse River. It seems not to be found there yet, despite recent and intense field-work (Meurisse-Genin *et al.*,in press). Thus, *H. bulgaromanorum* seems now extinct in Belgium, as it is thought to be in England (Botosaneanu 1981, Malicky 1984, Shirt in press).

On the other hand, the condition of *Hydropsyche fulvipes* (Curt) in Belgium may not be as bad as first claimed by Botosaneanu (1981) and as appearing in Stroot (1985a). Indeed, current researches reveal the presence in a few small streams of two larval forms which probably correspond to *H. fulvipes* and *H. saxonica* McL. The piece of information needs to be verified by collecting mature pupae and rearing larvae. In our opinion, it seems likely though. Besides, the same situation arose about *H. silfvenii* Ulmer, first only known from two old records (Stroot 1985a), but now proven to be abundant in several acidic streams of the Ardennes.

Beside the few more or less documented cases evoked above, there must be some other rare species now extinct or nearing extinction in Belgium. In doubt and awaiting circumstantial information, we think these rare species ought to be globally considered as at least "fragile or vulnerable". This can easily be conceived given the relative precariousness of small and isolated populations and of the particular environments inhabited by many rare species.

**More common species**

Some previously flourishing species now show a marked decrease and thus appear severely threatened in Belgium. This is expressed in the proportions of their records prior to and since 1950, considerably differing from the global proportion of data. The most restrictive and thus the most optimistic estimation points out 19 species now clearly decreasing. These 19 species (Table 1) have been proposed for the Red List of endangered species in the Belgian fauna (Stroot 1985b). About 30 additional species have shown a neat but less significant decline during the last decades.

Table 1. Red list of endangered Trichoptera of the Belgian fauna (previously common species recently showing a marked decrease throughout the country).

| | |
|---|---|
| *Chimarra marginata* (L.) | *Limnephilus stigma* Curt. |
| *Phryganea grandis* L. | *Colpotaulius incisus* (Curt.) |
| *Oligotricha striata* (L.) | *Grammotaulius nigropunctatus* (Retz.) |
| *Hagenella clathrata* (Kol.) | *Potamophylax luctuosus* Pill. & Mitt. |
| *Oligostomis reticulata* (L.) | *Goera pilosa* (Fabr.) |
| *Limnephilus affinis* Curt. | *Ceraclea nigronervosa* (Retz.) |
| *Limnephilus griseus* (L.) | *Triaenodes bicolor* (Curt.) |
| *Limnephilus marmoratus* Curt. | *Leptocerus interruptus* (Fabr.) |
| *Limnephilus nigriceps* (Zett.) | *Molanna angustata* Curt. |
| *Limnephilus politus* McL. | |

Most of these threatened species are lentic, as also observed for Nordrhein-Westfalen caddis flies (Wichard 1979), several being typical of small acidic and temporary pools (Hiley, 1976, 1978). Their decrease is clearly ascribable to the vanishing of such "useless" environments, regularly destroyed by drainage, as also reported from Scotland and England (Hiley 1978) and from north-western Germany (Wichard 1979). If not filled in, larger sheets of water are usually altered, mainly for purposes of intensive fishing and other recreational activities. Even where ponds still occur, very few can be seen as natural. A lot of them only shelter trite species as *Athripsodes aterrimus* (Steph.) or *Mystacides longicornis* (L.), recently showing clearly increasing numbers of records.

A few lotic species are also markedly decreasing. *Molanna angustata* Curt., still frequent in the Netherlands (Botosaneanu and Higler, pers. comm.), is one of the most threatened. Its larva is restricted to sandy bottoms, and thus to slowly running waters, sandy ponds being scarce in Belgium. Therefore, it could be particularly vulnerable to organic pollutions, via any drop of dissolved oxygen. The other clearly declining lotic species could also be the most sensitive to pollutions, as supposedly *Chimarra marginata* (L.), a potamophilic species. *Ch. marginata* is not only vanishing from Belgium but also from West-Germany (Wichard 1979 ; Tobias and Tobias 1984), where it is now almost impossible to find (Pitsch, pers. comm.), and from the Netherlands, where there is no recent record (Higler, pers. comm.). In Austria, its distribution area is progressively reduced towards the west (Malicky, pers. comm.).

Among the common species not showing a generalized decrease and thus likely to appear in a rather healthy condition, some still show a clear reduction of their distribution in Belgium, visible on the published maps (Stroot 1984b). The main reduction pattern concerns at least 10 species which disappeared from Middle Belgium to be now only found in High Belgium, or even in the Ardennes, and sometimes also in Kempenland (Fig. 1) : *Hydropsyche pellucidula* (Curt.) (Fig. 2a), *Polycentropus flavomaculatus* (Pict.), *Oligoplectrum maculatum* (Four.) (Fig. 2b), *Lasiocephala basalis* (Kol.), *Athripsodes cinereus* (Curt.)... Even species as *Hydropsyche siltalaï* Döhler and *Anabolia nervosa* (Curt.) (Fig. 2c.) almost completely disappeared from Middle Belgium where they used to be very common. Except this last one, all the concerned species are typically rheophilic. Their vanishing from Middle Belgium may certainly be ascribed to the increasing pollutions of the water courses there, due to intensive agricultural practices and to suburban housing development. The apparent vanishing of *Oligoplectrum maculatum* from Denmark (Svensson and Tjeder 1975) has been clearly attributed to organic pollution (Nielsen 1976a).

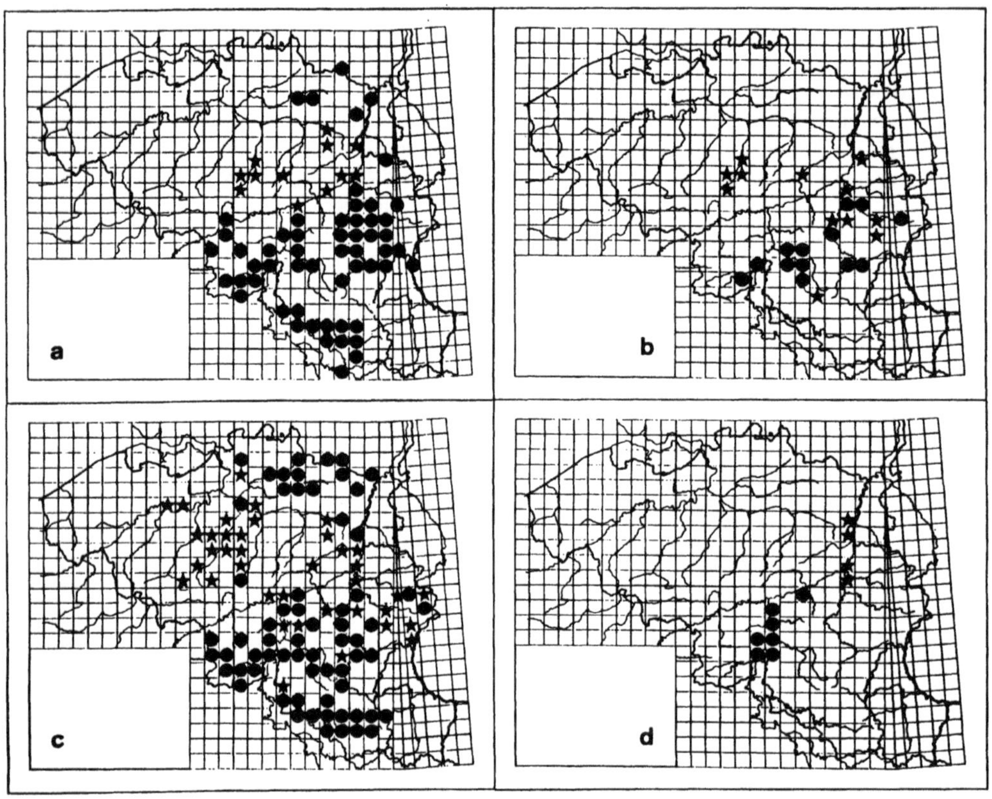


Figure 2. Distribution of (a) *Hydropsyche pellucidula*, (b) *Oligoplectrum maculatum*, (c) *Anabolia nervosa* and (d) *Hydropsyche exocellata* in Belgium, prior to (✱) and since (●) 1950 (after Stroot 1984b).

Some other species show a progressive withdrawal from the lowest parts of large rivers belonging to the potamal. It is the case in the Meuse River for species as *Hydropsyche exocellata* Dufour (Fig. 2d), *Psychomyia pusilla* (Fabr.), some Leptocerids, and to a lesser extent, for *Hydropsyche contubernalis* McL. and *Ecnomus tenellus* (Ramb.). Their situation in the Meuse, as well as the general state of the river itself, are described and discussed in Meurisse-Genin *et al.* (in press) : multiple pollutions including eutrophication, together with its progressive canalization, appear as the causes of the dramatic damaging of the lowest part of the Meuse River. The present-day project to "improve" the regulation of the upper reaches must be seen as a real threat towards the little potamon left in Belgium.

CONCLUSION

About half of the Belgian caddis fly fauna finally appears threatened at one or another extent (Table 2). The causes of its decrease are multiple. Nevertheless, they can be grouped in two categories :

(1) destruction of peculiar, limited and isolated environments, as springs and small acidic or temporary pools, particularly prejudicial to rare or specialized species ;
(2) alteration, including various forms of pollutions, of larger scale environments, as ponds, lakes and large rivers.

Mostly in consequence of its very high population density, Belgium has the pathetic privilege of concentrating on its small territory numerous examples of all the damages observed in different parts of the world and reviewed by Botosaneanu (1981) as regards Trichoptera. It is high time to react, in a country where speedways are built even in the most famous nature reserves.

Table 2. Summarizing table of the state of the caddis fly fauna in Belgium.The upper rhithral of the Ardennes remains the only quite healthy zone.

| State | Estimated number of species |
|---|---|
| Species considered as "fragile or vulnerable" because of their rareness | 48 (at least 3 extinct ?) |
| Previously common species now in a marked decrease throughout the country | 19 (Red List) + 30 ? |
| Species showing a clear reduction of their distribution area in Belgium | 20 |

## REFERENCES


Botosaneanu, L. 1981. Ordo Trichoptera et Homo insapiens. **Proc. 3rd Int. Symp. Trichoptera,** G.P. Moretti (ed.), Junk, The Hague : 11-19.

Heath, J. & Leclercq, J. 1981. Provisional Atlas of the Invertebrates of Europe, Maps 1-27, European Invertebrate Survey ed., Gembloux (Belgium) and Abbots Ripton (England).

Hiley, P.D. 1976. The identification of British limnephilid larvae (Trichoptera). **Syst. Ent.,** 1 : 147-167.

Hiley, P.D. 1978. Some aspects of the life histories of Limnephilidae (Trichoptera) related to the distribution of their larvae. **Proc. 2nd Int. Symp. Trichoptera,** M.I. Crichton (ed.), Junk, The Hague : 297-301.

Leclercq, J., Gaspar, C., Marchal, J.L., Verstraeten, C. & Wonville, C. 1980. Analyse des 1.600premières cartes de l'Atlas provisoire des Insectes de Belgique et première liste rouge d'Insectes menacés dans la faune belge. **Notes fauniques Gembloux,** 4 : 104 p.

Lécureuil, J.Y., Chovet, M., Bournaud, M. & Tachet, H. 1983. Description, répartition et cycle biologique de la larve d'*Hydropsyche bulgaromanorum* Malicky 1977 (Trichoptera, Hydropsychidae) dans la Basse Loire. **Annls Limnol.,** 19 : 17-24.

Malicky, H. 1977. Ein Beitrag zur Kenntnis der *Hydropsyche guttata*-Gruppe (Trichoptera, Hydropsychidae). **Z. Arbgem. öst. Ent.,** 29 : 1-28.

Malicky, H. 1984. The distribution of *Hydropsyche guttata* Pictet and *H. bulgaromanorum* Malicky (Trichoptera, Hydropsychidae) with notes on their bionomics. **Entomologist's Gaz.,** 35 : 257-264.

Marlier, G. 1949. Essai d'un catalogue des Trichoptères de Belgique. **Annls. Soc. ent. Belg.,** 85 : 108-134.

Meurisse-Genin, M., Reydams-Detollenaere, A., Stroot, P. & Micha, J.C. in press. Les macroinvertébrés benthiques de la Meuse belge : bilan de cinq années de recherches (1980 à 1984). **Arch. Hydrobiol.**

Nielsen, A. 1950. On the zoogeography of springs. **Hydrobiologia** 2 : 313-321.

Nielsen, A. 1976a. Pollution and caddis-fly fauna. **Proc. 1st Int. Symp.Trichoptera,** Junk, The Hague : 159-161.

Nielsen, A. 1976b. Revision of some opinions expressed in my 1942 paper. **Proc. 1st Int. Symp. Trichoptera,** H. Malicky (ed.), Junk, The Hague : 163-165.

Shirt, D.B. (ed.) in press. British Red Data Books : 2. Insects. Nature Conservancy Council, Peterborough.

Stroot, P. 1984a. Faunistique et répartition longitudinale des Trichoptères dans une rivière salmonicole de basse montagne, en Belgique. **Hydrobiologia,** 108 : 245-258.

Stroot, P. 1984b. Les Trichoptères de Belgique et des régions limitrophes (225 cartes). Atlas provisoire des Insectes de Belgique, I.R.S.N.B., Bruxelles, 75 p.

Stroot, P. 1985a. Actualisation du catalogue des Trichoptères de Belgique. **Soc. r. belg. Ent. (Suppl.),** 61 p.

Stroot, P. 1985b. Trichoptères menacés de la faune belge. Conférence-débat "Invertébrés menaçants, Invertébrés menacés", **Fac. Sci. Agro. Gembloux** : 55-57.

Stroot, P. & Depiereux, E. (in prep.). Proposition d'une méthologie pour établir des "Listes Rouges" d'invertébrés menacés. Exemple des Trichoptères de Belgique.

Svensson, B.W. & Tjeder, B. 1975. Check-list of the Trichoptera of North-Western Europe. **Ent. scand.,** 6 : 261-274.

Tobias, D. & Tobias, W. 1984. Rote Liste der Köcherfliegen (Trichoptera). Rote Liste der gefährdeten Tiere und Pflanzen in der Bundesrepublik Deutschland, Kilda, Greven :67-70.

Wichard, W. 1979. Rote Liste der in Nordrhein-Westfalen gerfährdeten Köcherfliegen (Trichoptera). Rote Liste der in Nordrhein-Westfalen gefährdeten Pflanzen und Tiere, Landesanstalt für Ökologie, Recklinghausen : 65-67.

# FAUNISTIC STUDIES ON IBERIAN TRICHOPTERA : A HISTORICAL SURVEY AND PRESENT STATE OF KNOWLEDGE

M.A. GONZALEZ, D. GARCIA DE JALON and L.W. Da TERRA

## SUMMARY

For more than a century from 1865 to 1972 there were only fragmentary and intermittent studies of Iberian caddis flies by a few entomologists. From 1972, the state of knowledge has been notably improved by continuing contributions from several authors, adding 58 species to the Iberian Trichoptera catalogue. Those species incorporated in the Iberian fauna after Limnofauna Europaea (2nd edition 1978) are reviewed and the distribution of all species in the Iberian peninsula is analyzed.

At present, we know 267 Iberian caddis fly species, distributed among 80 genera and 20 families. The Limnephilidae is dominant (56 spp.), followed by Leptoceridae (34 spp.) and Hydroptilidae (32 spp.).

In the Iberian fauna, even a though species of wide European distribution are predominant, the endemic species are an important component (30%) and are of a great zoogeographical value.

## INTRODUCTION

Together with some notes on dispersal on the Iberian Trichoptera fauna, the monograph by R. McLachlan (1874-84), and especially the First Additional Supplement represented an important first contribution to the knowledge of the caddis flies of the Iberian peninsula. This work which included many species from Portugal, recorded 63 species, 59 of which are still valid and 25 were new species.

During the first third of this century, Longino Navás was very active collecting and describing caddis flies from all over world, but especially from Spain, raising the number of catalogued Iberian species to 216 (Navás 1920).

However, this prolific work was unreliable because of the tendency of Navás for describing new species with inadequate diagnosis. Thus, from 129 Navás species revised by Schmid (1949) only 60 were valid. Nevertheless, Navás cited more than 170 caddis species described by other authors, from many localities in the Iberian peninsula, that together with his own valid ones gave a fairly good picture of the Trichoptera fauna of the North East of Spain, where he mainly collected and lived.

After Navás there was an interruption in Iberian Trichoptera studies, until 1952 when Schmid published the results of his travels collecting caddis flies all over Spain. One hundred species were cited, among them 25 were new for science (seven of these have since fallen to synonymy).

From 1952 to 1972, there were no notable contributions.

## STATE OF KNOWLEDGE

From 1972, and mainly in the last nine years (1978-86), we have seen a strong resurgence of research on Iberian Trichoptera. This period has been characterized by the publication of abundant information on the taxonomy and distribution of our Iberian caddis flies by indigenous and foreign authors. Our knowledge of the faunistic composition of some regions has improved and the Trichoptera catalogue of the Iberian peninsula has been considerably increased.

In order to gain an idea of this recent progress we will use as a reference the Limnofauna Europaea (2nd ed. 1978), where Botosaneanu and Malicky recorded 216 Iberian species in region 1. Some of these species should now be excluded in accord with present knowledge : *Rhyacophila guadarramica* Schm. as synonymous with *Rhyacophila meridionalis* Pict. (Garcia de Jalón 1983) ; *Hydropsyche dissimulata* Kum.& Bots. as synonymous with *Hydropsyche modesta* Navás (Malicky 1981) ; *Hydropsyche hibera* as synonymous with *Hydropsyche instabilis* Curtis (González & Botosaneanu 1985) ; *Tinodes antequerelus* Schm. as *T. foedellus* McL. (Malicky, personal communication) ; *Oecetis alexanderi* Kum. as *O. tripunctata* F. (Kumanski 1981) and *Anomalopteryx alacerrima* Schm. with seems clearly synonymous with *A. chauviniana* Stein.

*Synagapetus insons* McL. is, according to Malicky (1980), a Pyrenean species and its presence in the

Table 1. Additions to the Iberian Trichoptera fauna (region 1) since the revision of Botosaneanu & Malicky (1978), with references to the works where they are first cited and the Iberian provinces or regions where they have been found (see Figure 1 for location of the places).

| SPECIES | REFERENCES | IBERIAN LOCALITIES |
|---|---|---|
| *Ryacophila dorsalis* (Curtis, 1834) | García de Jalón, 1982 | BU LO SO GU HU TE |
| *Rhyacophila evoluta* McLachlan, 1879 | Puig *et al.*, 1981 | B L |
| *Rhyacophila mocsaryi tredosensis* Schm. 1952 | García de Jalón, 1982 | LE ZA BU LO SG AB M |
| *Rhyacophila obelix* Malicky | | NP |
| *Rhyacophila pascoei* MacLachlan, 1879 | García de Jalón, 1982 | CR MA |
| *Synagapetus lusitanicus* Malicky, 1980 | | LU S LE ZA P AV NP |
| *Synagapetus marlierorum* Botosaneanu, 1980 | | C PO LU OR LE NP CP |
| *Agapetus ochripes* Curtis, 1834 | González & Otero, 1984a | S |
| *Agapetus theischingeri* Malicky, 1980 | | CU |
| *Stactobia intermedia* González & Terra, 1981 | | LU CP |
| *Stactobiella risi* (Felber, 1908) | | NP: Moncao 25-VIII-83 1 ♂ |
| *Ithytrichia clavata* Morton, 1905 | González & Otero, 1981 | LU SA CC NP |
| *Oxyethira falcata* Morton, 1983 | Terra, 1981 | C M NP CP |
| *Oxyethira iglesiasi* González & Terra, 1982 | | NP CP |
| *Hydroptila cornuta* Mosely, 1922 | González & Terra, 1983 | C PO LU ZA SA AV GU J MA NP CP |
| *Hydroptila dejaloni* Botosaneanu, 1980 | | M NP |
| *Hydroptila espada* Malicky, 1981 | | CC CP SP |
| *Hydroptila giudicellorum* Botosaneanu, 1980 | | CC CP |
| *Hydroptila idefix* Malicky, 1979 | | C ZA AV NP CP |
| *Hydroptila tagus* García de Jalón & González, 1985 | | GU |
| *Hydroptila vichtaspa* Schmid, 1959 | | SP : PortoEspada 11-IV-84 2 ♂ |
| *Hydroptila vilaverde* Malicky & González, 1981 | | C LE P NP CP |
| *Allotrichia galaica* González & Malicky, 1980 | | LU LE ZA P SG M NP CP |
| *Philopotamus montanus caurelensis* González & Terra, 1979 | | LU OR O S LE ZA |
| *Wormaldia cantabrica* González & Botosaneanu, 1983 | | LU O S LE ZA CP |
| *Wormaldia lusitanica* González & Botosaneanu, 1983 | | C PO LU OR ZA CR NP CP |
| *Wormaldia subnigra* McLachlan, 1985 | García de Jalón, 1982 | J |
| *Hydropsyche contubernalis* McLachlan, 1865 | García de Jalón, 1983 | GU MU AB |
| *Hydropsyche* cf. *dinarica* Marinkovic, 1979 | García de Jalón, 1983 | LU LE GU HU AL CP |
| *Hydropsyche teruela* Malicky, 1980 | | TE |
| *Hydropsyche urgorrii* González & Malicky, 1980 | | C PO LU NP CP |
| *Pseudoneureclipsis lusitanicus* Malicky, 1980 | | NP |
| *Polycentropus terrai* Malicky, 1980 | | C LU NP |
| *Cyrnus monserrati* González & Otero, 1983 | | CC SP |
| *Paduniella vandeli* Décamps, 1965 | González & Otero, 1984b | HU CO SP |
| *Metalype fragilis* (Pictet, 1834) | García de Jalón, 1982 | GU J |
| *Tinodes algiricus* McLachlan, 1880 | González & Otero, 1984b | CO |
| *Tinodes baenai* González & Otero, 1984 | | CO |
| *Tinodes marrocanus* Mosely, 1938 | | MU:Alhama 2♂ leg. Lajonquiera |
| *Tinodes rostocki* MacLachlan, 1878 | González & Otero, 1984a | O |
| *Brachychentrus subnubilus* Curtis, 1834 | González, 1979 | C LE ZA VA BU SO SG CR GU NP |
| *Micrasema minimum* McLachlan, 1876 | González & Otero, 1981 | LU OR M TO |
| *Apatania eatoniana* McLachlan, 1880 | González & Otero, 1984a | S |
| *Apatania theischingerorum* Malicky, 1981 | | CU |
| *Drusus rectus* (McLachlan, 1868) | García de Jalón, 1982 | LE |
| *Limnephilus aistleitneri* Malicky, 1986 | | TE |
| *Limnephilus incisus* Curtis, 1834 | | ZA:Pedrazales 5-8-83 1♂+2♀ |
| *Stenophylax permistus* MacLachlan, 1895 | Filbá, 1974 | C LU O SO BI HU G |
| *Chaetopteryx villosa gonzalezi* Botosaneanu, 1980 | | M |
| *Psilopteryx sanabriensis* González & Otero, 1985 | | LU ZA |
| *Silonella* sp. | | MA: Igualeja 3-3-84 2P♂ + 3L |
| *Thremma tellae* González, 1978 | | C PO LU OR LE ZA SG AV M NP CP |
| *Athripsodes albifrons* (Linnaeus, 1758) | Terra, 1981 | C LU OR LE P SO HU GR NP |
| *Athripsodes bessae* Malicky & Terra, 1984 | | NP |
| *Ceraclea aurea* (Pictet, 1834) | González & Otero, 1981 | LU |
| *Ceraclea dissimilis* (Stephens, 1836) | García de Jalón, 1982 | VA P BU SO |
| *Adicella josephinae* González & Otero, 1980 | | LU |
| *Erotesis schachti* Malicky, 1982 | | H CP |

Iberian peninsula has not yet been verified.*Wormaldia beaumonti* Schmid is a good species (González & Botosaneanu 1983) that was considered as a subspecies in the Limnofauna Europaea. *Philopotamus amphilectus* McL. and *P. perversus* McL. are subspecies of *P. variegatus* Scop. and *P. montanus* Don. respectively (González & Terra 1979). *Sericostoma flavicorne* Schn. included provisionally a group of dubious species among which there were *S. selysii* Pict. and *S. pyrenaicum* Pict. but this is not possible, as *S. flavicorne* is a definite species that lives in Asia Minor and The Levant (Moubayed & Botosaneanu 1985).

In contrast fifty eight species must be added to the initial list in the Limnofauna ; they are enumerated in Table 1 with indications of their distribution in the Iberian peninsula.

It is remarkable that 29 of these species are new species described from Iberian specimens. Five species are first cited in this paper from Iberian localities : *Stactobiella risi* Felber and *Hydroptila vichtaspa* Schmid from Portugal, and *Tinodes marrocanus* Mosely, *Limnephilus incisus* Curtis and *Silonella* sp. from Spain. The specific determination of this last taxon is being studied.

As a synthesis of all the corrections and additions, we have listed 267 Iberian species, distributed in 80 genera and 20 families. Except for the Arctopsychidae and Molannidae, all European families are represented in the Iberian fauna. Limnephilidae is the most numerous family (56 species), followed by Hydroptilidae (34 species), Leptoceridae (32 species) and Rhyacophilidae (24 species). The composition of each family is presented in Table 2, which lists Iberian species and endemics.

Table 2. Number of endemics and total number of Iberian species distributed by families.

| Family | Known endemics from Iberian peninsula | | | Total number of species in Iberian peninsula |
|---|---|---|---|---|
| LIMNEPHILIDAE | 13 | - | 23 % | 56 |
| HYDROPTILIDAE | 12 | - | 35 % | 34 |
| LEPTOCERIDAE | 11 | - | 34 % | 32 |
| RHYACOPHILIDAE | 8 | - | 33 % | 24 |
| HYDROPSYCHIDAE | 7 | - | 37 % | 19 |
| PSYCHOMYIDAE | 3 | - | 18 % | 18 |
| GLOSSOSOMATIDAE | 8 | - | 50 % | 16 |
| POLYCENTROPODIDAE | 4 | - | 27 % | 15 |
| PHILOPOTAMIDAE | 6 | - | 50 % | 12 |
| OTHERS | 11 | - | 27 % | 41 |
| TOTAL | 83 | | | 267 |

It is our belief that the number of Trichoptera species inhabiting the Iberian peninsula could increase by at least 10-15% more. This belief is based on the incomplete knowledge of the caddis from some parts of the Iberian peninsula.

The map in Figure 1 shows the number of species found in each province, based on the bibliography and on unpublished recors by the authors.

The regions with a richer Trichoptera fauna are found in the northen half of the Iberian peninsula, especially in the northern-west quarter, which has been systematically explored by the authors since 1972. This includes the central and northern half of Portugal, Galicia (Coruña, Lugo, Orense and Pontevedra), the adjacent provinces (León and Zamora) and Sistema Central Mountains (Madrid, Segovia and Avila). Other provinces which have been little studied and need further research, are Asturias, Santander, Vizcaya and San Sebastián in the Cantabric region and to a lesser extent, the Spanish Pyrenees area.

Trichopterologists have paid little attention to the fauna of the meridional half of the Iberian peninsula, and this could be an explanation of the small number of species recorded in these southern provinces (less than 10 species in nine of them).

However, other factors determine the richer fauna of the north and west of the Iberian peninsula, in particular the rainfall and the geomorphology, that result in the higher density of swift streams with soft waters.

A FIRST ATTEMPT TO CHARACTERIZATE THE IBERIAN CADDIS FLY FAUNA

The Iberian Trichoptera fauna is characterized by a high number of endemics, nearly one third of the total of species, although the species widely distributed in Europe are predominant. This high proportion of endemic species is similar to that found by Moretti & Cianficconi (1981) in the Italian peninsula, and also coincides with the proportion of endemic Plecoptera for the Iberian peninsula (Berthélemy & Terra 1980).

Nearly all of these Iberian endemics have been described from caddis flies collected in the northern half of the peninsula and their known distributional area is more or less restricted to the north-west quarter of the peninsula.

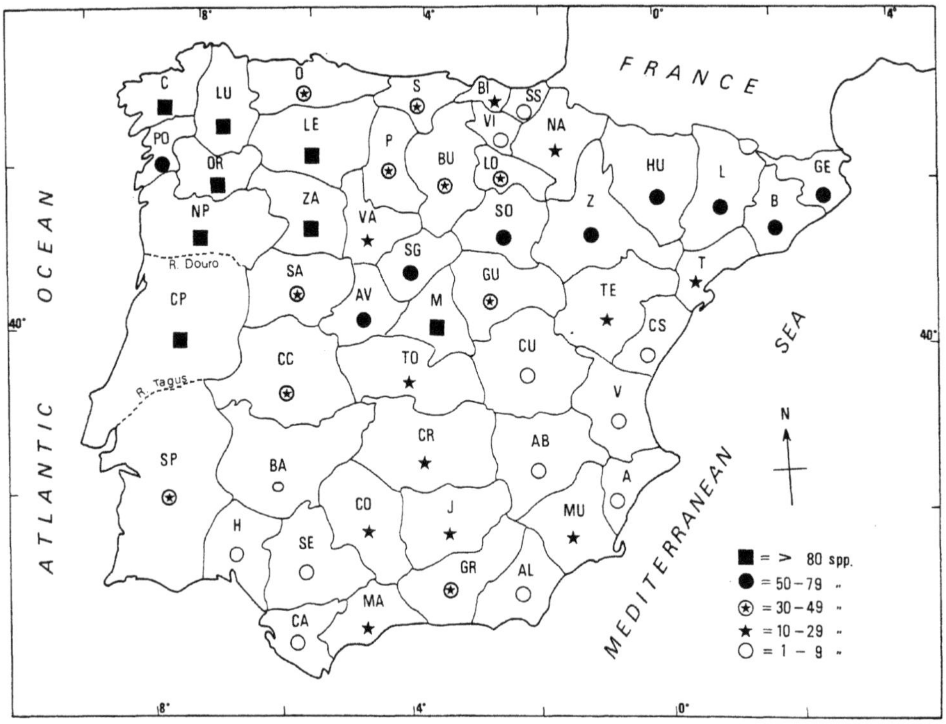


Figure 1. Numbers of Trichoptera species in the different provinces (Spain) and regions (Portugal) of the Iberian peninsula : A : Alicante, AB : Albacete, AL : Almería, AV : Avila, B : Barcelona, BA : Badajoz, BI : Vizcaya, BU : Burgos, C : Coruña, CA : Cádiz, CC : Cáceres, CO : Córdoba, CP : Central Portugal, CR : Ciudad Real, CS : Castellón, CU : Cuenca, GE : Gerona, GR : Granada, GU : Guadalajara, H : Huelva, HU : Huesca, J : Jaén, L : Lérida, LE : León, LO : Logroño, LU : Lugo, M : Madrid, MA : Málaga, MU : Murcia, NA : Navarra, NP : North Portugal, OR : Orense, O : Asturias, P : Palencia, PO : Pontevedra, S : Santander, SA : Salamanca, SE : Sevilla, SG : Segovia, SO : Soria, SP : South Portugal, SS : Guipúzcoa, T : Tarragona, TE : Teruel, TO : Toledo, V : Valencia, VA : Valladolid, VI : Alava, Z : Zaragoza, ZA : Zamora.

We suggest as an hypothesis for the explanation of this concentration of endemics that the Hesperico Massif has served as an evolutionary refuge for many cold running water species. From a geological point of view, the north-west corner of the Iberian peninsula belongs to the Hesperico Massif which constituted the old Iberian nucleus of archaic and paleozoic terrain, already emerged since then (Melendez 1958) and unconnected with the rest of Iberia until the end of the Miocene.

Another characteristic feature of the Iberian Hesperic fauna is that it contains elements in linking with Asia and Africa (e.g. gen. *Larcasia* ) together with American elements since West Europe was connected with North America in the Cretaceous period. The genera *Drunella* and *Serratella* (Ephemerellidae, Ephemeroptera) are well represented in the Nearctic region and are mainly confined to the north-west of Iberia (González del Tánago & Garcia de Jalón 1983).

Although it may be expected that some endemic species restricted to high mountains (e. g. *Adicella josephinae* Gonz. & Otero, *Chaetopteryx atlantica* Mal., *C. lusitanica* Mal., *Psilopteryx sanabriensis* Gonz. & Otero, *Potamophylax albergaria* Mal. and *Drusus cantabricus* Schm.) have a much limited distribution, probably other endemics living at lower altitudes would have a wider distribution than those at present known, in other regions not yet intensively studied .

The fauna of Southern Iberia is little known but it seems to be much poorer than the northern, although it has its own few endemics, e.g.,*Psilopteryx esparraguera* Schmid, *Rhyacophila nevada* Schmid, *Hydropsyche infernale* Schmid and *Tinodes baenai* Gonz. & Otero.

Besides these strictly Iberian endemics, there are about 20 species whose distribution area is confined to Iberia and the Pyrenees. Most of these species are limited to the northern of the peninsula, and some of them to the north-east quadrant, for instance, *Rhyacophila rupta* McL., *Notidobia sagarrai* Navas and *Schizopelex furcifera* McL.

Another characteristic of the Iberian Trichoptera fauna worthy of note is its connection with the northern Africa fauna. Nine species of caddis flies are spread over Iberia and Maghreb only. *Hydroptila campanulata* Morton, *H. cintrana* Morton, *Hydropsyche lobata* McL. and *Tinodes algiricus* McL. are examples of this group.

Finally, we must repeat that our present knowledge of the Iberian Trichoptera fauna is still incomplete, with many species only cited from one or two localities. Further research is needed to confirm the expressed ideas or to further the development of new explanations either.

## REFERENCES


Berthélemy, C. & da Terra, L.W. 1980. Plécoptères du Portugal (Insecta). **Annls. Limnol.** 16(2) : 159-182.

Botosaneanu, L. & Malicky, H. 1978. Trichoptera. *In* : J. Illies (ed.), Limnofauna Europaea 2nd ed., Fisher Verlag, Stuttgart : 333-359.

Filbá, E.L. 1974. Contribución al conocimiento de los Tricópteros cavernícolas de Cataluña. **IV. Simp. region. Bioespeol. Barcelona :** 21-35.

Garcia de Jalón, D. 1982. Contribución a la zoogeografía de los Tricópteros españoles. **SHILAP Rev. Lepid.,** 10(37, 38, 39) : 75-78, 157-165, 237-244.

Garcia de Jalón, D. 1983. Contribución al conocimiento de las larvas del género *Hydropsyche* (Trichoptera) ibéricas. **Act. I. Congr. iber. Entom. :** 275-285.

González, M.A. 1979. Observaciones sobre los Tricópteros de la Península Ibérica, I. **Bol. Asoc. esp. Entom.,** 3 : 219-223.

González, M.A. & Botosaneanu, L. 1983. Etude d'un groupe d'espèces nettement sympatriques de *Wormaldia* McL de l'ouest de la Péninsule Ibérique (Trichoptera). **Bull. Zool. Mus. Univ. Amsterdam,** 9(18) : 165-171.

González, M.A. & Botosaneanu, L. 1985. Etude de trois espèces d' *Hydropsyche* Pictet d'Espagne décrites par F. Schmid (Ins. : Trich.) **Bull. Zool. Mus. Univ. Amsterdam,** 10(13) : 89-95.

González, M.A. & Otero, J.C. 1981. Observaciones sobre los Tricópteros de la Península Ibérica, II. **Bol. Asoc. esp. Entom.,** 4 : 127-129.

González, M.A. & Otero, J.C. 1984a. Observaciones sobre los Tricópteros de la Península Ibérica : V, Tricópteros de los Picos de Europa (Norte de España). **Bol. asoc. esp. Entom.,** 8 : 47-52.

González, M.A. & Otero, J.C. 1984b. Observaciones sobre los Tricópteros de la Península Ibérica : VI, Tricópteros de Córdoba (Andalucía, Sur de España). **Nouv. Rev. Ent. (N.S.),** I(1) : 61-66.

González, M.A. & Da Terra, L.W. 1979. Contribución al conocimiento de los *Philopotamus* ibéricos (Trich., Philopotamidae). **Bol. Asoc. esp. Entom.,** 3 : 163-172.

González, M.A. & Da Terra, L.W. 1983. Observaciones sobre los Tricópteros de la Península Ibérica, III. **Bol. Asoc. esp. Entom.,** 6(2) : 245-248.

González del Tánago, M. & Garcia de Jalón, 1983. New Ephemerellidae from Spain (Ephemeroptera). **Aquatic Insects,** 5(3) : 147-156.

Kumanski, K. 1981. Faunistic investigations on Bulgarian Trichoptera to June 1980, with a revised check-list. **Proc. 3rd Int. Symp. Trichoptera,** G.P. Moretti (ed.), Junk, The Hague : 139-148.

Malicky, H. 1980. Beschreibungen von neuen mediterranean Köcherfliegen und bemerkungen zu bekannten (Trich.). **Z. Arbgem. Ost. Ent.,** 32(1/2) : 1-17.

Malicky, H. 1981. Neues über mediterrane, vorderasiatische und europäische Köcherfliegen (Trichoptera). **Zeits. für Entom.,** 16(2) : 175-190.

McLachlan, R. 1874-1884. A monographic revision and synopsis of the Trichoptera of the European fauna (reprint 1968). Classey. Hampton.

Meléndez, B. 1958. Mapa geológico de España y Portugal. Paraninfo Ed. Madrid.

Moretti, G.P.& Gianficconi, F. 1981. First list of Italian Trichoptera. **Proc. 3rd Int. Symp. Trichoptera.** G.P. Moretti (ed.), Junk, The Hague : 199-212.

Moubayed, Z. & Botosaneanu, L. 1985. Recherches sur les Trichoptères du Liban et principalement des bassins supérieurs de l'Oronte et du Litani (Insecta : Trichoptera). **Bull. Zool. Mus. Univ. Amsterdam,** 10(11) : 61-76.

Navás, L. 1920. Tricópteros (Ins.) de España. **Asoc. esp. progr. Cienc. (Sevilla),** 6 : 5-16

Puig, M.A., Bautista, I. Tort, M.J. & Prat, N. 1981. Les larves de Trichoptera de la rivière Llobregat (Catalogne, Espagne). Distribution longitudinale et relation avec la qualité del'eau. **Proc. 3rd Int. Symp. Trichoptera,** G.P. Moretti (ed.), Junk, The Hague : 303-310.

Schmid, F. 1949. Les Trichoptères de la collection Navás. **EOS, Madrid,** 25 : 305-426.

Schmid, F. 1952. Contribution à l'étude des Trichoptères d'Espagne. **Pirineos,** 8 : 627-695.

Terra, L.S.W. da. 1981. Lista faunistica de Tricópteros de Portugal. (Insecta, Trichoptera). **Bom. Soc. port. Entom.,** 12 : 1-42.

# ANALYSES OF TRICHOPTERAN FAUNAL SIMILARITIES IN NORTHERN PORTUGAL

L.S.W. TERRA and M.C. MOLLES, Jr.

## SUMMARY

Principal components analysis was used to explore similarities among light trap collections of adult Trichoptera made at nine sites in northern Portugal. Similarities among collections appeared to be strongly related to site altitude and the size and gradient of waters associated with collecting sites. The number, composition and ordering of site groupings were dependent on the taxon analyzed and as a consequence, it appears that in northern Portugal different taxa of Trichoptera have distinctive distributions along the same environmental gradient.

## INTRODUCTION

Terra and Gonzalez (1984) reported distributions of adult Trichoptera among six sites in northern Portugal. Their analysis mainly involved qualitative and semi-quantitative assessment of the Trichoptera taxa most characteristic of individual sites, with some discussion of site similarities. The present study focused on a quantitative exploration of similarities among these and three additional sites using principal components analysis (PCA). The goals were : 1) to quantitatively analyze similarities among trichopteran faunas in northern Portugal, and 2) to explore the influence of taxonomic level (order, family, tribe, etc.) and choice of taxon (e.g. Rhyacophilidae versus Hydropsychidae) on similarities indicated by PCA.

## METHODS

The nine sites were located at state fish hatcheries so that personnel of the Inland Fisheries Department of Portugal could operate the light traps. The sites were all located in northern Portugal (Fig. 1) and are associated with a wide variety of aquatic habitats (Table 1).

Table 1. Physical characteristics of waters associated with collecting sites.

| Site | Altitude (m) | Gradient (0/00) | Width (m) | Depth (cm) | Distance from Sea (km) |
|---|---|---|---|---|---|
| Mira (Mi) | 5 | 0 | * | * | 1 |
| Monção (Mo) | 10 | 0.2 | 175 | 200 | 35 |
| Vila do Conde (Vc) | 0 | 0.8 | 70 | 100 | 3 |
| Campelo (Cm) | 400 | 12 | 3 | 30 | 57 |
| Torno (To) | 605 | 130 | 1 | 20 | 65 |
| Castrelos (Cs) | 610 | 14 | 9 | 40 | 165 |
| Geres (Ge) | 660 | 41 | 2 | 30 | 60 |
| Relva (Re) | 720 | 4 | 9 | 50 | 90 |
| Fonte Santa (Fs) | 880 | 80 | 6 | 50 | 110 |

* Complex of lakes, ponds and canals

The light traps, previously described by Terra (1978), were operated at the sites for two to six years between 1976 and 1984. The number of collecting nights at sites varied from 33 to 728, depending on availability of personnel.

For analyses, all collections at a site were pooled to provide a basis for comparing relative species compositions. Because of differences in collecting effort between sites, species were represented as proportions of the total collection. Prior to analysis, these proportions were transformed (arcsine$\sqrt{\text{proportion}}$) to correct for bimodality (Zar 1974).

Figure 1. Map of Portugal showing the locations of the nine collecting sites included in the present study.

Principal components analysis was applied and interpreted as described by Molles and Terra (1986). A total of twelve analyses were performed : 1) all Trichoptera with individuals identified to species if possible, 2) all Trichoptera with individuals identified to family only, 3) each of the nine numerically dominant families, and 4) the leptocerid tribe Athripsodini. Sites were grouped according to their loadings on PCA factors (Molles and Terra 1986). Following Rotenberry and Wiens (1981), loadings were operationally judged to be significant if they exceeded 0.60, analogous to a correlation (n = 9) significant at the 0.05 level. In general, however, loadings either far exceeded or were far below this operational threshold. Species scores on factors exceeding 1.0 (±) were judged to contribute most significantly to factors and hence to be characteristic of sites.

## RESULTS AND DISCUSSION

The collections contained a total of over 145,000 individual Trichoptera belonging to 143 species. The number of individuals collected at each site varied from 3651 at Mira to 52,851 at Monção, the site for which there were the fewest collections. Thus, numbers had only weak correspondence to the number of collections at a site. Number of species collected at each site varied from 18 at Mira to 79 at Relva.

Twelve PCA's were performed (Table 2), each resulting in 3 to 6 site groupings, accounting for a total of 82.1 to 99.8% of the variation among sites.

The results of 4 of 12 PCA's (Table 2) will be discussed in detail : 1) all Trichoptera identified to species, 2) all Trichoptera identified to family only, 3) Rhyacophilidae, and 4) Hydropsychidae. The PCA including all Trichoptera, identified to species if possible, formed three groups of sites : 1) Torno, Fonte Santa, and Gerês, 2) Castrelos and Campelo, 3) Monção and Vila do Conde and separated two sites as unique : 1) Mira, and 2) Relva.

Table 2. Site groupings (indicated by underlining) based on principal components analysis of various subdivisions of the Trichoptera. Numbers indicate percent variance explained by each PCA factor forming a group. To = Torno, FS = Fonte Santa, Ge = Gerês, Cs = Castrelos, Cm = Campelo, Mo = Monção, Vc = Vila do Conde, Mi = Mira, Re = Relva.

| | | | | | | |
|---|---|---|---|---|---|---|
| Trichoptera (spp.) | To, Fs, Ge | Cs, Cm | Mo, Vc | Mi | Re | |
| Families | Ge, To | Cs, Cm | Mi | Vc, Mo | Fs | Re |
| Rhyacophilidae | Cm, Cs, Mo | To, Fs, Ge | Vc | Re | Mi | |
| Glossosomatidae | Cm, To, Mo | Cs, Re | Vc | Ge | Fs | Mi |
| Hydroptilidae | Cs, Cm | To, Fs | Re, Ge | Mi | Mo | Vc |
| Philopotamidae | Ge, Fs, To | Cs, Cm | Mo | Re | Vc, Mi | |
| Hydropsychidae | Ge, To, Fs, Cs | Re, Cm | Vc, Mi | Mo | | |
| Polycentropodidae | To, Fs, Ge, Cm, Cs | Vc, Mi, Mo | Re | | | |
| Psychomyidae | To, Fs, Vc | Cm, Ge | Cs, Re | Mi | Mo | |
| Limnephilidae | Fs, Cs | Vc, Mo, Mi | To, Ge | Re | Cm | |
| Leptoceridae | Mo, Mi, Vc | Ge, To | Re, Cs | Fs | Cm | |
| Athripsodini | Ge, To, Cm | Mo, Vc | Cs, Re | Fs, Mi | | |

The grouping of Torno, Fonte Santa, and Gerês was the result of a combination of high numbers of 16 taxa and low numbers of five others. The 16 common taxa (listed here and in all following presentations in order of decreasing importance or factor score) were : *Hydropsyche siltalai*, *Hydropsyche* spp. females, *Rhyacophila lusitanica*, *Tinodes assimilis*, *R. adjuncta*, *Psychomyia pusilla*, *Philopotamus montanus perversus*, *Sericostoma vittatum*, Polycentropodid females, *Chaetopteryx lusitanica*, *Allogamus laureatus*, *Tinodes waeneri*, *Polycentropus kingi*, *R. terpsichore*, *Plectrocnemia laetibilis*, and *Sericostoma* spp. females. The five ucommon taxa were : *Psychomyia* spp. females,*Psychomyia ctenophora*, *Cheumatopsyche lepida*, *Oxyethira archaica*, and *Athripsodes braueri*. The grouping of Torno, Fonte Santa, and Gerês seems appropriate since all are associated with small to medium sized, high gradient streams at medium to high altitude. It is interesting to note that the only difference between the results of this analysis and the second, where Trichoptera were identified to family only, was the separation of Fonte Santa from Gerês and Torno because of the greater abundance of Hydropsychidae at Fonte Santa.

The second group was comprised of Castrelos and Campelo largely as a consequence of a combination of eight abundant taxa : *Hydroptila cornuta*, *Psychomyia pusilla*, *Oxyethira frici*, *Hydroptila tineoides*, *Wormaldia variegata matheyi*, *Hydropsyche* spp. females,*Ithytrichia lamellaris* and *Hydroptila vilaverde*. It is apparent that this list is dominated by the family Hydroptilidae, the family which distinguished these two sites from the others in analysis two, the familial analysis. Physically, Castrelos and Campelo are similar since they are both associated with medium sized streams with moderate gradients at intermediate elevations.

The third group consisted of Monção and Vila do Conde, which were joined on the basis of a combination of high numbers of 13 taxa and low numbers of three others. The 13 abundant taxa joining Monção and Vila do Conde were : *Psychomyia ctenophora*, *Hydropsyche* spp. females, *Psychomyia* spp. females, *Mystacides azurea*, *Ceraclea sobradieli*, *Ecnomus deceptor*, *Tinodes assimilis*, *Cheumatopsyche lepida*, *Hydropsyche lobata*, *Polycentropus telifer*, *Tinodes waeneri*, *Hydropsyche exocellata*, and *Setodes punctatus*. The uncommon taxa were : *Rhyacophila lusitanica*, *Psychomyia pusilla*, and *Chaetopteryx lusitanica*. Grouping of Monção and Vila do Conde seems logical considering their physical characteristics, both are associated with the lower reaches of two of the larger rivers of northern Portugal. In analysis two, the families joining Monção and Vila do Conde were Psychomyidae and Leptoceridae.

Mira was the only site associated with lakes, ponds and slow flowing canals and was separated by PCA as distinctive mainly on the basis of a combination of seven common taxa and three uncommon taxa. Abundant species were : *Ecnomus tenellus*, *Oxyethira unidentata*, *Mystacides azurea*, *Agraylea sexmaculata*, *Tinodes waeneri*, *Hydroptila giudicellorum*, and *Limnephilus marmoratus*. Terra and González (1984) mentioned almost all these when listing the species characteristic of Mira. Taxa rare or absent were : *Psychomyia* spp. females, *Cheumatopsyche lepida*, *Psychomyia ctenophora*. The analysis of Trichoptera families also separated Mira from the rest, but mainly on the basis of the family Ecnomidae. Relva is physically distinctive from the remainder of the sites because, though it is at high altitude, the nearly stream is of low gradient. Principal components analysis also separated Relva as faunistically distinctive on the basis of a combination of 15 common taxa and five uncommon taxa. Common taxa were : *Oxyethira archaica, Athripsodes braueri, Psychomyia pusilla, Allogamus ligonifer, Psychomyia* spp. females, *Oxyethira frici,* Polycentropodid females, *Rhyacophila adjuncta, Cyrnus cintranus, Lepidostoma hirtum, Ceraclea sobradieli, Ithytrichia* spp. females, *Tinodes waeneri, Schizopelex festiva,* and *Psychomyia ctenophora.* Uncommon taxa were : *Tinodes assimilis, Hydroptila cornuta, Ecnomus deceptor, Hydropsyche lobata* and*Sericostoma* spp. females. In the family level analysis, Relva was separated from the other sites on the basis of a combination of three common families : Leptoceridae, Polycentropodidae, and Limnephilidae and two uncommon families : Ecnomidae and Psychomyidae.

Analyses of site similarities based on distributions of species of single families were considerably easier to interpret. The analysis of Rhyacophilidae produced two groups of sites and separated three sites as unique (Table 2). Campelo, Castrelos, and Monção formed the first group mainly because their collections contained substantial numbers of *Rhyacophila relicta.* Torno, Fonte Santa, and Gerês were grouped because of codominance of *R. lusitanica* and *R. adjuncta.* Vila do Conde was placed by itself because of dominance of *R. munda.* Relva was also separated from the rest, but as a result of dominance by *R. adjuncta.* Mira was unique because it was the only site where no Rhyacophilidae were collected. These results may be interpreted as follows : *Rhyacophila relicta* is found mainly in medium to large streams with cool temperatures such as those at Campelo, Castrelos, and Monção. *Rhyacophila lusitanica* and *R. adjuncta* are found in greatest abundance in small to medium sized, cool streams at high elevations such as those at Torno, Fonte Santa, and Gerês. However, *R. adjuncta* has a wider distribution than *R. lusitanica* and hence the former is dominant at Relva, associated with a high elevation stream with a low gradient. The only common species at Vila do Conde, *R. munda,* is most common in warm, low elevation streams and rivers of the southern Iberian peninsula. Mira is in an area of warm sluggish or standing waters and hence *Rhyacophila* are absent.

The pattern of distribution indicated by PCA for the Hydropsychidae is consistent with information in the literature (Badcock 1976 ; Boon 1978). Gerês, Torno, Fonte Santa, and Castrelos are grouped because they are characterized by an abundance of *Hydropsyche siltalai* and low numbers of *H. pellucidula.* These sites are at high altitude and *H. siltalai* is known to be characteristic of such places. Campelo and Relva are joined by an abundance of *H. pellucidula* wich is more common at lower elevations than *H. siltalai.* Again, as in the PCA of the Rhyacophilidae, Relva is separated from the other high elevation sites presumably because of the low-gradient stream at this site. Vila do Conde and Mira were joined because *H. lobata* was dominant at these sites. This species is native to Iberia and North Africa where it is mainly found in low elevation streams and rivers. Terra (1980) suggested that *H. lobata* is tolerant of elevated salinities, and perhaps it is physiologically tolerant in other ways as well, which would explain its being essentially the only hydropsychid in the polluted, lower Rio Ave. Monção is separated from the remainder of the sites because of numerical dominance of *Cheumatopsyche lepida* at this point in the Rio Minho.

The results of the 12 PCA's indicate considerable variation in number, composition, and ordering of site groupings (Table 2). It appears that site similarities were strongly influenced by the taxon analyzed, since no two analyses produced identical results. This seems particularly interesting when one compares the results for different families or for all Leptoceridae versus the Athripsodini. The unique result for each, demonstrates different distributions along the same environmental gradient and suggests different ecological requirements and evolutionary histories. Further, these results should encourage caution in references to trichopteran "communities". The role of chance as well as those of phylogenetic history, physico-chemical requirements and biotic interaction in determining trichopteran distributions and the composition of local assemblages needs further, careful investigation. The necessary complexity of such investigations argues for a focus on taxonomically well-known families, tribes, or genera having tractable larval and adult life histories.

## REFERENCES

Badcock, R. 1976. The distribution of the Hydropsychidae in Great Britain. **Proc. 1st Int. Symp. Trichoptera,** Malicky H (ed), The Hague, Junk : 49-58.

Boon, P.J.1978. The pre-impoundment distribution of certain Trichoptera larvae in the north Tyne river system (northern England), with particular reference to current speed. **Hydrobiologia,** 57 : 167-174.

Molles, M.C. Jr & Terra, L.S.W. In Press. Principal components analysis of temporal and spatial variation in trichopteran faunas of northern Portugal. **Proc 5th Int. Symp.Trichoptera**

Rotenberry, J.T. & Wiens, J.A. 1981. A synthetic approach to principal component analysis of bird/habitat relationships. Capen D.E. (ed) : **USDA Forest Serv. Gene. Tchn. Rep.,** RM-87, Rocky Mountain Forest and Range Exp. St., Fort Collins : 197-208.

Terra, L.S.W. 1978. First results of Trichoptera collecting with light traps at Vila do Conde (Portugal). **Proc. 2nd Int. Symp. Trichoptera,** M.I. Crichton (ed), Junk, The Hague : 75-82.

Terra, L.S.W. 1980. Notas sobre os Trichópteros do Rio Ave, em Vila do Conde, e seu relacionamento com a qualidade da agua. **Est. Inform.** no 287.

Terra, L.S.W. & Gonzalez, M.A. 1984. Notes on the distribution of caddisflies in Portugal. **Proc. 4th Int. Symp. Trichoptera,** J.C. Morse (ed.), Junk, The Hague : 391-396.

Zar, J.H. 1974. Biostatistical analysis. Prentice-Hall, Inc, New Jersey.

# THE CADDIS FLIES (TRICHOPTERA) OF THE PRIMEVAL FOREST IN BIAŁOWIEŻA

B. MOHAMMAD, J. MAJECKI and C. TOMASZEWSKI

## SUMMARY

During the years 1960-62, 2129 adult caddis flies, belonging to 72 species, were collected in the Białowieża Primeval Forest.*Holocentropus insignis* Martynov, *Lenarchus bicornis* (McLach.), *Ylodes detruncatus* (Martynov) and *Ylodes kawraiskii* (Martynov) are new records for Poland.

## INTRODUCTION

Białowieża Primeval Forest and Białowieża National Park constitute a unique nature reserve on the borders of central and eastern Europe. The preservation of many natural plant associations in varied habitats makes this area a special kind of refuge. Białowieża National Park, created in 1921, has the status of a special reserve covering the most natural and primitive part of the Forest (Faliński 1968). The central position of the National Park within the borders of Białowieża Primeval Forest (an area of 47sq.km) provides a natural zone which protects this area against the harmful influences of civilisation. The observations made in this part of the Forest show that the accumulation in the ecosystems of heavy metals from industrial emissions is not very high, and hence Białowieża National Park is one of the least contaminated National Parks in Poland (Grodzińska 1980). The investigations of Wróbel and Okołow (in press) indicate that water is formed here under almost natural condition. This is important, since it is a lowland area and the condition of the underground and surface water determines to a great extent the formation of characteristic plant associations (Faliński 1968). The water directly affects the animals living in the numerous rivers, old river beds, streams, canals, fens and temporary pools in the forest. It can be expected that these water ecosystems, little changed by man, are the last chance for the survival of many water invertebrates, including, of course, caddis flies.

Ulmer (1925) was the only author who has published on the caddis flies of Białowieża Forest. He listed 43 species, caught in 1918, mainly in the vicinity of the River Narewka. In order to add to this list, and carry out further studies, the authors of this paper have worked on Trichoptera caught in 1960-62 in the Białowieża Primeval Forest.

## THE CHARACTERISTICS OF THE INVESTIGATED AREA AND THE METHOD

Białowieża Primeval Forest, of 1250 sq.km., is the biggest forest complex in the Central European Plain. Within Polish borders it covers 580 sq.km. of the Podlasie Lowland. The area has a continental climate which restricts the vegetation and is associated with long-lasting snow. The average annual rainfall is about 600 mm, ranging from 426 to 857 mm (Faliński 1968). The area of Białowieża is almost wholly drained by the tributaries of the Rivers Narwia and Bug (the Vistula river basin), with only a small part by the River Niemen. The Narewka, a typical lowland river which crosses the centre of the Forest, with its tributaries in spring and after heavy rains floods the neighbouring areas. Lenitic habitats are represented by old river-beds, fens, pools and temporary pools. In some plant associations characteristic of the Forest such as marshy forests, ash-spruce-alder groves, hornbeam and oak-hornbeam clumps, the soil waters form temporary spring or spring and autumn pools. All the lakes are artificial (the dams on the Rivers Narewka and Perebel), with a minimal water current. The adult caddis flies from Białowieża, at present in the collection of the Evolutionary Biology Department of Łodz University were caught mainly by Prof. Adamczewski in May, June, July, August, September, and October in 20 localities (Fig. 1). The greatest number of specimens came from light traps. They were kept dry and not determined.

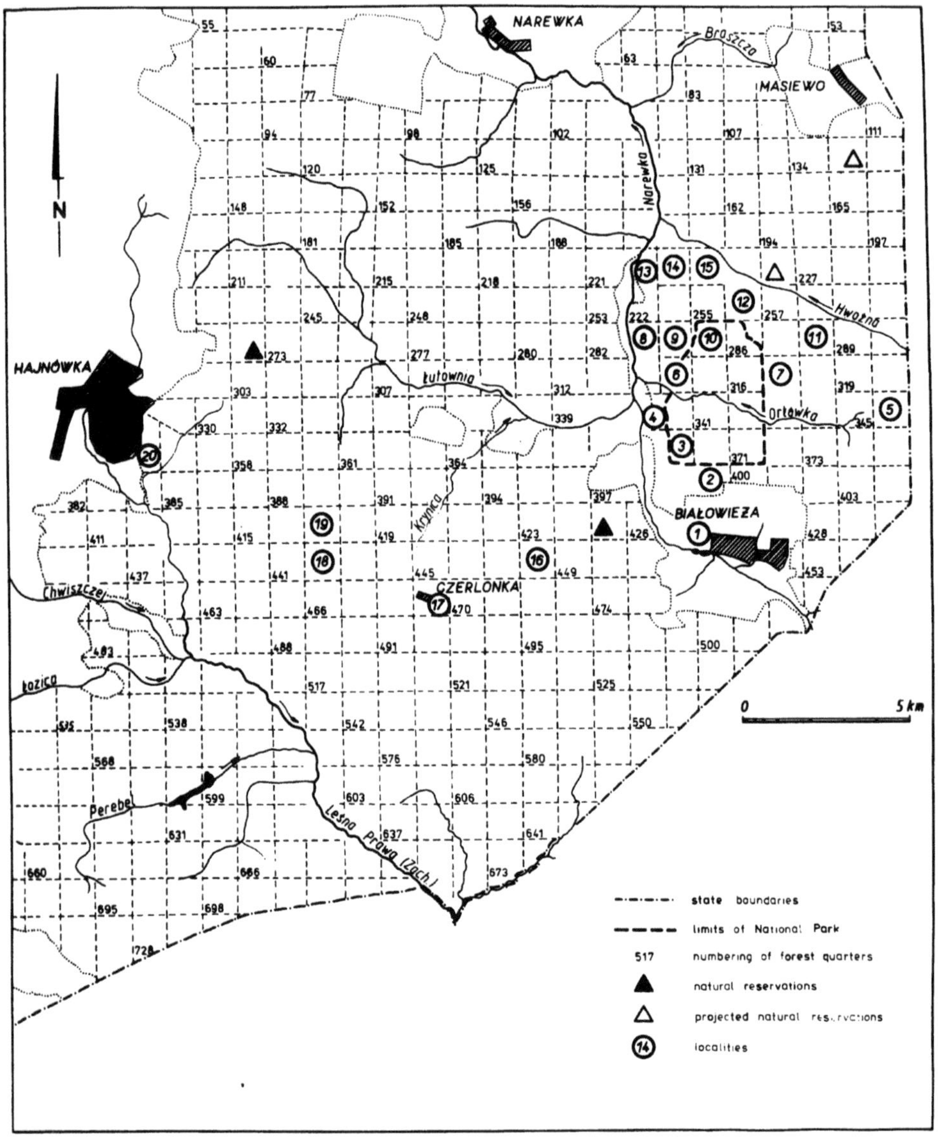


Figure 1. Location of the sampling sites in Białowieża Forest.

The localities marked on the map were grouped as follows :
I - locality 1, most systematic light trapping was carried out here : 57 samples, 1082 specimens.
II - localities 3, 6, 10, within the area of Białowieża National Park : 21 samples, 305 specimens.
III - localities 2, 4, 7, 8, 9, 12, lying close to the border of the National Park : 25 samples, 390 specimens.
IV - localities 5, 11, 13, 14, 15, between the Rivers Narewka and Hwoźna : 12 samples, 201 specimens.
V - localities 16, 17, 18, 19, 20, within the Forest : 13 samples, 151 specimens.

Table 1. List of Trichoptera collected in Białowieża Forest in the years 1960-62

| | LOCALITIES | | | | |
|---|---|---|---|---|---|
| | I | II | III | IV | V |
| *Hydropsyche angustipennis* (Curt.) * | | + | | | |
| *Hydropsyche bulgaromanorum* Mal. | + | | | + | |
| *Hydropsyche contubernalis masovica* Mal. | | | + | + | + |
| *Hydropsyche pellucidula* (Curt.) * | + | | + | + | |
| *Plectrocnemia conspersa* (Curt.) | + | | | | |
| *Polycentropus flavomaculatus* (Pict.) | | + | | | |
| *Holocentropus dubius* (Ramb.) | | + | | | |
| *Holocentropus insignis* Mart. | | + | + | | |
| *Holocentropus picicornis* (Steph.) | + | | + | + | |
| *Ecnomus tenellus* (Ramb.) | + | | | | |
| *Trichostegia minor* (Curt.) * | + | + | | | |
| *Agrypnia obsoleta* (Hag.) | | | + | | |
| *Agrypnia pagetana* Curt. * | + | + | + | + | + |
| *Agrypnia varia* (Fabr.) * | + | + | | + | |
| *Phryganea bipunctata* (Retz.) * | + | + | + | | |
| *Phryganea grandis* (L.) * | + | + | | | + |
| *Oligotrichia striata* (L.) * | | | + | | |
| *Hagenella clathrata* (Kol.) | | | | + | |
| *Oligostomis reticulata* (L.) * | | + | | | |
| *Ironoquia dubia* (Steph.) | | + | | + | |
| *Limnephilus auricula* Curt. * | + | + | + | | + |
| *Limnephilus bipunctatus* Curt. * | + | | | | |
| *Limnephilus dispar* McLach. | | | | + | |
| *Limnephilus elegans* Curt. * | | + | | | |
| *Limnephilus extricatus* McLach. * | + | | | + | |
| *Limnephilus flavicornis* (Fabr.) * | + | + | | + | + |
| *Limnephilus fuscicornis* Ramb. * | + | | | | |
| *Limnephilus griseus* (L.) * | + | + | + | + | + |
| *Limnephilus hirsutus* (Pict.) | + | | + | | |
| *Limnephilus ignavus* McLach. * | + | | + | | + |
| *Limnephilus lunatus* Curt. | + | | | | |
| *Limnephilus marmoratus* (Curt.) | + | + | + | + | |
| *Limnephilus nigriceps* (Zett.) | + | | | | |
| *Limnephilus politus* McLach. * | | + | | | |
| *Limnephilus rhombicus* (L.) * | + | | + | | |
| *Limnephilus sericeus* (Say) * | + | + | + | + | + |
| *Limnephilus sparsus* (Curt.) * | + | + | + | + | + |
| *Limnephilus stigma* Curt. * | + | + | + | + | + |
| *Limnephilus subcentralis* Brau. * | + | | | | + |
| *Limnephilus vittatus* (Fabr.) * | + | + | + | + | + |
| *Colpotaulius incisus* (Curt.) * | | + | | | |
| *Grammotaulius nigropunctatus* (Retz.) * | + | + | | | + |

| | LOCALITIES | | | | |
|---|---|---|---|---|---|
| | I | II | III | IV | V |
| *Grammotaulius nitidus* (Mull.) | + | | | | |
| *Glyphotaelius pellucidus* (Retz.) * | + | + | + | + | + |
| *Nemotaulius punctatolineatus* (Retz.) * | + | | | | |
| *Anabolia laevis* (Zett.) * | + | | | | |
| *Phacopteryx brevipennis* (Curt.) | + | + | + | + | |
| *Lenarchus bicornis* (MacLach.) | | | + | | |
| *Rhadicoleptus alpestris* (Kol.) * | | + | + | | |
| *Potamophylax nigricornis* (Pict.) | | + | | | |
| *Halesus digitatus* (Schrk.) | | + | + | | |
| *Halesus radiatus* (Curt.) | | + | + | | |
| *Halesus tessellatus* (Ramb.) * | | + | + | + | |
| *Micropterna sequax* McLach. | | | + | + | |
| *Chaetopteryx villosa* (Fabr.) | | + | | | |
| *Athripsodes albifrons* (L.) | | | + | | |
| *Athripsodes aterrimus* (Steph.) * | | + | + | | |
| *Athripsodes cinereus* (Curt.) * | + | + | | | |
| *Ceraclea dissimilis* (Steph.) | + | | | + | + |
| *Ceraclea fulva* (Ramb.) | + | + | | | |
| *Mystacides azurea* (L.) * | + | | | | |
| *Mystacides longicornis* (L.) * | + | | | | + |
| *Mystacides nigra* (L.)* | | | + | | |
| *Triaenodes bicolor* (Curt.) * | + | | | | |
| *Ylodes detruncatus* (Mart.) | | + | | | |
| *Ylodes kawraiskii* (Mart.) | + | | | | |
| *Erotesis baltica* (MacLach.) | | + | | | |
| *Oecetis furva* (Ramb.) | | | + | | |
| *Oecetis lacustris* (Pict.) * | + | | | + | |
| *Oecetis ochracea* (Curt.) | + | + | | | + |
| *Oecetis testacea* (Curt.) | + | | + | | |
| *Notidobia ciliaris* (L.) * | | | | + | |
| *Molanna angustata* (Curt.) | + | + | + | | + |
| *Molannodes tinctus* (Zett.) | + | | | | |
| Number of species | 45 | 38 | 32 | 24 | 18 |

* These species were determinated by Ulmer (1925

## RESULTS

There were 1291 male and 838 female caddis flies in the collection studied, and 83% of the specimens belonged to 34 species from the family Limnephilidae. According to Andersen (1978, 1979) this is a typical result in such investigation. Only 3 species, *Cyrnus trimaculatus* (Curt.), *Ceraclea senilis* (Burm.) and *Micropterna lateralis* (Steph.), reported by Ulmer (1925) for Białowieża Forest, were not found in this collection. Hence, it is clear that 32 species listed here (Tab. 1), are new for Podlasie and Białowieża Forest (Ulmer 1925, Tomaszewski 1965). Białowieża Primeval Forest is the second region in Poland where *Hydropsyche bulgaromanorum, Ylodes simulans* and *Oecetis testacea* are present. The four species, *Holocentropus insignis, Lenarchus bicornis, Ylodes detruncatus* and *Ylodes kawraiskii* are new records for Poland within its present borders. *Holocentropus insignis* and *Lenarchus bicornis* were previously found in the vicinity of Vilnius by Raciecka (1931).

Bialowieza Forest is the most southern locality for *L. bicornis* and *Y. detruncatus* which according to

Botosaneanu and Malicky (1978) and Tobias (1976) occur mainly in Scandinavia. In contrast, for *Y. kawraiskii* Bialowieza consitutes the most northern locality.

AKNOWLEDGEMENTS

The authors are grateful to Prof. S. Adamczewski for giving them access to his collection of caddis flies and to Dr. B. Szczesny for his assistance in the determination of some species.

REFERENCES

Andersen, I. 1978. Influence of Temperature on the Sex Ratio of Trichoptera in Light-Trap Catches in Western Norway. **Nor. J. Ent.,** 25 : 149-151.

Andersen, I. 1979. Some Caddis Flies (Trichoptera) in Western Norway, and their Arrival Pattern in Light Traps. **Fauna Norvegica,** 26 : 12-17.

Botosaneaunu, L. & Malicky, H. 1978. Trichoptera, in Illies J. (ed.) Limnofauna Europaea. Stuttgart, Fischer, 333-359.

Faliński, J.B. (ed). 1968. Park Narodowy w Puszczy Białowieskiej, Panstw. Wyd. Roln. Leśn.,Warszawa.

Grodzińska, K. 1980. Zanieczyszczenia Polskich Parków Narodowych Metalami Ciezkimi. **Ochrona Przyr.,** 43 : 9-27.

Raciecka, M. 1931. Chrósciki (Trichoptera) Północno-Waschodniej Polski ze Szczególnym Uwzględnieniem Obszaru Wileńsko-Trockiego. **Proc. Tow. Przyi. Nauk. w Wilnie,** 20 :1-34.

Tobias, W. 1976. Köcherfliegen und Steinfllegen einiger Gewasser in Sor Varanger (Nord-Norvegen) (Trichoptera, Plecoptera). IV. *Triaenodes detruncatus* Martynov neu für Norvegen. **Senckenbergiana biol.,** 57 : 49-53.

Tomaszewski, C. 1965. Chrusciki (Trichoptera), Katalog Fauny Polski.,Warszawa, 28 : 1-104.

Ulmer, G. 1925. Trichopteren und Ephemeropteren aus dem Bialowiser Wald. **Abh. Bayer. Akad. Wiss., Suppl.,** 6-9 : 339-342.

Wróbel, S. & Okolow, C. (in press). Białowieski Park Narodowy i Jego Wody.

# PRELIMINARY COMPARISON OF NEW GUINEA TRICHOPTERA WITH THE FAUNAS OF SULAWESI AND CAPE YORK PENINSULA.

A. NEBOISS


## SUMMARY

The Trichoptera of Sulawesi, New Guinea and Cape York Peninsula of North Eastern Australia are characterised and compared. Of the total of about 80 genera recognised, only about 20% (17) were found common to all three areas. On the species level only a few are common to these areas. The main differences between the Australian and Oriental caddis-fly species occur along the Weber's Line.


In 1980 a paper on caddis-fly distributions in Australia was presented to the third Symposium on Trichoptera (Neboiss 1981). Since then an attempt has been made to consolidate the published information on all species throughout the South-west Pacific region and to establish their distributions. Consequently the Atlas of Trichoptera of the region was assembled and published (Neboiss 1986). This publication exposed the incompleteness, inaccuracies and paucity of data from large sections of the region. In search for further information the existence of undescribed material held by several institutions became known, and this, together with recently accumulated expedition material from some unexplored districts considerably widens our knowledge and understanding of species distributions and relationships.

In this paper the Trichoptera faunas of Sulawesi, New Guinea and Cape York Peninsula of North-eastern Australia are characterised and compared (Figure 1). Work on most of the undescribed material is still at an early stage, but provisional lists of species are now compiled.

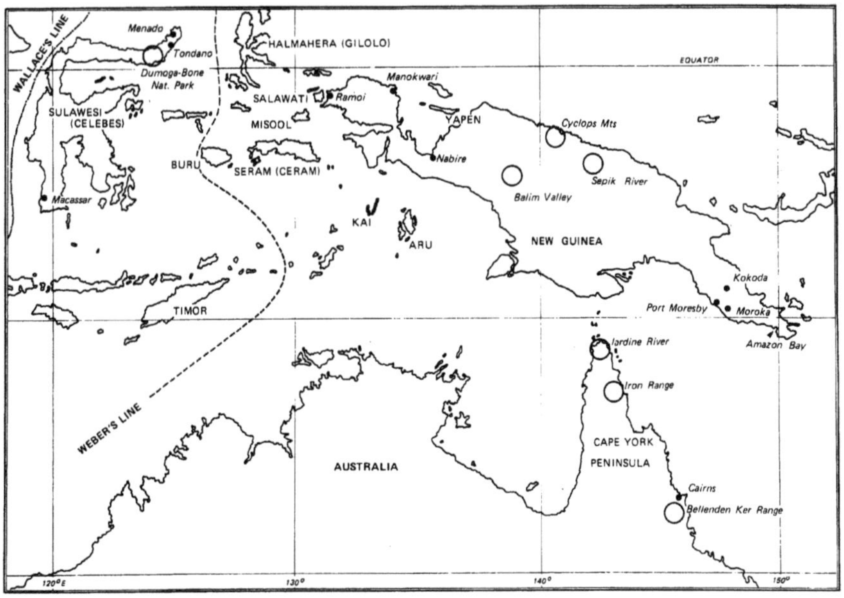


Figure 1. The Oriental Australian transition zone, showing the major caddis-fly collecting sites.

The three areas are located within the tropics and include more or less extensive mountain ranges reaching elevations above 1500 m, and having expanses of rainforest mingled with cultivated lowlands. Sulawesi is situated to the west of Weber's Line, which indicates a narrow zone of faunal balance between the contrasting Australian and Oriental species (Mackerras 1970, Powell *et al.* 1981). To the east of Weber's Line lies New Guinea, but Cape York Peninsula is situated further to the south.

According to recent information the islands of the Wallacea region (which includes Sulawesi) has not been connected to either Asian or Australian continents during Quaternary. Land connection between New Guinea and Cape York Peninsula (Australia) over Torres Strait, however, existed during most of Quaternary opening up only as recently as 6500-8000 years BP (Kikkawa *et al.* 1981).

Records of New Guinea Trichoptera are based mainly on specimens described in few publications. The first were collected by Wallace and described by McLachlan (1863) ; later collections were assembled by the italian naturalist Beccari and missionary Loria. Part of this material was later described by Navas (1923-33). The German Kaiserin-Augusta-Fluss (Sepik river) expedition material was described by Ulmer (1915, 1938) ; Miss Cheesman's collections mainly gathered at Cyclops Mts and Kokoda districts was described by Kimmins (1962) ; specimens from Central Highlands collected by Barrett was partly described by Korboot (1964, 1965) and British Speleological Expedition specimens described by Kumanski (1979). To this now added are data extracted from some 8000 specimens collected in the 1955-70 period and deposited in the Bishop Museum collection, Honolulu. The localities are widely distributed throughout the island from Manokwari in the north-west, Balim valley in the highlands of Irian Jaya to the Amazon Bay area at the far south-east, and therefore of considerable interest.

In his review of the Sunda Island fauna Ulmer (1951) recorded only 18 species from Sulawesi (Celebes). The information was based on extremely limited number of specimens from a few scattered localities. During the next 30 years only one species was added to the island fauna. Considering that about 100 species are known from Sumatra (Ulmer 1951) it is no surprise that among several thousand specimens collected by the Project Wallace 1985 expedition to the Dumoga-Bone national Park, North Sulawesi, more than 80, mostly undescribed species have been recognised.

Cape York Peninsula is the north-eastern part of the Australian continent. It has been regarded as a critical area for the interaction between the New Guinea and Australian species (Kikkawa *et al.* 1981). The northern most tip is situated only 150 km south of New Guinea, it gradually widens in southerly direction to about 590 km at the 18°S, thus forming a broad triangle.

Information on species present in the area is usually hidden in general publications. Although not very comprehensive, it nevertheless gives clear indications that the greatest species diversity is found in the few tracts of rainforest still existing in the area. The species number is much lower in the dry lowlands and deforested areas (Kikkawa *et al.* 1981). Several expeditions in the 1973-83 period to Cairns-Mossman, Iron Range and Jardine River areas by Max Moulds and to Bellenden Ker Range by the Queensland Museum, Earthwatch expedition, 1981 (Monteith, 1982) accumulated extensive material which is now being gradually sorted and studied.

The distribution of species in the region is rather irregular with only a very few species extending across the boundaries of the three areas. The genera, however, are represented more widely. A total of 80 genera are recognized as being present in the region ; of these 41 are known from Sulawesi, 44 from New Guinea and 47 from Cape York Peninsula. Only 17 genera are common to all three areas, 11 are shared by Sulawesi and New Guinea, but only 6 by New Guinea and Cape York Peninsulta. The interchange between the two latter areas has been considered as comparatively restricted because the moist adapted organisms of New Guinea rainforests and the dry adapted form from semi-arid savannah of Cape York Peninsula found the opposite environment hostile and unsuitable (Monteith 1974).

The three largest families are Hydroptilidae, Hydropsychidae and Leptoceridae, each with 14 to 16 genera and 78, 71 and 108 species respectively. The species numbers in Sulawesi are mostly based on material collected at Dumoga-Bone National Park, the data for New Guinea and Cape York Peninsula are extracted from the Atlas (Neboiss 1986) to which are added the somewhat incomplete data from a considerable number of specimens not yet positively identified or described. The following brief summaries on the families gives insight into the characteristics of each of the three areas.

The predominantly Palaearctic family Rhyacophilidae reaches the Sunda archipelago with 7 species (Ulmer 1951, Malicky 1978), and it is now for the first time recorded from Sulawesi. The family does not extend further eastward across Weber's Line into New Guinea. There it is replaced by the members of the family Hydrobiosidae, which has its centre of distribution in south-eastern Australia and New Zealand. The

Table 1. Genera and number of species recorded from each area (data for Sulawesi show literature records in brackets ; New Guinea and Cape York Peninsula data include undescribed species).

| FAMILY | GENUS | NUMBER OF SPECIES | | |
|---|---|---|---|---|
| | | Sulawesi | New Guinea | Cape York |
| RHYACOPHILIDAE | *Rhyacophila* | 1 | - | - |
| HYDROBIOSIDAE | *Apsilochorema* | 1 | 8 | 2 |
| | *Ulmerochorema* | - | - | 3 |
| | *Tanorus* | - | 10 | - |
| | *Ethochorema* | - | - | 2 |
| | *Ptychobiosis* | - | - | 1 |
| GLOSSOSOMATIDAE | *Agapetus* | - | 15 | 2 |
| | *Tagapetus* (?) | 3 | - | - |
| HYDROPTILIDAE | *Chrysotrichia* | 3 | 1 | - |
| | *Hydroptila* | 6 | 9 | 4 |
| | *Madioxyethira* | 1 | 1 | - |
| | *Orthotrichia* | 7 | 14 | 8 |
| | *Oxyethira* | 1 | 4 | 4 |
| | *Parastactobia* | 1 | - | - |
| | gen.indet. X | 1 | - | - |
| | *Hellyethira* | - | - | 4 |
| | *Maydenoptila* | - | - | 1 |
| | *Orphninotrichia* | - | - | 1 |
| | *Xuthotrichia* | - | - | 1 |
| | gen. indet. A | - | 2 | - |
| | gen. indet. B | - | 1 | - |
| | gen. indet. C | - | 1 | - |
| | *Plethus* (?) | - | 2 | - |
| PHILOPOTAMIDAE | *Chimarra* | 7(1) | 16 | 5 |
| | *Hydrobiosella* | - | - | 3 |
| HYDROPSYCHIDAE | *Oestropsyche* | 1(1) | 1 | - |
| | *Aethaloptera* | - | 1 | 1 |
| | *Polymorphanisus* | 1 | - | - |
| | *Macrostemum* | 1 | 5 | 1 |
| | *Baliomorpha* | - | 5 | 4 |
| | *Leptopsyche* | - | 1 | - |
| | *Diplectrona* | 1 | 4 | 8 |
| | *Sciops* | 1(2) | - | - |
| | *Hydropsyche* | 3(2) | 4 | - |
| | *Hydropsychodes* | 3 | 2 | - |
| | *Hydromanicus* (?) | (1) | - | - |
| | *Herbertorossia* | - | 7 | - |
| | *Abacaria* | - | 10 | - |
| | *Cheumatopsyche* | - | 1 | 2 |
| | *Smicrophylax* | - | - | 1 |
| | *Asmicridea* | - | - | 1 |
| STENOPSYCHIDAE | *Stenopsychodes* | - | - | 7 |

| FAMILY | GENUS | NUMBER OF SPECIES | | |
|---|---|---|---|---|
| | | Sulawesi | New Guinea | Cape York |
| POLYCENTROPODIDAE | *Hyalopsyche* | - | 1 | 1 |
| | *Hyalopsychella* | 1 | - | - |
| | *Nyctiophylax* | 1 | 1 | 8 |
| | *Polycentropus* | 4 | 17 | - |
| | *Polyplectropus* | 1(1) | 3 | 1 |
| | *Plectrocnemia* | - | - | 1 |
| DIPSEUDOPSIDAE | *Dipseudopsis* | (3) | - | - |
| ECNOMIDAE | *Ecnomus* | 3 | 4 | 10 |
| | *Ecnomina* | - | - | 5 |
| PSYCHOMYIIDAE | *Psychomyia* | 2 | - | - |
| | *Tinodes* | 4 | 1 | - |
| | *Paduniella* | 1 | - | - |
| GOERIDAE | *Goera* | 1 | - | - |
| MOLANNIDAE | *Molanna* | (1) | - | - |
| LEPIDOSTOMATIDAE | *Dinarthropsis* | - | 1 | - |
| | *Neolepidostoma* | 1 | 1 | - |
| | *Goerodes* | 2 | 1 | - |
| CONOESUCIDAE | *Coenoria* | - | - | 1 |
| CALOSIDAE | *Calocoides* | - | - | 1 |
| | *Caenota* | - | - | 3 |
| | *Pliocaloca* | - | - | 3 |
| HELICOPSYCHIDAE | *Helicopsyche* | - | - | 3 |
| PHILORHEITHRIDAE | *Tasmanthrus* | - | - | 1 |
| ATRIPLECTIDIDAE | *Atriplectides* | - | - | 1 |
| ODONTOCERIDAE | *Marilia* | - | - | 2 |
| CALAMOCERATIDAE | *Anisocentropus* | 2(3) | 12 | 3 |
| LEPTOCERIDAE | *Triplectides* | 1 | 1 | 6 |
| | *Westriplectes* | - | - | 1 |
| | *Notalina* | - | - | 1 |
| | *Lectrides* | - | 1 | - |
| | *Notoperata* | - | - | 1 |
| | *Symphitoneuria* | 1(1) | 2 | 1 |
| | *Symphitoneurina* | - | 1 | - |
| | *Adicella* | 1 | 1 | - |
| | *Triaenodes* | 2 | 10 | 14 |
| | *Leptocerus* | 1 | 1 | 3 |
| | *Setodes/Trichosetodes* | 3 | 3 | 1 |
| | *Oecetinella* | (1) | 1 | - |
| | *Oecetis* | 8(1) | 10 | 25 |

genus *Apsilochorema* is the only one shared by all three areas, of which the richest is New Guinea with about 10 species.

The Glossosomatidae of the region belong to the *Agapetus-Synagapetus-Tagapetus* group, which at present is insufficiently investigated, with numerous undescribed taxa. The three species found in Sulawesi appear to represent a species group distinct from those in New Guinea and Cape York Peninsula. The genus *Glossosoma* although known from Java and Sumatra, has not been recorded from Sulawesi nor east of Weber's Line.

Of the "micro" caddis-flies of the family Hydroptilidae of Sulawesi three species are already known from Northern Australia, two are in common with New Guinea and one elsewhere in Asia. Others are in 'oriental' genera that are unknown anywhere east of Sulawesi. Perhaps one of the most interesting finds is one of the immatures of a species of *Madioxyethira* , hither-to unknown, and unlike other Stactobiini larvae. These were collected from the faces of waterfalls. Four other genera are recorded from New Guinea and four from Cape York Peninsula but none of these are in common to the two areas.

The almost cosmopolitan genus *Chimarra* of the family Philopotamidae is well represented in the entire region, yet so far there are no records of any species in common to the three areas, or even to any two areas, each area appears to have its own species complex. A considerable number of species are undescribed and study of these should prove most rewarding exercice. Of this family only one other genus - *Hydrobiosella* occurs on Cape York Peninsula. The genus has numerous species in south-eastern Australia. Although regarded as subgenus of *Sortosa* by Ross (1956) it was reinstated to full generic level by Neboiss (1977).

The family Hydropsychidae with 16 recognised genera and more than 70 species is the second largest after Leptoceridae. It appears that only two, or probably three genera are common to all three areas, three are in common to Sulawesi and new Guinea, and three to New Guinea and Cape York Peninsula. The generic grouping of some New Guinea and Cape York Peninsula species is somewhat confused, but detailed study of the undescribed material no doubt will help to rectify the misinterpretations. It has been already established that the genus *Sciops* does not extend into the Australian fauna (Neboiss 1986), nor there are any records of it in New Guinea. The *Diplectrona* is another species complex which does not agree fully with the typical forms. The review of this group is already in progress. Disjunct distribution also rises questions of the genus *Cheumatopsyche* and forms similar to it.

The Australian genus *Stenopsychodes* of the family Stenopsychidae clearly differs from the Asian members of the family. The distribution in Australia is limited to the eastern coastal districts with most species (7) known from Cape York Peninsula. It is not known from New Guinea nor Sulawesi, however, one species of the typical oriental forms is recorded from Sumatra and Borneo.

The genus *Polycentropus* of the family Polycentropodidae in New Guinea has developed into a group of closely related species, all occuring in the highland streams. There is no direct resemblance to the species from Sulawesi and the genus is not recorded from Cape York Peninsula, where one species of the south-east Australian genus *Plectrocnemia* is recorded.

*Nyctiophylax,* a genus well known from the oriental region, is known from Sulawesi and New Guinea by one species in each, but from Cape York Peninsula at least 8 species could be referred to this genus.

*Hyalopsyche* and *Hyalopsychella* here are still retained in the family Polycentropodidae. Their distribution is very local but at least one species is known to occur in each area.

The family Dipseudopsidae has been recorded from Sulawesi, but its distribution does not extend eastward across the Weber's Line. Similarly also the Psychomyiidae distribution stops at this line, the only exception is genus *Tinodes* of which one species is recorded from New Guinea. The family Ecnomidae replaces Psychomyiidae in New Guinea and Australia with two described genera. At least 10 species of *Ecnomus* and 5 species of *Ecnomina* are known from Cape York Peninsula.

Of the tube-case-making caddis-flies of the superfamily Limnephiloidea three families - Goeridae, Molannidae and Lepidostomatidae are reaching Sulawesi from the west, with only the latter family marginally extending into the New Guinea area. In contrast to that, the Cape York Peninsula fauna includes six families not found in New Guinea or Sulawesi. Further two families - Calamoceratidae and Leptoceridae are well represented throughout the region. The family Leptoceridae with 14 genera and well over 100 species is the largest in the region. There are some genera which have not been recorded northwards beyond Cape York Peninsula, they are *Westriplectes, Notalina* and *Notoperata* ; on the other hand there are six genera common to all three areas.

ACKNOWLEDGMENTS

I am particularly indebted to Dr Alice Wells, Zoology Department, University of Adelaide, Mr David Cartwright, MMBW laboratories, Melbourne, Dr Gordon Nishida, Bishop Museum, Honolulu, Dr Geoff Monteith, Queensland Museum, Brisbane and Mr Max Moulds, Australian Museum, Sydney for the extensive material made available to me.

REFERENCES

Kikkawa,J., Monteith, G.B.& Ingram, G.1981. Cape York Peninsula : Major region of faunal interchange. *In* Ecological Biogeography of Australia (A. Keast ed.), Dr W. Junk, The Hague : 1697-1742.

Kimmins, D.E. 1962. Miss L.E. Cheesman's expedition to New Guinea : Trichoptera. **Bull. Br. Mus. Nat. Hist. (Entomol.)** , 11 : 99-187.

Korboot, K.1964. Eight new species of caddis-flies (Trichoptera) from the Australian region. **Pap. Dept. Ent. Univ. Qld.,** 2(2) : 47-56.

Korboot, K.1965. A new species of caddis-fly (Trichoptera) from New Guinea . **J. Entomol. Soc. Qld.,** 4 : 40.

Kumanski, K.1979. Trichoptera (Insecta) from New Guinea. **Aquatic Insects,** 1(4) : 193-219.

Malicky, H.1978. Beiträge zur Kenntnis der Insektenfauna Sumatras. 7. **Beitr. naturk. Forsh. SudwDtl.,** 37 : 159-173.

McLachlan, R.1863. On *Anisocentropus*, a new genus of exotic Trichoptera. **Trans. Entomol. Soc. Lond.,** (3) 1 : 492-496.

Mackerras, I.M.1970. Composition and distribution of the fauna (Chapter 9) *in* The Insects of Australia (I.M. Mackerras ed.) **Melb. Univ. Press,** Melbourne : 187-203.

Monteith, G.B. 1974. Focus on Cape York. Fauna subcommittee report. **Entomol. Soc. Qld. Brisbane,** 36 p.

Monteith, G.B.1982. Earthwach expedition to Bellenden Ker. **News Bull. Ent. Soc. Qld.,** 10(4) : 46-47.

Navás, L.1923. Estudio sobre Neuropteros. **Arx. Ins. Cienc. Barcelona,** 7 : 179.203.

Navás, L.1930. Insectos del Museo de Paris. **Broteria.Ser.Zool.**, 26 : 19-23.

Navás, L.1933. Insecta orientalia. **XII series.Men. Pont. Accad. Sci. Nuovi. Lin.**, 17 : 75-108.

Neboiss, A.1977. A taxonomic and zoogeographic study of Tasmanian caddis-flies (Insecta : Trichoptera). **Mem. Nat. Mus. Vict.,** 38 : 1-203.

Neboiss, A.1980. Distribution of Trichoptera families in Australia, with comments of fauna in the south-west. **Proc. 3rd Int. Symp. Trichoptera,** G.P. Moretti (ed.), Junk, The Hague : 265-272.

Neboiss, A.1986. Atlas of Trichoptera of the SW Pacific-Australian region. (Ser. Ent. 37), Junk Dordrecht, VIII + 286 p.

Neboiss, A.1986. Taxonomic changes in caddis-fly species from the south-west Pacific-Australian region with descriptions of new species (Insecta : Trichoptera). **Mem. Mus. Vict.,** 47 : 213-223.

Powell, C.McA., Johnson, B.D. & Veevers, J.J.1981. The early Cretaceous break-up of Eastern Gondwanaland, the separation of Australia and India, and their interaction with Southeast Asia. *In* Ecological Biogeography of Australia (A.Keast ed.), Dr W.Junk, The Hague : 17-29.

Ross, H.H.1956. Evolution and classification of the mountain caddis-flies. The University of Illinois Press. Urbana.

Ulmer, G.1915. Trichopteren des Ostens, besonders von Ceylon und New Guinea. **Deutsche Ent. Zeitschrift,** Berlin : 41-75.

Ulmer, G,1938. Einige neue Trichopteren von Neu Guinea aus dem Berliner Museum. Sitzber. Ges. naturf. Freunde, Berlin, 1937 : 398-403.

Ulmer, G.1951. Köcherfliegen (Trichopteren) von den Sunda Inseln (Teil 1). **Arch. Hydrobiol., Suppl.,** 19 : 1-528.

# THE CADDISFLIES OF SRI LANKA

P. CHANTARAMONGKOL

SUMMARY

The present state of faunistic knowledge of the caddisflies of Sri Lanka is discussed which is probably the best known caddis fauna of any tropical countries.

The caddisflies of Sri Lanka are probably the best and most completely known of any tropical countries. Many collections have been made and papers published, with Schmid's (1958) paper as the most important. Recently, the material collected by the Lund University Expedition of 1962 was studied by the author (Chantaramongkol 1985), and 29 species were described separately by Chantaramongkol and Malicky (1986).

The thesis by Chantaramongkol (1985) gave a summary of the fauna and included an atlas for the identification of all Ceylonese species, based on the principles proposed in the Atlas of European Trichoptera by Malicky (1983), where in a synoptic table the order is subdivided into families or genera by means of three characters : the spur formula, the presence or absence of ocelli, and the number of segments in the male maxillary palpi. Further identification to species is done by comparison of figures.

In the present state of knowledge, 222 species of Trichoptera are known from Sri Lanka, only 26 of them being found outside the island. One of them, *Ecnomus tenellus*, has a wide distribution including Europe. The percentage of endemics is probably not as high as these figures indicate because the fauna of adjacent southern India is poorly known.

The work by Chantaramongkol (1985) included an ecological analysis of all species as far as it was possible from the available data. A clear distinction was found between the faunas of waters at different altitudes. Rhyacophilidae, Glossosomatidae, Goeridae, Lepidostomatidae, Ceylanopsyche and Helicopsychidae clearly prefer the higher altitudes of the mountains in the southern part of the island. In Hydroptilidae, Philopotamidae, Ecnomidae, Psychomyidae, Hydropsychidae and Leptoceridae many species were found at lower altitudes, but some were also present at higher altitudes. In the lowlands the faunas of standing waters and of large rivers are similar. Standing waters at intermediate altitudes (between 150 and 1500 m) are relatively poor in numbers of species. There are indications that in Sri Lanka a longitudinal zonation in river systems exists, but the present data did not permit a clear definition of the zones.

Schmid (1984) has presented, by extrapolation from his collections in several parts of India, a hypothesis of an extraordinary richness of Trichoptera in southern Asia, much higher than in any other regions of the world. He estimated the total number of southern Asian caddisflies to be about 40.000 species. The caddis fauna of Sri Lanka is now well enough known to say that this island does not belong to this postulated extremely rich region. The total of existing species in Sri Lanka may be 250 - 300, judging from the 222 species now known and from the high percentage of species found at repeated occasions by many collectors. This is a figure comparable with the numbers in Europe which range in most cases between 200 and 300 per country (e.g. in Great Britain, Sweden, Finland, Austria, Italy, Bulgaria, Greece, Turkey).

The human influence on the caddis fauna in Sri Lanka is certainly important. For hundreds, or even thousands of years, many man-made impoundments were constructed which changed the hydrology of natural waters, but this influence has not been studied. Pollution may also be important in the densely populated parts of the country, but only the striking abundance of *Hydropsyche katugahakanda* in a heavely polluted stream indicated a high pollution tolerance of this species.

AKNOWLEDGEMENT

I thank Dr. M. I. Crichton for the correction of the English text.

REFERENCES

Chantaramongkol, P. 1985. Die Köcherfliegen von Sri Lanka. Taxonomische Grundlagen für ihre Verwendung als Gewässer-Indikatoren. Dissertation zur Erlangung des Doktorgrades, Formal - und Naturwissenschaltiliche Fakultät, Universität Wien. 399 pp.

Chantaramongkol, P. & Malicky, H. 1986. Beschreibung von neuen Köcherfliegen (Trichoptera, Insecta) aus Sri Lanka. - **Ann. Naturhist. Mus. Wien** (in press).

Malicky, H. 1983. Atlas of European Trichoptera. Junk, The Hague, Boston, London, X + 298 pp.

Schmid, F. 1958. Trichoptères de Ceylan. - **Arch. Hydrobiol.** 54 : 1-173.

Schmid, F. 1984. Un essai d'évaluation de la faune mondiale des Trichoptères (Abstract). - **Proc. 4th Int. Symp. Trich.** J.C. Morse (Ed.), Junk, The Hague : 337.

# PRELIMINARY INVESTIGATIONS ON THE FAUNISTICS AND ECOLOGY OF HONG KONG TRICHOPTERA

D. DUDGEON

SUMMARY

An analysis of the composition and geographic affinities of Hong Kong Trichoptera has not previously been attempted. This is no doubt a direct consequence of the paucity of available information. The data reviewed herein are derived from the results of collections in Hong Kong over the last 5 years, in addition to a survey of the Asian Trichoptera literature. The results indicate that Hong Kong caddisflies comprise a predominance of warm-adapted taxa with a close affinity to tropical Oriental faunas. However, the potential for very high trichopteran diversity in the territory is not realized as a consequence of habitat limitation brought about by environmental degration.

## INTRODUCTION

The caddisflies of Hong Kong are poorly known but include species which are widespread in China and/or the Oriental region, as well as species recorded only from within the territory. Information on the ecology of these animals is scant. This review collates information on the systematics of Hong Kong Trichoptera and indicates their relationships to other Asian faunas. These findings are related to characteristics of freshwater habitats in Hong Kong.

## RESULTS

Hong Kong is situated 250 km south of the Tropic of Cancer, between latitudies 22°7' - 22°9'N and longitudes 113°52' - 114°30'E, on the southern coast of the People's Republic of China adjoining Guangdong Province. The landscape comprises the southern end of an eroded mountain chain of Jurassic origin, made up of folded and metamorphosed volcanic and granitic rocks with local sedimentary outcrops. A rise in sea level following the last glaciation has resulted in a drowned coastline with former mountain tops being represented by small offshore islands, the highest peak attaining 958 m. Consequently, Hong Kong lacks an extensive coastal plain and much of the land area comprises steep hillsides supporting grasses and low scrub on thin acidic soils. Only in sheltered valleys does forest obtain, the restricted lowland being given over to market gardening (13 % of total land area) or to urbanization (10% of land area and increasing).

The climate is tropical monsoon with mild dry winters (February mean minimum - 15.2°C) and hot wet summers (July mean maximum - 27.9°C). Most of Hong Kong's average precipitation of 2170 mm falls in summer resulting in increased stream discharge and sediment loads.

Little is known of the biology of Hong Kong freshwaters. This is surprising in view of a reliance upon rivers and streams draining into reservoirs as the major source of water supplies. The potential diversity of freshwater fauna is high as Hong Kong lies in an area of transition between the tropical Oriental fauna and the temperate faunas of China and Japan. Against this background the composition of the local Trichoptera fauna can be discussed.

Hong Kong caddisflies have been known for over 130 years (Walker, 1852; McLachlan, 1862) but are not well represented in the literature. Indeed, prior to 1986, only 6 species representing 5 families had been recorded from the territory : *Stenopsyche angustata* Martynov (Stenopsychidae), *Dipseudopsis collaris* McLachlan (Dipseudopsidae), *Melanotrichia serica* Barnard & Dudgeon (Xiphocentronidae), *Macrostemum fastosum* (Walker), *Macrostemum lautum* (McLachlan), (Hydropsychidae) and *Ascalaphomerus humeralis* Walker (Calamoceratidae) (Ulmer 1926, 1932 ; Wu 1937 ; Dudgeon 1983 ; Barnard & Dudgeon 1984). *Melanotrichia serica* is known from nowhere else but is unlikely to be endemic, *Macrostemum fastosum* is widespread in the Oriental region, whilst the remaining species have been recorded from China.

Given Hong Kong's location between the rich Chinese fauna and the diverse but poorly known faunas of the Oriental tropics, to what extent has the local faunal composition been influenced on the one hand by the territory's tropical climate, whilst on the other by its proximity to China ? Table 1, compiled from published studies, indicates the relative generic and specific richness of trichopteran families at 7 Asian sites along latitudinal gradient from the equator to the Palaearctic. To present a more complete account of the Hong Kong Trichoptera, published records of adult caddisflies are supplemented by data from studies of the larvae (Dudgeon 1982, 1983, 1984). At least 65 species have been found as larvae in the territory (Table 1) and, although their description awaits establishment of larval-adult associations, their inclusion raises the Hong Kong species total almost 10 fold. In making this list I assumed that all species already known from Hong Kong were represented as larvae in my collections. As this may not in fact be the case, the species total can be regarded as slightly conservative. Note that 14 families known as larvae have not previously been recorded from Hong Kong. However, only 5 of these (Hydroptilidae, Xiphocentronidae, Psychomyiidae, Sericostomatidae and Helicopsychidae) are not recorded from Guangdong Province (Dudgeon, unpublished observations). Their diminutive size indicates that they may have been overlooked by past collectors.

Despite the different land areas of the regions considered, some interesting trends emerge. It is striking that the tropical Oriental sites (Sumatra, Peninsula Malaysia, Java and Sri Lanka) are similar with respect to total numbers of genera (55-61) and overall faunal composition. The Chinese Trichoptera are clearly different and richer, both in terms of species and genera, reflecting the region's large total area and latitudinal spread. The caddisfly fauna of North Korea is poorer (comparable in richness to the much smaller area of Hong Kong) but exhibits clear similarities to the Chinese fauna in terms of family representation. These comments must, however, be tempered by the fact that tropical Asian Trichoptera are very poorly known and the overall similarity of Sri Lankan, Malaysian, Sumatran and Javan faunas may be more apparent than real. For example, examination of a small collection of Trichoptera from Sumatra increased the Rhyacophilidae of that island from 1 species to 5 (Malicky 1978a, 1978b). Significantly, Schmid (1984) estimates the trichopteran fauna of South-west Asia at 40,000 species, out of a world total of at least 50,000.

Ranking the families in each region on the basis of generic (Table 2) and specific (Table 3) richness revealed similarities/dissimilarities in their caddisfly faunas. Based on generic richness, the first 4 rankings at the tropical Asian sites were a robust group of only 5 families (Hydroptilidae, Hydropsychidae, Leptoceridae, Psychomyiidae and Polycentropodidae), with the 5th rank adding a further 4 taxa (Lepidostomatidae, Calamoceratidae, Philopotamidae and Glossosomatidae) (Table 2). The Hydroptilidae, Hydropsychidae and Leptoceridae are about equally rich in these tropical Asian countries and differences in ranking probably reflect the prevailing state of knowledge (the Hydroptilidae have hardly been studied in Peninsula Malaysia). Generic rankings of families in China and North Korea emphasized the importance of hydropsychids and leptocerids, with polycentropodids and lepidostomatids attaining secondary status. By contrast with tropical Asia however, the Phryganeidae and especially the Limnephilidae emerge as high-ranking diverse taxa and there is a northwards increase in representation of the Limnephiloidea.

Marked deviations of Chinese and North Korean faunas from those of the Oriental tropics are also apparent from family rankings on the basis of species richness (Table 3). Limnephilidae and Phryganeidae once more attain high ranks and are joined by Stenopsychidae and Rhyacophilidae, although the Hydropsychidae and Leptoceridae are still well represented as they are in the tropical Asian group rankings. The emergence of Rhyacophilidae and Stenopsychidae as important northern groups reflects their high specific but low generic diversity in the Holarctic. However with the exception of these two families, specific richness generally paralleled generic richness as an indicator of family ranking/importance.

Despite Hong Kong's transitional position, the local Trichoptera are clearly affiliated to the tropical Asian grouping. A distinctive characteristic is seen in the relative importance of the Calamoceratidae, but 4 of the 5 major tropical families are in top ranking positions whilst phryganeids and limnephilids have not been recorded from the territory. Significantly, a recent study of Mongolian caddisflies (Mey and Dulmaa 1985) emphasizes the importance of Rhyacophilidae, Limnephilidae and Phryganeidae as primarily northern taxa. Some similarity with more northern regions is indicated by the elevated specific ranking of Rhyacophilidae, but overall the Hong Kong caddisfly fauna is clearly affiliated with the tropical realm, differing only in detail in the importance of the Calamoceratidae. The Hydropsychidae are evident as the most diverse Hong Kong group regardless of specific or generic ranking criteria.

The close resemblance of the Hong Kong caddisfly fauna to that of the Oriental tropics can be attributed to climatic similarity, while habitat characteristics could account for the predominance of Calamoceratidae among the Hong Kong Limnephiloidea. An examination of the distribution of Trichoptera in China (Dudgeon, unpublished) reveals that Phryganeidae and Limnephilidae are largely absent from areas of China normally considered to be included in the Oriental Region. Only *Eubasilissa regina* McLachlan (Phryganeidae) extends

TABLE 1. Numbers of genera (and species) making up Trichoptera families in 7 regions of Asia.

| | Sumatra 5.5°N-5.9°S | Peninsula Malaysia 1.5-6.5°N | Java 6.0-9.0°S | Sri Lanka 6.5-10.3°N | Hong Kong 22°N | China 19-50°N | North Korea 38-43°N |
|---|---|---|---|---|---|---|---|
| Rhyacophilidae | 1 (5) | 2 (6) | 1 (3) | 1 (1) | 2 (4) | 2 (45) | 1 (13) |
| Hydrobiosidae | 1 (2) | 1 (3) | 1 (1) | 1 (1) | 0 | 2 (3) | 1 (1) |
| Glossosomatidae | 3 (5) | 2 (3) | 3 (3) | 1 (10) | 1 (2) | 2 (14) | 2 (3) |
| Hydroptilidae | 8 (8) | 2 (4) | 15 (16) | 13 (35) | 3 (3) | 4 (4) | 2 (7) |
| Philopotamidae | 1 (5) | 1 (8) | 2 (7) | 2 (20) | 1 (4) | 3 (18) | 1 (3) |
| Stenopsychidae | 1 (1) | 1 (2) | 0 | 0 | 1 (2) | 1 (44) | 1 (4) |
| Psychomyiidae | 4 (6) | 7 (14) | 4 (6) | 4 (15) | 1 (1) | 2 (3) | 1 (2) |
| Xiphocentronidae | 0 | 0 | 0 | 1 (1) | 1 (1) | 1 (2) | 0 |
| Polycentropodidae | 4 (4) | 5 (9) | 2 (3) | 4 (20) | 2 (5) | 6 (8) | 3 (5) |
| Dipseudopsidae | 1 (2) | 1 (4) | 1 (2) | 1 (3) | 1 (1) | 1 (3) | 0 |
| Ecnomidae | 1 (8) | 1 (6) | 1 (3) | 1 (13) | 1 (1) | 1 (2) | 1 (2) |
| Arctopsychidae | 0 | 0 | 0 | 0 | 0 | 2 (3) | 1 (1) |
| Hydropsychidae | 12 (29) | 9 (35) | 12 (36) | 11 (29) | 8 (27) | 11 (45) | 4 (7) |
| Phryganeidae | 0 | 0 | 0 | 0 | 0 | 10 (19) | 2 (2) |
| Limnocentropodidae | 0 | 1 (1) | 0 | 0 | 0 | 1 (1) | 0 |
| Molannidae | 0 | 0 | 0 | 1 (1) | 1 (1) | 1 (2) | 1 (1) |
| Calamoceratidae | 2 (4) | 2 (5) | 2 (4) | 2 (8) | 4 (4) | 3 (7) | 0 (0) |
| Odontoceridae | 1 (1) | 1 (1) | 1 (2) | 1 (1) | 1 (1) | 2 (5) | 1 (3) |
| Leptoceridae | 9 (18) | 14 (27) | 7 (10) | 12 (39) | 4 (4) | 10 (22) | 4 (8) |
| Goeridae | 3 (7) | 1 (3) | 1 (2) | 1 (6) | 1 (1) | 1 (5) | 1 (2) |
| Limnephilidae | 0 | 0 | 0 | 0 | 0 | 18 (44) | 4 (4) |
| Lepidostomatidae | 2 (2) | 5 (10) | 3 (4) | 1 (4) | 1 (1) | 7 (10) | 2 (3) |
| Brachycentridae | 0 | 0 | 0 | 0 | 0 | 2 (2) | 0 |
| Sericostomatidae | 0 | 0 | 0 | 1 (7) | 1 (1) | 1 (1) | 1 (1) |
| Uenoidae | 0 | 0 | 0 | 0 | 0 | 1 (1) | 0 |
| Helicopsychidae | 1 (2) | 1 (1) | 1 (1) | 2 (9) | 1 (1) | 0 | 0 |
| Totals | 55 (109) | 57 (142) | 57 (103) | 61 (223) | 36 (65) | 95 (313) | 34 (72) |

TABLE 2. Major trichopteran families in Asia, arranged in decreasing rank order according to generic richness. Tied ranks are assigned equal scores.

| Sumatra | Peninsula Malaysia | Java | Sri Lanka | Hong Kong | China | North Korea | |
|---|---|---|---|---|---|---|---|
| Hydropsychidae | Leptoceridae | Hydroptilidae | Hydroptilidae | Hydropsychidae | Limnephilidae | Leptoceridae | I |
| Leptoceridae | Hydropsychidae | Hydropsychidae | Leptoceridae | Calamoceratidae | Hydropsychidae | Limnephilidae | II |
| Hydroptilidae | Psychomyiidae | Leptoceridae | Hydropsychidae | Leptoceridae | Phryganeidae | Hydropsychidae | III |
| Polycentropodidae | Polycentropodidae | Psychomyiidae | Polycentropodidae | Hydroptilidae | Leptoceridae | Polycentropodidae | IV |
| Psychomyiidae | Lepidostomatidae | Lepidostomatidae | Psychomyiidae | Polycentropodidae | Lepidostomatidae | Phryganeidae | V |
| | | Glossosomatidae | Calamoceratidae | Rhyacophilidae | | Lepidostomatidae | |
| | | | Philopotamidae | | | | |
| | | | Helicopsychidae | | | | |

TABLE 3. Major trichopteran families in Asia, arranged in decreasing rank order according to species richness. Tied ranks are assigned equal scores.

| Sumatra | Peninsula Malaysia | Java | Sri Lanka | Hong Kong | China | North Korea | |
|---|---|---|---|---|---|---|---|
| Hydropsychidae | Hydropsychidae | Hydropsychidae | Leptoceridae | Hydropsychidae | Hydropsychidae | Rhyacophilidae | I |
| Leptoceridae | Leptoceridae | Hydroptilidae | Hydroptilidae | Polycentropodidae | Rhyacophilidae | Leptoceridae | II |
| Ecnomidae | Psychomyiidae | Leptoceridae | Hydropsychidae | Calamoceratidae | Limephilidae | Hydropsychidae | III |
| Hydroptilidae | Lepidostomatidae | Philopotamidae | Philopotamidae | Leptoceridae | Stenopsychidae | Hydroptilidae | IV |
| Goeridae | Polycentropodidae | Psychomyiidae | Polycentropodidae | Philopotamidae | Leptoceridae | Polycentropodidae | V |
| | | | | Rhyacophilidae | | | |

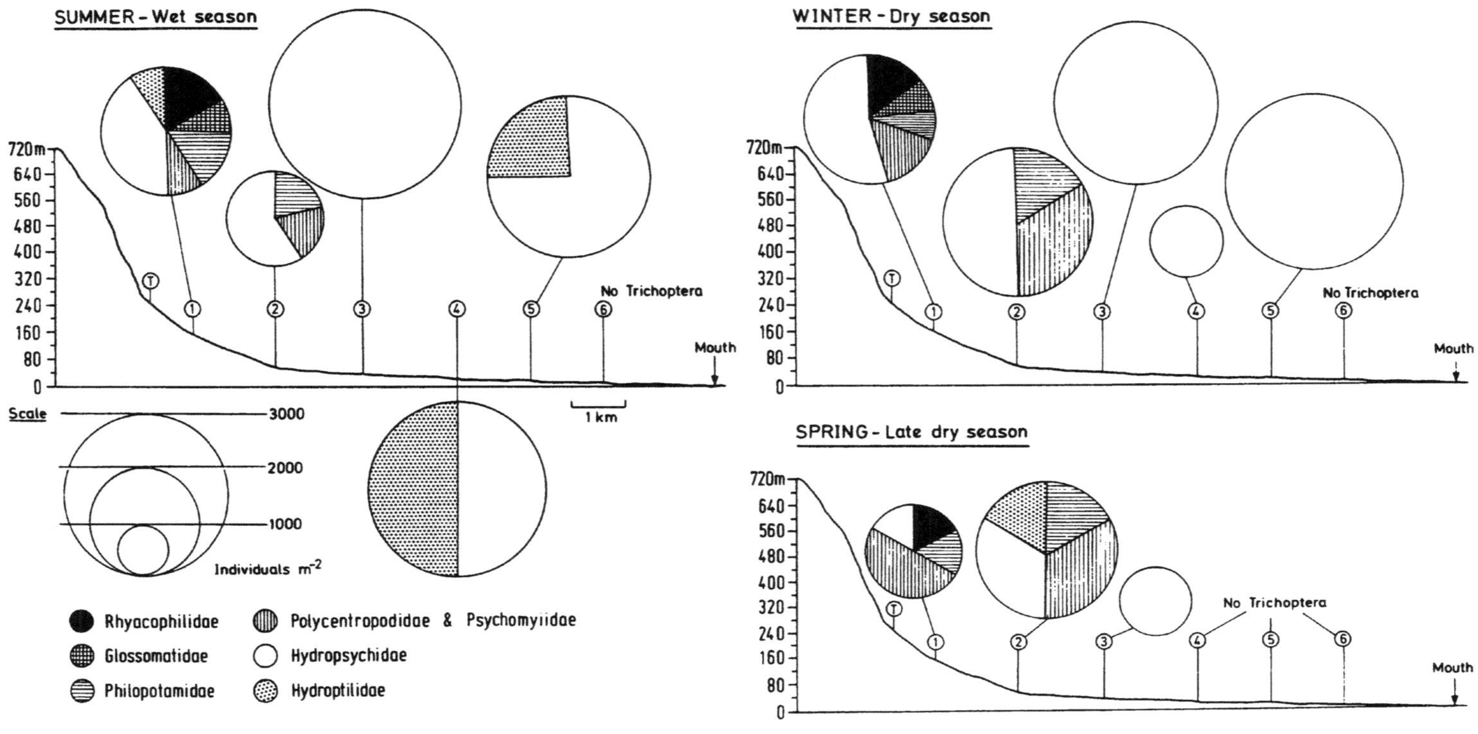


Figure 1. Seasonal changes in the distribution and abundance of trichopteran families along the Lam Tsuen River, New Territories, Hong Kong.

south into the Oriental Region (the northern limits of which are marked by latitude 30°N, extending eastwards from the southern border of Tibet across to Shanghai), and although three Limnephilidae (*Limnephilus externus* Hagen, *Nothopsyche intermedia* Martynov and *Platyphylax lanuginosus* McLachlan) have been recorded from the Shanghai environs they are not found further south. Similar latitudinal/climatic constraints probably limit the southerly spread of Rhyacophilidae and Stenopsychidae ; both families are cool-adapted (Ross 1956) and exhibit a marked decline in species richness towards the equator. Overall, those elements of the Asian caddisfly fauna which are confined to China and North Korea are those which Ross (1956) has termed cool-adapted, while most elements of the Hong Kong and Oriental fauna are warm-adapted. It is perhaps worth emphasizing that use of the terms cool- and warm adapted does not imply an understanding of the physiological mechanisms determining the distribution of caddisfly taxa. Indeed it should be noted that the cool adapted Rhyacophilidae and Stenopsychidae do include certain species restricted to tropical regions.

While Hong Kong has the potential to support a diverse, tropical, caddisfly fauna, the realization of this potential depends upon habitat availability. Lotic and lentic habitats are present , but diverse communities are largely confined to streams and the upper course of rivers. However, many hill streams flow only after the summer rains and do not support permanent caddisfly populations. These streams drain into storage reservoirs whose pristine waters are heavily stocked with fish (Cyprinidae, Ophicephalidae and introduced Cichlidae) ; predation prevents the establishment of Trichoptera in such sites. Lentic waters associated with agricultural habitats are often used to raise fish with dire consequences for aquatic insects - where aquaculture is not carried out contamination of standing waters by fertilizers and/or pesticides has long since eliminated the caddisfly inhabitants. The remaining hillstreams and rivers include habitats of varying suitability according to their flow schedules, pollution load, and inclusion/exclusion in Government mosquito control procedures. Intensive land use in Hong Kong (a consequence of an increase in population from 600,000 to over 5 million since World War II) has resulted in gross pollution of lowland running waters by wastes from pig and fowl farms as well as by human sewage and industrial effluents. In only a few localities does the annual flushing of rivers associated with the summer rains permit establishment of Trichoptera in the lower reaches. Such a seasonal pattern is very apparent in the Lam Tsuen River, New Territories, Hong Kong which is polluted by agricultural wastes. Species richness declines downstream (Fig.1), although the population densities of certain Hydropsychidae in the lower course increase in response to high food availability associated with organic pollution. Seasonal effects influence this downstream trend as pollution loads increase in parallel with declining discharge volume in the dry season. Trichopteran diversity in the middle and lower course is thereupon further limited, in extreme cases leading to the elimination of all caddisflies.

Calamoceratidae, Odontoceridae and Lepidostomatidae are only abundant in woodland streams in sheltered valleys. Potentially diverse groups such as Leptoceridae are unable to attain the richness seen in the Oriental tropics due to a lack of suitable habitat. Similarly, the relative paucity of Hydroptilidae in Hong Kong can be attributed to low habitat diversity as a result of environmental degradation. Significantly, I have been unable to collect the genera *Dipseudopsis* and *Ascalophomerus* in Hong Kong, despite past records of their presence.

In summary, diverse caddisfly assemblages are found only in those relatively unpolluted lotic habitats with year-round (albeit fluctuating) flows which are not regularly sprayed to control mosquitoes. Caddisflies of standing waters have long since been exterminated. As a consequence, the Hydropsychoidea are well represented in Hong Kong, along with some Rhyacophiloidea, but the Limnephiloidea are more poorly represented reflecting habitat limitation.

To conclude : the problems faced by biologists studying Trichoptera in Hong Kong mirror, in miniature, those prevailing elsewhere in Asia. The increasing pressure for rural development leads to inevitable habitat degradation and destruction, and we in Hong Kong are witnessing the disappearance of many caddisfly species, a phenomenon occurring throughout much of China and the Oriental tropics. It is indeed a tragedy that without prompt remedial action the potential for many caddisflies to delight and instruct will be lost forever.

ACKNOWLEDGEMENTS

I am grateful to Dr Hans Malicky (Lunz, Austria) for constructive comments on a draft of this manuscript.

REFERENCES

Barnard, P.C. & Dudgeon, D. 1984. The larval morphology and ecology of a new species of *Melanotrichia* from Hong Kong (Trichoptera : Xiphocentronidae). **Aquatic Insects**, 6 : 245-252.

Dudgeon, D. 1982. Aspects of the microdistribution of insect macrobenthos in a forest stream in Hong Kong. **Arch.Hydrobiol.Suppl.**, 64 : 221-239.

Dudgeon, D.1983. An investigation of the drift of aquatic insects in Tai Po Kau. Forest stream, New Territories, Hong Kong. **Arch.Hydrobiol.** , 96 : 434-447.

Dudgeon, D.1984. Longitudinal and temporal changes in functional organization of macroinvertebrate communities in the Lam Tsuen River, Hong Kong. **Hydrobiologia,** 111 : 207-217.

Malicky, H. 1978a. Beitrage zur Kenntnis der Insekten fauna Sumatra : Teil 7 : Köcherfliegen (Trichoptera) aus Sumatra und West-Neuguinea. 1. Rhyacophilidae, Glossosomatidae, Stenopsychidae Goeridae. **Beitr. Naturk. Forsch. SudwDtl.** , 37 : 159-173.

Malicky, H. 1978b. *Rhyacophila krauskasseggae* n. sp., eine neue Köcherfliege (Trichoptera) aus Sumatra. **Z. Arbeitsgemeinschaft Osterr. Entomologen,** 30 : 41-42.

McLachlan, R.1862. Characters of new species of exotic Trichoptera ; also of one new species inhabiting Britain. **Trans. R. ent. Soc. Lond.,** 3rd Series, 3 : 301-311.

Mey, W.C. & Dulmaa, A. 1985. Die Köcherfliegenfauna der Mongolei (Insecta, Trichoptera). **Mitt. Zool. Mus. Berlin.** , 61 : 79-104,

Ross, H.H. 1956. Evolution and classification of the mountain caddisflies. Univ.Illinois Press, Urbana.

Schmid, F. 1984. Essai d'évaluation de la faune mondiale des Trichoptères. **Proc 4th Int. Symp. Trichoptera,** J.C. Morse (ed.), Junk, The Hague, Ser.Ent., 30 :337.

Ulmer, G. 1926. Beitrage zur Fauna sinica. III. Trichopteren und Ephemeropteren. **Arch.Natg.Berlin,** 91(A5) : 19-110.

Ulmer, G. 1932. Aquatic insects of China, Article IV. Neue Chinesische Trichopteren, nebst Ubersicht über die Bisher aus China bekannten Arten. Teil II. Liste der Chinesischen Trichopteren, mit Angaben über Synonymie, Litteratur und Geographische Verbreitung. **Peking Nat. Hist.Bull.,** 7 : 135-157.'

Walker, F. 1952. Catalogue of the specimens of neuropterous insects in the collection of the British Museum. p 76.

Wu,C.E. 1937. Trichoptera. Catalogue Insectorum Sinensium, III. The Fan Memorial Institute of Biology, Beijing, China : 1269-1312.

# AN INTRODUCTION TO THE TRICHOPTERA FAUNA OF JAPAN, WITH A TENTATIVE CHECKLIST OF GENERA.

K. TANIDA

SUMMARY

Mainly based on the iconographical work of the larvae from Japan (Tanida 1985) which containded 66 genera and 107 + species of caddis larvae with keys and illustrations, the Trichoptera fauna of Japan is examined and a tentative checklist of the genera is provided, where 98 genera are listed with synonyms. The Trichoptera fauna of Japan is characterized by a mixture or Oriental, Holarctic, Asian Palaearctic and Nearctic genera along ones common to almost all faunal regions. The Japanese genera of Trichoptera are classified into 10 groups according to their patterns of zoogeographical distribution.

## INTRODUCTION

Not a small number of researches have been conducted on the Trichoptera fauna of Japan and some revisional or iconographical studies on adults (Tsuda 1942, Kuwata 1965, Tani 1977) and larvae (Tsuda 1959, Tsuda & Akagi 1962) have been also provided, and recently 8 genera of larvae were newly described from Japan and some genera were transferred to appropriate genera based on larval taxonomy (Tanida 1985). However, the classification of Trichoptera of Japan has remained unsatisfactory. One of the main reasons seems to be the insufficient revisional studies on most genera, which includes larvae as well as adults.

Considering the present situation mentioned above, I intend to present a tentative but up-to-date checklist of Trichoptera genera of Japan and to analyze the generic composition from a regional zoogeographical view point to clarify the status of the Japanese Islands in Trichoptera evolution and dispersal.

## CHECKLIST OF GENERA

Tsuda (1942) presented a list of Trichoptera species, which contained 194 species belonging to 74 genera from Japan, Korea, Taiwan and Chisima, but he failed to include some larval genera recorded by Iwata (1927a, b, c, 1928a, b, 1930), and some of the genera presented in Tsuda's work should be changed according to the current knowledge of Trichoptera taxonomy. Until then no checklist has been published for the Japanese species. Although further examinations on publications and type series specimens of the Japanese Trichoptera are indispensable to complete the species checklist, I intend to make a tentative but up-to-date checklist of the genera, based mainly upon my own study of the larvae and the examination of publications including Fischer's (1960-73) Trichoptera Catalogue and other recent studies including some revisional and iconographical works on Trichoptera from Japan.

### A Tentative Checklist of Trichoptera Genera of Japan.

Family Stenopsychidae

1.*Stenopsyche* McLachlan, 1866. =*Philopotamopsis* Iwata, 1927. (Syn. : Kuwayama, 1930.) ; *Parastenopsyche* Kuwayama, 1930. (Syn. and revision of adults : Schmid, 1969.)

Family Philopotamidae

2. *Chimarra* Stephens, 1829. (Revision of adults : Kobayashi, 1980).

3. *Wormaldia* McLachlan, 1865. =*Nanagapetus* Tsuda, 1942. (Syn.: Ross, 1956. Revision of adults : Kobayashi, 1980.)

4. *Dolophilodes* Ulmer, 1909. =*Sortosa* Navás, 1918 (Syn. : Wiggins, 1977 ; Revision of adults : Kobayashi, 1980.)

Family Psychomyiidae

5. *Melanotrichia* Ulmer, 1906. =*Kibuneopsychomyia* Tsuda, 1942 ; *Tsukushitrichia* Kobayashi, 1964. (Syn. : Barnard & Dudgeon, 1984.)

6. *Psychomyia* Latreille, 1829.

7. *Tinodes* Curtis, 1834.
8. *Psychomyiella* Ulmer, 1908. (No larval description for Japanese species.)

Family Polycentropodidae

9. *Ecnomus* McLachlan, 1864.
10. *Dipseudopsis* Walker, 1852. =*Bathytinodes* Iwata, 1927. (Syn. : Tsuda, 1937.)
11. *Nyctiophylax* Brauer, 1865.
12. *Kyopsyche* Tsuda, 1942. (No larval description for Japanese species.)
13. *Plectrocnemia* Stephens, 1836.
14. *Phylocentropus* Banks, 1907. (No larval description for Japanese species.)
15. *Polyplectropus* Ulmer, 1905. (No larval description for Japanese species.)
16. *Cyrnus* Stephens, 1836. (No larval description for Japanese species.)
17. *Paranyctiophylax* Tsuda, 1942. (No larval description.)
18. *Neureclipsis* McLachlan, 1864. (Recorded by Iwata, 1927a, doubtful in idenfication)
19. *Holocentropus* McLachlan, 1878. (Recorded by Iwata, 1927a, doubtful in identification.)
20. *Polycentropus* Curtis, 1835. (Further examination needed for Japanese species.)

Family Hydropsychidae

21. *Diplectrona* Westwood, 1840.
22. *Macrostemum* Kolenati, 1859. =*Macronema* Pictet, 1836 (part). (Transfer : Flint & Bueno-Soria, 1982.)
23. *Parapsyche* Betten, 1934. =*Arctopsyche* (part). (Revision of adults : Schmid, 1968)
24. *Arctopsyche* McLachlan, 1868. (Revision of adults : Schmid, 1968.)
25. *Hydropsyche* Pictet, 1834. (Revision of adults and larvae : Tanida, in press.)
26. *Cheumatopsyche* Wallengren, 1891. =*Hydropsychodes* Ulmer, 1905. (Syn. :Kimmins, 1963.) ; *Hydropsyche* (part). (Transfer : Tanida, in press.)
27. *Hydromanicus* Brauer, 1865. (No larval description for Japanese species.)

Family Rhyacophilidae

28. *Apsilochorema* Ulmer, 1907. =*Psilochorema* McLachlan, 1866 (part.). (Transfer and revision of adults : Ross, 1956.)
29. *Himalopsyche* Banks, 1940. (Revision of adults : Schmid & Botosaneanu, 1966.)
30. *Rhyacophila* Picket, 1834. (Revision of adults : Schmid, 1970.)

Family Glossosomatidae

31. *Agapetus* Curtis, 1834. =*Synagapetus* McLachlan, 1879. (Syn. and revision of adults : Ross, 1956.)
32. *Glossosoma* Curtis, 1834. =*Mystrophora* Klapalek, 1892. (Syn. and revision of adults : Ross, 1956.)
33. *Electragapetus* Martynov, 1934 (Revision of adults : Ross, 1956. No larval description for Japanese species.)
34. *Padunia* Martynov, 1910. =*Uenotrichia* Tsuda, 1942. (Syn. and transferred from Hydroptilidae : Marshall, 1979.)

Family Hydroptilidae

35. *Stactobia* McLachlan, 1880.
36. *Orthotrichia* Eaton, 1873.
37. *Oxyethira* Eaton, 1873.
38. *Hydroptila* Dalman, 1819.
39. *Palaeagapetus* Ulmer, 1912. (Adults & larvae : Ito & Hattori, 1986.)

Family Limnocentropodidae

40. *Limnocentropus* Ulmer, 1907. =*Kitagamia* Iwata, 1927. (Syn. Tsuda, 1942. Revision of adults and larvae : Wiggins, 1969a.)

Family Phryganopsychidae

41. *Phryganopsyche* Wiggins, 1959. =*Phryganea* Linnaeus, 1758 (part). (Transfer and revision of adults and larvae : Wiggins, 1959, 1969b.)

Family Phryganeidae

42. *Oligotricha* Rambur, 1842. =*Neuronia* Stephens, 1829 (part). (Transfer and revision of adults : Wiggins & Kuwayama, 1971.)
43. *Dasystegia* Wallengren, 1880. (No larval description for Japanese species.)
44. *Eubasilissa* Martynov, 1930. =*Neuronia* (part). (Transfer : Tanida, 1985.)

45. *Semblis* Fabricius, 1775. =*Neuronia* (part). (No larval description for Japanese species.)
46. *Colpomera* McLachlan, 1862. = *Phryganea* Linnaeus, 1758 (part). (No larval description for Japanese species.)

Family Brachycentridae

47. *Brachycentrus* Curtis, 1834. =*Oligoplectrodes* Martynov, 1909 ; *Brachycentriella* Iwata, 1927. (Syn. and revision of adults and larvae : Flint, 1984 : Revision of adults : Wiggins, Tani & Tanida, 1985.)
48. *Eobrachycentrus* Wiggins, 1965. =*Brachycentrus* (part) ; *Micrasema* (part). (Transfer and revision of adults : Wiggins, Tani & Tanida, 1985.)
49. *Micrasema* McLachlan, 1876.

Family Limnephilidae

50. *Dicosmoecus* McLachlan, 1875. =*Kogurea* Kobayashi, 1962 ; *Stenophylax* Kolenati,1848 (part). (Syn. and revision of adults and larvae : Wiggins & Richardson, 1982.)
51. *Ironoquia* Banks, 1916. =*Allophylax* Banks, 1907. (Recorded as Allophylax by Iwata, 1927a, doubtful in identification.)
52. *Pseudostenophylax* Martynov, 1909. (Revision of adults : Schmid, 1955.)
53. *Drusus* Stephens, 1837. =*Potamorites* McLachlan, 1867. (Recorded as *Potamorites* by Iwata, 1927a & 1928a, doubtful in identification.)
54. *Metanoea* McLachlan, 1880. (Recorded by Iwata, 1928a, doubtful in identification.)
55. *Hydatophylax* Wallengren, 1891. =*Platyphylax* McLachlan, 1871 (part). (Transfer and revision of adults : Schmid, 1950.)
56. *Nemotaulius* Banks, 1906. =*Glyphotaelius* Stephens, 1837, (part) ; *Grammotaulius* Kolenati, 1848, (part). (Transfer and revision of adults : Schmid, 1952a.)
57. *Asynarchus* McLachlan, 1880. =*Limnophilus* (part); (Transfer and revision of adults Schmid, 1954b.)
58. *Lenarchus* Martynov, 1914.
59. *Limnephilus* Leach, 1815. =? *Grammotaulius* (part).
60. *Anabolia* Stephens, 1837. (Recorded by Iwata, 1927a, doubtful in identification.)
61. *Nothopsyche* Ulmer, 1905. (Revision of adults : Schmid, 1952b.)
62. *Halesus* Stephens, 1836. (No larval description for Japanese species.)
63. *Moropsyche* Banks, 1906. (Revision of adults : Schmid, 1954a.)
64. *Apatania* Kolenati, 1848. (Revision of adults : Schmid, 1953, 1954.)
65. *Neophylax* McLachlan, 1871. =*Halesinus* Ulmer, 1907. (Syn. : Schmid, 1955.)
66. *Goera* Stephens , 1829.
67. ?*Imania* Martynov, 1935. (No larval description for Japanese species.)

Family Uenoidae

68. *Uenoa* Iwata, 1927. =*Eothremma* Martynov, 1933. (Syn. : Tsuda, 1937. Revision of adults and larvae : Wiggins, Weaver & Unzicker, 1985.)

Family Lepidostomatidae

69. *Neoseverinia* Ulmer, 1908. (Revision of adults and larvae : Ito, 1983.)
70. *Goerodes* Ulmer, 1907. =*Atomyiella* Tsuda, 1936. (Syn. : Tsuda, 1942.) ;*Yamatopsyche* Tani, 1971. (Syn. : Ito, 1948b) = *Dinarthrodes* Ulmer 1907 (part) ; *Crunobiodes* Martynov, 1928 (part) ; *Crunoeciella* Ulmer, 1905 (part). (Transfer : Ito, 1984b ; Revision of adults and larvae : Ito, 1978.)
71. *Dinarthrodes* Ulmer, 1905. (No larval description for Japanese species.)
72. *Ayabeopsyche* Tsuda, 1942. (No larval description.)
73. *Paradinarthrodes* Tani, 1971. (No larval description.)
74. *Dinarthrum* McLachlan, 1871. (Revision of adults and larvae : Ito, 1984a.)
75. *Lepidostoma* Rambur, 1842. (Recorded by Iwata, 1927a, doubtful in identification.)
76. *Lasiocephala* Costa, 1857. (Recorded by Iwata, 1927a, doubtful in identification.)

Family Sericostomatidae

77. *Gumaga* Tsuda 1938.

Family Odontoceridae

78. *Psilotreta* Banks, 1899.
79. *Perissoneura* McLachlan, 1871.
80. *Odontocerum* Leach, 1815. (Recorded by Iwata, 1928a, doubtful in identification.)

Family Molannidae

81. *Molanna* Curtis, 1834. (Revision of adults : Wiggins, 1968.)

Family Helicopsychidae

82. *Helicopsyche* Siebold, 1856.

Family Calamoceratidae

83. *Anisocentropus* McLachlan, 1863. =*Kizakia* Iwata, 1927. (Syn. : Ulmer, 1951.)
84. *Georgium*Fischer, 1964. =*Rhabdoceras* Ulmer, 1905. (Changed as preoccupied. No larval description for Japanese species.)
85. *Ganonema* McLachlan, 1866.
86. *Asotocerus* McLachlan, 1866. (No larval description for Japanese species.)
87. *Heteroplectron* McLachlan, 1871. (Recorded by Tsuda, 1942, doubtful in identification. No larval description for Japanese species.)

Family Leptoceridae

88. *Triplectides* Kolenati, 1859. (No larval description for Japanese species.)
89. *Setodes* Rambur, 1842.
90. *Trichosetodes* Ulmer, 1915.
91. *Oecetis* McLachlan, 1877.
92. *Ceraclea* Stephens, 1829. =*Leptocerus* (part). (Transfer : Tanida, 1985.)
93. *Triaenodes* McLachlan, 1865.
94. *Mystacides* Berthold, 1827.
95. *Leptocerus* Leach, 1815. (No larval description for Japanese species.)
96. *Parasetodes* McLachlan, 1880. (No larval description for Japanese species.)
97. *Triaenodella* Mosely, 1932. (No larval description for J apanese species.)
98. *Tobikera* Matsumura, 1931. (Probably =*Triplectides* : Tsuda, 1942.)

ZOOGEOGRAPHICAL CONSIDERATIONS OF TRICHOPTERA GENERA

In the preceding section, 98 Trichoptera genera were recognized from Japan, however, 12 genera are doubtful in generic identification and more than 17 genera are in need of further investigation on their generic status.

Other 66 genera can be classified into the following groups according to their patterns of distribution in zoogeographical regions. Most of the information on their regional distribution are derived from Wiggins (1977) and/or Fischer (1960-1973). Figures in the parentheses following generic names show the species number of larvae described in Tanida (1985) and the number of species recorded from Japan until now.

**Classification of the Japanese genera according to their patterns of distribution in faunal region**

1. Genera endemic to Japan (3 genera)
   *Neoseverinia* (1-1), *Perissoneura* (1-2), *Georgium* (1-1).
2. Genera common to the Asian Palaearctic region (4 genera).
   *Stactobia* (1-?), *Dasystegia* (0-1), *Semblis* (0-2), *Dinarthrum* (0-3).
3. Genera common to the Oriental region (9 genera).
   *Melanotrichia* (1-1) , *Dipseudopsis* (1-1), *Apsilochrorema* (1-1), *Electragapetus* (0-1) *Limnocentropus* (1-1), *Phryganopsyche* (1-2), *Uenoa* (1-1), *Nothopsyche* (3-7+),*Ganonema* (1-?).
4. Genera common to the Holarctic region ( 8 genera).
   *Ecnomus* (1-2), *Micrasema* (3-5), *Hydatophylax* (2-4), *Nemotaulius* (1-3), *Asynarchus* ((2-2), *Limnephilus* (3-6+), *Lenarchus* (1-1), *Ceraclea* (0-7).
5. Genera common to the Nearctic region (2 genera).
   *Palaeagapetus* (0-1), *Eobrachycentrus* (1-4).
6. Genera common to the oriental and Holarctic regions (9 genera).
   *Rhyacophila* (19-44+), *Glossosoma* (1-5+), *Oligotricha* (1-3), *Eubasilissa* (1-2?), *Goera* (6-7), *Moropsyche* (1-3+), *Apatania* (1-9+), *Molanna* (1-1), *Setodes* (1-5).
7. Genera common to the Oriental and Nearctic regions (6 genera).
   *Parapsyche* (3-3), *Arctopsyche* (3-3), *Himalopsyche* (1-1), *Imania* (0-1+),*Psilotreta* (1-2), *Gumaga* (1-1).

8. Genera common to the Oriental and Ethiopian regions (3 genera).
*Stenopsyche* (2-3), *Goerodes* (4-10), *Trichosetodes* (0-2).
9. Genera common to the Asian Palaearctic, Oriental and Nearctic regions (3 genera).
*Neophylax* (3-3), *Pseudostenophylax* (1-1), *Dicosmoecus* (1-1).
10. Genera common to almost all faunal regions (19 genera).
*Chimarra* (1-2), *Wormaldia* (1-10), *Dolophilodes* (4-9), *Psychomyia* (1-2), *Tinodes* (0-1), *Nyctiophylax* (1-0), *Plectrocnemia* (1-?), *Diplectrona* (3-4), *Macrostemum* (1-1),*Hydropsyche* (8-9), *Cheumatopsyche* (2-2), *Agapetus* (1-4+), *Orthotrichia* (1-?), *Oxyethira* (1-4), *Hydroptila* (1-4), *Helicopsyche* (1-1), *Anisocentropus* (1-1), *Oecetis* (1-4), *Triaenodes* (1-2).

At least, 70 genera of Trichoptera have been recorded from Japan. Wiggins (1977) listed 143 genera from North America, the number of the Japanese ones reaches nearly half of the latter. Considering the area of Japanese islands and the present insufficient situation of Trichoptera taxonomy in Japan, it is doubtless that the Japanese islands have a rich Trichoptera fauna. The genera include Oriental, Holarctic and Asian Palaearctic elements and also some genera common to almost all faunal regions. The recent discoveries of genera *Eobrachycentrus* (Wiggins, Tani & Tanida 1985) and *Palaeagapetus* (Ito & Hattori 1986) in Japan show an interesting biogeographical relationship between the Japanese Islands and the North American Continent, along with the other genera common to the Oriental and Nearctic regions.

With the development of Trichoptera taxonomy in Japan, certainly, we will have to add more genera than listed here. The future checklist of Trichoptera species or genera of eastern Asia based on up-to-date taxonomical studies is indispensable to understand the zoogeography of Japanese Trichoptera.

REFERENCES


Barnard, P.C & Dudgeon, D.1984. The larval morphology and ecology of a new species of *Melanotrichia* from Hong Kong (Trichoptera : Xiphocentronidae). **Aquatic Insects,** 6 : 245-252.

Fischer, F.C.J.1960-1973. Trichopterum catalogus. Amsterdam : Nederlandse Entomologische Vereeniging. Vol. 1-15.

Flint, O.S.1984. The genus *Brachycentrus* in north America, with a proposed phylogeny of the genera of Brachycentridae (Trichoptera). **Smithsonian Contr. Zool.,** 398 : 1-58.

Flint, O.S. & Bueno-Soria, J. 1982. Studies of Neotropical caddisflies, XXXII : The immature stages of *Macronema* by the resurrection of *Macrostemum* (Trichoptera : Hydropsychidae). **Proc. biol. Soc. Wash.,** 95 : 358-370.

Ito, T.1978. Morphological and ecological studies on the caddisfly genus *Dinarthrodes* in Hokkaido, Japan (Trichoptera, Lepidostomatidae) I. The larval development and the cases on four species of *Dinarthrodes*. **Kontyû, Tokyo,** 46 : 574-584.

Ito, T.1983. Morphology and bionomicus of *Neoseverinia crassicornis* (Ulmer) (Trichoptera, Lepidostomatidae) II. Larvae, egg and bionomics. **Kontyû, Tokyo,** 51 : 322-329.

Ito, T.1984a. Three new species of *Dinarthrum* (Trichoptera, Lepidostomatidae). **Kontyû, Tokyo,** 52 : 1-20.

Ito, T.1984b. On the genus *Goerodes*(Trichoptera, Lepidostomatidae) in Japan. **Kontyû, Tokyo,** 52 : 506-515.

Ito, T. & Hattori, T. 1986. Description of a new species of *Palaeagapetus* (Trichoptera, Hydroptilidae) from Northern Japan, with notes on bionomics. **Kontyû, Tokyo.,** 54 : 143-151.

Iwata, M.1927a. Trichopterous larvae from Japan. **Zool. Mag., Tokyo,** 39 : 209-272. (In Japanese).

Iwata, M.1927b. Trichopterous larvae from Japan. **Annot. Zool. Japan,** 11 : 203-233.

Iwata, M.1927c. Trichopterous larvae from Japan. II. **Zool. Mag., Tokyo,** 39 : 389-394. (In Japanese).

Iwata, M.1928a. Trichopterous larvae from Japan. III. **Zool. Mag., Tokyo,** 40 : 115-130. (In Japanese).

Iwata, M.1928b. Trichopterous larvae from Japan. IV. **Zool. Mag., Tokyo,** 40 : 237-241. (In Japanese).

Iwata, M.1930. Trichopterous larvae from Japan. V. **Zool. Mag., Tokyo,** 42 : 59-66. (In Japanese)

Kimmins, D.E.1963. On the Trichoptera of Ethiopia. **Bull.Br. Mus. nat. Hist. (Ent).,** 13 : 9-170.

Kobayashi, M.1964. A new genus and a new species of Hydroptilidae from Japan (Trichoptera). **Kontyû, Tokyo** 32 : 211-213.

Kobayashi, M.1980. A revision of the family Philopotamidae from Japan. **Bull. Kanagawa Pref. Mus.** 8 : 85-104 + 8pls.

Kuwata, K.1965. trichoptera. *Iconographia Insectorum japonicorm naturali edita*, ed. Asahina, S. *et al.*. Tokyo : Hokuryukan, 3 : 163-166 +pls. 82-83. (In Japanese)

Kuwayama, S.1930. The Stenopsychidae of Nippon. **Ins. Matsumurana,** 4 : 109-120.

Marshall, J.E.1979. A review of the genera of the Hydroptilidae (Trichoptera). **Bull.Br. Mus. nat. Hist. (Ent).,** 39 : 135-239.

Ross, H.H.1956. Evolution and classification of the mountain caddisflies. Urbana. Univ. Illinois Pr. 213 p.

Schmid, F.1950. Monographie du genre *Grammotaulius* Kolenati (Trichoptera, Limnophilidae). **Rev. Suisse Zool.,** 57 : 317-352.

Schmid, F.1952a. Les generes *Glyphotaelius* Steph. et *Nemotaulius* Bks. (Trichopt. Limnophil.). **Bull.soc. Vaud. Sci. Nat.,** 65 : 213-244.

Schmid, F.1952b. Le groupe de *Chilostigma*. **Arch. Hydrobiol.,** 47 : 75-163.

Schmid, F.1953. Contribution à l'étude de la sous-famille des Apataniinae, I. **Tijdschr.v.Entomol.,** 96 : 109-167.

Schmid, F. 1954a. Contribution à l'étude de la sous-famille des Apataniinae, II.**Tijdshr.v.Entomol.,** 97 : 1-74.

Schmid, F.1954b. Le genre *Asynarchus* McL. **Mitt.Schweiz. Ent.Ges.,** 27 : 57-96.

Schmid, F.1955. Contribution à l'étude des Limnophilidae (Trichoptera). **Mitt. Schweiz. Ent. Gez.,** 28 : 1-245.

Schmid, F.1968. La famille des Arctopsychides (Trichoptera). **Mem. Soc. entomol. Quebec,**1: 1-84 +2 pls.

Schmid, F.1969. La famille des Stenopsychides (Trichoptera). **Can. Ent.,** 101 : 187-224 + pls.1-7.

Schmid, F.1970. Le genre *Rhyacophila* et la famille des Rhyacophilidae (Trichoptera). **Mem. ent. Soc.Can.,** 66 : 1-230.

Schmid, F. & Botosaneau, L. 1966. Le genre *Himalopsyche* Banks (Trichoptera, Rhyacophilidae). **Ann. Soc. entomol. Quebec,** 11 : 123-176.

Tani, K.1977. Trichoptera. Coloured illustrations of the insects of Japan, II, eds. Ito, S., Okutani,T. & Hiura, I., Osaka : Hoikusha : 184-206+pls. 43-44. (In Japanese).

Tanida, K.1985. Trichoptera. An illustrated book of aquatic insects of Japan, ed. T.Kawai, Tokyo, Tokai Univ. Pr., 167-215. (In Japanese)

Tsuda, M.1937. Benthonic animals on muddy bottom of Lake Iniwa-naiko, Hikone, Shiga. **Jap.J.Limnol.** 7 : 127-128. (In Japanese).

Tsuda, M.1942. Japanische Trichopteren, I. Systematik. **Mem. Coll. Sci. Kyoto Imp. Univ.B,** 17 : 239-339.

Tsuda, M.1959. Trichoptera. Illustrated insect larvae of Japan, ed. Ezaki, T. *et al.,* Tokyo : Hokuryukan, 126-153. (In Japanese).

Tsuda, M. & Akagi, I. 1962. Trichoptera. Aquatic Entomology, ed. Tsuda, M., Tokyo : Hokuryukan : 112- 148. (In Japanese).

Ulmer, G.1951. Köcherfliegen (Trichopteren) von den Sunda-Inseln. Teil I. **Arch. Hydrobiol. Suppl.,** 19 : 1-528.

Wiggins, G.W.1959. A new family of the Trichoptera from Asia. **Can. Ent.,** 91 : 745-757.

Wiggins, G.W.1968. Contributions to the systematics of the caddisfly family Molannidae in Asia (Trichoptera). **Life Sci. Contr. Royal Ontario Mus.,** 72 : 1-26.

Wiggins, G.W.1969a. Contributions to the biology of the Asian caddisfly family Limnocentropodidae (Trichoptera). **Life Sci. Contr. Royal Ontario Mus.,** 74 : 1-29.

Wiggins, G.W.1969b. A new species of the caddisfly family Phryganopsychidae from Japan (Insecta : Trichoptera). **Bull. nat. Sci. Mus. Tokyo.,** 12 : 211-215.

Wiggins, G.W.1977. Larvae of the North American caddisfly genera (Trichoptera). Toronto and Buffalo : Univ. Toronto Pr., 401 p.

Wiggins, G.W. & Kuwayama, S.1971. A new species of the caddisfly genus *Oligotricha* from northern Japan and Sakhalin, with a key to the adults of the genus (Trichoptera : Phryganeidae). **Kontyû, Tokyo,** 39 : 340-346.

Wiggins, G.W. & Richardson, J.S. 1982. Revision and synopsis of the caddisfly genus *Dicosmoecus* (Trichoptera : Limnephilidae ; Dicosmoecinae). **Aquatic Insects,** 4 : 181-217.

Wiggins, G.W. Tani, K. & Tanida, K. 1985. *Eobrachycentrus,* a genus new to Japan, with a review of the Japanese Brachycentridae. **Kontyû, Tokyo,** 53 : 59-74.

Wiggins, G.W. Waever III, J.S. & Unzicker, J.D.1985. Revision of the caddisfly family Uenoidae (Trichoptera). **Can. Ent.,** 117 : 763-800.

# A PRELIMINARY STUDY OF THE SUBFAMILY HYDROPSYCHINAE (TRICHOPTERA : HYDROPSYCHIDAE) IN CHINA

L. X. TIAN and Y. W. LI

## SUMMARY

This paper proposes a phylogenetic tree of the subfamily Hydropsychinae by comparing the morphology of male genitalia and analysing geotropical distribution of all genera of the subfamily. This subfamily seems to originate from the Australian region and spread through the Indo-Malayan region northwards to the Palearctic then to the Nearctic region westwards to the Ethiopan region. Genus *Smicridea* may be the most primitive type survival. The genus *Hydropsyche* developed directly from it. Other genera developed from a 'median body', then split into two branches : *Ceratopsyche* branch and *Cheumatopsyche* branch. A list of 46 species (23 species new to Science) of Chinese Hydropsychinae belonging to 8 genera (one genus new to Science) is also presented.

## INTRODUCTION

Banks (1940), Martynov (1930, 1931), Ulmer (1926, 1928, 1933), Mosely (1942) and Schmid (1959, 1963, 1965, 1968) described 53 species of Hydropsychinae from China. In recent years, Kimmins (1957), McFarlane (1976), Ross (1977) used characters of wing venation, tibiae and male genitalia for better definitions of the genera. Ross and Unzicker (1977) discussed the evolution of the male genitalia of *Hydropsyche* and divided this genus into two genera : *Symphitopsyche* and *Hydropsyche*. The present paper is based on the study of Ross (1977). It gives the list of the Chinese species (the figures and the descriptions of these species will be published later on) and proposes a phyletic outline of the evolution of the genera. We prefered Ross and Unzicker's terminology to Schmid' s (1980).

## MATERIAL STUDIED

In 1957-1985, more than 3000 specimens were collected with light traps placed near streams in Anhui, Fujian, Guangdong (include Hainan island), Guangxi, Guizhou, Hebei, Heilongjiang, Hunan, Jiangsu, Shanxi, Sichuan, Xizang, Yunnan and Zhejiang provinces. The material is preserved dry. Some specimens have been spread and their genitalia treated in 10% KOH.

## DISTRIBUTION AND RELATIONSHIP OF THE GENERA

The Hydropsychinae include 24 genera and about 550 species. Its distribution is ubiquitous. None of its single genus has such a wide distribution.

*Cheumatopsyche* is ubiquitous except for the Neotropical region and especially well represented in the Indo-Malayan region.
*Smicridea* and *Asmicridea* are Neotropical and Australian ; they seem to be geographical substitutes for *Cheumatopsyche*.
*Ceratopsyche* and *Potamyia* are Palearctic, Nearctic and Indo-Malayan.
*Hydropsyche* is ubiquitous except for the Neotropical region.
*Mexipsyche* is Chinese, Mexican and Javanese.
*Hydromanicus, Hydatopsyche, Abacaria, Ecnopsyche, Occutanspsyche* and *Hydatomanicus* are Indo-Malayan.
*Calopsyche, Aeteapsyche, Orthopsyche, Herbertorossia* and *Austropsyche* are Australian.
*Symphitopsyche* and *Plectropsyche* are Ethiopian.
*Oropsyche* and *Streptopsyche* are Nearctic.
*Hydronema* is Central Asiatic.

The subfamily originally split into two lineages, one containing *Hydropsyche* and *Smicridea* and the other one constituting a 'median body' as defined by Ross and Unzicker. This taxon is characterized by the presence of a bud on the ventral endothecal processes. From this 'median body' two branches developed. *Ceratopsyche* is typical of the first branch. Two or three lobelets arising from a large tubular endotheca, a pair of long membranous lobes and the phallotremal sclerites sunk into the endotheca characterize *Occutanspsyche*. *Symphitopsyche* shows a very small endothecal process. *Mexipsyche* is provided with bare phallotremal sclerites, a sclerotized ventral endothecal lobe cleft dorsally and the endothecal processes sclerotized. The second branches includes most of the genera of the subfamily, in which the ventral endothecal lobe is lost or transformed into a small membranous knob or two sclerites and the endothecal processes are mobile or immobile. The most primitive member is *Orthopsyche* in which the membranous ventral endothecal lobe is split into three parts. *Streptopsyche, Plectropsyche* and *Calosopsyche* have membranous valves at the lower part of the phallobase. *Cheumatopsyche* and *Potamyia* are not considered here. Because of the shape of their X$^{th}$ segment, *Hydromanicus,Hydatomanicus* and *Hydatopsyche* form a group and have free preanal appendages. *Hydatomanicus* may have evolved directly from *Hydromanicus*.

The interpretations above are summarized in Figure 1.

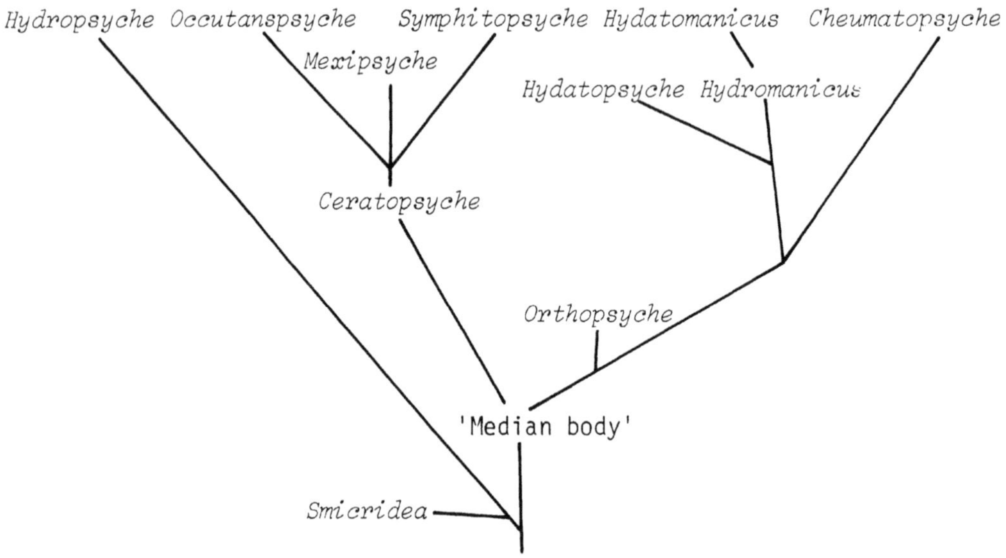


Figure 1. Phylogeny of the subfamily Hydropsychinae.

LIST OF THE CHINESE HYDROPSYCHINAE

This family is richly represented in China, especially in the meridional part. Nine genera and 77 species have been recorded up to the present. Eight genera and 46 species are listed below. All the type specimens are deposited in the Institute of Zoology, Academia Sinica, Beijing and in the Nanjing Agricultural University.

1 - *Hydropsyche hedini* Forsslund
Guangyuan and Qinchennshan (Sichuan)

2 - *Hydropsyche ornatula* McLachlan (Rec.nov)
Wulingshan (Hebei)

3 - *Ceratopsyche columnata* (Martynov) (comb.nov.)
Xixinang (Henan), Guangyuan and Qinchennshan (Sichuan), Huangguoshu (Guizhou), Fupin (Shanxi).

4 - *Ceratopsyche compressa* (sp.nov.)
Holotype ♂ : Bijie (Guizhou), 22 Sept 1984 ; paratype 4♀♀, allotype ♀collected with holotype.

5 - *Ceratopsyche conoidea* (sp.nov.)
Holotype ♂: Miaoersnan (Guangxi), 10 July 1984.

6 - *Ceratopsyche fukiensis* (Schmid) (comb.nov.)
Wuyishan (Fujian).
7 - *Ceratopsyche gautamittra* (Schmid) (comb.nov & rec.nov.)
Huangguoshu (Guizhou).
8 - *Ceratopsyche hainanensis* (sp.nov.)
Holotype ♂ : Jianfenglin in Hainan island (Guangdong), 18 May 1984; allotype ♀ and paratypes ♂♂ collected with holotype.
9 - *Ceratopsyche nevae* (Kolenati) (comb.nov., rec.nov.)
Maodan river (Heilongjiang).
10 -*Ceratopsyche penicillata* (Martynov) (comb.nov.)
Wuyishan (Fujian) ; Emeishan (Sichuan) ; Yinjian (Yunnan).
11 - *Ceratopsyche simulata* Mosely (comb. nov. & syn. nov.
West-Tianmushan (Zhejiang) ; Jinxiu and Miaoershan (Guangxi) ; Wuyishan (Fujian).
12 - *Ceratopsyche tetrachotoma* (sp. nov.)
Holotype ♂ : Huaping (Guangxi) 7 July 1984.
13 - *Ceratopsyche trifora* (sp. nov.)
Holotype ♂ : Huangshan (Anhui) 8 July 1984 ; Allotype ♀ and paratypes 2♂♂ collected with holotype; ♂ Fanjinshan (Guizhou).
14 - *Ceratopsyche kozhantschikovi* (Martynov) (comb. nov. & rec. nov.)
Maodan river (Helongjiang).
15 - *Mexipsyche rhomboana* (Martynov) (comb.nov.)
Jitang, Jiangba, Changdu, Suiba, Cuolane, Nuiwuqisangdo, Zogong (Xizang) ; Markang (Sichuan).
16 - *Mexipsyche furcula* (sp.nov.)
Holotype ♂ and allotype ♀, Wuyishan and Guadun (Fujian) ; paratypes 3♂♂ and 10♀♀ collected with holotype.
17 - *Mexipsyche ditalon* (sp.nov.)
Holotype ♂: Meto and Gedang (Xizang).
18 -*Mexipsyche hoenei* (Schmid) (comb. nov.).
Wuyishan (Fujian) ; Emeishan (Sichuan).
19 - *Mexipsyche kaznakovi* (Martynov) (comb. nov., rec. nov.).
Zadadiya, Maxian, Lielamu, Zhangmu and Jilong (Xizang).
20 - *Mexipsyche malformatio* (sp. nov.).
Holotype ♂ : Litang (Sichuan) 2 June 1982 ; allotype ♀ collected with holotype ; paratype ♂ Litang (Sichuan) 16 Aug. 1982 and 18 ♂♂ : Jianda military depot (Xizang) 26-31 July 1976.
21 - *Mexipsyche pungentis* (sp. nov.)
Holotype ♂ : Pianma and Lushui (Yunnan) ; allotype ♀ and paratypes 1 ♂ and 2♀♀ collected with holotype.
22 - *Mexipsyche polyphylla* (sp. nov.)
Holotype ♂: Renbu and Shada (Xizang), 9 Aug. 1983 ; Paratype ♂ : Renbu and Shada (Xizang) collected with holotype ; ♂ Jilong (Xizang) 5 June 1984, 2♂♂ Linzhi (Xizang), 9 Aug. 1983, Meto and Beipong (Xizang) 29 Aug. 1983 and ♂ Meto and Jialasa (Xizang) 2 July 1983.
23 - *Mexipsyche tibetana* (Schmid) (comb. nov.)
Mankang and Haitong (Xizang) 12 Aug. 1982, Noushui and Yaojaping (Yunnan) 2 June 1982.
24 - *Mexipsyche dolosa* (Banks) (comb. nov.)
Wuyishan, Shaowu and Guiling (Fujian).
25 - *Mexipsyche fryeri* (Ulmer) (comb. nov., rec. nov.)
Yinjian and Liuku (Yunnan).
26 - *Mexipsyche retronsis* (sp. nov.)
Holotype ♂: Gershiquan river, Ali (Xizang) 6 June 1975 ; paratypes 2♂♂ collected with holotype and Gui lake, Pulan (Xizang) 21 July 1976.
27 - *Occutanspsyche polyacantha* (gen. nov., sp. nov.)
Holotype ♂: Nankuenshan (Guangdong) 20 June 1981.
28 - *Hydromanicus canaliculatus* (sp. nov.)
Holotype ♂ : Wuyishan and Guandun (Fujian) 1 Aug. 1983 ; paratypes ♀ and 4♂♂ collected with holotype.

29 - *Hydromanicus deceptus* (Banks) (comb.nov) & rec. nov.)
Xinnong in Hainan island (Guangdong) and Dazhou & Shaowu (Fujian).

30 - *Hydromanicus emeius* (sp. nov.)
Holotype♂ : Emeishan and Jinshui (Sichuan) 27 Aug. 1984.

31 - *Hydromanicus fissus* (sp. nov.)
Holotype♂ : Emeishan (Sichuan) 23 Apr. 1957.

32 - *Hydromanicus umbonatus* (sp. nov.)
Holotype ♂: Fugong and Maji (Yunnan).

33 - *Hydromanicus ovitus* (sp. nov.)
Holotype ♂: Wuhua County (Guangdong) 25 June 1978.

34 - *Cheumatopsyche dubitans* Mosely
Wuyishan (Fujian), Fanjingshan (Guizhou) and Sangzhi (Hunan).

35 - *Cheumatopsyche longiclasper* (sp. nov.)
Holotype♂ : Jianyang and Masha (Fujian) 6 Aug. 1983.

36 - *Cheumatopsyche spinosa* Schmid
Huanggoshu (Guizhou).

37 - *Cheumatopsyche trifascia* (sp.nov.)
Holotype ♂ : Wuyishan and Guadong (Fujian) 11 Aug. 1983 ; paratype ♀ collected with holotype ; paratypes 10 ♂♂ & 1♀ collected with holotype, 1 ♂ Shaowu Nakou 4 Aug. 1984, 2 ♂♂ Jianyang Masha (Fujian) 6 Aug. 1983 and 1♂ Tianmushan (Zhejiang) July 1984.

38 - *Cheumatopsyche uenoi* (Tsuda) (comb. nov. & Syn. nov.)
Harbin (Helongjiang), Nanjing (Jiangsu), Hanshu (Hunan), Simao, Jinhong, Shilin and Kummin (Yunnan), Biije (Guizhou).

39 - *Cheumatopsyche ventricosa* (sp. nov.)
Holotype ♂ : Guadun and Wuyishan (Fujian) 11 Aug. 1983 ; allotype ♀ collected with holotype Paratypes 10 ♂♂ and 1 ♀ Guadun (Wuyishan), Sangang (Wuyishan) 8-11 Aug. 1983, 1 ♂ Nakou (Shaowu county) 4 Aug. 1983, 2♂♂ and 4 ♀♀ Masha (Jianyang) 6 Aug. 1983 and 1 ♂ Tianmushan (Zhejiang) 12 Aug. 1984.

40 - *Potamyia bicornis* (sp.nov.)
Holotype ♂: Tongchong (Yunnan) 22 May 1983 ; allotype ♀ and paratype ♂ collected with holotype.

41 - *Potamyia chinensis* (Ulmer) (comb. nov.)
Wuhua county and Jianfengling in Hainan island (Guangdong) ; Jianyang Masha, Shaowu Dazhou and Shaowu Nakou (Fujian) ; Hualian Farm (Yunnan).

42 - *Potamyia jinhongensis* (sp. nov.)
Holotype♂ : Jinhong (Yunnan) ; allotype ♀ and paratype♂ collected with holotype.

43 - *Potamyia parva* (homonym.nov.)
*Hydropsychodes chinensis* Martynov 1930
Jinhong (Yunnan).

44 - *Potamyia proboscida* (sp. nov.)
Holotype ♂ : Bawangling in Hainan Island (Guangdong) 23-28 May 1983 ; allotype ♀ and paratypes 3 ♂♂collected with holotype.

45 - *Potamyia straminea* (McLachlan) (comb. nov., rec. nov.)
Aksu (Xinjiang autonomous region).

46 - *Potamyia yunnanica* (Schmid) (comb. nov.)
Mangkang and Yianjin (Xizang) ; Liuku, Gongshan, Xiangyang (Yunnan).

REFERENCES

Banks, N. 1940. Reports on certain groups of Neuropteroid Insects from Szechwan, China. **Proc. U.S. Nat. Mus. Wash.** 88 : 173-220.

Forsslund, K.H. 1935. Swedisch-Chinesiche Wissenschaftliche Expedition nach den Nordwestlichen Provinzen Chinas. **Arklv. Zool.,** 27 A(31) : 1-21.

Kimmins, D.E. 1957. Notes of some british species of the genus *Hydropsyche* (Trichoptera). **Ent. Gaz.,** London, 8 : 199-210.

Mc Farlane, A.G. 1976. A generic revision of New Zealand Hydropsychinae (Trichoptera). **J. R. Soc. New. Zealand.,** 6 : 361-369.

Martynov, A.V. 1930. On the Trichopterous fauna of China and eastern Tibet. **Proc. Zool. Soc.**, London : 65-112.

Martynov, A.V. 1931. Reports on a collection of insects of the order Trichoptera from Siamand China. **Proc. U. S. Nat. Mus.**, 79(25) : 1-20.

Mosely, M.E. 1942. Chinese Trichoptera : a collection made by Mr. M.S. Yang in Foochow.**Trans. R. ent. Soc.**, London, 92 : 342-362.

Ross, H.H & Unzicker, J.D. 1977. The relationship of the genera of american Hydropsychidae as indicated by phallic structures (Trichoptera, Hydropsychidae). **J. Georgia ent. Soc.**, 12 : 298-312.

Schmid, F. 1959. Quelques Trichoptères de Chine. **Mitt. Zool. Mus.**, Berlin, 35 : 317-345.

Schmid, F. 1963. Quelques Trichoptères de Chine II. **Bonn. zool. Beitrag**, 16 : 127-154.

Schmid, F. 1965. Ergebnisse der zoologischen Forschungen von Dr. Z. Kaszab in der Mongolei, **Reichenbachia**, Dresden, 7 : 201-203.

Schmid, F. 1968. Trichoptera 2. Ergebnisse der zoologischen Forschungen von Dr. Z. Kaszab in der Mongolei. **Reichenbachia**. Dresden, 11 : 5-16.

Schmid, F. 1980. Les Insectes et Arachnides du Canada 7. Genera des Trichoptères du Canada et des Etats adjacents. Agric. Can. Ottawa. Publ. 1962 : 1-296.

Ulmer, G. 1926. (1925). Beiträge zur Fauna sinica. III. Trichopteren und Ephemeropteren. **Arch. Natg. Berlin.** 91 : 19-110.

Ulmer, G. 1928. Über die Trichopterengattung *Hydromanicus* und Verwandte. **Konorvia Z. Syst. Ins.**, 7 : 314-318.

Ulmer, G. 1933. Aquatic Insects of China Article IV Neue Chinesiche Trichopteren, nebst Uebersicht über die bischet aus China bekannte Arten. Teil II. **Peking Nat. Hist. Bull., 7 : 135-157 & 195-218.**

# 5. PHYLOGENY

# ON THE BIOGEOGRAPHY OF THE *OXYETHIRA* GROUP, TRIBE HYDROPTILINI (HYDROPTILINAE, HYDROPTILIDAE, TRICHOPTERA)

A.WELLS

## SUMMARY

The biogeography of the *Oxyethira* group, tribe Hydroptilini (Hydroptilinae, Hydroptilidae) is examined using an analytical approach. Distributions are substituted for taxa in a dendrogram of generic relationships of the three Hydroptilini groups. The area in common for the *Oxyethira* group is Australasia and for the *Agraylea* group, the Palaeartic, while little can be deduced from the cosmopolitan distribution of the Hydroptila group. The ecology of extant *Oxyethira* group members is considered and an origin in warm, slow to still waters of Gondwanaland is postulated. Problems arising from this hypothesis are discussed.

Six tribes are recognised in the hydroptilid sub-family Hydroptilinae, and of these, the Hydroptilini is considered to be one of the more highly derived (Marshall 1979). Within this tribe, Marshall designated three generic groups, the *Agraylea, Hydroptila* and *Oxyethira* groups. She considered the *Hydroptila* and *Oxyethira* groups to be sister groups and to be more recently derived than the *Agraylea* group (Fig. 1). *Hydroptila,* the sole member of the *Hydroptila* group, is almost cosmopolitan in distribution, while the *Agraylea* group has members in the Nearctic, Palaearctic, Africa, and SE Asia (Fig. 1).

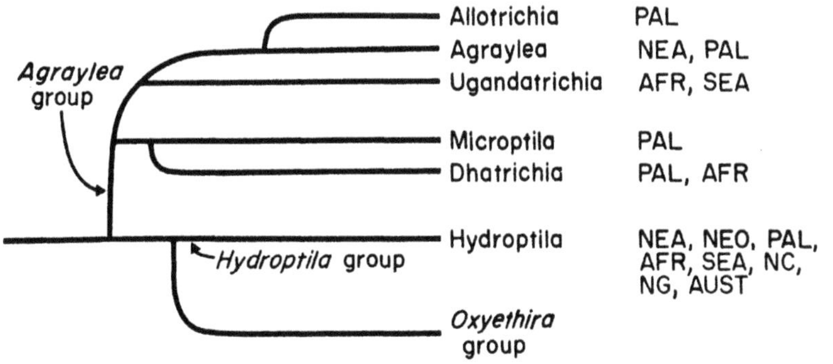


Figure 1. A phylogeny for the three generic groups in the tribe Hydroptilini. Genera and their distributions are given for the *Agraylea* and *Hydroptila* groups. (Adapted from Marshall 1979). Abbreviations are as follows :- AFR, Africa ; AUST, Australia ; NC, New Caledonia ; NEA, Nearctic ; NEO, Neotropics ; NG, New Guinea ; NZ, New Zealand ; PAL, Palaearctic ; SEA, SE.Asia.

This paper examines the biogeography of the *Oxyethira* group, which is presently considered to be comprised of 10 genera (Wells 1986), all with representatives in Australasia, and only two or perhaps three of which extend beyond that region (Table 1).

A list of taxa and their distributions is only a starting point in a biogeographical study. The analytical approach to biogeography, advocated by Platnick and Nelson (1978), requires examination of phylogenetic relationships and patterns of distribution in relation to the geological history of the regions involved. In addition, Endler (1982) and Davis (1982) argue that neglect of ecology can lead to spurious interpretations in biogeography. Thus, all of these aspects should be considered.

Table 1. *Oxyethira* group genera and their distributions.

| GENUS | DISTRIBUTION |
|---|---|
| *Acanthotrichia* | Australia |
| *Acritoptila* | Australia, New Caledonia |
| *Austratrichia* | Australia |
| *Hellyethira* | Australia, New Caledonia, Japan (?) |
| *Mulgravia* | Australia |
| *Oxyethira* | Cosmopolitan |
| *Orphninotrichia* | Australia |
| *Paroxyethira* | New Caledonia, New Zealand |
| *Tricholeiochiton* | Australia, Palaearctic, SE Asia |
| *Xuthotrichia* | Australia |

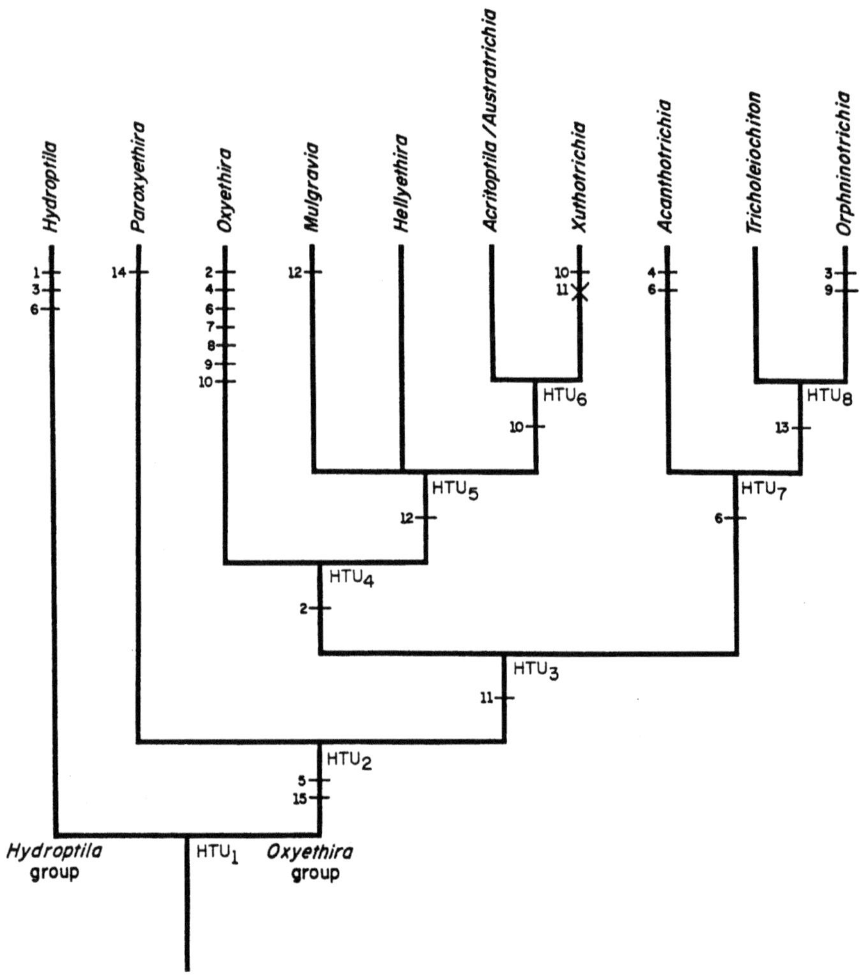


Figure 2. Cladogram of relationships of genera in the Australasian Hydroptilini, derived by the WISS method (Farris *et al.* 1970). HTU's are hypothetical ancestors which share derived steps ; bars indicate derived steps ; a cross indicates a reversal in a transformation series ; numbers refer to coded adult characters from Wells (1986).

In a recent study of the systematics and biogeography of Australian Hydroptilidae (Wells 1986), the phylogeny of *Oxyethira* group genera was analysed using several cladistic methods. Cladograms of relationships, based on adult characters and using *Hydroptila* as the out-group, were compared and tested against data on larvae. Arguments were advanced in favour of an hypothesis of relationships, derived by using the Weighted Invariant Step Strategy (WISS method) of Farris *et al.* (1970), and represented here in Figure 2. (This method is basically Hennigian in approach but proceeds by clustering of taxa which share **steps** leading to derived states, not necessarily the actual **states.**)

According to this cladogram, *Paroxyethira* was the first off-shoot of a lineage that later split to form a clade consisting of *Oxyethira* and *Hellyethira-Acritoptila-Austratrichia-Mulgravia-Xuthotrichia,* and its sister clade, in which *Acanthotrichia* is the sister group of *Tricholeiochiton-Orphninotrichia.* The cladogram is not entirely satisfactory as three taxa are indistinguishable from their immediate ancestors, there is one trichotomy, a reversal in a character state (the cross below *Xuthotrichia* ), and several parallelisms [see (Wells 1986) for details].

The next step in the exercise is to substitute distributions for taxa in the cladogram, as in Figure 3, which is extended to include Marshall's dendrogram for *Agraylea* group genera, and their distributions. Distributions shared by all or most members of a monophyletic group are postulated to be ancestral while novel distributions must be explained as resulting from dispersals or extinctions.

According to Figure 3, the distribution in common for *Oxyethira* group genera, excluding *Paroxyethira,* is Australia. If *Paroxyethira* is included, the distribution expands to Australasia. Among *Agraylea* group members, all but one is Palaeartic. Thus, these data are consistent with a Gondwanaland origin for the *Oxyethira* group and a Laurasian origin for the *Agraylea* group. The origin of the *Hydroptila* group is, as yet, undetermined. Future careful analysis of intrageneric relationships may help to decide its place of origin.

It seems plausible, then, to explain the origins of these groups in terms of vicariance resulting from fragmentation of Pangaea. Similar hypotheses have been invoked to explain relationships of other unrelated Australasian taxa ( Humphries 1981).

There are problems, though, with a putative Gondwanan origin for the *Oxyethira* group. The distribution of *Paroxyethira* is best explained as a result of vicariance : *Paroxyethira* diverged in isolation in New Zealand/New Caledonia after the rafting of these land masses from Antarctica (c. 80-90 million years ago (Crook 1981)). Hence, $HTU_3$, the ancestor of the other genera, must have been present on the land mass comprised of Antarctica-South America-Australia. Yet only *Oxyethira* is known from South America and is represented there by members of a relatively apomorphous sub-genus, not closely related to either of the two sub-genera found in Australia (Kelley 1984). On the other hand, all but two, or possibly three, of the genera descended from $HTU_3$ are Australian endemics.

One of these two genera, *Tricholeiochiton,* is the sister group of an Australian endemic, *Orphninotrichia,* and therefore probably originated in Australia (where it is now restricted to the north) and subsequently dispersed to SE Asia and the Palaearctic. This invokes dispersal against prevailing winds, but is not impossible and has been suggested for other groups, e.g. for seven species in the Carabidae by Darlington (1971).

Greater difficulties are encountered though when attempting to explain the present wide distribution of *Oxyethira. Oxyethira (Dampfitrichia),* which Kelley (1984) considers to be one of the more highly derived sub-genera, occurs in northern Australia and in the Neotropics, Palaearctic and SE Asia. The Australian representatives can be postulated to be relatively recent arrivals from the Oriental region (especially since one species extends from Burma to N Australia), or descendents of Oriental forms.

Sub-genus *Trichoglene,* to which the species of *Oxyethira* in southern and eastern Australia belong, is restricted to Australasia and is less easily accounted for :-According to Kelley (1984) it is the most primitive of all *Oxyethira* sub-genera. Its presence in New Zealand and New Caledonia suggests that it had evolved before their separation from Gondwanaland, yet the cladogram in Figure 2 depicts the dichotomy which led to the evolution of *Oxyethira* as a more recent event than that which gave rise to *Paroxyethira.*

An alternative explanation is that *Oxyethira (Trichoglene)* dispersed from Australia to New Zealand and New Caledonia. Some species in other genera have reached New Caledonia from Australia relatively recently (Wells 1986), but there is no evidence of similar dispersals to New Zealand.

The sub-genus *Oxyethira* , the next off-shoot after *Trichoglene,* is Holarctic and SE Asian in its present distribution which is consistent with it having a Laurasian origin, but then raises the problem of the absence of *Oxyethira (Trichoglene)* and other group members from South America (and Africa). If it is taken further back to a Pangaean origin, the absence of *Agraylea* group genera from Australasia must be explained.

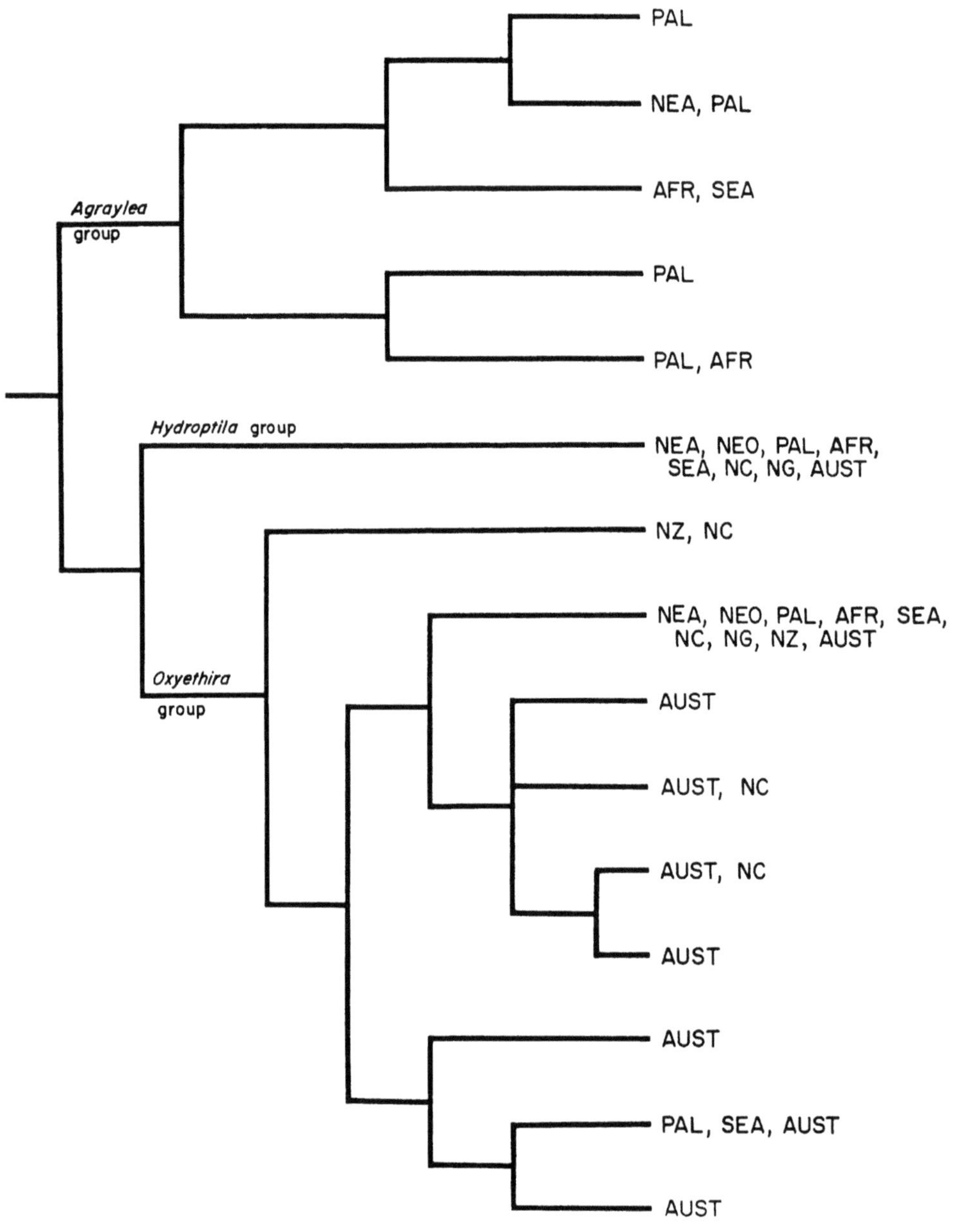


Figure 3. Area cladogram for taxa in Figures 1 and 2, i.e. for genera in the tribe Hydroptilini. Abbreviations as for Figure 1.

Perhaps as Endler (1982) and Davis (1982) insist, the answer is to be found in ecology. Unfortunately, at present only limited ecological data are available for this group. Lepneva (1964) describes habitats of several northern hemisphere *Oxyethira* species and of *Tricholeiochiton fagesii* Guinard as slow to stagnant waters, amongst algae and plant thickets. Wiggin's (1977) description of habitats of North American *Oxyethira* is similar, and many Australian *Oxyethira* group species are found in slow to still waters (Wells 1985). Known larvae of all but the torrenticolous genus, *(Orphninotrichia)*, are of a long-legged form purported to be adapted to feeding on, and clambering amongst, filamentous algae. Their cases are flat and many are made of secretion only.

It is likely, then, that the ancestor of the *Oxyethira* group was a long-legged filamentous alga feeder, living in warm, slow to still waters. Such habitats, were probably abundant in relatively low-relief, high-rainfall eastern Gondwanaland during the Cretaceous, when temperatures were equable (Kemp 1981), and still abound on the Australian continent where most *Oxyethira* group genera are now found.

Edmunds (1982) claims that life history patterns can be informative of past histories, and that in southern North America, Ephemeroptera of austral origin show patterns of late summer emergence. However, this pattern does not seem to occur in these hydroptilids :-Nielsen (1948) notes that there are two generations per year in *Oxyethira* species in "southerly countries" (of the northern hemisphere), and from studies of the Australian fauna Wells (1986) deduced that many *Oxyethira* group members are multivoltine, with their life cycles being strongly temperature dependent. This small amount of information, although differing from Edmund's findings, is consistent with an origin in warmish waters.

In conclusion, the *Oxyethira* group is perceived as a relatively apomorphous sector of the Hydroptilidae. It probably originated in the warm, slow to still waters of eastern Gondwanaland although an earlier Pangaean origin cannot be discounted entirely. Subsequent vicariance and adaptive radiation resulted in splitting and divergence of members, some of which adapted to the cooler, faster waters of the newly uplifted ranges of eastern Australia [early Tertiary, (Nix 1981)]. Thus, while some have persisted in their ancestral habitat of warm, slow waters, others have undergone a reversal of the usual trichopteran trend of radiation from cool montane streams to warmer waters, and are now found in faster flowing, cooler rivers.

In many groups with Gondwanan origins, Australasian representatives tend to be relatively primitive (Mackerras 1970, Winterbourne 1982), although Brundin (1965) found that some Australasian Chironomidae are more apomorphous than their close allies in South America. It is interesting to speculate whether new *Oxyethira* group members are yet to be found in any warm, slow waters in southern South America. If they are, will they be more plesiomorphic or more highly derived than the known forms ? Whatever they are, if found, they could finally settle some of the questions about the history of the *Oxyethira* group.

## REFERENCES


Brundin, L. 1965. On the real nature of transantarctic relationships. **Evolution**, 19 : 496-505.

Crook, K.A.W. 1981. The break-up of the Australian Antarctic segment of Gondwanaland. *In* Keast, A.(ed.) Ecological biogeography of Australia. Junk, The Hague, **Monogr. biol.**, 41 : 1-14.

Davis, G.M. 1982. Historical and ecological factors in the evolution, adaptive radiation, and biogeography of freshwater mollusks. **Amer.Zool.**, 22 : 375-395.

Darlington, P.J. 1971. The carabid beetles of New Guinea. Part IV. General considerations ; analysis and history of fauna ; taxonomic supplement. **Bull.Mus. Comp. Zool.**, 142 : 129-337.

Edmunds, G.F. 1982. Historical and life history factors in the biogeography of mayflies. **Amer. Zool.**, 22 : 371-374.

Endler,J.A.1982. Problems in distinguishing historical from ecological factors in biogeography. **Amer. Zool.**, 22 441-452.

Farris, J.S., Kluge, A.G. & Eckardt, M.J. 1970. A numerical approach to phylogenetic systematics. **Syst. Zool.**, 19 : 172-189.

Humphries, C.J. 1981. Biogeographical methods and the southern beeches (Fagaceae : *Nothofagus*).*In* Funk, V.A. and Brooks, D.R. (eds.) Advances in cladistics. N.Y. Botanical Gardens, Bronx, N.Y. :177-207.

Kelley, R.W. 1984. Phylogeny, morphology and classification of the micro-caddis fly genus *Oxyethira* (Trichoptera : Hydroptilidae). **Trans. Am. Ent. Soc.**, 110 : 435-463.

Kemp, E.M. 1981. Tertiary palaeogeography and the evolution of Australian climate. *In* Keast, A.(ed.) Ecological biogeography of Australia. Junk, The Hague, **Monogr. biol.**, 41 : 31-49.

Lepneva, S.G. 1964. Larvae and pupae of the Suborder Annulipalpia Trichoptera. Fauna U.S.S.R. 2(1). Trans., Israel Program Sci. Trans., Inc., 1970. Zool. Inst. Akad. Nauk. S.S.S.R., (N.S.) 88 : 1-566.

Mackerras, I.M. 1970. Composition and distribution of the fauna. *In* Mackerras, I.M. (ed.) The insects of Australia. A textbook for students and research workers. CSIRO, **Melbourne Univ.Press.**, Victoria : 187-204.

Marshall, J.E. 1979. A review of the genera of the Hydroptilidae (Trichoptera). **Bull. Br. Mus.(Nat. Hist.) Entomol.**, 39 : 135-239.

Nielsen, A. 1948. Postembryonic development and biology of the Hydroptilidae. **Biol. Skr. Dan. Vid. Selsk.,** 4(1), 1-200.

Nix, H.A. 1981. The environment of Terra Australis. *In* Keast, A. (ed.) Ecological biogeography of Australia. Junk, The Hague, **Monogr. biol.,** 41 : 103-133.

Platnick, N.I. & Nelson, G. 1978. A method of analysis for historical biogeography. **Syst. Zool.,** 27 : 1-16.

Wells, A. 1985. Larvae and pupae of Australian Hydroptilidae (Trichoptera) with observations on general biology and relationships. **Aust. J. Zool. Suppl.** 113 : 1-69.

Wells, A.1986. The systematics and biogeography of Australian Hydroptilidae (Trichoptera). Unpublished Ph.D. thesis, Univ. Adalaide.

Wiggins, G.B. 1977. Larvae of the North American caddisfly genera (Trichoptera). **Univ. Toronto Press,** Toronto & Buffalo, 401 p.

Winterbourne, M.J. 1980. The freshwater insects of Australasia and their affinities. **Palaeogeogr., Palaeoclim. Palaeoecol.,** 31 : 235-249.

# HIGHER CLASSIFICATION OF TRIPLECTIDINAE (TRICHOPTERA : LEPTOCERIDAE)

J.C. MORSE and R.W. HOLZENTHAL

## SUMMARY

A phylogenetic analysis of 29 morphological characters of larvae, pupae, and adults for the 22 nominal genus-group taxa of Triplectidinae implies that the three tribes Grumichellini (including *Atanatolica, Grumichella,* and *Triplexa* ), Hudsonemini (including *Condocerus, Hudsonema,* and *Notalina*) and Triplectidini (including *Notoperata, Triplectides,* and *Westriplectes*) are each monophyletic and the latter two are sister tribes. The other 13 nominal genus-group taxa are probably synonyms of these genera. At least 69 of the 124 known species of the subfamily are included in *Triplectides*, suggesting a need to recognize 17 or more subgenera within it for conceptual and predictive purposes.

## INTRODUCTION

Comprehensive revisions of the genera of the leptocerid subfamily Triplectidinae Ulmer, 1906, have been published by Ulmer (1907) and Mosely (1936). Regional studies by Mosely and Kimmins (1953 for Australia and New Zealand), Ulmer (1951, 1955 for the Sunda Islands), Neboiss (1977 for Tasmania, 1986 for the Southwest Pacific and Australian Region), and Holzenthal (1985, 1986a, 1986b for the Neotropical Region) have added considerably to our understanding of this primarily Australasian-Neotropical subfamily. The works of these authors, along with many smaller papers, have increased the subfamily to 124 known species in 22 nominal genus-group taxa. Nine genus-group taxa have been considered synonyms in previous publications and three have been transferred to Triplectidinae from the sister subfamily Leptocerinae (Table 1).

Morse (1981) grouped the triplectidine genera into three tribes (Grumichellini, Hudsonemini, and Triplectidini) based on the phylogenetic evidence known to him at that time. Since then, Holzenthal (1985, 1986b) presented new information refining Morse's hypotheses. The purpose of this paper is to review these data along with additional information and to use them to propose a more complete phylogeny and classification for the subfamily. We believe this classification will be useful for predicting yet-poorly-known characteristics of species in this subfamily and for preparing future revisions of its natural taxa.

## PHYLOGENETIC PHILOSOPHY AND METHOD

According to Hennig (1966), only monophyletic groups should be given formal taxonomic status. A monophyletic group is most readily recognized by its possession of one or more synapomorphies or homologues (uniquely shared characters). According to Remane (1956) and Ruppert (1982), homologues are characters recognized, in practice, by unique resemblance because of (a) correspondence in external spatio-temporal connections, (b) correspondence in internal spatio-temporal connections, and (c) similarity by these criteria with intermediate characters in other organisms or ontogenetic or paleontologic stages. Any two or more intrinsic characters may be homologous, including those of anatomy, physiology, biochemistry, behavior, genetics, and development (Eldredge and Cracraft 1980). The limits on useful character selection are the entire organism at the outermost extreme and the level of uniformity at the innermost extreme (Ruppert 1982). The probability of inferrring erroneously that characters are, or are not, homologous is greater for characters nearer bland uniformity (such as "reduced" or "lost" characters). The hierarchic distribution of homologues among organisms provides the basis for the hierarchical ordering of taxa employed in modern biological classification.

Table 1. Nominal genus-group taxa of Triplectidinae Ulmer 1906.

| | |
|---|---|
| *Atanatolica* Mosely 1936 | *Symphitoneurina* Schmid 1950 |
| *Condocerus* * Neboiss 1977 | *Triplectides* Kolenati 1859 |
| *Grumichella* *Müller 1879 | *Notanatolica* Mac Lachlan 1866 |
| *Leptocellodes* *Ulmer 1911 | synonymized by Mosely 1936 |
| synonymized by Ulmer 1955 | *Pseudonema* Mac Lachlan 1862 |
| *Hudsonema* Mosely 1986 | synonymized by Ulmer 1905 |
| *Lectrides* Mosely 1953 | *Tetracentron* Brauer 1865 |
| *Notalina* Mosely 1936 | synonymized by Ulmer 1907 |
| *Neonotalina* Holzenthal 1936, as subgenus | *Tobikera* Matsumura 1931 |
| *Triplexina* Mosely 1953, | synonymized by Tsuda 1942 |
| synonymized by Neboiss 1977 | *Triplectidina* Mosely 1936 |
| *Notoperata* Neboiss 1977 | *Triplexa* Mosely 1953 |
| *Symphitoneuria* Ulmer 1906 | *Gracilipsodes* Sykora 1967 |
| *Loticana* Mosely 1936 | synonymized by Morse 1981 |
| synonymized by Mosely and Kimmins 1953 | *Westriplectes* Neboiss 1977 |

* originally described in Leptocerinae Leach 1815

## HOMOLOGUES

Homologues (or synapomorphies) suggesting a hierarchical ordering of the genera of Triplectidinae are listed in Table 2 and the resulting phylogeny is depicted in Figure 1. In previous works, Weaver (1984) identified homologues numbered 1 and 2 in Table 2 ; Morse (1981), those numbered 9, 10, 11, 13, 14, and 18; and Holzenthal (1985, 1986b), those numbered 20, 21, 22, 23, and 24. Other homologues in Table 2 are inferred here for the first time, although several were first suggested by other authors as diagnostic characters.

Illustrations have been published for most of these homologues and ancestral states or else their nature is readily understandable from published prose. Five homologues deserve special mention, however. (a).The partially desclerotized fifth segment of the maxillary palp in adult leptocerids (homologue #5) superficially resembles that in Curvipalpia of Annulipalpia (Weaver 1984 and this volume). (b) Only in Leptoceridae and Macronematinae (Hydropsychidae) are the males typically larger and more colorful than the females, the male antennae and forewings are longer and the hindwings are broader (homologue #7). Based on numerous homologues at the family and higher levels of universality, we think these most likely are independent evolutionary developments. (c) Separate pre-pronotal sclerites (homologue #16, Figures 2 and 3) have been observed only in adults of Hudsonemini and Triplectidini. (d) Anterolateral pronotal lobes consistently overhang and conceal the dorsal apex of the proanepisternum in adult Hudsonemini and Triplectidini (homologue #17, Figure 3). These lobes in other leptocerids are smaller (Figure 4) ; leptocerid posterolateral pronotal lobes always partially obscure the proanepisternum. (e) Lateral pronotal warts are absent (homologue #19) in all Grumichellini and Hudsonemini examined, suggesting that they might be sister lineages. However, we think that such a simple or uniform character is far more likely to have evolved twice independently than are the pre-pronotal sclerites or broad anterolateral pronotal lobes, which conversely imply that Triplectidini and Hudsonemini are sister lineages.

## TRIPLECTIDINAE GENERA

Evidence for the independent monophyly of each of the six grumichelline and hudsonemine genera was provided by Holzenthal (1985, 1986b). Uniquely shared homologues are not yet known for the species of either *Notoperata* or *Westriplectes* ; each of these two genera thus may be paraphyletic. Within Triplectidinae, the long thyridial cell of the male forewing (homologue #29) is unique to *Triplectides sensu lato,* implying monophyly for its species.

The classification of the 69 species of *Triplectides s.l.* is under study and the details are, for the most part, outside the scope of this paper. Eventually, however, one dilemma almost certainly must be decided : whether to recognize 17 or more genera for the various monophyletic lineages already identified (Figure 1) or only subgenera within the more traditional *Triplectides.* At the moment, assuming that remaining phylogenetic problems can be resolved, we are inclined toward the latter, more conservative choice because of the additional phylogenetic information provided by the more broadly defined genus *Triplectides.* The alternative of leaving *Triplectides* is it now is known (paraphyletic, with *Triplectidina, Symphitoneurina, Lectrides,* and *Symphitoneuria* as excluded genera) is incompatible with Hennig's (1966) phylogenetic

Table 2. Homologues and ancestral states for taxa and sister groups of Leptoceridae and Triplectinidae as shown in Figure 1.

| # | Taxon | Sister Group | Form[1] | Structure | Homologue | References[2] | Ancestral State | References[2] |
|---|---|---|---|---|---|---|---|---|
| 1 | Leptoceridae | Molannidae | F | lateral lamellae | present | Nn 61 | absent | Nn 57 |
| 2 | " | " | L P A | antennae | long | L 765A, Hi 853, 855 | short | L 682, L p.522, My p.140 |
| 3 | " | " | A | pronotum | deeply incised | Wi 16.141 | nearly entire | Wi 16.143 |
| 4 | " | " | A | lateral sulci of vertex | present | Wi 16.141, MH 2 | absent | Wi 16.143 |
| 5 | " | " | A | maxillary palp segment 5 | flexible | Mr 23 | inflexible | My p.140 |
| 6 | " | " | A | wing cell 2; nigma | appar. absent; appar. before M | B 21, S 660, 661 | present; in cell 2 | B 20, S 748 |
| 7 | " | " | A | size male: size female | > 1 | MH text | < 1 | MH text |
| 8 | " | " | L | posterior part of lateral sclerite of abd. segment I | narrow and spineless | Wg 9.1 | broad and spinuous or absent | U 14, 25; Wg 11.1 |
| 9 | Leptocerinae | Triplectidinae | A | apparent hindwing cell 3 | absent | MK 191 | present | MK 134 |
| 10 | " | " | A | hindwing sectoral crossvein | absent | MK 191 | present | MK 134 |
| 11 | " | " | A | mesotibial preapical spurs | absent | MK p.198 | present | MK p.198 |
| 12 | " | " | L | foreleg trochantin | obtuse, broad | U 99, 108 | sinuous, long-tapering | U 87, 88 |
| 13 | Triplectidinae | Leptocerinae | M | phallic parameres | absent | MN 77c, 77e | present | Mr 31 |
| 14 | " | " | M | phalicata | indistinct, possibly plate-like | MN 77c, 77e | distinct, tubular | Mr 31 |
| 15 | " | " | P | large tooth on mandible | present | U 94 | absent | U 113 |
| 16 | Hudsonemini + Triplectidini | Grumichellini | A | pre-pronotal sclerites | present | MH 2, MH 3 | absent | MH text |
| 17 | " | " | A | anterolateral pronotal lobes | broad | MH 3 | narrow | MH 4 |
| 18 | Grumichellini | Hudsonemini + Triplectidini | A | mesotibial preapical spurs | absent | MK p.198 | present | MK p.198 |
| 19 | " | " | A | lateral pronotal warts | absent | MH text | present | MH 2 |
| 20 | Atanatolica + Triplexa | Grumichella | M | apical processes of abd. segment X | present | Ho 7.5E, MK 155 | absent | Ho 7.54B |
| 21 | " | " | M | mesal setae of apex of infer. appendage | present | Ho 7.5C, MK 155f | absent | Ho 7.54C |
| 22 | Hudsonemini | Triplectidini | M | 2nd article of inf. appendage | absent | MK 167 | present | MK 134 |
| 23 | " | " | L | 2ndary setae of anal proleg | present | Hz 1E | absent | Ho 8.1A |
| 24 | Condocerus + Hudsonema | Notalina | M | inferior appendage | short | MK 167 | long | MK 134 |
| 25 | Triplectidini | Hudsonemini | M | basal lobe of inf. appendage | articulated on narrow strap | MN 71b | not articulated | MK 172e |
| 26 | Westriplectes + Triplectides | Notoperata | A | forewing discal cell apex | broad | Nb 688, 689 | narrow | Nb 718, 720 |
| 27 | " | " | M | phallotremal sclerite | trident-shaped | MN 77e | U-shaped | MH 5 |
| 28 | " | " | M | basal lobe of inf. appendage | long | Nb 683, 693 | short | Nb 712, 717 |
| 29 | Triplectides | Westriplectes | M | forewing thyridial cell | long | Nb 690 | short | Nb 689 |

[1] A = adult, F = female, L = larva, M = male, P = pupa.

[2] B = Betten 1934, Hi = Hickin 1967, Ho = Holzenthal 1985, Hz = Holzenthal 1986b, L = Lepneva 1971, MH = this paper, MK = Mosely and Kimmins 1953, MN = Morse and Neboiss 1982, Mr = Morse 1975, My = Mosely 1939, Nb = Neboiss 1977, Nn = Nielsen 1980, S = Schmid 1980, U = Ulmer 1955, Wi = Wiggins 1984, Wg = Wiggins 1977. Numbers refer to figures unless indicated otherwise.

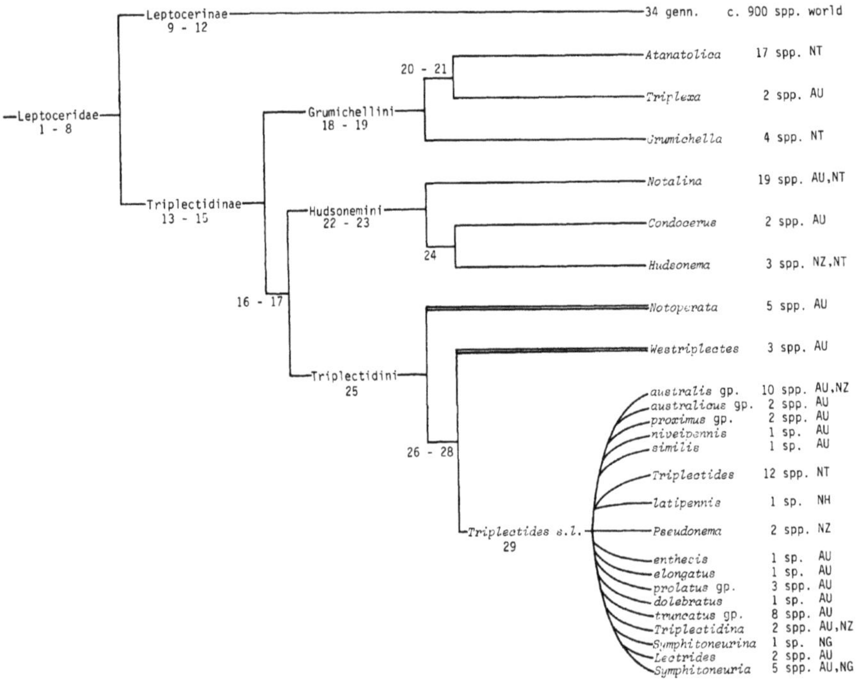


Figure 1. Phylogeny, classification, and distribution of Triplectidinae genera. Numbers with each subterminal node refer to homologues in Table 2. Double lines indicate possible paraphyly. AU = Australia, NG = New Guinea, NH = New Hebrides, NT = Neotropical Region, NZ = New Zealand. Teh 14 species of *Triplectides s.l.* from Asia and Baltic Amber are not included.

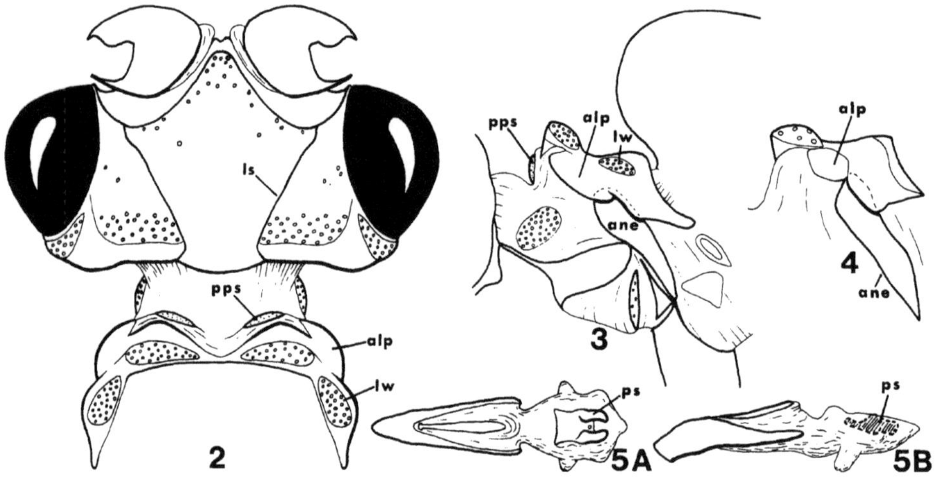


Figure 2-5. Characters of Triplectidinae. 2, *Notoperata maculata* (Mosely), head and pronotum, dorsal ; 3, same, prothorax, lateral ; 4, *Atanatolica* sp., prothorax, lateral ; 5A-B, *N. maculata,* phallus, A - dorsal, B - lateral. alp = anterolateral lobe of pronotum, ane = proanepisternum, Is = lateral sulcus of vertex, Iw = lateral wart of pronotum, pps = pre-pronotal sclerite, ps = phallotremal sclerite.

systematics philosophy and is unacceptable to us for practical reasons discussed by Morse (1981) in the Third International Symposium on Trichoptera.

Regardless of the classification scheme ultimately used for Triplectidini genera and subgenera, we think that the foregoing evidence provides a useful tool for predicting ecological and behavioral features and other structural characters in Triplectidinae, a focus for future research.

## ACKNOWLEDGEMENTS

Discussions with Drs. O. S. Flint, Jr., and A. Neboiss stimulated much of this research and loans of specimens through them from the U.S. National Museum of Natural History and the Museum of Victoria, respectively, made it possible. We are grateful to both.

This is Technical Contribution No. 2595 of the South Carolina Agricultural Experiment Station, Clemson University.

## REFERENCES

Betten, C. 1934. The caddis flies or Trichoptera of New York State. **Bull. New York State Mus.,** 292 : 1-576, pls. 1-67.

Eldredge, N. & Cracraft, J. 1980. Phylogenetic patterns and the evolutionary process. **Columbia Univ. Press,** New York.

Hennig, W. 1966. Phylogenetic systematics. **Univ. Illinois Press,** Urbana.

Hickin, N.E. 1967. Caddis larvae, larvae of the British Trichoptera. **Fairleigh Dickinson Univ. Press,** Rutherford, New Jersey.

Holzenthal, R.W. 1985. Studies in Neotropical Leptoceridae (Trichoptera) : Their diversity, evolution, and biogeography, with revisions of selected genera. Ph. D. dissertation, Clemson University.

Holzenthal, R.W. 1986a. The Neotropical species of *Notalina,* a southern group of long-horned caddiflies(Trichoptera : Leptoceridae). **Syst. Ent.** 11 : 61-73.

Holzenthal, R.W. 1986b. Studies in Neotropical Leptoceridae (Trichoptera), VI : Immature stages of *Hudsonema flamenii* (Navás) and the evolution and historical biogeography of Hudsonemini (Triplectidinae). **Proc. Ent. Soc. Washington,** 88 : 268-279.

Lepneva, S.G. 1971. Larvae and pupae of *Integripalpia,* Trichoptera. Fauna of the U.S.S.R., 2(2). Zool. Inst Akad. Nauk S.S.S.R., new Ser., (1966) 95 : 1-560. Trans. Israel Prog. Sci. Trans., Inc.

Morse, J.C. 1975. A phylogeny and revision of the caddisfly genus *Ceraclea* (Trichoptera, Leptoceridae). **Contrib. Amer. Ent. Inst.,** 11(2) : 1-97.

Morse, J.C. 1981. A phylogeny and classification of family-group taxa of Leptoceridae (Trichoptera). **Proc. 3rd Int. Symp. Trichoptera,** G.P. Moretti (ed.), Junk, The Hague : 257-264.

Morse, J.C. & Neboiss, A. 1982. *Triplectides* of Australia (Insecta : Trichoptera : Leptoceridae). **Mem. Natn. Mus. Victoria,** 43 : 61-98.

Mosely, M.E. 1936. A revision of the Triplectidinae, a subfamily of the Leptoceridae (Trichoptera). **Trans. R. Ent. Soc. London,** 85(3) : 91-129.

Mosely, M.E. 1939. The British caddis flies, a collector's handbook. George Routledge & Sons, Ltd., London.

Mosely, M.E. & Kimmins, D.E. 1953. The Trichoptera (caddis-flies) of Australia and New Zealand. Adlard and Son, Ltd. , Dorking, Surrey.

Neboiss, A. 1977. A taxonomic and zoogeographic study of Tasmanian caddis-flies. **Mem. Natn. Mus. Victoria** 38 : 1-208, pls. 1-3.

Neboiss, A. 1986. Atlas of Trichoptera of the SW Pacific - Australian Region. Dr W.Junk Pub., Dordrecht.

Nielsen, A.1980. A comparative study of the genital segments and the genital chamber in female Trichoptera. **Det Kong. Danske Vidensk. Selsk. Biol. Skr.,** 23 : 1-200.

Remane, A. 1956. Die Grundlagen des natürlichen Systems, der vergleichenden Anatomie und der Phylogenetik. Akad. Verlags., Geest & Portig K. -G., Leipzig.

Ruppert, E.E. 1982. Homology recognition as a basis for comparative biology. **Proc. 3rd Int. Symp. Tardigrada,** ed. D.Nelson, E. Tennessee State Univ. Press, Johnson City : 31-54.

Schmid, F. 1980. Genera des Trichoptères du Canada et des Etats adjacents. Les insectes et arachnides du Canada, partie 7. Ministre des Approvisionnements et Services Canada, Hull, Québec.

Ulmer, G. 1907. Trichoptera. Genera Insectorum, 60a, ed. P.Wytsman.

Ulmer, G. 1951. Köcherfliegen (Trichopteren) von den Sunda-Inseln (Teil I). **Arch. Hydrobiol. Suppl.,** 19 : 1-528.

Ulmer, G.1955. Köcherfliegen (Trichopteren) von den Sunda-Inseln (Teil II). **Arch. Hydrobiol. Suppl.,** 21: 408-608.

Weaver, J.S. III. 1984. The evolution and classification of Trichoptera, part I : The groundplan of Trichoptera. **Proc. 4th Int. Symp. Trichoptera,** J.C. Morse (ed.), Junk, The Hague : 413-419.

Wiggins, G.B. 1977. Larvae of the North American caddisfly genera (Trichoptera). **Univ. Toronto Press**, Toronto.

Wiggins, G.B. 1984. Trichoptera. An introduction to the aquatic insects of North America, 2nd edition, R.W. Merritt and K.W.Cummins (ed.), Kendall/Hunt Pub. Co., Dubuque, Iowa : 271-311.

# PHYLOGENY OF THE *POLYCENTROPUS INSULARIS* SPECIES-GROUP (TRICHOPTERA : POLYCENTROPODIDAE)

S. W. HAMILTON

## SUMMARY

The *Polycentropus insularis* species-group comprises five Neotropical species, one occuring in the Lesser Antilles and the others variously distributed in northern South America and southern Central America. Based on homologous structures of the male terminalia, a phylogeny is proposed for these species.

## INTRODUCTION

In a recent study of Caribbean *Polycentropus sensu stricto* (Hamilton in press), the phylogenetic relationship of the *P. insularis* species-group to the other New World *Polycentropus sensu stricto* was discussed. The five species constituting the *P. insularis* group are variously distributed in northern South America, southern Central America, and the Lesser Antilles. The *P. insularis* group has been proposed as a sister group to the large, Neotropical *P. gertschi* group (Flint 1974). There are difficulties with this hypothesis owing to the questionable monophyly of the *P. gertschi* group. Although the species of both groups together form a monophyletic unit, the questioned monophyly of the *P. gertschi* group casts doubt on whether the *P. insularis* group is an ingroup or a sister group to the *P. gertschi* group (Hamilton, in press). I have shown elsewhere (Hamilton, in press) that the *P. insularis* group is a member of the South American - Caribbean track (Rosen 1976). Geologic data indicate the northern South America - Antilles component of this track must be the result of dispersal to the recently emergent Lesser Antilles (Coney 1983, Hamilton in press).

The purpose of this paper is to present a completely resolved phylogeny of the *P.. insularis* group, as well as brief diagnoses and illustrations of male terminalia for each of the five species, expanding the earlier work of Hamilton (in press).

The phylogenetic methods and terminology are those of Henning (1966), Remane (1956), and Ruppert (1982) as presented by Morse and Holzenthal elsewhere in these proceedings.

## TAXONOMIC DATA

### The *Polycentropus insularis* Group, Flint, 1974

The *P. insularis* group has been diagnosed by Flint (198O) and is supported by homologues (synapomorphies) 3-5 in the phylogenetic analysis (Fig. 6). These three homologues serve to separate the species of the *P. insularis* group from other species of *Polycentropus s. s.*

### *Polycentropus altamani* Yamamoto, 1967

Figure 1

Each preanal appendage is relatively short and broad apically . Each inferior appendage has a relatively long, broad, ventral portion with an apical point, and a dorsolateral, setose portion that is broad and well developed. The apicoventral lip of the phallobase is long and evenly curved, and the endophallic processes are united mesally, lightly sclerotized, and only slightly decurved at the apices. The subphallic support (Fig. 2f) is a large, narrow sclerite below the phallus.

Distribution : I have seen records or specimens from Panama, Costa Rica, Nicaragua, Ecuador, and Venezuela.

### *Polycentropus biappendiculatus* Flint, 1974

Figure 2

Each preanal appendage is relatively short and broadly pointed apically. The setose, dorsolateral portion of each inferior appendage is narrow and truncate apically. The ventral portion of each inferior appendage is devoid of setae laterally and the apical portion is very acute, more narrow than in *P. cuspidatus*, and has a ventral shoulder. The apicoventral lip is more strongly curved than in *P. cuspidatus* and the endophallic processes are slightly shorter. The subphallic support is smaller than in *P. altamani.*

Distribution : This species was described from west-central Suriname and has been collected recently at the base of Cerro de la Neblina, Venezuela.

*Polycentropus cuspidatus* Flint, 198O

Figure 3

Each preanal appendage is relatively elongate, with a deep posterior emargination. In this feature this species somewhat resembles *P. surinamensis.* The dorsolateral portion of each inferior appendage is narrow, elongate, thumb-like, being more rounded apically than in *P. biappendiculatus.* The ventral portion of each inferior appendage is devoid of setae laterally and narrows evenly to an acute apex. The apicoventral lip of the phallobase is more evenly curved than in *P. biappendiculatus* and the endophallic processes are longer. The subphallic support is very lightly sclerotized and poorly developed.

Distribution : This species is only known from central Ecuador.

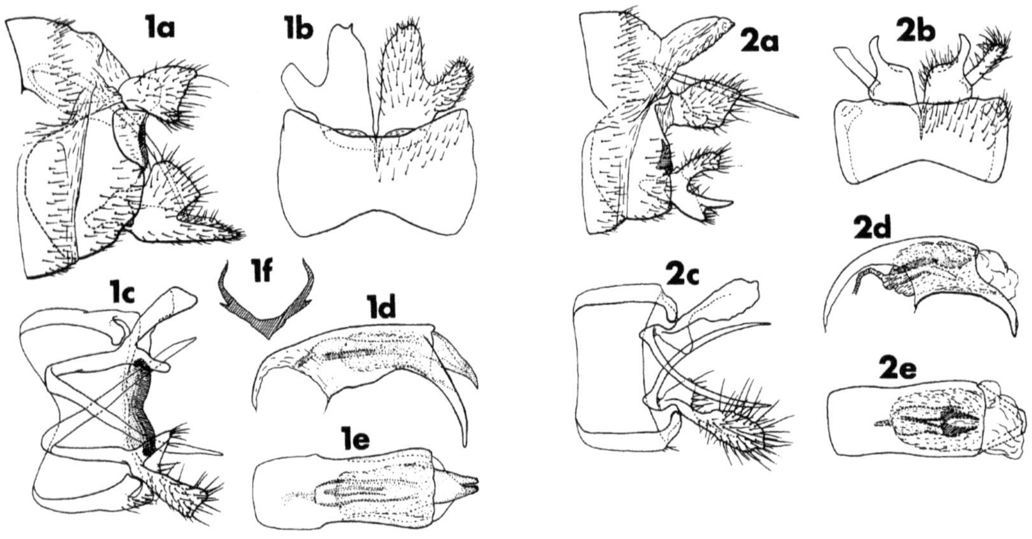


Figures 1-2. Male terminalia. 1, *Polycentropus altmani.* 2, *Polycentropus biappendiculatus.* a, left lateral ; b, ventral ; c, dorsal (terga VIII-X removed) ; d, phallus left lateral ; e, phallus, dorsal ; f, phallic support sclerite, posterior.

*Polycentropus surinamensis* Flint, 1974

Figure 4

Each preanal appendage is relatively elongate on the dorsal margin, but is not so produced on the ventral margin as in *P. cuspidatus.* The dorsolateral portion of each inferior appendage is relatively broad and developed, while the ventral portion is very short and acute laterally, and bears spine-like setae on the mesal ridge. The apicoventral lip of the phallobase is relatively straight and the endophallic processes are large and strongly upcurved at the apices. The subphallic support is a broad, well developed U-shaped sclerite.

Distribution : This species is known from the type series collected in east-central Suriname.

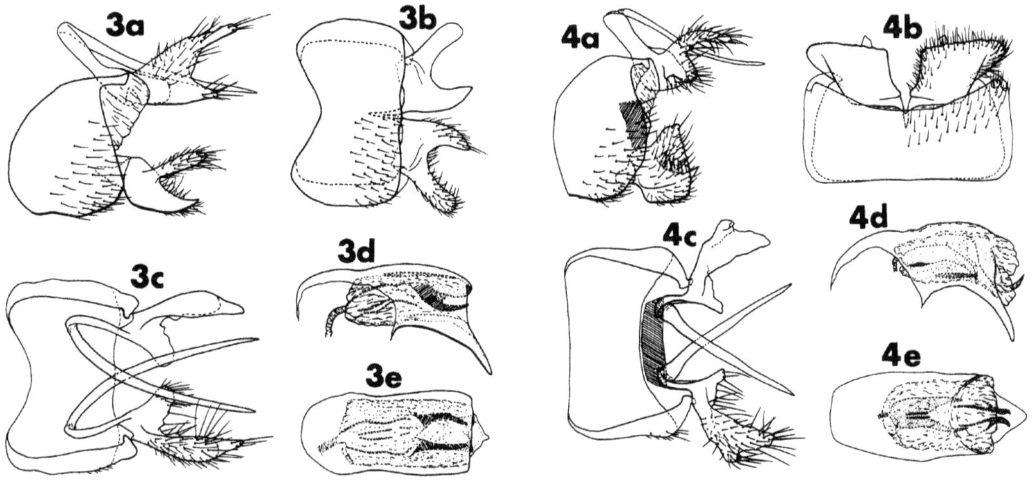


Figures 3-4. Male terminalia. 3, *Polycentropus cuspidatus*. 4, *Polycentropus surinamensis*. a, left lateral ; b, ventral ; c, dorsal (terga VIII-X removed) ; d, phallus, left lateral ; e, phallus, dorsal.

*Polycentropus insularis* Banks, 1938
Figure 5

Each preanal appendage is somewhat short, broad, and rounded apically. Each inferior appendage has the dorsolateral portion broad, similar to *P. cuspidatus*, and the ventral portion is setose and moderately elongate, but not so long as in *P. altmani* ; the ventral portion is narrow and acute in ventral aspect (Fig. 5b). The apicoventral lip of the phallobase is strongly decurved and the endophallic processes are broad and strongly curved at the base and straight apically. The subphallic support is apparently not present.

Distribution : This species was described from Grenada and has been collected recently in Dominica.

PHYLOGENETIC ANALYSIS

A phylogenetic hypothesis for the *P. insularis* group is presented in Figure 6. Monophyly of the *P. insularis* + *P. gertschi* group is established by two homologues. The (1) apicoventral lip of the phallobase is attenuated into a long, acute, ventrally directed process and (2) the narrow intermediate appendages have a distinct flexion area or articulation at or near their bases. Monophyly of the *P. gertschi* group was hypothesized by Flint (1980), but this is in doubt for reasons presented more fully eslewhere (Hamilton, in press). The *gertschi* group and other New World *Polycentropus* are used as outgroups for the phylogenetic analysis of the *P. insularis* group.

Monophyly of the *P. insularis* group is proposed based on three homologues. Each (3) intermediate appendage extends anteriorly as a long loop into the abdomen before extending posteriorly below the membranous segment X, (4) there is a pair of heavy, sclerotized endophallic processes which are dorsally positioned in the retracted phallus, and (5) each dorsolateral corner of sternite IX is produced posteriorly forming narrow pedicels on which each appendage articulates.

The (6) endophallic processes being more heavily sclerotized and curved establishes the monophyly of four of the species, exluding *P. altmani* Yamamoto. The sister-species relation ship of *P. biappendiculatus* Flint and *P. cuspidatus* Flint is establised by two homologues, (7) the dorsolateral setose portion of each inferior appendage is reduced to a narrow, digitate process and (8) the ventral portion of each inferior appendage is narrow, acute, and devoid of lateral setae.

The sister-species relationship of *P. surinamensis* and *P. insularis* is indicated by (9) the spine-like setae of the mesal ridge of each inferior appendage and (1O) the apical region of the ventral portion being reduced in length.

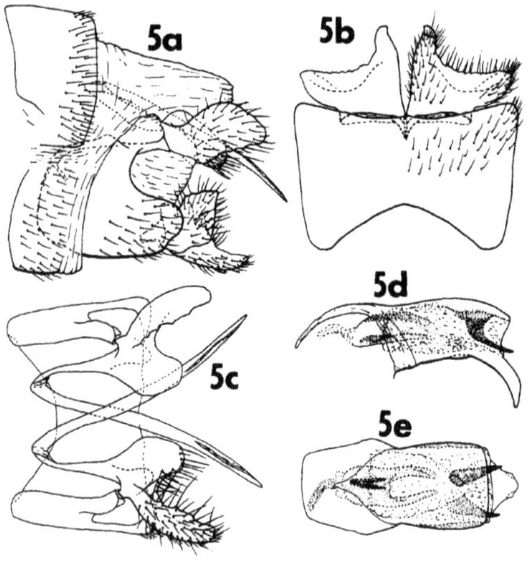


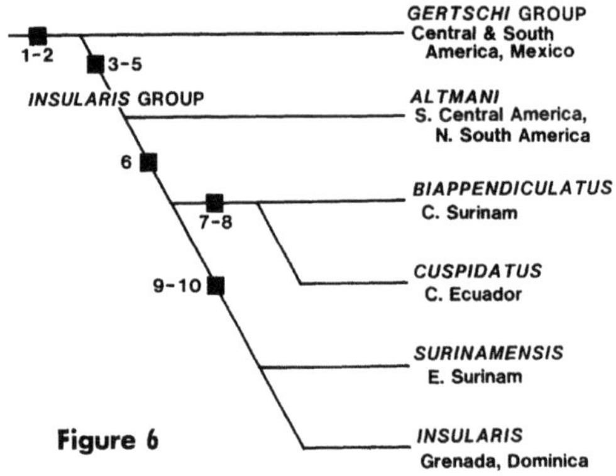


Figures 5-6. 5, male terminalia, *Polycentropus insularis.* a, left lateral ; b, ventral ; c, dorsal (terga VIII-X removed) ; d, phallus, left lateral ; e, phallus, dorsal. 6. Phylogeny of the *P. insularis* group.

ACKNOWLEDGMENTS

For reviewing the manuscript of this paper I am indebted to M. Rothschild, J. Morse, and P. Adler, Clemson University. Specimens examined for this research were borrowed through the kindness of O. Flint, Jr., Smithsonian Institution, Washington, D. C. and S. Shaw, Museum of Comparative Zoology, Harvard University.

This is Technical Contribution No. 2597 of the South Carolina Agricultural Experiment Station, Clemson University.

REFERENCES

Banks, N. 1958. New West Indian neuropteroid insects. **Rev. Entomol.**, 9 : 285-304.

Coney, P. J. 1983. Plate tectonic constraints on the biogeography of Middle America and the Caribbean Region. **Ann. Missouri Bot. Gard.**, 69 : 432-443.

Flint, O. S., Jr. 1974. Studies of Neotropical caddisflies, XV : The Trichoptera of Surinam. Studies of the Fauna of Suriname and other Guyanas, 14 : 1-155.

Flint, O. S. Jr. 1980. Studies of Neotropical caddisflies, XXIX : The genus *Polycentropus* (Trichoptera : Polycentropodidae). **J. Washington Acad. Sci.**, 70 : 148-160.

Hamilton, S. W. (in press). Historical biogeography of two groups of Caribbean *Polycentropus* (Trichoptera : Polycentropodidae). *In*, Liebherr, J. K. (ed.). Zoogeography of West Indian Insects. **Cornell Univ. Press**, Ithaca, New York.

Hennig, W. 1966. Phylogenetic systematics . Univ. Illinois Press, Urbana.

Remane, A. 1956. Die Grundlagen des natürlichen Systems, der vergleichenden Anatomie und der Phylogenetik. Akad. Verlags., Geest & Portig K.-G., Leipzig.

Rosen, D. E. 1976. A vicariance model of Caribbean biogeography. **Syst. Zool.**, 24 : 431-464.

Ruppert, E. E. 1982. Homology recognition as a basis for comparative biology. *In*, Nelson, D. (ed). **Proc. 3rd Int. Symp. Tardigrada, E. Tennessee State Univ. Press**, Johnson City : 31-54.

Yamamoto, T. 1967. New species of the caddisfly genus *Polycentropus* from Central America (Trichoptera : Polycentropodidae). **J. Kansas Entomol. Soc.**, 40 : 127-132.

# ECOLOGICAL AND EIDONOMIC TRENDS IN MEDITERRANEAN STENOPHYLACINI

H. MALICKY

SUMMARY

1. The limnephilid caddisflies of the *Stenophylax* group which had a Central European ancestor evolved in the Mediterranean Region in connection with the summer-dry conditions of streams.
2. Later several species of this group were isolated in high mountains and adapted their life cycles accordingly, and some of them became brachypterous.
3. Other species of the group extended their areas northward and lost their summer dormancy stepwise with the latitude.
4. Some isolated populations of the *Mesophylax aspersus* complex reduced their spur number from 134 to 122.
5. Some southern populations of *Potamophylax* species which are widespread in Europe changed their wing coloration independently from each other from dark brown to yellow.

According to Schmid (1957), the genera *Stenophylax, Micropterna* and *Mesophylax* form a homogenous group. *Mesophylax* stands somewhat aside, but *Stenophylax* and *Micropterna* could well be united in one genus. There is only one difference : in *Micropterna* the first tarsal segment of the male foreleg is shorter than the second ; in *Stenophylax* it is longer, but this is not always distinct. In *Micropterna muehleni* McL. 1884 the segments are equal. Thus, the two generic names are kept mainly for continuity in nomenclature.

At present we know about 35 species of the group, 12 of which are also found in Europe outside the true Mediterranean Region, and only one of them (*Micropterna lateralis* Stephens 1837) is not found in the South. It is the only monophyletic group of limnephilids whose centre of distribution and evolution is in the Mediterranean Region. The Stenophylacini are widespread in northern temperate regions, with a clear centre in the mountains of Central Europe. Only one genus (*Hydatophylax*) occurs also in North America where there are four more genera closely related to it (Schmid 1955). It is therefore very probable that the *Stenophylax* group also had its ancestor in Central Europe. They evolved in the Mediterranean Region as an adaptation to streams which dry out in summer. The larvae develop quickly during the winter under rainy conditions ; the adults emerge before the drought, and sometimes hide in caves where they mey be found in large numbers. Oviposition takes place in autumn when the streams have water again after the start of the rainy season (Bouvet 1971, 1977, Denis 1974, Malicky 1981). Some of the few species occurring in Central or Northern Europe kept this cycle also in permanent streams, but in other species the summer dormancy is lacking and has probably been lost. In other species the reduction of dormancy is correlated with the latitude (Malicky 1981).

Another striking tendancy is the occurrence of certain species in high mountains in positions marginal to the Mediterranean Region. *Mesophylax aethiopicus* Malicky 1976 lives in standing waters in high elevations (3600 m) in Ethiopia, and is the only limnephilid in Africa south of the Sahara (Malicky 1976, 1978). The adults are fully winged. The life cycle is unknown. *Micropterna baduka* Mey & Müller 1979 lives in the Caucasus at elevations of 1000-2000 m mainly in lakes, and rarely in streams. *Micropterna thaleri* Malicky 1985 is found in small brooklets in the High Atlas at 3400 m. Both species are brachypterous offspring of the widespread *M. nycterobia* McL. Their life cycles are unknown, but the fact of brachypterism means that they cannot be the same as in *M. nycterobia* in which long-distance flights are necessary for the spring emigration from breeding sites to caves and for the autumn migration in search of oviposition sites (Malicky and Winkler 1974). It is possible that other species occasionally or regularly have a similar biology, e.g. *Micropterna wageneri* Malicky 1971 which is found in high elevations in Central Italy (Moretti, pers. comm.).

Another trend is found in *Mesophylax aspersus* Rambur 1842 which is perhaps a mixture of species or subspecies. The normal spur formula of the adults is 134. In some regions, however, it appears that all specimens have a spur formula of 122. I found this in all specimens I have seen from the Hoggar Mts. in Central Sahara (36 specimens), the Pamir Mts. in Afghanistan (1), Saudi Arabia (1), and Sardinia (over 100 specimens) (Malicky in press 1, 2). The ecological and evolutionary meaning of this reduction is unknown.

*Potamophylax cingulatus* Stephens 1837 and *P. latipennis* Curtis 1834 are widespread in Central and Northern Europe and occur also in some parts of the southern peninsulas. Normally they have more or less dark brown wing pattern, but in some places in the south they are yellow. The yellow animals from Calabria were described as *P. cingulatus gambaricus* Malicky 1971. All the *cingulatus* specimens I have collected in the Sierra de Montseny near Barcelona (Catalonia, Spain) are yellow also. On the other hand, *P. latipennis* which is rare in Turkey appears to be constantly yellow in this country (Sipahiler, pers. comm.). The yellow coloration could be a parallelism with the normally pale reddish or yellowish *Stenophylax* and *Micropterna,* but an ecological meaning is not clear. In Calabria and in the Sierra de Montseny they live in cool streams in wooded areas and have apparently no summer dormancy or any kind of diapause.

ACKNOWLEDGEMENT

I thank Dr. M. I. Crichton for the correction of the English text.

REFERENCES

Bouvet, Y. 1971. La diapause des Trichoptères cavernicoles. **Bull. Soc. zool. Fr.**, 96 : 375-384.

Bouvet, Y. 1977. Conditions de vie des Trichoptères subtroglophiles (Insectes, Limnephilidae) ; leurs réactions aux variations des facteurs du milieu. Thèse, Univ. Claude Bernard Lyon, 82 p.

Denis, C. 1974. Données sur la chronologie de la maturation ovarienne et sur la diapause de *Micropterna sequax* McLachlan (Trichoptera, Limnephilidae). **Bull. Soc. Sci. Bretagne**, 49 : 125-129.

Malicky, H. & Winkler, G. 1974. Untersuchungen über die Höhlenimmigration von *Micropterna nycterobia* (Trichoptera, Limnephilidae). **Oecologia (Berl.)**, 15 : 375-382.

Malicky, H. 1976. Ein neuer *Mesophylax* (Trichoptera, Limnephilidae) aus Äthiopien. **Ent. Z. (Stuttg.)**, 86 : 43-45.

Malicky, H. 1978. Das Männchen von *Mesophylax aethiopicus* Mal. (Trichoptera, Limnephilidae). **Sitzber. Öst. Akad. Wiss., Math. natw. Kl.**, Abt. I, 186 : 279-284.

Malicky, H. 1981. The phenology of dispersal of several caddisfly (Trichoptera) species in the island of Crete. - **Proc. 3rd Int. Symp. Trichoptera**, G.P. Moretti (ed.), Junk, The Hague, : 157-163.

Malicky, H. 1985. Neue Beiträge über mediterrane *Micropterna*-Arten (Trichoptera, Limnephilidae). **Mitt. Ent. Ges. Basel**, 35 : 27-35.

Malicky, H. (in press). 1). The caddisflies of Saudi Arabia and adjacent regions. - Fauna of Saudi Arabia.

Malicky, H. (in press) 2) Die Köcherfliegen (Trichoptera) des Iran und Afghanistans. **Z. Arbgem. Öst. Ent.**

Mey, W. & Müller, A. 1979. Neue Köcherfliegen aus dem Kaukasus. **Reichenbachia** 17 : 175-182.

Schmid, F. 1955. Contribution à l'étude des Limnophilidae (Trichoptera) **Mitt. Schweiz. Ent. Ges. Beih.**, 28 : 1-245.

Schmid, F. 1957. Les genres *Stenophylax* Kol., *Micropterna* St. et *Mesophylax* McL. (Trichopt. Limnoph.). **Trab. Mus. Zool. Barcelona, N.S. Zool.**, 2(2) : 1-51.

# 6. BIOLOGY AND PHYSIOLOGY

# EFFET DE LA PHOTOPERIODE SUR LA MATURATION GENITALE DE FEMELLES DE *LIMNEPHILUS RHOMBICUS* D'EUROPE CENTRALE

C. DENIS


## SUMMARY

The central European population under study was influenced by photoperiod like other populations of this species. Long photophases momentanly arrest female genital development. However a given summer photoperiod induces in central European females a shorter genital rest than in other conspecific females.


## INTRODUCTION

Plusieurs espèces de Limnéphilidés, dont *Limnephilus rhombicus*, pondent en automne bien que les adultes apparaissent au printemps, principalement en Mai et en Juin. A l'émergence les femelles sont immatures et le demeurent pendant la période estivale. Novak et Sehnal (1963, 1965) ont, les premiers, mentionné que les femelles de certaines espèces de Trichoptères présentent ainsi un repos génital et montré que ceci est dû à l'action des longues photophases d'été. Cependant les mêmes auteurs ont constaté que dans les Hauts Tatras des femelles de *L. rhombicus* avaient des oeufs mûrs dès la fin Juillet, ce qui semble impliquer dans leur cas un développement génital ininterrompu en été. Certaines observations ont conduit le Dr. Malicky à penser que la population de *L. rhombicus* de Lunz am See (Autriche) ne présentait pas, elle, non plus, de repos génital en été.

Le Dr. Malichy m'ayant aimablement adressé des pontes de *L. rhombicus* provenant de Lunz am See, j'ai cherché à déterminer si cette population possédait des caractéristiques biologiques particulières. Pour cela j'ai effectué deux séries parallèles d'expériences, l'une sur des femelles originaires de Lunz am See et l'autre sur des femelles de la Région de Rennes (Bretagne) qui, elles, présentent un net repos génital en été (Denis 1973 et 1981).

## MATERIEL ET METHODES

La population de Lunz am See utilisée dans ce travail est issue de deux pontes récoltées au début de Septembre 1984. Les expériences ont été réalisées sur les animaux provenant de ces deux pontes et sur leurs descendants de première génération. La population de Bretagne provient de pontes récoltées au début d'Octobre dans l'étang des Maffrais près de Rennes.

Dans la nature, le développement embryonnaire et larvaire de *L. rhombicus* s'effectue entre l'automne et le printemps suivant, à des photophases comprises entre 8 et 14 heures environ. Lorsque les femelles deviennent adultes, leurs ovocytes sont en prévitellogenèse. Si elles sont maintenues artificiellement à une photophase assez courte, elles entrent assez rapidement en vitellogenèse. Mais dans la nature elles sont exposées à une photopériode de type estival et la vitellogenèse est retardée. De plus le délai entre l'émergence et le début de la vitellogenèse est proportionnel à la durée de la photophase.

Dans le but de mettre en évidence d'éventuelles différences entre les femelles de Lunz am See et celles de Rennes, j'ai exposé des adultes des deux populations à différentes photopériodes. Les animaux utilisés dans les expériences ont été maintenus à la température de 15°C pendant tout leur cycle biologique. Les larves se sont développées à une photopériode de type hivernal (photophase de 12 heures). Les adultes ont été répartis en lots soumis respectivement aux photophase suivantes : 12, 16 et 18 heures. Chaque lot comprenait de 30 à 40 femelles accompagnées de mâles. Pour chaque lot j'ai cherché à déterminer le délai entre l'émergence et le début de la vitellogenèse (cette dernière est repérable par la dissection ou par l'apparition des premiers accouplements), ainsi que le délai entre l'émergence et l'oviposition.

RESULTATS

**Délai entre l'émergence et le début de la vitellogenèse**

Les résultats des expériences sont présentés dans le tableau 1.

Tableau 1 : Délai entre l'émergence et le début de la vitellogenèse pour des femelles de *Limnephilus rhombicus* originaires de Lunz am See (Autriche) et de Rennes (Bretagne) à une température de 15°C et à différentes photophases.

| Larves | Adultes | Lunz am See | Rennes |
|---|---|---|---|
| 12h | 12h | 2 à 3 semaines | 3 à 4 semaines |
| | 16h | 1 à 2 mois | 2,5 à 3,5 mois |
| | 18h | 1,5 à 2,5 mois | 3 à 4 mois |

Le tableau 1 montre qu'à une photophase donnée le délai entre l'émergence et le début de la vitellogenèse est toujours plus court pour les femelles de Lunz am See que pour celles de Rennes. Mais le tableau 1 met également en évidence une augmentation de ce délai en fonction de la durée de la photophase pour les femelles des deux populations. Donc les photopériodes estivales induisent chez les femelles de Lunz am See un repos génital comme chez celles de Rennes, mais pendant moins longtemps.

Les résultats concernant la population de Lunz am See permettent de comprendre pourquoi Novak et Sehnal (1963, 1965) ont capturé des femelles mûres dès le mois de Juillet. Des femelles émergeant en Mai se trouvent exposées à une photophase qui croît jusqu'à environ 16 heures. Dans ces conditions la vitellogenèse peut commencer un mois plus tard, c'est-à-dire en Juin. De très rares femelles (4 en tout) sont d'ailleurs entrées en vitellogenèse après des délais inférieurs à ceux indiqués dans le tableau 1. La vitellogenèse dure de 2 à 3 semaines (Le Lannic, 1975, 1976). Certaines femelles d'Europe Centrale peuvent donc avoir des oeufs mûrs en Juillet.

**Délai entre l'émergence et l'oviposition**

Les résultats des expériences sont présentés dans le tableau 2.

Tableau 2 : Délai entre l'émergence et l'oviposition pour des femelles de *Limnephilus rhombicus* originaires de Lunz am See (Autriche) et de Rennes (Bretagne) à une température de 15°C et à différentes photophases.

| Larves | Adultes | Lunz am See | Rennes |
|---|---|---|---|
| 12h | 12h | 1 à 1,5 mois | 1,5 à 2 mois |
| | 16h | 2,5 à 3,5 mois | 4 à 5 mois |
| | 18h | 3 à 4 mois | 5 à 6 mois |

Les données présentées dans le tableau 2 montrent que :

- A une photophase donnée les femelles de Lunz am See pondent plus tôt que celles de Rennes. En effet, les premières pondent en Septembre (certaines même un peu avant), comme celles de Tchécoslovaquie (Novak et Sehnal, 1963), tandis que les secondes ne pondent qu'en Octobre.
- Lorsque les femelles sont maintenues à une photophase de 12 heures elles pondent toutes en une quinzaine de jours, tandis que sous une photophase estivale la période de ponte s'étend sur un mois au moins. Il est vraisemblable que la diminution de la durée de la photophase à la fin de l'été et en automne a notamment

pour effet de raccourcir la période de ponte. Dans la région de Rennes les femelles de *L. rhombicus* pondent pendant la première moitié d'Octobre. Par ailleurs, Novak et Sehnal (1963) précisent que les masses ovigères de cette espèce sont toutes déposées durant la première moitié de Septembre.

Sachant que la vitellogenèse dure de 2 à 3 semaines, la comparaison entre les données du tableau 1 et celles du tableau 2 montre que les femelles maintenues à la photophase de 12 heures pondent dès que leurs oeufs sont mûrs, tandis que celles qui sont soumises à une photophase estivale retiennent leurs oeufs pendant un certain temps. Ce fait s'observe aisément en élevage. Il est donc probable que les femelles trouvées mûres en juillet dans les Hauts Tatras ne pondent qu'au début de Septembre comme les autres femelles d'Europe Centrale.

## CONCLUSION

Cette étude montre que les deux populations étudiées sont l'une et l'autre sensibles à l'action de la photopériode. Dans les deux cas les photophases estivales provoquent un arrêt temporaire du développement génital des femelles. Mais une photophase estivale donnée induit chez les femelles de Lunz am See un repos génital moins long que chez celles de Rennes. Les différences quantitatives ainsi mises en évidence sont vraisemblablement dûes à des différences génétiques entre les populations de Bretagne et d'Europe centrale.

## REMERCIEMENTS

Le Dr Malicky (Biologische Station, Lunz am See, Autriche) a eu la gentillesse de m'adresser des pontes provenant d'Europe Centrale grâce auxquelles ce travail a pu être réalisé. Je lui en suis très reconnaissant.

Je tiens également à remercier Madame Rault et Monsieur Dedelot pour l'aide technique qu'ils m'ont apportée.

## REFERENCES

Denis, C. 1973. Obtention d'un cycle biologique sans diapause chez *Limnephilus rhombicus* et *Anabolia nervosa* (Trichoptera, Limnephilidae). **Bull. Soc. Sci. Bretagne** 48 : 197-207.

Denis, C. 1981. Action de la photopériode sur la maturation génitale des femelles de quelques Limnéphilidés. **Proc. 3rd Int. Symp. Trichoptera,** G.P. Moretti (ed.), Junk, The Hague : 57-66.

Le Lannic, J. 1975. Contribution à l'étude du développement et de la maturation de l'appareil reproducteur de *Limnephilus rhombicus* L. **Bull. Soc. Zool. Fr.,** 100(4) : 539-550.

Le Lannic, J. 1976. Développement de l'appareil reproducteur de quelques Trichoptères Limnéphilidés et premières données expérimentales sur son fonctionnement. Thèse 3ème cycle, Rennes, 117 p.

Novak, K & Sehnal, F. 1963. The development cycle of some species of the genus *Limnephilus* (Trichoptera). **Cas. csl. spol. ent.,** 60 : 68-80.

Novak, K & Sehnal, F. 1965. Imaginal diapause bei den in periodischen gewässern lebenden Trichopteren. **Proc.12. Int. Congr. Ent. London,** 434.

# MALE MATE RECOGNITION IN LEPTOCERIDAE

E. PETERSSON and J.O. SOLEM

## SUMMARY

Field studies were carried out on male swarms of *Mystacides azurea* L. and *Athripsodes cinereus* Curtis (Leptoceridae) at Lake Erken in south Sweden. It was found that : (1) A dummy conducted in the swarm attracted the swarming males, which flew up to it for a short investigation. (2) The males were more attracted to a dummy pendulating over or in the swarm (resembling an approaching receptive female) than to any other type of movement. (3) A dummy infused with female extract (dichloromethane) placed just above the swarm did not attract the males more than the control. These experiments support the hypothesis that the male mate recognition in leptocerids mainly is visual. A sexual dimorphism in eye proportions was also found, which significance is briefly discussed.

## INTRODUCTION

More than half of the Scandinavian species of Leptoceridae males have been observed to aggregate in swarms, usually over the water surface or the reed belt. Pair formations in the swarms have been reported by several authors (Wesenberg-Lund 1943, Hickin 1953, Solem 1984). In southern localities (e.g. south Sweden), swarms of leptocerids occur generally at dawn and dusk , but in montane and high latitude localities, the species are more likely to perform swarming behaviour at any time of the day, when conditions like wind and temperature are favourable (c.f. Solem 1978).

A conspicuous feature in *Mystacides* is that males have distinctly larger eyes than females. This sexual difference is similar to what is well known among many species of Ephemeroptera. The eye dimorphism, and that swarming does not occur in the dark, made us hypothesize that male mate recognition in North European Leptoceridae mainly is visually based.

The purpose of this study was to test this hypothesis and to quantify the eye-proportions of both sexes, as well as to investigate whether pheromones are involved in mate recognition.

## MATERIAL AND METHODS

The field studies were carried out at an oligotrophic lake (Lake Erken) in south Sweden. To investigate the visual perception of swarming males of *Mystacides azurea* L. and *Athripsodes cinereus* Curtis, dummies were presented to the males. The dummies (dead mayflies, dead leptocerid females, rubber septa, etc.) were fastened to a line on a fishing rod and were conveyed in different manners in and over the swarms (Fig. 1). Numbers of males flying to the dummy during 1 min were counted.

Five females of *M. azurea* were caught when flying tandem with a male. The females were placed in 0.5 - 1.0 ml dichloromethane for 1.5 h extraction. The tissue was removed and the extracts were infused to rubber septa (F-138 Septa from Supelco Inc.) . The rubber septa were fastened to a line on a fishing rod and suspended 3 - 5 cm above the water surface (i. e. type A in Fig. 1) at the swarming site. Males which inspected the rubber septa were video recorded. Afterwards their time spent at the rubber septa were measured to the nearest 0.03 s.

The eye-proportions were measured using the method of Malicky (1978), 13 males and 14 females of *M. azurea* were measured. The corresponding numbers for *A. cinereus* were 12 and 11, and for *Rhyacophila nubila* Zetterstedt (Rhyacophilidae) 16 and 17. Identification was made by use of Macan (1973) and Malicky (1984).

## RESULTS

The various types of movements of the dummy (Fig. 1) were differently attractive to the males. A higher number of males inspected the dummy when it was moved in the horizontal level (type C) than when it was moved up and down (type B), or kept quiet (type A) (Tab. 1). We could not detect any difference in the attraction of males of *M. azurea* and *A. cinereus* when using a dead mayfly (*Ephemera danica*), dead females of the studied species, or circular rubber septa. *M. azurea* was more easily attracted than *A. cinereus*.

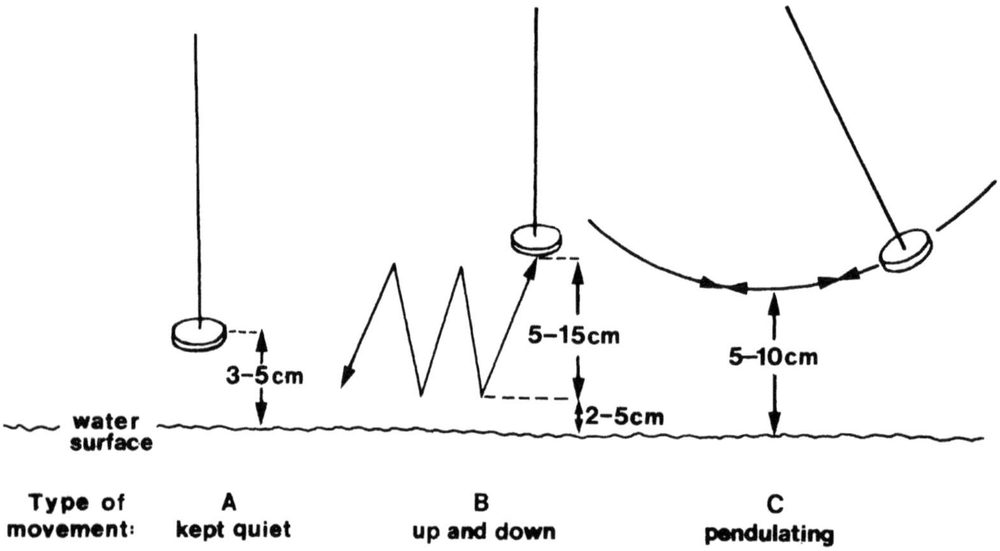


Figure 1. The different types of movements of the dummies.

Table 1. No. males per min of *M. azurea* and *A. cinereus* inspecting the dummy performing the A, B and C type of movement. See Fig. 1 for further explanation. Statistical analysis : Kruskal-Wallis test.

| species | type of movement | $\bar{x} \pm$ S.D. | N | |
|---|---|---|---|---|
| | A | 2.33 ± 1.63 | 6 | |
| *M. azurea* | B | 1.71 ± 1.25 | 7 | H = 15.13, p < 0.001 |
| | C | 5.64 ± 2.40 | 14 | |
| | A | 0.13 ± 0.35 | 8 | |
| *A. cinereus* | B | 0 | 8 | H = 6.42, p < 0.05 |
| | C | 0.67 ± 0.71 | 8 | |

Males of *M. azurea* showed no difference in "inspection time" between a rubber septum infused with female extract and the control (Tab. 2). The mean distance between the rubber septum and the head of the animal was 0.75 ± 0.12 cm (N = 7), i.e. about half the length of the male antenna (1.70 ± 0.25 cm, N=8).

The studied leptocerid species were found to have a sexual dimorphism in eye-proportions. In both species the males had larger eyes than the females (Fig. 2a and 2b). The dimorphism was more expressed in *M. azurea* than in *A. cinereus*. *R. nubila* showed no such dimorphism (Fig. 2c).

DISCUSSION

It is obvious from the experimental data that different patterns of movements initiate reactions among the males. Males are able to discriminate between various types of movements. The significance of this is that males can discriminate between females that are receptive for mating and those that are not. Females receptive for mating may show this by flying in the horizontal level and will be intercepted more often by males, while females showing other types of flight movements, e.g. up and down movements as when egg-laying, will more rarely be intercepted. Also, the patterns of movements will discriminate between females and males entering the swarms of *M. azurea*, because males are mostly doing up and down movements.

Table 2. "Inspection time" (s) for males of *M. azurea* at a rubber septum infused with female extract and at a control.

| | $\bar{x} \pm$ S.D. | N | |
|---|---|---|---|
| Female extract | 0.16 ± 0.15 | 10 | $p < 0.129$ (Mann-Whitney U-test) |
| Control | 0.20 ± 0.11 | 14 | |

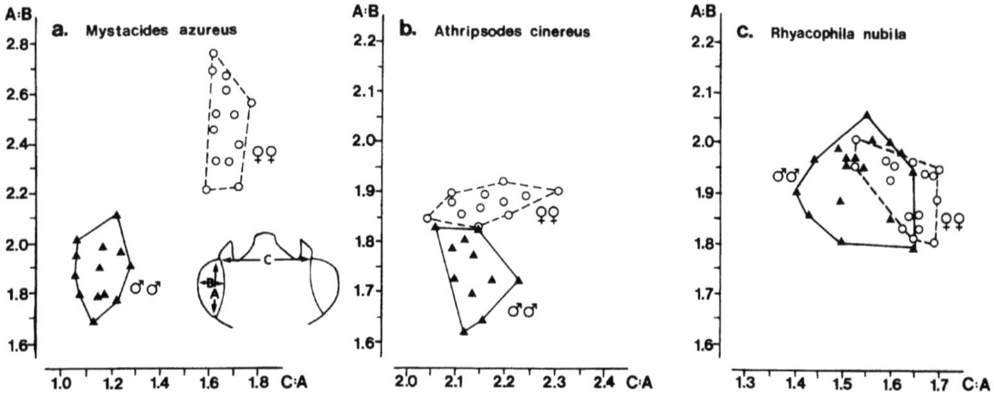


Figure 2. Sexual dimorphism in eye-proportions in *M. azurea, A. cinereus* and *R. nubila*. The scale are not the same in the diagrams. The outline drawing in the figure is in dorsal view.

Judging from the flight patterns of males and females, we conclude that long distance female sex pheromones are not involved in Leptoceridae. However, it might be possible that short distance pheromones were involved when the female approached the swarming males. If short distance female sex pheromones were present, the males should be expected to examine the extract-dummy more carefully than the control, but they did not, which suggests that short distance female sex pheromones do not exist in *M. azurea*.

The sexual eye-dimorphism in *M. azurea* and *A. cinereus* might be seen as an adaptation for mate finding (Thornhill and Alcock 1983). Males apparently compete for arriving females and male mating success then is, at least partially, correlated to the detection ability. *R. nubila* which have been shown to have sex pheromones (Solem 1985), showed no sexual eye-dimorphism. This indicates that the mate finding strategies in Leptoceridae and Rhyacophilidae are differently based.

Sullivan (1980) suggested that orientation in insect swarms mainly is visually based. This seems true even for swarming leptocerids, since all data in this study indicate visual perception and none that chemical cues were involved. Since the distance between the dummy and the head of the animals was less than the length of the male antenna, the antenna might be used as a tactile receptor, which was suggested by Tozer *et al.* (1981) for the North American leptocerid *Nectopsyche albida* (Walker). The male mate recognition might then be divided into two steps : (1) Visual perception ; the males inspect any object which have a "proper" size and behave in a "proper" way, and (2) the final and definitive identification of the object : this step is presumably not only visually based, but chemical, tactile and behavioural cues might be involved as well.

In which way the females of *M. azurea* and *A. cinereus* are attracted to the swarming males, (the far distant attraction), is not known, but even here a visual orientation may be included. If the swarm itself, and not the swarming site, is the main attraction for the females, male swarms can presumably attract females more easily than can single males. The swarms will also serve as a species-specific mating aggregation which prevents interspecific mating.

## REFERENCES


Esben-Petersen, P. 1916. Vaarfluer. Danmarks fauna, no 19. G.E.C.. Gads Forlag, Copenhagen.

Hickin, N.E. 1953. *Mystacides nigra* L. Courtship of a Caddis Fly. **Entomol. mon. Mag.**, 89 : 71.

Macan, T.T. 1973. A key to the adults of the British Trichoptera. **Freshwater Biol. Ass. Sc. Publ.**, 28 : 1-151.

Malicky, H. 1978. The taxonomic significance of eye proportions in adult caddisflies. **Proc. 2nd Int. Symp. Trichoptera**, M. I. Crichton (ed.), Junk, The Hague : 166-157.

Malicky, H. 1983. Atlas of European Trichoptera. Junk, The Hague, 298 p.

Solem, J.O. 1978. Swarming and habitat segregation in the family Leptoceridae (Trichoptera). **Norw. J. Ent.**, 25 : 145-148.

Solem, J.O. 1984. Adult behavior of North European caddisflies. **Proc. 4 th Int. Symp. Trichoptera**, J.C. Morse (ed.), Junk, The Hague : 375-382.

Solem, J.O. 1985. Female sex pheromones in *Rhyacophila nubila* (Zetterstedt) (Trichoptera, Rhyacophilidae) and arrival pattern to sticky traps. **Fauna norv. Ser. B**, 32 : 80-82.

Sullivan, R.T. 1980. Insect swarming and mating. **Florida Ent.**, 64(1) : 44-65.

Thornhill, R. & Alcock, J. 1983. The evolution of insect mating systems. Harvard University Press.

Tozer, W.E., Resh, V.H. & Solem, J.O. 1981. Bionomics and adult behavior of a lentic caddisfly *Nectopsyche albida* (Walker). **Am. Midl. Nat.**, 106 : 133-144.

Wesenberg-Lund, C. 1943. Biologie der Süsswasserinsekten. Gyldenalske Boghandel, Copenhagen.

# TECHNIQUES FOR DEMONSTRATING SEX PHEROMONES IN TRICHOPTERA

V. H. RESH, J. K. JACKSON and J. R. WOOD

## SUMMARY

Sex pheromones have recently been demonstrated for several species of Trichoptera. This paper presents an overview of techniques that have been used for studying long-range sex attractants of adult caddisflies. Experimental techniques for demonstrating the presence of sex pheromones, identifying the site of pheromone production, and analyzing and interpreting data from field bioassays are presented. Experimental design, adult flight behavior(s), environmental factors, and population phenology may confound interpretation of field data. Sex pheromones are useful in elucidating systematic, ecological, and behavioral relationships among the Trichoptera.

## INTRODUCTION

The presence of sex pheromones in Trichoptera was experimentally demonstrated by Wood and Resh (1984), although earlier anecdotal evidence had indicated that caddisflies use some type of chemically mediated communication (e.g. Kelner-Pillault 1975, Anderson 1978). Sex pheromones have now been demonstrated for caddisflies in several families : Rhyacophilidae [*Rhyacophila nubila* (Zetterstedt) (Solem 1985)] ; Glossosomatidae [*Agapetus marlo* Milne (Resh, unpublished data)] ; Limnephilidae [*Dicosmoecus gilvipes* (Hagen) (Resh and Wood 1985)] ; Molannidae [*Molanna angustata* Curtis (Solem, unpublished data)] ; Sericostomatidae [*Gumaga griseola* McL. and *G. nigricula* McL. (Wood and Resh 1984)] ; and presumably (see below) Helicopsychidae [*Helicopsyche borealis* (Hagen) (Resh *et al.* 1984)] . The importance of sex pheromone research in the study of systematic, ecological , and behavioral relationships has been well documented for several insect orders (Bell and Cardé 1984). Since several other researchers are now beginning to examine sex pheromones in caddisflies, this paper presents an overview of techniques that we have found useful in studying long-range sex attractants of species in this order. For a detailed treatment of techniques in pheromone research as applied to other insect groups, see Hummel and Muller (1984).

## DEMONSTRATION OF SEX PHEROMONES

We have used the following experimental treatments in demonstrating the presence of sex pheromones in caddisflies : live caged males, live caged females, chemical extracts of males, chemical extracts of females, and controls. For live organism treatments, pupae have been collected from study sites, reared in the laboratory, and these adults have been used in experimentation. Resh and Solem (unpublished data) found that sweep net-collected, light trap-collected, and laboratory-reared specimens yielded similar test responses.

In conducting pheromones tests, males and females are placed separately into cylindrical plastic, screened cages (made from 30 X 40 mm film containers) and then added to pheromone traps. We have used commercially available sticky traps in all tests done in California (Pherocon 1C, manufactured by Zoecon Corporation, Palo Alto, California ), but Solem (1985, and personal comm.) has used sticky traps made from flat pieces of cardboard (10 cm x 30 cm) covered with white paper and an adhesive. Since birds were attracted to the insects caught in these traps, Solem surrounded the sticky traps with a pillow cage of chicken wire that had a mesh coarse enough for insects to enter but fine enough to exclude birds.

The definitive proof of the presence of pheromones is attraction to crude extracts of live specimens. Extracts of males and females have been prepared by placing four to five specimens of the same sex in a glass vial with from 2 ml (for small species such as *G. nigricula*) to 5 ml (for larger species such as *D. gilvipes*) of dichloromethane. After one hour, the bodies are removed and discarded . This extract is then added to the dispensing agent (see below) in the pheromone trap. Other investigators that have studied exocrine secretions of caddisflies (e.g. Duffield et al. 1977, Duffield 1981) have obtained extracts by briefly dipping the specimen into the solvent. We compared the rate of male captures using extracts of female *G. griseola* obtained by soaking whole bodies (1hour exposure) with extracts obtained by dipping whole bodies (2 second exposure). Field tests for this comparison were done in Las Trampas Creek (Contra Costa Co., California, USA) on June 16-17, 1986. Six times as many caddisflies were collected with the whole body extracts as with whole body dips.

When preparing the extract treatment, the choice of extracting agent and the extract dispensing device are important considerations. The success of the extract treatment depends on the solubility of the pheromone in the solvent. In all tests thus far published (i.e. Wood and Resh 1984, Resh and Wood 1985, Solem 1985), dichloromethane has been used as the solvent. In earlier studies of caddisfly exocrine glands, Duffield *et al*. (1977) and Duffield (1981) used *n* -pentane and dichloromethane, respectively. Benzene, diethyl ether, ethanol, and hexane have also been used as solvents in studies of insect pheromones (Golub and Weatherston 1984). We have found three different pheromone dispensing devices to be effective : rubber septa, cotten balls, and foam stoppers (e.g. Indenti-plug manufactured by Gaymar Industries, Buffalo, NY). Although we have obtained similar results with all three devices, we have found foam stoppers to be the most convenient to use.

In setting live caged adults and extracts in pheromone traps, we have made several observations that may be useful to others considering such studies. We have found that extract poured onto a foam stopper at dusk will be effective through the night, even to the point of attracting not only dawn active males (e.g. *Gumaga* spp.) but sometimes attracting males over subsequent days. In conducting the field bioassay, the traps comprising the different treatments should be randomly assigned to the different trap locations. For statistical purposes, we have found that no fewer than five replicates of each treatment and the control should be run. Controls for live organism treatments should contain an empty cage ; in extract controls, traps should contain the solvent and the dispensing device. In our experiments, traps have been placed in riparian vegetation, preferably within 10-15 m of the stream edge. We placed traps at least 10 m apart to avoid interactions among treatments. We generally place them at dusk and retrieve them the following morning. At this time, the number and sex of all caddisflies collected in the "stickum" of each pheromone trap is recorded. Experiments using *D. gilvipes* were allowed to run at least 2 days, after it was discovered that transporting the females to the study site inhibited pheromone release for 1 day. *G. nigricula* and *G. griseola* are not adversely affected by transportation.

## DATA ANALYSIS AND INTERPRETATION

We have compared the results of different treatments using standard statistical procedures : transformations followed by t-test or analysis of variance and Dunnett's test (e.g. Wood and Resh 1984). Other approaches are equally valid within the limits of the experimental design.

Several factors may complicate data analysis and interpretation. Random captures of adults on control traps commonly occur (e.g. Wood and Resh 1984, Table 1 ; Solem 1985, Table 1). If the random capture is a female, and she continues to release sex pheromones, the control trap then becomes an unplanned additional treatment, in that males may be attracted by her pheromone. If the random capture is a male, distinguishing between treatments and controls may become difficult. For example, Resh *et al.* (1984) attempted to demonstrate whether *H. borealis* uses sex pheromones. Females of this species have calling positions that are similar to those of Lepidoptera species that release sex pheromones (Resh *et al.* 1984), and similar movements have been found in other caddisfly families as well (e.g. Resh and Houp 1986). However, field trials have been unable to demonstrate consistently that statistically higher numbers of *H. borealis* males are captured at caged female and female extract treatments than are captured at control treatments. In pheromone tests with *G. nigricula*, high numbers of *H. borealis* are routinely captured in both control and treatment traps (range = 0 to 11 individuals per trap). In part, we believe this results from some aspects of the flight behavior of *H. borealis* that causes high numbers to appear in the control traps.

Several factors are associated with the release of and response to sex pheromones, including the location of the individual releasing and the individual(s) receiving the pheromone concentration, and visual cues to which the receiver may also respond. The importance of these factors has been well demonstrated ; the shape, color, and location of traps in the habitat have been found to strongly influence captures (Cardé and Elkinton 1984). This reflects the significance of interactions between the releaser and the receiver. Although pheromone attraction has been found to increase with increasing pheromone concentration, extremely high concentrations can repel insects (Cardé and Baker 1984, Cardé and Elkinton 1984). Two other factors can interfere with the experiment : meteorological (e.g. wind, temperature) or phenological (e.g. end of flight season) conditions. Thus, a lack of captures in a pheromone study does not mean that pheromones are not present in that population.

## SITE OF PHEROMONE PRODUCTION

Resh and Wood (1985) recently examined the site of sex pheromone production in three species of caddisflies and reported that the female fifth abdominal sternite contains a substance that attracts conspecific males and is most likely produced by the paired exocrine glands located on that segment. The site of sex pheromone production was examined using extracts obtained from different body parts. First, head, thorax, and abdomen were dissected for 25 females ; five replicates, each containing parts of five individuals were prepared. The resulting biossay (from Resh and Wood 1985) indicated that only the abdomen attracted males. Following this, other females were dissected and each abdominal segment was used as a treatment. When only abdominal segment 5 attracted males, the next biossay consisted of examining the fifth abdominal tergit compared to the fifth sternite. Since the fifth abdominal sternite attracted males, the final biossay consisted of the fifth abdominal sternite as one treatment and the remainder of the body (i.e. the whole body minus fifth abdominal sternite) as the other treatment. This last bioassay confirmed the results of the three previous bioassays. In contrast to our studies with *D. gilvipes*, *G. griseola*, and *G. nigricula*, Solem (1985) found that with *R. nubila*, extracts of the fourth sternite from females attracted males but that no glands were visible on the fourth segment. Thus far, only one of the large number of presumed pheromone organs that have been described in caddisflies (see Wood and Resh 1984 for a partial review) has been positively associated with reproductive behavior (Resh and Wood 1985). Additional experimental evidence is needed to establish the precise function (e.g. pheromone, kairomone, allomone ; Birch and Haynes 1982) of other exocrine glands in Trichoptera.

Two additional considerations should also be noted. In doing the individual body part bioassays mentioned above, we always ran an additional treatment : a whole body extract. This treatment provided us with a way of knowing whether males were responding to pheromones the test night ; this is obviously useful in interpreting zero counts. Second, in dissecting body parts, especially sternites from tergites, we found that using different forceps for each body part avoided contamination.

## DIRECTIONS FOR CURRENT AND FUTURE RESEARCH

Resh and Wood (1985) have discussed how pheromone research may provide important systematic and ecological information about the insect order Trichoptera, the relationships among the families in this order, and interordinal relationships between the Trichoptera and Lepidoptera. We are currently conducting research on caddisfly pheromones involving the chemical isolation and identification of the sex pheromones of *D. gilvipes*, diel studies of pheromone release and response in *D. gilvipes* and *Gumaga* spp., and an examination of the role that pheromones have in mate selection. A multitude of other questions involving caddisfly pheromones remain unanswered, such as the presence of sex pheromones in other families of caddisflies, the anatomy and physiology of pheromone-producing glands in females and pheromone receptors in males, the presence of aphrodisiac (i.e. short distance) pheromones in caddiflies, and the precise nature of the pheromone releasing behavior in females. Presumably the Trichoptera possess a rich diversity of reproductive behaviors. However, so little research has been done on adult behavior that a recent comprehensive review of insect mating systems (Thornhill and Alcock 1983) contained only four references to caddisflies.

## ACKNOWLEDGEMENTS

We thank J. O. Solem for permission to cite unpublished data. The research leading to this report was supported by the office of Water Research and Technology, U.S.D.I., project No. A-063-CAL, and the University of California Water Resources Center, Project UCAL-WRC-W-519.

## REFERENCES

Anderson, N.H. 1978. Continuous rearing of the limnephilid caddisfly, *Clistoronia magnifica* (Banks).*In* M.I. Crichton (ed), **Proc. 2nd Int. Symp. Trichoptera.**, The Hague, Netherlands, Junk : 317-329.

Bell, W.J. & Cardé, R.T. (eds). 1984. Chemical ecology of insects. Sunderland, MA, Sinauer Associates, Inc., 524p.

Birch, M.C. & Haynes, K.F. 1982. Insect pheromones. London, Edward Arnold Ltd., 60p.

Cardé, R.T. & Baker, T.C. 1984. Sexual communication with pheromones. pp. 355-383. *In* W.J. Bell and R.T. Cardé (eds). Chemical ecology of insects, Sunderland, MA, Sinauer Associates, Inc., 524p.

Cardé, R.T. & Elkinton, J.S. 1984. Field trapping with attractants : methods and interpretation. pp. 111-129. *In* H.E. Hummel and T.A. Miller (eds). Techniques in pheromone research, New York : Springer-Verlag, Inc. 464p.

Duffield, R.M. 1981. 2-Nonanol in the exocrine secretion of the nearctic caddisfly, *Rhyacophila fuscula* (Walker) (Rhyacophilidae : Trichoptera). **Proc. ent. Soc. Wash.,** 83 : 60-63.

Duffield, R.M., Blum, M.S., Wallace, J.B., Lloyd, H.A. & Regnier, F.E. 1977. Chemistry of the defensive secretion of the caddisfly *Pycnopsyche scabripennis* (Trichoptera : Limnephilidae). **J. chem. Ecol.,** 3 : 649-656.

Golub, M.A. & Weatherston, I. 1984. Techniques for extracting and collecting sex pheromones from live insects and from artificial sources. pp. 223-285. *In* H.E. Hummel and T.A. Miller (eds). Techniques in pheromone research, New York, Springer-Verlag, Inc., 464p.

Hummel, H.E. & Miller, T.A. (eds). 1984. Techniques in pheromone research. New York, Springer-Verlag, Inc., 464p.

Kelner-Pillault, S. 1975. Attirance sexuelle chez un Trichoptère : *Enoicyla pusilla* Burmeister. **Bull. Soc. ent. Fr.,** 80 : 252-257.

Resh, V.H. & Houp, R.E. 1986. Life history of the caddisfly *Dibusa angata* and its association with the red alga *Lemanea australis*. **J. N. Am. Benth. Soc.,** 5 : 28-40.

Resh, V.H. & Wood, J.R. 1985. Site of sex pheromone production in three species of Trichoptera. **Aquatic Insects,** 7 : 65-71.

Resh, V.H., Lamberti, G.A. & Wood, J.R. 1984. Biology of the caddisfly *Helicopsyche borealis* (Hagen) : a comparison of North American populations. **Freshwat. Invert. Biol.,** 3 : 172-180.

Solem, J.O. 1985. Female sex pheromones in *Rhyacophila nubila* (Zetterstedt) (Trichoptera, Rhyacophilidae) and arrival pattern to sticky traps. **Fauna norv., Ser. B.,** 32 : 80-82.

Thornhill, R. & Alcock, J. 1983. The evolution of insect mating systems. Cambridge, Massachusetts : Harvard University Press, 547p.

Wood, J.R. & Resh, V.H. 1984. Demonstration of sex pheromones in caddisflies (Trichoptera). **J. chem. Ecol.,** 10 : 171-175.

# A STUDY OF EGG MASSES OF *GLYPHOTAELIUS PELLUCIDUS* (RETZIUS), (TRICHOPTERA : LIMNEPHILIDAE)

M. I. CRICHTON

## SUMMARY

Regular field observations were made on egg masses laid on leaves over the outflow from Millbarn Pond in Berkshire in autumn 1984. More egg masses were laid on wet nights, and many were in contact with others. The egg masses are described. Larvae, reared in the laboratory from 9 egg masses, gave rise to 40 male and 44 female adults of *Glyphotaelius pellucidus* in March and April 1985. The suggestion by another worker that similar egg masses, collected at the same site in 1974, had been laid by *Nemotaulius punctatolineatus* is rejected.

## INTRODUCTION

Over the years since 1951 I have collected and observed caddis flies at Millbarn Pond, which is sited between my home in Mortimer and the University of Reading. Millbarn Pond (Grid ref. SU 671663) was formed in the 18th century by building a dam across the valley of the Lockram Brook, which has its source about 3 km away on the plateau gravel of Burghfield. This eutrophic pond has an area of 3. 4h, a maximum depth of about 3 m, and is about 60 m above sea level. The water runs out through a sluice at the south end of the dam and over seven brick-built steps down into a deeply-shaded valley. Only in wet weather is there any considerable flow of water.

A Robinson light trap (Robinson and Robinson 1950), using a MV lamp of 80 W, was operated on the dam close to the outflow, from May to November in 1953-55, and also through the autumn of later years. Thus a lot of information has been accumulated, and a total of 71 species of caddis flies recorded (Crichton 1960, 1965). Particular attention was directed to the Limnephilidae, of which 24 of the 59 British species were caught. The common species are listed in Table 1.

Table 1. Light trap catches of the common species of Limnephilidae from Millbarn Pond, 1953-82, listed in order of abundance.

| | |
|---|---|
| *Limnephilus lunatus* Curt. | 4206 |
| *Limnephilus flavicornis* (F.) | 1124 |
| *Halesus digitatus* (Schr.) | 1006 |
| *Limnephilus auricula* Curt. | 872 |
| *Glyphotaelius pellucidus* (Retz.) | 541 |
| *Anabolia nervosa* (Curt.) | 233 |
| *Limnephilus sparsus* Curt. | 233 |
| *Stenophylax permistus* McL. | 231 |
| *Limnephilus vittatus* (F.) | 215 |

Analysis of the autumn catches showed that the largest numbers of limnephilids were caught on warmer nights when there was rain. It was suggested that rain might act as a releasing stimulus to mating and egg-laying.

## THE EGG MASSES OF LIMNEPHILIDAE

Most limnephilids lay their eggs in a mass of jelly on leaves or other surfaces above water. This habit distinguishes them from other caddis which lay their eggs at the water surface, or underneath by crawling down vegetation or rocks. Among the British limnephilids it is normal for egg masses to be found above water. Known exceptions to this habit are *Drusus annulatus* Steph. (Gower 1966) and *Ecclisopteryx guttulata* (Pict.) (Nielsen 1942). Wiggins (1973) gave a useful summary of oviposition data with particular reference to diapause and the exploitation of temporary pools by caddis. He showed that oviposition above water has become most widely established in the Limnephilidae.

The egg masses of only a few Limnephilidae have been described ; there is some information in Silfvenius (1906), but other accounts are scattered and fragmentary. Thus, there is no available publication to help the collector to identify egg masses in the field. The only certain way is to observe the female in the act of egg-laying, or to rear larvae until they are identifiable or become adults.

In autumn 1980 I started to make observations on limnephilid egg masses at Millbarn Pond, and I suggested to R. J. James that he made them the subject of his final year undergraduate project at the University of Reading. He confined females in containers in the laboratory, and was successful in obtaining egg masses of *H. digitatus, S. permistus, Chaetopteryx villosa* (F.), *A. nervosa, L. flavicornis* and *L. lunatus.* He related these masses to ones he found at Millbarn Pond and elsewhere ; colour photographs and descriptions were included in his project (James 1981). There would appear to have been no published accounts of the egg masses of *H. digitatus, S. permistus, C. villosa* or *A. nervosa.* Thus, we now had information on five of the common species from Millbarn Pond in Table 1.

In further searches at Millbarn Pond in autumn 1981 and 1982 I found large egg masses, characteristically ribbed transversely, on the leaves of bur-reed, *Sparganium erectum* L. growing near the dam, and also on laurel, *Prunus laurocerasus* L. and ivy, *Hedera helix* L., over the outflow. These masses were not laid by any of the five species described above by James. They were too large to have come from *L. auricula* or *L. vittatus,* and probably also not from *L. sparsus.* This left*Glyphotaelius pellucidus* as the most likely source.

## THE EGG MASSES OF *GLYPHOTAELIUS PELLUCIDUS*

On 20 September 1974, M. I. Middleton, a research student at the University of Bristol, who was investigating the egg jelly of the Trichoptera, visited Millbarn Pond and collected egg masses from exposed roots and the brickwork of dam and outflow. He decided that they had been laid by *G. pellucidus,* although he did not find adults or rear through any larvae. From the observations of James, referred to above, and the description of Gower (1967), they were probably laid by *L. lunatus.* Middleton also found, on the leaves of ivy and sycamore, *Acer pseudoplatanus* L. above the outflow, eleven egg masses with transverse lines of eggs giving them a ribbed appearance. He decided that they had been laid by *Nemotaulius punctatolineatus* (Retz.) because they, and the 1st instar larvae which hatched from them, resembled the drawings and descriptions of this species by Wesenberg-Lund (1910). Again, he found no adults and did not rear through any of the larvae (Middleton 1977).

The only confirmed records of *N. punctatolineatus* in Britain are one male on 2 July 1965 and one female on 18 June 1967 from near Aviemore in Scotland by Pelham-Clinton (1966, and *in litt.* 25 February 1970). These dates are comparable with Denmark, where Wesenberg-Lund recorded adults in June and egg masses from mid-June to early August. The ribbed egg masses which I had found at Millbarn Pond in autumn 1981 and 1982 resembled very closely the photographs in Middleton's paper of 1977 which he claimed were of *N. punctatolineatus.* I could not accept this identification, but did not then have the opportunity to rear the larvae. Hence the following detailed study in the field and laboratory from September 1984 to the emergence of adults of *Glyphotaelius pellucidus* in March and April 1985.

Visits were made to Millbarn Pond nearly every day from 3 September to 4 November. I examined all leaves hanging over the 8 m of outflow steps and a further 12 m of the stream beyond. Any leaf which carried a ribbed egg mass was marked with a numbered Dymo label, and a sketch made. Thus the date when each egg mass was first seen was recorded, and notes were kept of its development up to the disappearance of the larvae into the stream. Some photographs were also taken. The leaves selected by the caddis flies were of the dominant plants, laurel, ivy and a few elder, *Sambucus nigra* L. As a consequence of a long dry spell in July and August the level of water in Millbarn Pond was so low that there was no outflow of water at the beginning of September.

A careful search on 3 September revealed no egg masses. On the following night there was rain, the first for many weeks, and the next morning 7 egg masses were found. A further 18 masses were found after the next wet night, which was 13/14 September. Thereafter, eggs were laid on both wet and dry nights up to 17/18 October ; out of a total of 134 egg masses, 80 were laid on wet nights. Their distribution is given in Fig. 1 in weeks corresponding to light trap catches of *G. pellucidus* in three earlier seasons. This species belongs to the group of limnephilids with an extended flight period and a summer diapause, followed by maximum activity in the autumn when eggs are laid (Crichton and Fisher 1981).

Egg masses were found on the upper leaf surfaces, normally close to the lowest point, which was where, in wet weather, water collected before dropping off the leaf. Only 33 remained isolated throughout their development; the remaining 101 were in contact with others. Examples are illustrated in Figure 2. Thus, there is a tendency for females to lay their egg masses touching ones laid earlier. A group of four adjacent laurel leaves attracted a total of 17 egg masses, of which 12 had been laid on two wet nights, suggesting that a number of females had been stimulated to oviposit in the proximity of others. Eggs were laid between 0.3 and 1.9 m above the water, with 69% between 0.5 and 1.5 m. Because of the summer drought it was 18 September before any water came over the outflow, and yet 38 egg masses had already been laid over the dry bed of the stream. During the whole period no eggs were found on leaves which were not directly over the course of the outflow stream.

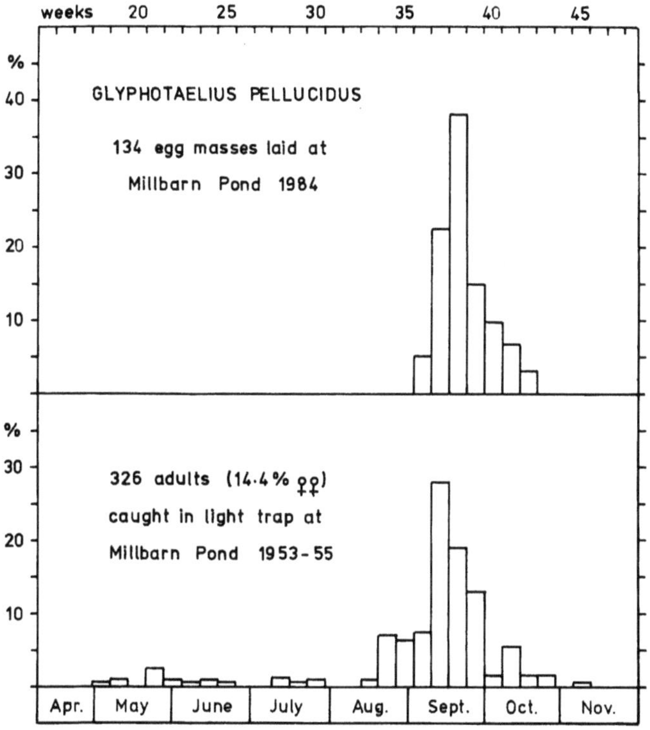


Figure 1. Numbers of egg masses of *Glyphotaelius pellucidus* laid each week over the outflow from Millbarn Pond in autumn 1984. Catches of adult *G. pellucidus* in a Robinson light trap on the dam, 1953-55.

When first laid, the egg masses were opaque white and 7-10 mm in length. Within a day or two they had become translucent and swollen to 10-15 mm, and the eggs had acquired a yellow tinge. In a sample of 18 masses the average number of eggs was 245 (range 156-349). The one shown in Figure 2E contained 349, while that in Figure 2F contained 243 eggs; both illustrate clearly the transverse rows of eggs giving a ribbed effect. (A convenient method for counting the eggs was found to be soaking the egg mass in 5% KOH in a watch glass for 1-2 hours, thus dissolving the jelly. After washing gently in water which was mainly drained away the eggs were counted with the help of a fine brush.) Within 2-3 weeks the larvae were moving freely in the jelly, which now became pendulous, and after a few days they dropped into the stream below. The clustering of larvae is shown in mass 1 Figure 2B, and a final empty pendulous mass of jelly in Figure 2D.

Nine egg masses, laid between 3 September and 9 October, were brought into an aquarium in the laboratory in October and November, and kept at a temperature of about 15°C. The larvae were reared in plastic containers, with a supply of fallen leaves of the silver maple, *Acer saccharinum* L. Leaves and water were changed about twice a week ; no aeration was needed. Numbers of larvae were reduced to prevent overcrowding. From these larvae, 40 male and 44 female caddis flies emerged between 26 March and 23 April 1985. They were all *Glyphotaelius pellucidus*. The suggestion of Middleton (1977) that *Nemotaulius punctatolineatus* had laid egg masses at Millbarn Pond must be rejected.

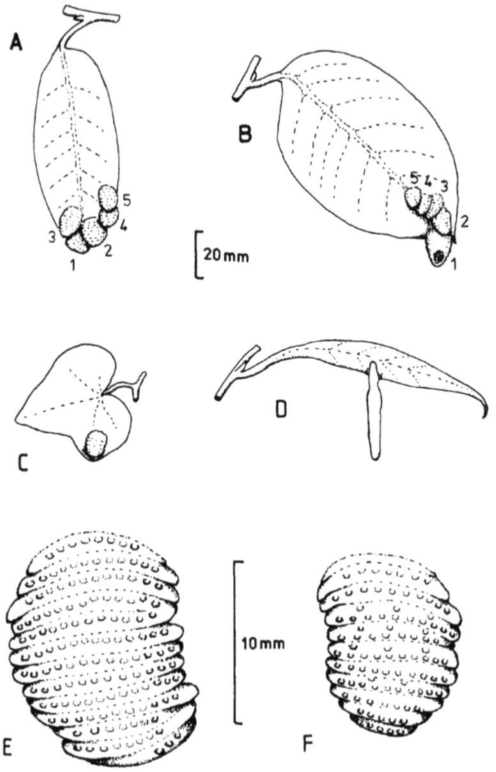


Figure 2. Egg masses of *Glyphotaelius pellucidus* on leaves over the outflow from Millbarn Pond.
A. laurel leaf on 3.X.84 with egg masses laid on (1) 13.IX., (2) 15.IX., (3) 23.IX., (4) & (5) 27.IX.
B. laurel leaf on 3.X.84 with egg masses laid on (1) 13.IX., with mass of larvae, (2) 14.IX., (3) 23.IX., (4) 25.IX., (5) 2.X.
C. ivy leaf on 21.IX.84 with egg mass laid on 3.IX.
D. laurel leaf on 3.X.82, the egg mass was laid on 7.IX., all larvae had left by 28.IX.
E. egg mass from ivy leaf on 25.IX.84, containing 349 eggs.
F. egg mass from laurel leaf on 3.X.84, containing 243 eggs.

ACKNOWLEDGEMENTS

I am indebted to Sir Michael Milne-Watson for letting me work over many years at Millbarn Pond in the grounds of Oakfield. I am grateful to Professor K. Simkiss for giving me facilities and support, after my retirement, in the Department of Pure and Applied Zoology. I also acknowledge a grant from The Royal Society towards the cost of attending the Symposium.

REFERENCES

Crichton, M.I. 1960. A study of captures of Trichoptera in a light trap near Reading, Berkshire. **Trans. R. ent. Soc. Lon.**, 112 : 319-344.

Crichton, M.I. 1965. Observations on captures of Trichoptera in suction and light traps near Reading, Berkshire. **Proc. R. ent. Soc. Lond.(A)**; 40 : 101-108.

Crichton, M.I. & Fisher, D.B. 1981. Further observations on limnephilid life histories, based on the Rothamsted Insect Survey. **Proc. 3rd Int. Symp. Trichoptera,** G.P. Moretti (ed.), Junk, The Hague : 47-55.

Gower, A.M. 1966. The life cycle of *Drusus annulatus* Steph. (Trich. Limnephilidae) in watercress beds. **Entomologist' s mon. Mag.,** 101 : 133-141.

Gower, A.M. 1967. A study of *Limnephilus lunatus* Curtis (Trichoptera : Limnephilidae) with reference to its life cycle in watercress beds.**Trans. R. ent. Soc. Lond.,** 119 : 283-302.

James, R.J. 1981. Observations on the reproduction of some limnephilid caddis flies (Trichoptera). Univ. Reading, Dep. Zool. Unpublished Undergraduate Project No.173.

Middleton, M.I. 1977. The possible discovery of the egg masses of *Nemotaulius punctatolineatus* (Retzius) in Britain (Trichoptera : Limnephilidae). **Entomologist's Gaz.,** 28 : 45-50.

Nielsen, A. 1942. Über die Entwicklung und Biologie der Trichopteren mit besonderer Berücksichtigung der Quelltrichopteren Himmerlands. **Arch. Hydrobiol., Suppl.,** 17 : 255-631.

Pelham-Clinton, E.C. 1966. *Nemotaulius punctatolineatus* (Retzius), a caddis fly new to the British Isles (Trichoptera, Limnephilidae). **Entomologist's Gaz.,** 17 : 5-8.

Robinson, H.S. & Robinson, P.J.M. 1950. Some notes on the observed behaviour of Lepidoptera in flight in the vicinity of light sources, together with a description of a light trap designed to take entomological samples. **Entomologist's Gaz.,** 1 : 3-20.

Silfvenius, A.J. 1906. Trichopterologische Untersuchungen I. Über den Laich der Trichopteren. **Acta Soc. Fauna Flora Fenn.,** 28(4) : 1-128.

Wesenberg-Lund, C. 1910. Über die Biologie von *Glyphotaelius punctatolineatus* Retz. nebst Bemerkungen über das freilebende Puppenstadium der Wasserinsekten. **Int. Rev. ges. Hydrobiol.,** 3 : 93-114.

Wiggins, G.B. 1973. A contribution to the biology of caddisflies (Trichoptera) in temporary pools. **Life Sci. Contr., R. Ont. Mus.,** 88 : 1-28.

# ADAPTIVE EGG LAYING BEHAVIOUR IN TWO SPECIES OF CADDISFLIES

C. OTTO


## SUMMARY

*Glyphotaelius pellucidus* and *Nemotaulius punctatolineatus* deposited their egg masses on leaves about 2 and 1 m above the ground/water, respectively. In both species about half the number of egg masses observed (n=112 and 122, respectively) were deposited in groups. In all but two of the leaves carrying more than one egg mass physical contact was established between the egg masses. Desiccation reduced egg survival under natural conditions. After desiccation in the laboratory hatching from egg masses deposited singly or in pairs differed and was more successful in the latter. A low surface/volume ratio of jointly laid egg masses apparently reduced dessication and hence increases survivorship of eggs. It is suggested that this is the adaptive significance of communal egg laying behaviour. Caddisfly species ovipositing in damp areas lay egg masses singly.


## INTRODUCTION

Egg laying behaviour in caddisflies is very diverse. Some species deposit their egg masses under the water surface while others oviposit while flying above or resting on it. Still others oviposit on objects above the water (Silfenius 1906, Hanna 1961, Elliott 1971). Among those species laying eggs above water, oviposition usually takes place just above the water surface. Where water is absent, egg masses may be found on the ground in damp vegetation, beneath stones, bark etc. (Novak and Sehnal 1963, Wiggins 1973, Hiley 1978). However, at least two limnephilid species deposit their egg masses on leaves at considerable heights, namely *Glyphotaelius pellucidus* (Retz.) and *Nemotaulius punctatolineatus* (Retz.) (Silfenius 1906, Wesenberg-Lund 1908, 1911). *G. pellucidus* is often found in temporary pools, while *N. punctatolineatus* usually inhabits lake shores (Wichard and Reichel 1969, Lepneva 1971, Otto 1976). The eggs of both species are surrounded by a gelatinous matrix which, after swelling, liquefies when the eggs hatch (Middleton 1977, Berté and Pritchard 1982). Wiggins (1973) pointed out, that the gelatinous matrix of caddisfly species normally ovipositing in water is much more susceptible to desiccation than that in the egg masses of species habitually laying out of water. However, even the more resistant eggs may suffer from desiccation (Novak and Sehnal 1963) and the egg laying behaviour of *G. pellucidus* and *N. punctatolineatus* has therefore been investigated to see if there are any further features which might reduce mortality due to this factor.

## MATERIAL AND METHODS

Leaves on which egg masses of *G. pellucidus* and *N. punctatolineatus* had been deposited were inspected in September 1983. The egg masses of *G. pellucidus* were found at a temporary pool which was dry during August and September (Otto 1976), whilst those of *N. punctatolineatus* were found at a lake shore (Otto 1979) in South Sweden. The height above the ground or water surface of leaves carrying egg masses was measured, and the number and position of egg masses on each leaf were noted. For each species the number of eggs in each of seven masses was counted.

Egg masses appearing partly or totally desiccated were collected in the field and immersed in tap water in the laboratory and the number of hatchlings counted. Fresh egg masses which appeared moist were treated in the same way as a control. For both species seven leaves bearing single egg masses and seven more bearing two egg masses were taken to the laboratory, air dried at 2O° C for one week. After two days of severe drying they were moistened daily for the final five days and then immersed in tap water in separate glass jars. The number of hatched larvae was noted at weekly intervals for seven weeks, and the I instar larvae were successively sorted out.

RESULTS

*G. pellucidus* egg masses were found on beech (*Fagus sylvatica* L.), alder (*Alnus glutinosa* L.) hazel (*Corylus avellana* L.) and mountainash (*Sorbus aucuparia* L.) leaves. The mean height at which the egg masses were deposited was 2.1. m (S.D.=0.5, n=80). *N. punctatolineatus* egg masses were deposited on alder (*Alnus glutinosa* L.) willow (*Salix* sp.) and black alder (*Rhamnus frangula* L.) leaves at mean height of 1.2 m (S.D. = 0.4, n = 82) above the water surface of the lake. Some egg masses were also found above the ground about 4 m away from the water margin. For the purpose of egg laying conveniently situated leaves were superabundant. A total of 112 *G. pullucidus* egg masses was found, of which 44% were deposited in contact with at least one other mass on the same leaf. From 122 *N. punctatolineatus* egg masses 54% were deposited together with other masses. Between egg masses deposited on the same leaf physical contact was established on all leaves but two bearing egg masses of *G. pellucidus*. The maximum number of egg masses per leaf was 8 in *G. pellucidus*, while the corresponding figure in *N. punctatolineatus* was 4 (Table 1). A $X^2$ test for "goodness-of-fit" showed that the egg masses of *G. pellucidus* confirmed to a Poisson distribution ($P > 0.05$). *N. punctatolineatus* on the other hand deviated markedly from a Poisson distribution ($P < 0.01$) indicating a non random distribution here.

Table 1. Distribution of number of egg masses per leaf in *G. pellucidus* and *N. punctatolineatus*, respectively.

| | Number of egg masses per leaf | | | | |
|---|---|---|---|---|---|
| | 1 | 2 | 3 | 4 | 8 |
| *G. pellucidus* | 62 | 11 | 4 | 2 | 1 |
| *N. punctatolineatus* | 56 | 16 | 6 | 4 | |

In both species, on leaves bearing two egg masses, one egg mass was typically deposited vertically above the other one partly covering the older mass. The leaves were bent downwards because of the extra burden. Several days may have elapsed between the two depositions because larvulae were sometimes seen in the old egg mass, while the upper (young) one only contained eggs.

In *G. pellucidus* susceptibility to desiccation seemed unrelated to the number of egg masses per leaf as judged from occurrence of desiccated masses in the field ($X^2$, n.s.) (Table 2). In *N. punctatolineatus*, on the other hand, significantly more singly deposited egg masses showed signs of desiccation (25%) compared to those laid jointly (6%) ($X^2$, $P < 0.01$) (Table 2).

Table 2. Relationship between the number of egg masses per leaf and the incidence of desiccation in *G. pellucidus*and *P. punctatolineatus* .

| | *G. pellucidus* | | *N. punctatolineatus* | |
|---|---|---|---|---|
| | Number of egg masses | | | |
| | Desiccated | not desiccated | desiccated | not desiccated |
| 1 egg mass leaf $^{-1}$ | 16 | 45 | 14 | 42 |
| > 1 egg mass leaf $^{-1}$ | 14 | 35 | 4 | 62 |

In *G. pellucidus* and *N. punctatolineatus* the mean number of eggs per mass was 180 and 167, respectively (Table 3). When fresh egg masses were immersed in water almost all eggs hatched over a period of three weeks, and no eggs hatched after four weeks of submergence (Table 3). In heavily desiccated egg masses collected in the field the majority of the eggs had not hatched after seven weeks, and most egg masses were then affected by fungi (Table 3). Thus desiccation decreases egg survival under natural conditions.

Table 3. Number of eggs per mass and number of eggs failing to hatch from fresh and desiccated masses of *G. pellucidus* and *N. punctaıolineatus* collected in the field. For all values given n = 7.

| | *G. pellucidus* | | | *N. punctatolineatus* | | |
|---|---|---|---|---|---|---|
| | No of eggs $mass^{-1}$ | No of eggs not hatched | | No of eggs $mass^{-1}$ | No of eggs not hatched | |
| | | Fresh mass | Desiccated mass | | Fresh mass | Desiccated mass |
| $\overline{X}$ | 179.9 | 1.3 | 101.0 | 167.3 | 10.1 | 129.4 |
| S.D. | 64.8 | 1.8 | 33.8 | 39.3 | 17.9 | 34.4 |

Egg masses desiccated in the laboratory hatched for six weeks. Assuming equal hatching success in both egg masses deposited jointly on the same leaf on the average 42 (S.D. = 25, n = 14) and 60 (S. D. = 43, n = 14) eggs hatched from each of the masses of *G. pellucidus* and *N. punctatolineatus*, respectively. In those egg masses deposited singly the corresponding figures were 29 (S.D. = 44, n = 7) and 32 (S.D. = 46, n = 7), respectively. Significantly more eggs hatched from egg masses deposited jointly in *G. pellucidus* (Mann-Whitney U test, $P < 0.05$), while the difference was not significant in *N. punctatolineatus* $P < 0.10$).

DISCUSSION

Desiccation of egg masses deposited jointly is likely to be slower than from single masses because of a reduced surface volume ratio and this factor would be expected to increase egg survival. By ovipositing above existing egg masses females could increase their fitness. However, not all egg masses are laid jointly suggesting a trade off between different types of egg laying behaviour. First, prolonged searching for previously deposited egg masses may increase the risk of being taken by predators. Second, as a consequence of jointly laid egg masses a large number of small larvae with poor locomotory powers will appear in a restricted area. This may eventually lead to local depletion of high quality food. (Otto 1974, 1975, Richardson and Mackay 1984), in turn leading to retarded larval growth and hence lower competitive ability. As a result small imagines, being less fecund (Svensson 1975), will appear from areas of high larval density. Third, in patches where a high number of larvae hatch as a result of jointly laid egg masses mortality from cannibalism is likely to be increased, even among leaf shredding species (Novak and Sehnal 1963, Erman 1981). Support for this hypothesis, that cannibalism may counterbalance the advantage of jointly deposited egg masses, was obtained from the fresh egg masses hatched in the laboratory. Approximately 7 x 180 and 7 x 167 *G. pellucidus* and *N. punctatolineatus* larvae, respectively, were placed in two separate cages (20 x 20 cm) through which water circulated. In spite of superabundance of food in the form of shed alder leaves only 434 and 11 *G. pellucidus* and *N. punctatolineatus* larvae, respectively, survived six weeks after submergence of egg masses. Many empty cases with remnants (sclerties etc.) of small larvae were observed. The three factors listed above as well as others may contribute in making communal egg laying hazardous.

In butterflies eggs deposited jointly (in clusters) often contrast sharply with the leaf surface on which they are laid (Owen 1971). In contrast to eggs laid singly those laid jointly are brightly colored (yellow, orange or red), and aposematic colouring of eggs of egg-clustering species signals egg toxicity to avoid predation and parasitism (Stamp 1980). Since the matqrix is transparent and the eggs are brownish, egg masses of caddisflies are not aposematically colored but rather cryptic.

Slow rate of desiccation is proposed as an important reason for egg masses being laid jointly. If so, this egg laying behaviour should be found also in other species ovipositing in dry places such as on leaves at considerable distance from the ground. Unfortunately only the two species here investigated have conclusively been shown to use leaves situated high above the ground (Silfenius 1906, Klingstedt 1926, Wiggins 1973). Conversely, in species ovipositing in damp places on or near the ground or just above the water surface egg masses would be expected to be deposited singly. In fact observations by Silfenius (1906), Gower (1967), Anderson (1976), B.W. Svensson (pers. comm.), K. Novák (pers. comm.) and myself show that caddisfly species such as *Limnephilus politus* Mcl., *Limnephilus lunatus* (Curt.), *Pseudostenophylax edwardsi* (Banks.), *Limnephilus griseus* (L.), *Limnephilus rhombicus* (L.), *Limnephilus stigma* Curt. and *Limnephilus centralis* (Curt.) which oviposit in damp places, all deposit their egg masses singly. These observations support the hypothesis that joint egg-laying might serve to reduce desiccation.

Egg predators such as dipterans may substantially contribute to egg mortality in caddisflies (Wisseman and Anderson 1984). Egg masses might be deposited jointly in order to swamp the predators (Wilson 1975), particularly if egg masses deposited in trees are more easily found by invertebrate predators than those on the ground. This could represent an additional advantage in communal egg laying by *G. pellucidus* and *N. punctatolineatus.*

ACKOWNLEDGEMENTS

I thank Mr. S. Otto for assistance in the field, I also thank Dr B. Malmqvist, University of Lund, for comments on the manuscript. Support was received from the Swedish Natural Science Research Council.

REFERENCES

Anderson, N.H. 1976. The distribution and biology of the Oregon Trichoptera. **Oregon State Univ., Techn. Bull.**, 134 : 1-152.

Berté, S.B. & Pritchard, G. 1982. The phenomenon of egg mass liquefaction in *Nemotaulius hostilis* (Hagen) (Trichoptera : Limnophilidae). **Freshwater Invertebrate Biol.**, 1, 49-51.

Elliott, J. M. 1971. The life history and biology of *Apatania muliebris* Mclachlan (Trichoptera). **Entomologist's Gaz.**, 22 : 245-251.

Erman, N.A. 1981. Terrestrial feeding migration and life history of the stream-dwelling caddisfly, *Desomna bethula* (Trichoptera : Limnephilidae). **Can. J. Zool.**, 59 : 1658-1665.

Gower, A.M. 1967. A study of *Limnephilus lunatus* Curtis (Trichoptera : Limnephilidae) with reference to its life cycle in watercress beds. **Trans. R. Entomol. Soc., London**, 119 : 283-302.

Hanna, H.M. 1961. Observations on the egg-laying of some British caddis flies and on case-building by newly hatched larvae. **Proc. R. Entomol. Soc., London** (A), 36 : 57-62.

Hiley, P.D. 1978. Some aspects of the life histories of Limnephilidae (Trichoptera) related to the distribution of the larvae. **Proc. 2nd Int. Symp. Trichoptera**, M.I. Crichton(ed.), Junk, The Hague : 297-301.

Klingstedt, H. 1926. Beobachtungen über die Biologie inbesondere das Eilegen von *Limnophilus decipiens* Kol. (Trich.). **Notulae Entomol.**, 6 : 118-120.

Lepneva, S.G. 1971. Larvae and pupae of Intergripalpia. Fauna of USSR, Trichoptera, Vol. II A.A. Strelkov (ed.), Israel Program for Scientific Translations, Jerusalem : 1-700.

Middleton, M. I. 1977. The possible discovery of the egg masses of *Nemotaulius punctatolineatus* (Retzius) in Britain (Trichoptera : Limnephilidae). **Entomologist's Gaz.**, 28 : 45-50.

Novak, K. & Sehnal, F. 1963. The development cycle of some species of the genus *Limnephilus* (Trichoptera). **Cas. Ces. Spol. Entomol.**, 60: 68-80.

Otto, C. 1974. Growth and energetics in a larval population of *Potamophylax cingulatus* (Steph.) (Trichoptera) in South Swedish stream. **J. Anim. Ecol.**, 43 : 339-361.

Otto, C. 1975. Energetic relationships of the larval population of *Potamophylax cingulatus* (Trichoptera) in a South Swedish stream. **Oikos**, 26 : 159-169.

Otto, C. 1976. Habitat relationship in the larvae of three Trichoptera species. **Arch. Hydrobiol.**, 77 : 505-517.

Otto, C. 1979. The effects on a pike (*Esox lucius* L.) population of intensive fishing in a South Swedish lake. **J. Fish. Biol.**, 15 : 461-468.

Owen, D. 1971. Tropical butterflies. Clarendon, Oxford.

Richardson, J.S. & Mackay, R.J. 1984. A comparison of the life history and growth of *Limnephilus indivisus* (Trichoptera : Limnephilidae) in three temporary pools. **Arch. Hydrobiol.**, 99 : 515-528.

Silfenius, A.J. 1906. Über den Laich der Trichopteren. **Acta Soc. Fauna Flora Fenn.**, 28 : 1-128.

Stamp, N.E. 1980. Egg deposition patterns in butterflies : Why do some species cluster their eggs rather than deposit them singly ? **Am. Nat.** , 115 : 367-380.

Svensson, B. W. 1975. Morphometric variation of adult *Potamophylax cingulatus* (Trichoptera) reflecting environmental heterogeneity in a South Swedish stream. **Oikos**, 26 : 365-377.

Wesenberg-Lund, C. 1908. Über tropfende Laichmassen. **Internat. Rev. gesamt. Hydrobiol. Hydrogr.**, 1 : 869-871.

Wesenberg-Lund, C. 1911. Über die Biologie von *Glyphotoelius punctatolineatus* Retz. nebst Bemerkungen über das freilebende Puppenstadium der Wasserinsekten. **Internat. Rev. gesamt. Hydrobiol. Hydrogr.**, 3 : 93-114.

Wichard, W & Reichel, H. 1969. Zur Trichopterenfauna periodischer Gewässer. **Nachr. Bayer. Entomol.**, 18 : 57-58.

Wiggins, G.B. 1973. A contribution to the biology of caddisflies (Trichoptera) in temporary pools. **Life Sc. Contrib., R. Ontario Museum**, 88 : 1-28.

Wilson, E.O. 1975. Sociobiology - The New Synthesis. Harvard University Press. Cambridge, Massuchusetts.

Wisseman, R.W. & Anderson, N.H. 1984. Mortality factors affecting Trichoptera eggs and pupae in an Oregon coast range watershed. **Proc. 4th Int. Symp. Trichoptera**, J.C. Morse(ed.), Junk, The Hague : 455-460.

# OBSERVATIONS ON GILL NUMBER AND RESPIRATORY ADAPTATION IN CADDIS LARVAE

R.M. BADCOCK, M.T. BALES and J.D. HARRISON

## SUMMARY

The numbers of gill filaments were counted on caddis larvae and pupae from sites with different thermal regimes and on larvae maintained at different temperatures in the laboratory.

Intraspecific differences in gill number were found for both larvae and pupae, individuals from warmer conditions having more gills.

In species having single gill filaments, the increase occurs by the production of gills at new sites. Species with multiple gill filaments may increase the number by both an increase in the number of sites and an increase in the number of filaments per tuft.

The relationship between gill number and dry weight of larvae is discussed.

## INTRODUCTION

In our laboratory, we have been looking at acclimation to temperature in caddis larvae, interpreting acclimation as acclimatization to a single factor (Vernberg and Vernberg 1972). Bales has extended the work begun by Harrison (Harrison and Badcock 1981) and will present his own work on respiratory rates as an acclimatory response to temperature, discussing its significance in relation to distribution. This paper considers intraspecific variation in the number of gill filaments as an acclimatory response to temperature and consequently to dissolved oxygen concentration.

Firstly, Wichard (1974, 1978) found that the size of gills in some limnephilid larvae and in larvae of *Molanna angustata* Curt. was related to the size of the larvae but that the numbers of gills in these forms were determined by the environmental oxygen tension. He demonstrated it by keeping larvae in standing water and in flowing water conditions.

## FIELD OBSERVATIONS

Harrison, working in Staffordshire, England, has compared the numbers of gill filaments of larvae of the limnephilid, *Chaetopteryx villosa* (F.) from adjacent streams with differing temperature regimes. Tumbling Brook, the head stream, is a tributary of the trout beck, Coombes brook. Tumbling Brook has more equable temperatures than the main stream, which is more extreme and warmer in summer (Harrison and Badcock 1981, Fig. 1). In May and in July, both fifth instar and fourth instar larvae of *Chaetopteryx* from Coombes Brook, then the warmer stream, had significantly more gill filaments than those from Tumbling Brook (Table 1.). For fifth instar larvae, the results for each or the two months taken separately were just as significant as the combined results.

There was a trend towards a greater number of filaments in July than in May for fifth instar larvae but the figures were not significant.

Bales has counted the gill filaments of fifth instar larvae and of pupae of the limnephilid *Potamophylax cingulatus* (Stephens) in the same two streams during June 1985.

There was a highly significant difference between the two streams in the numbers of filaments present in fifth instar larvae (Table 2a) and in pupae (Table 2b), those present in Coombes Brook, then the warmer stream, having more gill filaments and the numbers were nearer to the maximum found, whereas those in Tumbling Brook were more variable. There was a significant difference in the dry weights of fifth instar larvae from the two streams, some of those from Coombes Brook being slightly heavier, but the two sites are so close together that the populations are unlikely to be genetically distinct.

Bales studied the same species, *Potamophylax cingulatus*, in two streams in the grounds of Keele University, Staffordshire, the Dell stream being warmer in summer and cooler in winter than Springpool stream. Although the difference in summer temperatures between the two streams was often only 2°C or less, it was sufficient for a population of *Hydropsyche instabilis* (Curt.) to dwell in Springpool stream and *H. angustipennis* (Curt.) in Dell stream. Larvae of *Potamophylax cingulatus* were collected from both streams in June 1985 for gill counts and again fifth instar larvae from the warmer stream had more gill filaments and the variation (as shown by the standard deviation) for larvae from the cooler stream was greater (Table 3a).

Table 1. Number of gill filaments in larvae of *Chaetopteryx villosa* (F.) in Coombes Brook (CB) and Tumbling Brook (TB).

| | Date | Stream | Sample size | Mean no. gf/larva | Signiificance (p) |
|---|---|---|---|---|---|
| 5th instar | | | | | |
| | 22. v .79) | TB | 41 | 42.39 | <0.001 |
| | + 9.vii.79) | CB | 23 | 46.87 | |
| 4th instar | | | | | |
| | 22. v .79) | TB | 26 | 25.15 | <0.001 |
| | + 9.vii.79) | CB | 13 | 35.62 | |

Table 2. Number of gill filaments in (a) larvae and (b) pupae of *Potamophylax cingulatus* (Steph.) in Coombes Brook (CB) and Tumbling Brook (TB), (June 1985).

a) 5th instar larvae

| Stream | Sample size (n) | Mean no. gf/larva | Trans* | SD ($\delta$n-1) | Significance (p) |
|---|---|---|---|---|---|
| TB | 29 | 51.6 | 69.8 | ± 7.22 | <0.001 |
| CB | 22 | 57.6 | 85.4 | ± 5.57 | |

Trans* = Arc sin transformation of % of maximum number of gill filaments.

b) Pupae

| Stream | Sample size (n) | Mean no. gf/pupa | SD ($\delta$n-1) | Significance (p) |
|---|---|---|---|---|
| TB | 14 | 48.4 | ± 4.77 | <0.001 |
| CB | 14 | 57.1 | ± 2.50 | |

However for fourth instar larvae from these two streams, it was the other way round, those from Springpool stream having the greater number of filaments and the lower variation (Table 3b). The values were less significant, however, only being significant at the 5% level. We suspect that these larvae may have over-wintered as fourth instars and not yet moulted to fifth instars. There was no significant difference in the size of the fourth instar larvae in the two streams, as measured by head capsule width, and regressions of gill filament number on head capsule width were not significant for either stream i.e. there was no relation between gill number and head capsule width.

The regression of gill filament number on dry weight in fifth instar larvae did not show any correlation but there was a significant correlation at the 1% level for this in the fourth instar larvae from the Dell stream in July 1985. We think that this difference between instars may have been due to the fact that almost all the fifth instar larvae from the Dell stream in July had close to the maximum number of gill filaments found (58), the mean being 55.6 ± 2.48. On the other hand, July work showed that the maximum number of gill filaments found in the fourth instar larvae was 51, the range for Dell fourth instar *P. cingulatus* then being 37-51 and the mean 43.2 ±3.54.

When the mean numbers of gill filaments in fifth instar larvae of *Potamophylax cingulatus* from the four streams are compared, analysing results for the streams in pairs and considering all the various combinations, there is a significant difference in the numbers of gill filaments of larvae from all the pairs of streams (Table 4).

Table 3. Number of gill filaments in *Potamophylax cingulatus* (Steph.) larvae in Dell and Springpool streams (June 1985).

a) 5th instar larvae

| Stream | Sample size (n) | Mean no. gf/larva | Trans* SD (δn) | Significance (p) |
|---|---|---|---|---|
| Springpool | 66 | 54.8 | 76.4 ± 8.6 | <0.001 |
| Dell | 66 | 56.7 | 81.5 ± 6.3 | |

Trans* = Arc sin transformation of maximum number of gill filaments.

b) 4th instar larvae

| Stream | Sample size (n) | Mean no. gf/larva SD (δn) | Significance (p) |
|---|---|---|---|
| Springpool | 21 | 42.8 ± 2.99 | <0.05 |
| Dell | 20 | 40.2 ± 4.08 | |

Table 4. 5th instar larvae of *Potamophylax cingulatus*. Statistical comparison of mean numbers of gill filaments in pairs of streams.

| Stream | Actual mean no. of gf/larva | t | Significance (p) |
|---|---|---|---|
| TB v | 51.6 | 3.24 | <0.002 |
| Springpool | 54.8 | | |
| TB v | 51.6 | 7.52 | <0.001 |
| Dell | 56.7 | | |
| TB v | 51.6 | 8.05 | <0.001 |
| CB | 57.6 | | |
| Springpool v | 54.8 | 3.85 | <0.001 |
| Dell | 56.7 | | |
| Springpool v | 54.8 | 4.58 | <0.001 |
| CB | 57.6 | | |
| Dell v | 56.7 | 2.58 | <0.02 |
| CB | 57.6 | | |

However, as might be expected, the mean numbers of gill filaments that are closest together are those from the two streams that are warmer in summer, namely Coombes Brook and Dell stream, both with high mean numbers of filaments. The means for larvae from the two streams that are cooler in summer- Tumbling brook and Springpool stream are less close but both at the lower end of the spectrum. It is worth noting that

Tumbling Brook, with the lowest mean number of gill filaments per larva, is a typical spring-fed head stream near its source and supports a population of *Diplectrona felix* McL., whereas *Hydropsyche instabilis* is the hydropsychid associated with Springpool stream.

## EXPERIMENTAL REARING

Fourth instar larvae of *Potamophylax cingulatus* were collected from the Dell stream in July 1985 and divided into two groups : one was maintained in the laboratory at 6°C and the other at 18°C. At the end of 33 days, they were still in the fourth instar and there was no significant difference in the number of gill filaments between the two groups.

However, when a population of fourth instar *Sericostoma personatum* (Spence) was similarly treated, after 40 days there was a significant difference at the 1% level in the number of gill filaments between larvae kept at 6°C and those at 18°C, larvae in both groups having moulted to fifth instars, but those at the warmer temperature had more gill filaments. There was a significant correlation between gill number and dry weight in this instance.

## DISCUSSION

It looks as if morphological adaptation occurs at moults and conspecific larvae in streams with warmer temperatures (and therefore less dissolved oxygen) have more gill filaments.

However, there is no change from single filaments to multiple filaments. In *Potamophylax cingulatus*, new single filaments are added further down the body. In *Sericostoma personatum* the increase occurs by the development of filaments at new sites and also by some increase in the number of filaments per tuft.

Within a given larval instar, apparently there may sometimes be a correlation between size, as measured by dry weight, and number of gill filaments, larger larvae having more filaments, or the presence of more filaments enabling them to grow larger. However, there is not always a correlation if the filament number is near to the maximum any way, as may occur particularly in fifth instar larvae in warmer water. This may be why Wichard (1974) did not find a correlation between number of gills and size of larva. We have not examined size of gills and we have not yet progressed to possible correlations between number of filaments and weight specific respiration, but Bales discusses respiratory rates in the next paper.

## ACKNOWLEDGEMENTS

We are grateful to the Natural Environment Research Council for the Training Award that is currently supporting M.T.B. in his Trichoptera work and to the Department of Biological Sciences, University of Keele, for the research studentship that supported J.D.H. for three years. Also, we thank the Royal Society for the Protection of Birds, the Staffordshire Trust for Nature Conservation and Mr. M. Waterhouse for permission to collect in Coombes Valley Nature Reserve.

## REFERENCES

Harrison, J.D. & Badcock , R.M. 1981. Effects of acclimation to temperature on the respiratory rate of some limnephilid larvae. **Proc. 3rd Int. Symp. Trichoptera,** G.P. Moretti, (ed.), Junk, The Hague : 111-116.

Vernberg, W.B. & Vernberg, F.J. 1972. Environmental Physiology of Marine Animals. Springer-Verlag, Berlin, Heidelberg, New York,

Wichard, W. 1974. Zur morphologischen Anpassung von Tracheenkiemen bei Larven der Limnephilini Kol. (Insecta, Trichoptera) I. Autökologische Untersuchungen im Eggstatter Seegetbiet im Chiemgau. **Oecologia,** 15 : 150-167.

Wichard, W. 1974. Zur morphologischen Anpassung von Tracheenkiemen bei Larven der Limnephilini Kol. (Insecta, Trichoptera) II. Adaptationsversuche unter verschiedenen $0_2$ - Bedingungen wahrend der larvalen Entwicklung. **Oecologia,** 15 : 169-175.

Wichard, W. 1978. Structure and function of the trachael gills of *Molanna angustata* Curt. **Proc. 2nd Int. Symp. Trichoptera,** M.I. Crichton (ed.), Junk, The Hague : 293-296.

# RESPIRATION RATES AND DISTRIBUTION OF CADDIS LARVAE IN RELATION TO ACCLIMATION TO TEMPERATURE

M.T. BALES and R.M. BADCOCK

## SUMMARY

The effect of temperature acclimation on the respiratory rate of a range of caddis larvae was studied using a micro-winkler method.

In all five experimental species, acclimation of larvae to a range of constant temperatures in the laboratory is shown to affect their subsequent respiratory ability when tested at other temperatures. All larvae maintained at the warmer temperature showed a lower weight specific respiration than larvae maintained at the cooler temperature.

The results are discussed with reference to the distributions of the species. There is some evidence that species with a lower ability to acclimate have a more restricted ecological distribution.

## INTRODUCTION

Water temperature is one of the more important environmental parameters, having a profound influence on life cycles and, as suggested by a number of authors, including Kamler (1965) and Philipson and Moorhouse (1976), probably being an important factor affecting the distribution of species.

As aquatic poikilotherms are not able to regulate their body temperature and thermal diffusion is more rapid than molecular diffusion (Hazel and Prosser 1974) the body temperature will at all times be close to the ambient water temperature. Therefore the ability to acclimate to temperature is likely to be of great importance for individuals of a species to function efficiently at widely different habitat temperatures.

Acclimation in this study is defined as a compensatory response of an organism to an alteration of one environmental factor over a period of days or weeks. Some doubts exist about the occurrence and importance of temperature acclimation in the invertebrates. A number of authors have demonstrated a change in respiratory rates following temperature acclimation (e.g. Vernberg and Vernberg 1972 ; Bulnheim 1979). Kulkarni (1978) states that there is irrefutable evidence that poikilothermic invertebrates undergo metabolic compensation in response to thermal variations. However Keister and Buck (1974) state that insects in general have little ability to compensate for environmental temperature.

The aim of this study was to compare the ability of a number of species of caddis larvae to acclimate to temperature and to relate this to their distributions.

## MATERIALS AND METHODS

Fourth and fifth instar larvae were collected from a single site, split into two groups matched approximately for size, and maintained in aerated pond water at the required constant temperature for 4-5 weeks with suitable food provided.

Immediately before the determination of the oxygen consumption the larvae were all maintained at an intermediate temperature for 24 hours, without food.

The oxygen consumption was measured using the closed bottle method, with one larva per bottle. The oxygen content of the water was determined by micro-winkler analysis using the azide modification, the iodine level being measured with a spectrophotometer (Rees and Hilton 1977). The larvae were weighed after oven drying at 60°C for 48 hours.

Five species were used in this study. *Hydropsyche angustipennis* (Curtis), *Hydropsyche pellucidula* (Curtis), *Hydropsyche instabilis* (Curtis) and *Sericostoma personatum* (Spence) were acclimated to 6 and 18°C and *Potamophylax cingulatus* (Stephens) was acclimated to 5 and 10°C.

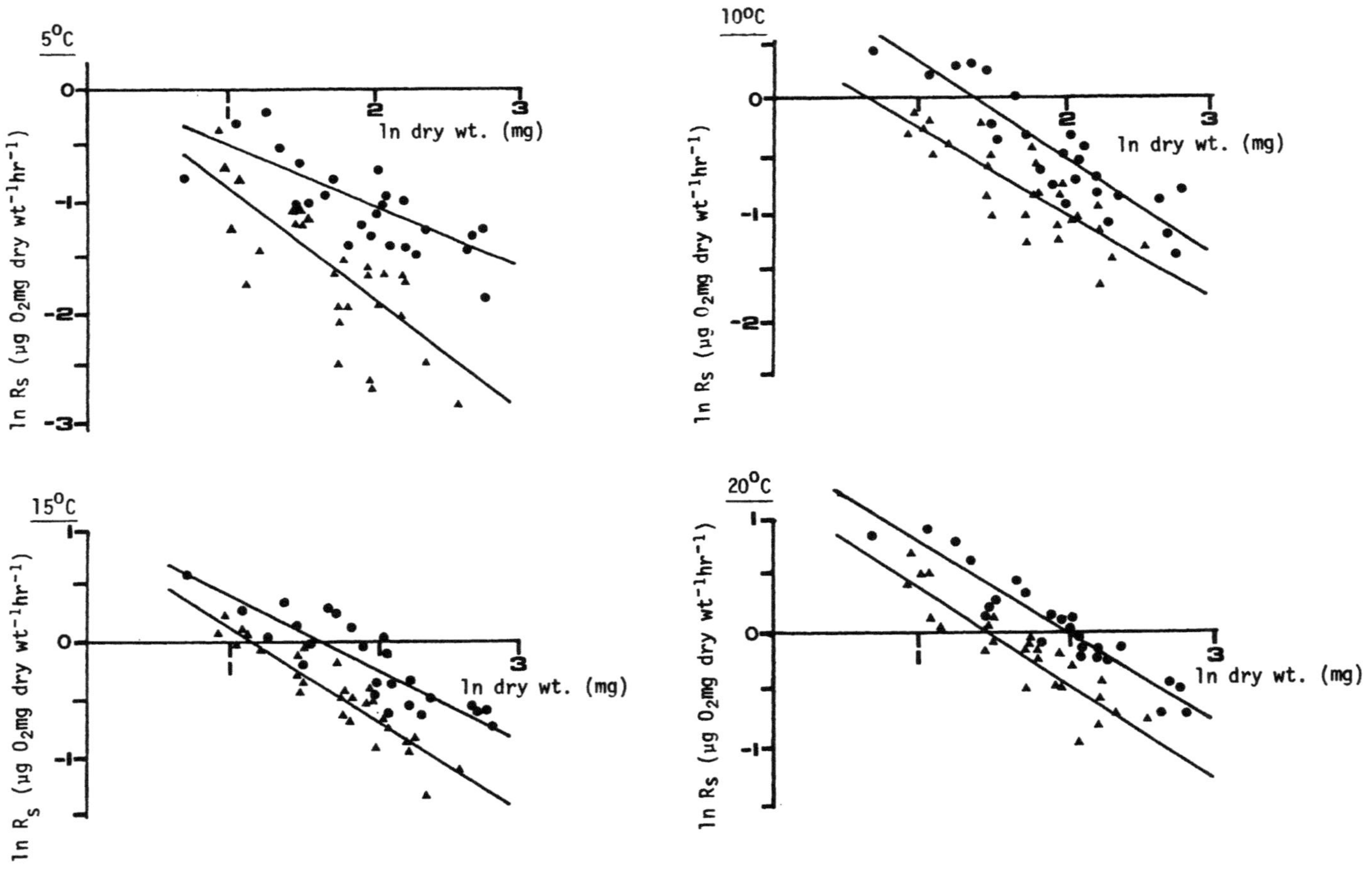


Figure 1. Ln-ln regressions of weight specific oxygen consumption ($R_S$) in *Seriscostoma personatum* against dry weights, at 5, 10, 15 and 20°C. ●, 6°C acclimated larvae ; ▲, 18°C acclimated larvae.

RESULTS

The weight specific respiration, i.e. the oxygen consumption per unit body weight, was determined for *Sericostoma personatum* at 5, 10, 15 and 20°C. The weight specific respiration (Rs) is plotted against dry weight on a ln-ln scale in Fig.1. In all four cases the linear regressions of Rs on dry weight are significant at the 0.1% level. Comparison of the pairs of regression lines using analysis of covariance shows no significant difference in the slopes of the lines, with the exception of the data of 5°C, which are significantly different at the 5% level. However in all cases there is a significant difference at the 0.1% level in the elevation of the lines.

The weight specific respiration of the larvae acclimated to 18°C was lower than that for the larvae acclimated to 6°C at all experimental temperatures and for all body weights.

Similar results were obtained for *Hydropsyche angustipennis*, the respiration being determined at 8, 12, 16 and 20°C. Again a ln-ln plot of Rs against dry weight, shown in Fig. 2, produces linear regressions which

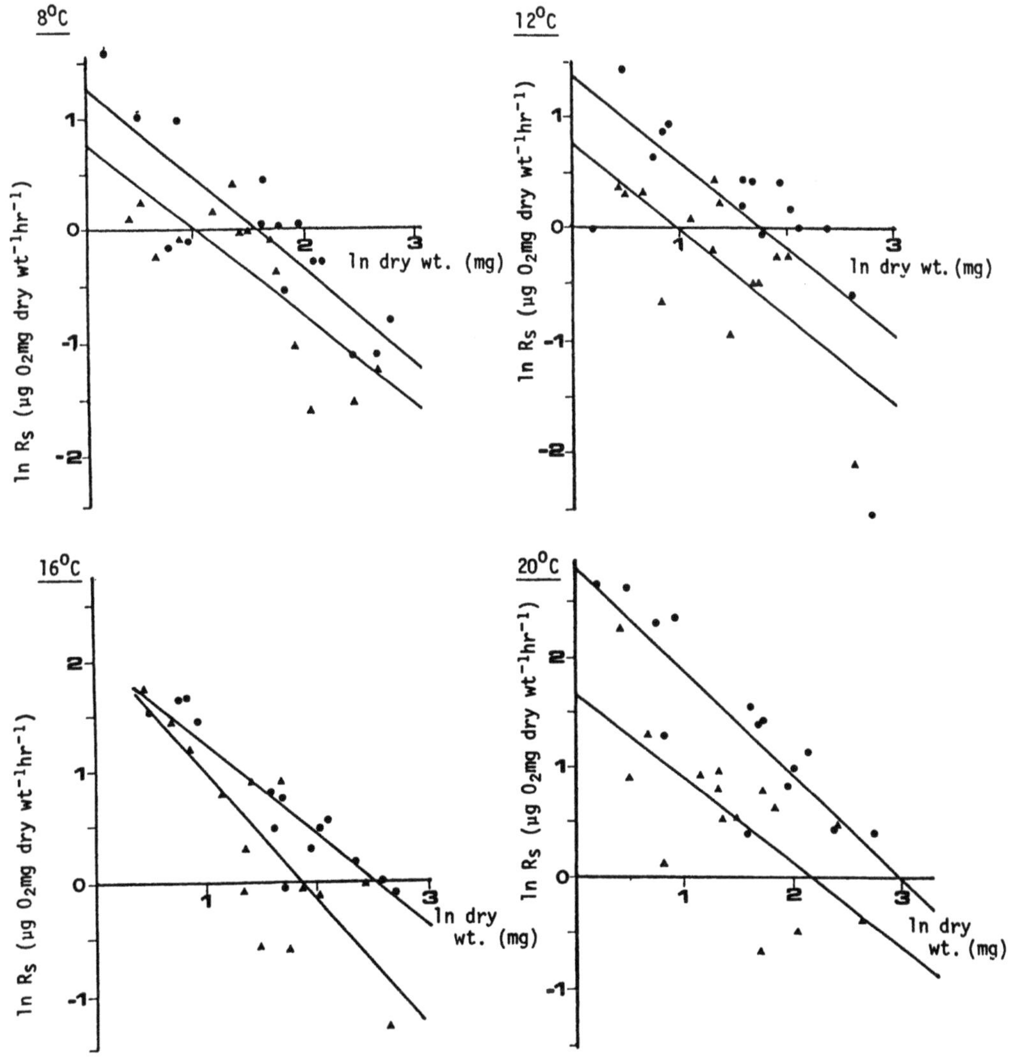


Figure 2. Ln-ln regressions of weight specific oxygen consumption ($R_S$) in *Hydropsyche angustipennis* against dry weights, at 8, 12, 16 and 20°C.●, 6°C acclimated larvae ; ▲, 18°C acclimated larvae.

are all signficant at the 0.1% level. No significant difference was found between the slopes of the pairs of lines, with the exception of the 16°C data which were significantly different at the 1% level. However, the elevations of the lines are all significantly different at levels ranging from 5% for the 12°C data to 0.1% for the 20°C data.

As with *S. personatum*, the larvae maintained at the higher acclimation temperature have a lower weight specific respiration for all body weights and at all experimental temperatures.

DISCUSSION

The data presented so far confirm that the temperature acclimation does occur in some species of caddis larvae over a period of four to five weeks. The main aim of this study is to compare the acclimation ability of a range of species, and to see if this has any relationship with their distribution. In order to compare species the measured weight specific respirations are converted to a standard body weight of 5mg, a weight falling within the range of weights for all of the species.

A measure of the extent to which acclimation has occurred is calculated as the percentage decrease in the weight specific respiration of the 18°C larvae compared with the 6°C larvae. Table 1 shows the mean weight specific respiration and the extent to which acclimation has occurred at 20°C for four species. For the three Hydropsychids there was a large variation in their weight specific respiration ; it ranged, for the 6°C acclimated larvae, from 3.85 for *H. angustipennis* down to 1.524 for *H. instabilis*, with *H. pellucidula* intermediate at 2.485. The percentage decrease in the weight specific respiration of the 18°C larvae compared with the 6°C larvae also showed a large variation, in the same order, ranging from 57% for *H. angustipennis* down to 25% for *H. instabilis*.

Table 1. Comparison of weight specific respiration (Rs) at 20°C standardized to a body weight of 5mg, of four species following acclimation to 6 and 18°C.

| | 6°C acclimated larvae | | 18°C acclimated larvae | | PD |
|---|---|---|---|---|---|
| | n | $R_S$ | n | $R_S$ | |
| *Sericostoma personatum* | 25 | 1.395 ± 0.038 | 28 | 0.943 ± 0.03 | 37.4 |
| *Hydropsyche angustipennis* | 14 | 3.851 ± 0.304 | 15 | 1.653 ± 0.213 | 57.1 |
| *Hydropsyche pellucidula* | 13 | 2.485 ± 0.092 | 14 | 1.446 ± 0.07 | 41.8 |
| *Hydropsyche instabilis* | 14 | 1.524 ± 0.075 | 14 | 1.139 ± 0.078 | 25.3 |

$R_S$ - weight specific respiration ($\mu g\ O_2$ Mg dry $wt^{-1}hr^{-1}$) mean ± S.E.

PD - percentage decrease in $R_S$ of the 18°C acclimated larvae compared to the 6°C acclimated larvae.

The weight specific respiration of *Sericostoma personatum* at 1.39 for the 6°C larvae is lower than that found for the three Hydropsychids, while the percentage decrease in the weight specific respiration of the 18°C larvae compared with the 6°C larvae is simalar to that found for *H. pellucidula*.

Linear regressions of ln weight specific respiration on ln dry weight for larvae of *Potamophylax cingulatus* acclimated to 5 and 10°C were again significant (5°C larvae, r = -0.831, 10°C larvae, r = -0.79). Analysis of covariance shows no significant difference in the slope of the lines, but a significant difference at the 0.1% level for the elevation of the lines. The larvae acclimated to the higher temperature again have a lower weight specific respiration at all body weights.

Although sufficient data for an adequate range of different species are not yet available for drawing firm conclusions, a number of tentative points can be made.

In all five species studied so far the ability to acclimate to temperature has been demonstrated.

Differences in the extent to which acclimation occurs indicate a possible relationship between ability to acclimate and the distribution of a species. *Hydropsyche instabilis*, which has a rather restricted distribution in cooler water has the lowest ability to acclimate of the five species studied, while *H. angustipennis* which has a more general southerly distribution in Britain (Badcock 1976), and is well known as being tolerant of high temperature and lower oxygen concentrations, has the greatest ability to acclimate. *H. pellucidula* may be considered intermediate in both distribution and its acclimation ability. The trend for increasing weight specific respiration from *H. instabilis* to *H. angustipennis* is the reverse of that found by Hildrew and Edington (1979). This may reflect interspecific differences in the response to water flow. Respiration in this experiment was determined in still water while in the experiment of Hildrew and Edington the water was stirred.

As data are only available as yet for one cased species, comparisons with other species are rather speculative. However the weight specific respiration of *S. personatum* is lower than the values obtained for the three hydropsychids. This may represent an inherent difference between cased and net spinning caddis, although alternatively it may be the result of differential activities, differences in the amount of non-metabolising material or possible respiratory benefits of the case. The percentage decrease in the weight specific respiration of the 18°C larvae compared with the 6°C larvae is similar to that found for *H. pellucidula*. Like *H. pellucidula* it has a fairly widespread distribution.

The occurrence of partial compensation in *P. cingulatus* following acclimation to 5 and 10°C was in contrast to results reported by Harrison and Badcock (1981) who found reverse acclimation. At the time the authors pointed out that this was a preliminary study and that the data base was low.

In conclusion, all species studied so far have exhibited differences in the metabolism of groups of larvae acclimated to two different constant temperatures over a period of weeks. Some evidence points towards a relationship between the ability of a species to acclimate to temperature and its distribution. This suggests that a greater acclimation ability may allow a species to extend its distribution.

It is important to appreciate that, although it is simple to demonstrate the occurrence of acclimation in a single species, comparisons of different species are more problematic as other factors, such as current velocity, will also affect the oxygen consumption to varying degrees. It should also be remembered that, although the differences are considered to be a response to temperature, they may in reality be a response to the different oxygen solubilities at the two temperatures.

## ACKNOWLEDGEMENTS

We are greteful to the Natural Environment Research Council for the training award currently supporting M.T.B.

## REFERENCES

Badcock, R.M. 1976. The distribution of the Hydropsychidae in Great Britain. **Proc. 1st Int. Symp. Trichoptera**, H. Malicky (ed.), Junk, The Hague : 48-58.

Bulnheim, H.P. 1979. Comparative studies on the physiological ecology of five euryhaline *Gammarus* species. **Oecologia**, 44 : 80-86.

Harrison, J.D. & Badcock, R.M. 1981. Effects of acclimation to temperature on the respiratory rate of some limnephilid larvae. **Proc. 3rd Int. Symp. Trichoptera**, G.P. Moretti (ed.), Junk, The Hague : 111-116.

Hazel, J.P. & Prosser, C. L. 1974. Molecular mechanisms of temperature compensation in Poikilotherms. **Physiol. Rev.**, 54, 3 : 620-677.

Hildrew, A.G. & Edington, J.M. 1979. Factors facilitating the coexistence of Hydropsychid caddis larvae (Trichoptera) in the same river system. **J. An. Ecol.**, 48 : 557-576.

Kamler, E. 1965. Thermal conditions in mountain waters and their influence on the distribution of Plecoptera and Ephemeroptera larvae. **Ekol. Pol., Ser. A.**, 13 : 377-414.

Keister, M. & Buck, J. 1974. Respiration : some exogenous and endogenous effects on rate of respiration. *In* M. Rockstein (ed.). The physiology of Insecta, 6, Acad. Press : 469-509.

Kulkarni, G.K. & Nagabhushanam, R. 1978. Thermal acclimation induced compensation in biochemical constituents of the Indian freshwater Leech, *Poecilobdèlla viridis* (Blanchard). **Hydrobiologia**, 58 (1) : 3-6.

Philipson, G.N. & Moorhouse, B.H.S. 1976. Respiratory behaviour of larvae of four species of the family Polycentropodidae (Trichoptera). **Freshwater Biol.** 6 : 347-353.

Rees, T.D. & Hilton, J. 1977. Improved efficiency in the Winkler method for the BOD test. Laboratory Practice Feb. : 91-93

Vernberg, W.B. & Vernberg, F.J. 1972. Environmental physiology of marine animals. Springer-Verlag, Berlin.

# CONTRIBUTION A LA CONNAISSANCE DES EPIBIONTES ET PARASITES DES LARVES DES TRICHOPTERES ITALIENS

C. CORALLINI SORCETTI and G.P. MORETTI

## SUMMARY

This study represents an updated review of the epibionts and parasites that have been found in the larvae of different species of Trichoptera coming from several regions of Italy. Various forms of both plant and animal life were studied : Algae-Fungi-Protozoa-Trematoda-Rotatoria- Nematoda-Nematomorpha-Annelida-Insecta.

## INTRODUCTION

Nous nous proposons dans cette étude d'exposer et de mettre à jour les observations et les informations que l'on possède aujourd'hui sur les épibiontes et les parasites des Trichoptères italiens. Depuis plusieurs années déjà, les auteurs s'occupent de ce problème, avec une attention particulière pour les protozoaires Ciliés (Péritriches) et les Sporozoaires (Grégarines) (Moretti et Corallini Sorcetti 1976-1980 ; Corallini Sorcetti 1981).

## EPIBIONTES

La cuticule des larves des insectes Trichoptères est l'hôte de colonies souvent très nombreuses de bactéries. En 1976 et en 1980, Moretti *et al.* ont étudié la présence de colonies de bactéries sur l'épicuticule des larves d'*Hydropsyche pellucidula* Curtis et d' *Hydropsyche modesta* Navas. On a constaté que la riche pubescence de leur tégument favorise l'établissement de ces micro-organismes qui peuvent atteindre des concentrations beaucoup plus élevées que dans les eaux fluviales. Les genres suivants ont été reconnus : *Aeromonas, Pseudomonas, Alcaligenes* ainsi que des formes pathogènes pour l'homme telles que *Escherichia coli, Shighella* sp., *Klebsiella* sp..

Chez les larves d'*H. modesta, H. pellucidula, Cheumatopsyche lepida* Pictet, *Limnephilus rhombicus* L., *Halesus appenninus* Mor. Spin, *Allogamus antennatus* Mcl, on a souvent observé la présence de filaments de *Phormidium tenue* (Cyanophyceae) qui se fixent sur la tête, l'abdomen, les pygopodes et les soies. Ces peuplements provoquent souvent des zones d'une couleur brunâtre sur le tégument. Un examen minutieux des morceaux de cuticule, effectué aussi avec des sections en paraffine, n'a révélé aucune forme d'approfondissement d'hyphes ou de filaments suceurs sous l'épicuticule, excluant la possibilité d'une action parasitaire.

Des stades aquatiques d'*Allogamus antennatus* de la rivière Nera sont fortement colonisés par une algue épizootique qui s'inscrit dans le genre *Gloxidium* (Euclorophyciae) et est principalement localisée sur les trachéo-branchies où elle forme de véritables manchons.

Il faut aussi rappeler que les espèces de Bacillariophyceae sont très nombreuses, celles-ci se trouvant, comme les bactéries, sur le tégument des larves des phryganes.

Les représentants des protozoaires ciliés sont très abondants parmi les épibiontes. Sur la cuticule d' *Odontocerum albicorne* Scop. on a trouvé des individus appartenant à l'ordre des Acinétiens localisés sur les sclérites abdominaux et sur les filaments trachéo-branchiaux. Le court pédoncule, la forme en calice et le gros noyau sont les éléments qui permettent d'attribuer ces individus au genre *Metacineta* (Moretti-Corallini 1985).

Plus fréquents et très répandus sont les Péritriches sessiles, les espèces de Trichoptères qui présentent des colonies de ces Protozoaires sont très nombreuses ; seules les larves carnivores et prédatrices semblent, en général, rester indemnes. La distribution des épibiontes est différente selon qu'ils colonisent des espèces avec ou sans fourreaux. En ce qui concerne les premières, ils s'installent principalement sur les sclérites thoraciques, la tête, les membres, les pygopodes, les trachéo-branchies, tandis qu'ils s'établissent sur le corps entier, en ce qui concerne les secondes, avec une plus grande prédilection pour les segments thoraciques et abdominaux, formant souvent des feutrages denses. De plus, l'absence de fourreaux favorise l'installation d'espèces à pédoncule long et ramifié. Au total on a reconnu quatre genres (Moretti *et al.* 1978) parmi lesquels *Epistylis* est le plus fréquent, alors que*Vorticella* est le plus rare ; et les espèces suivantes : *Vorticella convallaria* (L.) Noland - *Vorticella constricta* Fromentel - *Epistylis breviramosa* Stiller - *Epistylis rotans* Sveç - *Epistylis* sp. - *Opercularia phryganeae* Kaal - *Opercularia* sp - *Carchesium epistylis* Clap. u. L. - *Carchesium polipinum* Linné. Ces Ciliés sont aussi importants sous l'aspect de la saprobicité (Curds C.R., B Sc, Ph. D. 1969) et par rapport aux peuplements de Protozoaires présents dans le cours d'eau (Badcock 1986). (Pl. I).

La découverte des Rotifères est fréquente surtout chez les larves de la famille des Drusinés et des Limnéphilidés. Ils se multiplient en grande quantité, particulièrement dans les élevages en laboratoire et ils s'établissent principalement autour de l'orifice anal, exploitant comme aliment les matières fécales libérées par les larves. On les remarque aussi au niveau des sclérites thoraciques, au point d'attache des membres et plus rarement au niveau de l'abdomen. Les oviductes et germovitellaria pairs indiquent que ces formes appartiennent aux Bdelloïdes, la forme du mastax fort dont chaque uncus est pourvu d'un certain nombre de côtes parallèles, permet de les attribuer à la famille des Philodinés, genres *Philodina* et *Rotaria*. Ils vivent presque toujours en commun avec des colonies de Péritriches qui représentent ainsi pour eux un *pabulum* facile.

La présence des Ciliés favorise d'autres phénomènes d' ectosymbiose entre des larves de Trichoptères (*L. rhombicus, Potamophylax cingulatus* Steph) et des Oligochètes du genre *Naïs* qui vivent cachés sous leur pédoncule.

Les petites larves des Chironomidés Orthocladinés trouvées sur l'abdomen de *H. pellucidula*, ont le même comportement. Celles-ci construisent une sorte de tunnel de soie entre les colonies d'*Epistylis* et *Carchesium*, et elles en sortent pour s'alimenter de ces mêmes Protozoaires. Steffan, en 1967, cite déjà des cas d'associations entre les Chironomidés et des larves de *Phryganea* sp. et de Limnephilidés.

PARASITES

Dans les élevages en laboratoire, les nymphes des Trichoptères sont souvent envahies par les moisissures (*Saprolegnia* sp). Celles-ci se forment à cause de la faible oxygénation ou dans le cas de phénomènes de putréfaction du matériel présent dans l'aquarium. La *Saprolegnia* sp. est, en effet, un saprophyte, mais ses hyphes parviennent à pénétrer à travers les petites loges et les cocons nymphaux, ainsi qu'à travers la cuticule, envahissant le corps de l'insecte et provoquant la mort.

Comme nous l'avons déjà mentionné, de nombreuses études sont consacrées aux Grégarines ; en 1980, Moretti et Corallini avaient signalé une affection de grégarinose pour 38 taxa de Trichoptères, alors qu'on avait trouvé 29 espèces d'Eugrégarines en Italie. Actuellement, à la liste des Trichoptères intéressés par des phénomènes d'infestation, il faut ajouter *Polycentropus flavomaculatus* Pictet. et *P. mortoni* Mosely qui sont les hôtes de la Grégarine *Pileocephalus agilis* Geus, tandis que pour les Protozoaires il faut signaler la découverte d'*Asterophora caliciformis* Baudoin dans des larves de *Cheumatopsyche lepida*, et deux nouvelles espèces de *Pomania : P. morettii* (Corallini 1981) et *P. cianficconii* (Corallini, sous presse Actes SIB 1984) qui sont toutes les deux les hôtes des larves de *Potomophylax cingulatus* récoltées dans la rivière Nera (Marches) et dans l'émissaire du Lac Sirino (Basilicate). les Grégarines sont présentes dans des infestations plus ou moins massives mais, en général, elles sont bien supportées par l'hôte. Le gamonte est le stade de développement le plus fréquent, les dimensions varient selon l'espèce et oscillent d'une dizaine de µm dans le genre *Gregarina*, jusqu'à des dimensions visibles à l'oeil nu dans le genre *Pomania*. La syzygie s'observe facilement seulement pour les espèces du genre *Gregarina* qui présente aussi un dimorphisme sexuel ; dans les autres taxa que nous avons trouvés, elle représente un stade de développement plutôt bref qui se transforme rapidement en gamétocystes. Une autre phase très importante du cycle biologique pour la classification de ces Protozoaires est celle du céphalin, à cause des différentes formes que peut assumer l'appareil de fixation aux cellules de la paroi intestinale ou épimérite. Il faut souligner que les céphalins de *Globulocephalus hydropsyches* Baudoin, qui présente une évolution intracellulaire, (Planche 1) forment des niches dans les parois du mésentère. Il est probable, par conséquent, que dans le cas d'une infestation massive se produisent des dégats aux larves. Des altérations se produisent sans aucun doute par l'appareil de fixation de *Pomania* qui est constitué par une ventouse tellement robuste qu'elle détache les cellules de la paroi intestinale. Dans la zone abdominale qui correspond à la localisation intestinale des trophocytes, on remarque une nécrose des tissus sous-jacents. Ces deux cas, où une véritable action spoliatrice des parasites est mise en évidence, sont les seuls que nous avons observés.

Chez *H. modesta, H. pellucidula, Drusus improvisus* McL, on a trouvé des kystes de Trématodes. On les a remarqués aussi bien dans la région thoracique que dans la région abdominale, sous la cuticule, dans l'épaisseur comprise entre la musculature et l'intestin. Ils ont une forme ovalaire de 350 µm, montrent une enveloppe pseudochitineuse à double paroi et un amas interne moins consistant qui a révélé, seulement dans des cas rares, la présence d'un embryon (Boria 1977-1978). L'attribution du parasite n'est pas encore bien définie, mais il semble certain, selon l'étude de Vojtek (1969), qu'il s'agit de *Prohemistobolum anacetabolum.*

Dans les couches hypodermales et dans les corps gras d'*Ecnomus tenellus* Ramb., *Potamophylax cingulatus* et *Allogamus antennatus,* on a trouvé de nombreux kystes qui ont une consistance gélatineuse et qui, après fixation, montrent une surface riche en spinules et un embryon interne. Ils sont entourés (T1) d'un tissu de réaction de l'hôte, mesurent 200-250 µm et on en compte jusqu'à 10 dans un même individu. Il s'agit probablement de Nématodes, mais l'absence de sujets à un stade de maturation plus avancé ne nous a pas permis de les déterminer. Les larves végétariennes des Limnephilidés et des Sténophylacinés ont souvent dans l'intestin des Nématodes adultes qui peuvent s'inscrire dans le genre *Rhabditis* à cause de la forme du pharynx. Ils sont probablement ingérés en même temps que l'aliment.

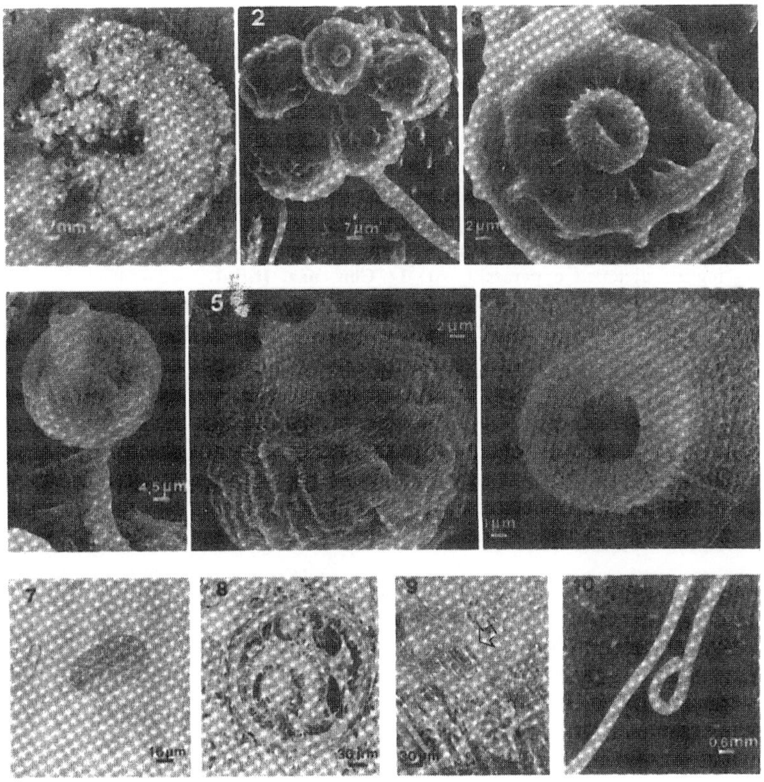


Planche 1 : 1) La larve d' *Hydropsyche modesta* avec le feutrage de *Carchesium polipinum.* 2) *Epistylis breviramosa* : colonie. 3) détail de *E. breviramosa.* 4) *Episylis* sp. 5) détail d' *Epistylis* sp. 6) détail du péristome d'*Epistylis* sp. 7) *Globulocephalus hydrophyches* . 8) kyste de Nématode. 9) Larve échinodéroïde. 10) Extrémité à deux lobes du mâle de *Gordius aquaticus.*

Sur la paroi intestinale de *Rhyacophila hartigi* Mal. et de *R. rougemonti* McL, on a trouvé des kystes de Gordiacés plutôt nombreux, 40 à 60 par intestin avec des dimension variables entre 70 et 120 µm. La larve échinodéroïde est visible et a un corps très distinct dans la partie céphalique et le tronc. La partie céphalique est constituée par une trompe qui peut s'invaginer et qui porte des épines et des styles. *Rhyacophila* représente probablement un hôte impropre puisque les kystes ne parviennent pas à se développer. Au contraire, les larves des Stenophylacinés et particulièrement *Potamophylax cingulatus,* hébergent des *Gordius* déjà développés. On a compté jusqu'à 5 vers dans un même individu. Ils se disposent dans la cavité coelomique, longitudinalement à l'intestin, comme les glandes de la soie, on peut aussi les apercevoir à travers la paroi du corps puisque les corps gras se réduisent considérablement. Nous avons pu observer clairement l'extrémité postérieure à deux lobes du mâle de *Gordius aquaticus*. Souvent dans les fourreaux vides on remarque encore des vers. (Pl. I)

En revanche, le parasite le mieux connu en Europe pour les Trichoptères, *Agryotypus armatus*, n'est pas très fréquent. On a trouvé peu de fourreaux d'*Odontocerum albicorne* dans lesquels cependant, on peut très bien voir le tube noir, d'une longueur d'environ 2mm, qui sert à la respiration de la nymphe de *Agryotypus*.

REFERENCES


Badcock, R.M. 1986. Peritrich ciliates epizoic on larvae of *Brachycentrus subnubilus* (Trichoptera) importance in relation to the total protozoan population in streams. **Hydrobiologia**, 132 : 125-131.

Boria, L. 1977-1978. Ecologia di un Ecclisopterigino della fauna italiana nella sorgente del F. Aso. Thèse Dip. Et. sup., Perugia.

Corallini Sorcetti, C. 1981. *Pomania morettii* sp.n. mihi in una larva di *Potamophylax cingulatus* Steph (F. Nera, Marche, Macerata). **Riv. Idrob.**, 20(3) : 561-570.

Curds, C.R. 1969. An illustrated key to the British freshwater Ciliated Protozoa commonly found in activated sludge. Water Centre London.

Moretti, G.P. & Corallini Sorcetti, C. 1976. Eugregarines infesting the digestive tract of Trichopteran larvae from some italian regions. **Boll. Zool.**, 43 : 69-73.

Moretti, G.P., Corallini Sorcetti, C & Montini, R. 1978. Epibiosi e parassitismo negli insetti Tricotteri. **Riv. Idrob.**, 17(1) : 3-24.

Moretti, G.P. & Corallini Sorcetti, C. 1981. Gregarines in Trichoptera larvae. **Proc. 3rd Symp. Trichoptera**, G.P. Moretti, (ed), Junk, The Hague.: 213-217.

Moretti, G.P. & Corallini Sorcetti, C. 1985. *Odontocerum albicorne* Scop. (Trichoptera). La larva : morfologia, anatomia, regime dietetico e parassiti. **At. 14 Con. naz. It.Ent.** : 233-240.

Moretti, G.P., Rossi, J., Clementi,F., Costamagna, L., Ingi, M & Spinelli, G. 1976. Fisionomia della colonizzazione batterica reperibile sul tegumento delle larve di *Hydropsyche dissimulata* e *Hydropsyche pellucidula*. **Riv. Idrob.**, 15(2) : 237-254.

Moretti, G.P., Rossi, J., Spinelli, G. & Clementi, F. 1980. Aggiornamento sulla fisionomia della colonizzazione batterica reperibile sul tegumento delle larve di *Hydropsyche pellucidula*e *Hydropsyche dissimulata*. **At 12 Con. naz. It. Ent.** 189-199.

Stefan, A.W. 1967. Ectosymbiosis in Aquatic Insect. *In* Symbiosis II. Ed. Academic Press : 207-289.

Vojtek, J. 1969. Die Bedentung der Wirbellosen fur die Levenszik der Unteroranung *Cyathocotylata*. **Sudarikov, Folia Un. Purkin Brunensis** : 135-146.

# INCIDENCE DES ELEMENTS TRICHOPTEROLOGIQUES DANS LE CONTENU INTESTINAL DES POISSONS DU COURS MOYEN DU PO (ITALIE)

P. GATTAPONI and C. CORALLINI SORCETTI

## SUMMARY

This research looks at 572 fish specimens belonging to twenty species taken from monthly catches over a two-year period (June 1983- May 1985) along the mid-stretch of the Pô River in the area of Piacenza. The objective of the study is to determine the presence of Trichoptera in the natural diet of fish living in eupotamic waters, the species most predator of these insects, the Trichoptera species most affected by this predation and seasonal trends.

## INTRODUCTION

Une enquête d'une durée de deux ans a été faite (Juin 1983-Mai 1985) portant sur la présence d'éléments trichoptérologiques dans le régime des poissons du cours moyen du Pô. La partie du fleuve ayant fait l'objet de l'enquête comprend les provinces de Piacenza et de Cremona (Fig. 1). Nous avons choisi ce secteur du fleuve étant donné qu'il représente une zone plutôt riche en espèces de poissons ; Gandolfi et Le Moli (1977) signalent la présence de 39 espèces pour le cours moyen-bas du Pô. Il y a, en outre, de nombreuses informations concernant la caractérisation de ce milieu aquatique, illustrées dans les actes du séminaire "Ecosistema del Medio Pô" (Piacenza, Laboratoire Central Enel-DCO) publié dans la Riv. di Idrobiologia 20(1), 1981.

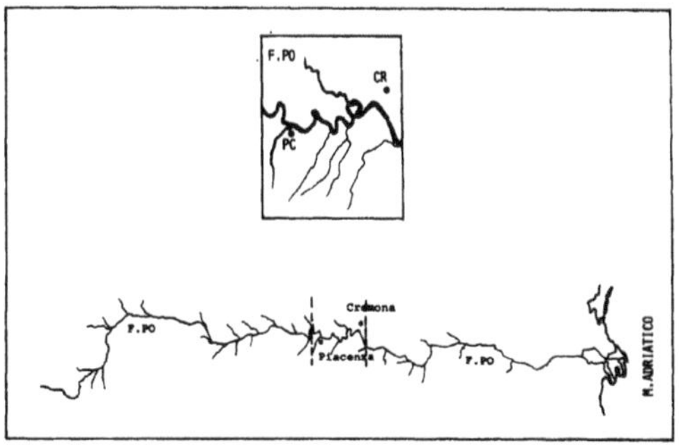


Figure 1.
Le cours du Pô et la partie examinée

## METHODOLOGIE

Les sujets ont été capturés au moyen de techniques locales en usage chez les pêcheurs : le trémail fixe, le trémail mobile et le "curiosa". En laboratoire, nous avons procédé à l'analyse des sujets, c'est-à-dire à l'identification de l'espèce, de l'âge et du sexe. Nous avons ensuite extrait le tube digestif, séparé les différentes composantes du contenu alimentaire et nous avons calculé l'importance relative, exprimée en pourcentage volumétrique, de tous les aliments repérés.

## RESULTATS

La partie du fleuve qui est l'objet de notre enquête est caractérisée par des méandres, des bras morts, des bas-fond, des îles fluviales, des confluences, des barrages, des peuplements hydrophytiques émergés et submergés (Moretti 1981). Le peuplement macrobenthique est constitué par les groupes suivants : les Turbellariés, les Oligochètes, les Hirudinés, les Mollusques, les Isopodes, les Amphipodes, les Décapodes, les Ephéméroptères, les Odonates, les Hémiptères, les Coléoptères, les Trichoptères et les Diptères (Citroni-Ruffo 1981). Parmi les insectes, les Diptères sont les plus nombreux et, au total, 109 taxons ont été individualisés (Ferrarese 1981).

On a distingué 14 espèces de Trichoptères dont la moitié seulement au stade d'*imago* (Tabl. 1) Les larves se trouvent dans le benthos, associées aux macrophytes et sur des substrats artificiels. Dans le benthos et sur les substrats artificiels prédominent les Hydropsychidae parmi lesquelles *Hydropsyche modesta* constitue l'espèce la plus abondante. (Moretti *et al.* 1981)

Au total, 572 poissons, appartenant à 20 taxons, ont été examinés (Tabl. 2). Les exemplaires les plus nombreux appartiennent à *Barbus barbus plebejus* et à *Leuciscus cephalus cabeda.* Les espèces qui contenaient des éléments trichoptèrologiques dans leur alimentation sont au nombre de 10. Le contenu alimentaire de la plus grande partie des sujets examinés se compose de déchets organiques digérés et de limon très fin pétri avec du mucus (Fig. 2) ; ensuite viennent les végétaux constitués par des fragments d'hydrophytes, d'algues, de graines et de feuilles. Le restant, présent en pourcentages différents, appartient au peuplement macrobenthique dont nous avons déjà parlé. Nous signalons, d'autre part, la découverte, dans le régime de nombreux poissons, de statoblastes et de restes de colonies de Bryozoaires parfois très abondants, tout comme des restes de poissons dans les espèces omnivores et prédatrices

Tableau 1. Trichoptères trouvés dans le cours moyen du Pô.

| TAXA | IMAGO | LARVES | | |
|---|---|---|---|---|
| | | Benthos | substrats artificiels | macrophytes |
| 1/*Hydroptila sparsa* Curtis | x | | | |
| 2/ *Agraylea sexmaculata* Curtis | x | | | |
| 3/ *Allotrichia pallicornis* Eaton | x | | | |
| 4/ *Hydropsyche modesta* Navas | x | x | x | |
| 5/ *Hydropsyche pellucidula* Curtis | | x | x | |
| 6/ *Polycentropus flavomaculatus* Pictet | x | x | | x |
| 7/ *Holocentropus dubius* Ramb | | | | x |
| 8/ *Holocentropus picicornis* Steph | x | | | x |
| 9/ *Psychomyia pusilla* Fabr | | x | x | |
| 10/*Ecnomus tenellus* Ramb | x | | | x |
| 11/ *Athripsodes cinereus* Curtis | x | | | |
| 12/ *Oecetis furva* Ramb. | x | | | |
| 13/ *Oecetis lacustris* Pictet | x | | | |
| 14/ *Leptocerus tineiformis* Curt | x | | | |

L'âge moyen des individus appartenant aux 10 taxa qui présentaient des éléments trichoptèrologiques, oscille entre un minimum de 3,71 années et un maximum de 5,4 années (Tabl. 3). On n'enregistre donc pas de valeurs très variées entre les différentes espèces. La sex-ratio est nettement en faveur des femelles à l'exception de *Tinca tinca* et de *Lepomis gibbosus.* Il y a beaucoup d'intestins vidés de leur contenu alimentaire et la fréquence de ceux qui présentent des restes de phryganides est faible. Les espèces de Trichoptères qui ont été découvertes sont : *Hydropsyche modesta, H. pellucidula, Hydropsyche* sp., *Athripsodes dissimilis* et *Limnephilus* sp.

Ce sont surtout les Hydropsychidae qui sont les proies préférées des prédateurs et qui, par ailleurs, sont les Trichoptères les plus nombreux dans le benthos. Il est intéressant d'observer que les restes de *H. pellucidula* sont plus fréquents que ceux de *H. modesta* qui est le Trichoptère prédominant dans cette partie du fleuve. Dans certains cas nous ne sommes pas parvenus à une détermination spécifique des pièces, celles-ci étant trop petites et en mauvais état.

*Ictalurus melas* est le poisson qui, selon les pourcentages, s'alimente le plus de Trichoptères, il est suivi par *Scardinius erythrophthalmus* et par *Barbus barbus plebejus.*

Tableau 2. Espèces de Poissons examinés, nombre de Poissons, composantes alimentaires

| Composantes alimentaires / Taxa | N. POISSONS EXAMINES | VEGETAUX | BRYOZOA | MOLLUSCA | ANNELIDA | CRUSTACEA | INSECTA | TRICHOPTERA | POISSONS |
|---|---|---|---|---|---|---|---|---|---|
| *Alburnus albidus alborella* Defil | 4 | | | | | | + | | |
| *Barbus barbus plebejus* Va | 132 | + | + | + | + | + | + | + | + |
| *Carassius carassius* L | 35 | + | + | + | | + | + | + | |
| *Chondrostoma genei* Bon | 22 | + | + | + | | + | + | | |
| *Chondrostoma soëtta* Bp | 47 | + | + | + | | | + | + | + |
| *Cyprinius carpio* L. | 13 | + | + | + | | | + | | + |
| *Cyprinius varpio* v. *specularis* Bp | 10 | + | + | + | | | + | | + |
| *Leuciscus cephalus cabeda* Risso | 126 | + | + | + | + | | + | + | + |
| *Rutilus pigus* Lac | 5 | + | | + | | + | + | + | |
| *Rutilus rubilio* Bp | 39 | + | + | + | | | + | + | + |
| *Scardinius erythrophthalmus* L | 44 | + | + | + | | + | + | + | |
| *Tinca tinca* L | 22 | + | | + | | + | + | + | |
| *Lepomis gibbosus* L | 30 | + | | + | + | + | + | + | + |
| *Micropterus salmoides* Lacép | 8 | + | + | | | + | + | | + |
| *Esox lucius* L | 6 | | | | | | | | + |
| *Perca fluviatilis* L | 18 | | | | | + | | | + |
| *Ictalurus melas* Raf | 6 | + | | | | + | + | + | |
| *Alosa fallax nilotica I.* Geoffr. | 1 | | + | | | | | | |
| *Silurus glanis* L | 2 | | | | | | + | | |
| *Liza ramada* Riss | 2 | | | | | | | | |

On trouve des restes de Trichoptères pendant toutes les saisons (Tabl. 4). *Barbus barbus plebejus* est l'espèce qui présente des restes pendant toute l'année, avec une plus grande fréquence en hiver, au printemps et en automne. Dans d'autres espèces telles que : *T. tinca, L. gibbosus*, on trouve des Trichoptères seulement en été.

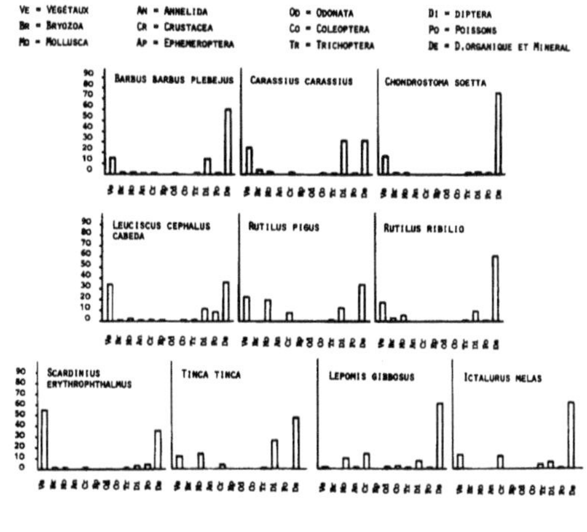


Figure 2. Contenus alimentaires : % des différents éléments par rapport au volume total.

A partir des données enregistrées, il résulte, en réalité, que le taux des Trichoptères est très modeste, pratiquement insignifiant par rapport aux autres aliments.

Il est cependant intéressant de comparer leur pourcentage à celui des autres insectes (Fig. 3). Après les Diptères, qui sont nettement prédominants, il apparaît que les insectes qui constituent la proie préférée sont les Trichoptères, suivis par les Coléoptères, les Odonates et les Ephéméroptères. De plus, il faut souligner que parmi les Diptères, on trouve souvent de grosses larves employées comme appât pour capturer les poissons. Ce sont pour la plupart des stades larvaires qui constituent la proie, sauf pour un seul exemplaire ailé femelle d'*Hydropsyche* sp., trouvé dans *B. barbus plebejus*, avec des ailes et un abdomen encore en bon état. Les proies ne sont pas ingérées entièrement, mais sous forme de fragments, et dans la plupart des cas elles sont complètement digérées, tandis que la cuticule et les parties chitineuses subistent telles quelles. Dans les *Hydropsyche*, par exemple, on trouve souvent des abdomens presque vides, mais la pubescence caractéristique de l'abdomen reste intacte.

Tableau 3. Espèces de poissons contenant des éléments trichoptérologiques, nombre, sexe, âge, Trichoptères trouvés.

| Taxa | Intestins examinés | vides | avec Trichoptères | sexe ♂ | ♀ | age | Trichoptères trouvés |
|---|---|---|---|---|---|---|---|
| *Barbus barbus plebejus* | 132 | 18 | 8 (6,06 %) | 21 | 111 | 5,01 | *H. pellucidula* - *H. modesta* *H.* sp. |
| *Carassius carassius* | 35 | 8 | 2 (5.71 %) | 1 | 34 | 3,87 | *H. pellucidula* |
| *Chondrostoma soëtta* | 47 | 24 | 1 (2,12 %) | 14 | 33 | 4,76 | *H.* sp. |
| *Leuciscus cephalus cabeda* | 126 | 46 | 5 (3,96 %) | 32 | 94 | 4,63 | *H. pellucidula* |
| *Rutilus pigus* | 5 | | 1 (20 %) | 2 | 3 | 5,4 | *H. modesta* |
| *Rutilus rubilio* | 39 | 13 | 1 (2,56 %) | 3 | 36 | 3,71 | *Athripsodes dissimilis* |
| *Scardinius erythrophthalmus* | 44 | 15 | 3 (6,81 %) | 8 | 36 | 4,95 | *H. pellucidula* |
| *Tinca tinca* | 22 | 8 | 1 (4,54 %) | 12 | 10 | 4,27 | *Limnephilus* sp. |
| *Lepomis gibbosus* | 30 | 7 | 1 (3,33 %) | 18 | 12 | 4,4 | *H. modesta* |
| *Ictalurus melas* | 6 | | 2 (33,33 %) | 2 | 4 | 3,85 | *H. pellucidula* - *H. modesta* |

Tableau 4. Contenu alimentaire : éléments trichoptèrologiques dans les différentes saisons.

| TAXA | SAISONS | | | |
|---|---|---|---|---|
| | PRINTEMPS | ETE | AUTOMNE | HIVER |
| *Barbus barbus plebejus* | x | x | x | x |
| *Carassius carassius* | | | x | |
| *Chondrostoma soëtta* | x | | | |
| *Leuciscus cephalus cabeda* | x | | | x |
| *Rutilus pigus* | | | | x |
| *Rutilus rubilio* | x | | | |
| *Scardinius erythrophthalmus* | x | x | x | |
| *Tinca tinca* | | x | | |
| *Lepomis gibbosus* | | x | | |
| *Ictalurus melas* | | x | x | |

Moretti *et al.* (1981), se référant au rapport du Dr. R. Vitali, avaient fait allusion à la présence des stades aquatiques des Trichoptères dans *B. barbus plebejus* et dans *L. cephalus cabeda.* Il résulte que ce dernier montre un taux plus élevé de restes trichoptérologiques, même avec des proportions plus faibles tandis que *B. barbus plebejus* révèle une destruction moins avancée des larves ingérées et des individus presque intacts sans dépigmentation du dessin de la tête ou du thorax. Les pourcentages saisonniers des présences diffèrent de nos observations, puisqu'une plus grande incidence des restes à été mise en évidence pendant l'hiver ainsi que leur absence en été et en automne.

En ce qui concerne la découverte et la destruction des restes, on peut remarquer que :

1) *Barbus barbus plebejus* : présente des restes d' *H. modesta* et d' *H. pellucidula* consistant en parties chitineuses sclérifiées, en têtes, en champs denticulés de genae en pro-meso-méta-notum, en membres plus ou moins fragmentés, en abdomens au corps adipeux effiloché mais non détruit ; dans le contenu alimentaire on relève aussi la présence de tubes malpighiens avec des granules d'urates.

2) *Carassius carassius* : met en évidence des fragments de larves d' *H. pellucidula* qui sont des morceaux de capsules céphaliques avec un clypeus, des mandibules et des segments thoraciques.

3) *Chondrostoma soëtta* : présente des fragments plutôt abîmés de pleures, de capsules céphaliques, de pygopodes d'*Hydropsyche* sp.

4) *Leuciscus cephalus cabeda* : montre des restes de larves d' *H. pellucidula*, des têtes entières, des mandibules denticulées et des soies intactes, des segments thoraciques, des coxa, des fémurs, des pattes antérieures à petit trochanter, des abdomens à pubescence intacte ; on aperçoit encore le contenu alimentaire.

5) *Rutilus pigus* : nous avons trouvé des têtes de jeunes *H. modesta.*

6) *Rutilus rubilio* : de petits fourreaux d'*Athripsodes dissimilis* ont été trouvés avec des statoblastes de Bryozoaires.

7) *Scardinius erythrophthalmus* : présente des clypeus, des mandibules, des sclérites thoraciques avec une pigmentation et une pubescence intactes, des fémurs, et des morceaux de tissus d'*H. pellucidula.*

8) *Tinca tinca* : on a trouvé des sclérites thoraciques ainsi que des fragments de clypeus de *Limnephilus* sp.

9) *Lepomis gibbosus* : met en évidence des clypeus, des fragments de capsules céphaliques d'*H. modesta.*

10) *Ictalurus melas* : montre des restes de *H. modesta* avec des capsules céphaliques, des clypeus, des labres, des mandibules, des champs denticulés des genae, tandis qu' on a des clypeus, des sclérites thoraciques, des fémurs et des tibias d'*H. pellucidula.*

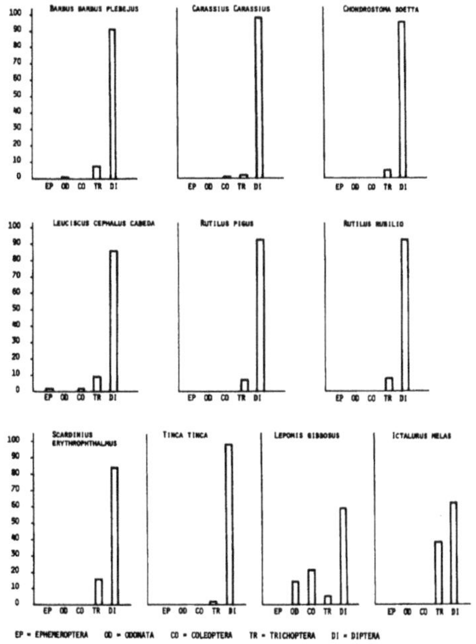


Figure 3. Contenus alimentaires : % des différents ordres par rapport au volume total des insectes.

## CONCLUSIONS

Deux années d'enquête sur 20 espèces de poissons du cours moyen du Pô démontrent que seulement 10 taxa montrent des restes de Trichoptères avec un taux de présence et un pourcentage très réduit par rapport aux autres aliments. Il faut cependant souligner qu'après les Diptères les Trichoptères sont les proies préférées.

*H. pellucidula* est le Trichoptère le plus fréquemment ingéré. Bien qu'étant peu nombreux, les restes que nous avons trouvés ont permis de constater la présence de deux espèces de Trichoptères qui s'ajoutent à la liste de celles qui ont été signalées pour le cours moyen du fleuve : *Limnephilus* sp.- *Athripsodes dissimilis.*

REMERCIEMENTS

Nous remercions l'Enel de Piacenza qui nous a fourni les poissons pour notre recherche.

REFERENCES

Cironi, R. & Ruffo, S. 1981. Alcune considerazioni sul popolamento a macroinvertebrati del fiume Po a Caorso. **Riv. Idrob.,** 20(1) : 47-76.

Ferrarese, U. 1981. Nota sui Ditteri del medio corso del Fiume Po a Caorso (Piacenza). **Riv. Idrob.,** 20(1) : 245-253.

Gandolfi, G & Le Moli, F. 1977. Distribuzione della fauna ittica nel Po. **Quad. I.R.S.A.,** 32 : 723-758.

Moretti, G.P. 1981. Impostazione e significato dell ' indagine : Caratteristiche generali dell 'ecosistema del medio Po. **Riv. Idrob.,** 20(1) : 15-45.

Moretti, G.P., Tucciarelli, F. & Cianfiocconi, F. 1981. Composizione e consistenza del popolamento Tricotterologico nell ' ecosistema fluviale del Medio Po (Caorso-Piacenza). **Riv. Idrob.,** 20(1) : 231-244.

# 7. LIGHT TRAPS AND FLIGHT PATTERNS

# RESULTS OF FOUR YEARS' INTENSIVE LIGHT TRAPPING IN WESTERN SWITZERLAND

C. SIEGENTHALER

## SUMMARY

Between 1980 and 1983, 12 light traps were operated from April to November in locations in Western Switzerland chosen for their diverse geographical and meteorological conditions.

Large captures (70,000 insects belonging to 15 families) allowed an extensive study of geographical distribution and relative abundance in correlation with local variables such as altitude, surroundings and seasonal changes.

## INTRODUCTION

From 1980 to 1983, light traps were operated from April to November in 12 locations in Western Switzerland. Light traps are known to be very effective in attractiving Trichoptera (Malicky1973), there have been detailed studies such as those of Crichton (1960, 1971, 1974). The dead insects were collected every few days : approximately 70,000 caddis flies comprising 117 species were caught.

## DESCRIPTION OF THE PROSPECTED AREA

Western Switzerland presents a very wide range of biotopes due to its exceptionally varied altitudes. The choice of the locations was made to allow the study of populations in as many different biotopes as possible. The light traps were mostly installed on river banks in order to study their Trichoptera fauna (8 traps) : others were set up some distance away from water so as to examine the flying capabilities of the different species (4 traps). This phenomenon has been studied in more detail in my earlier report (Siegenthaler 1985).

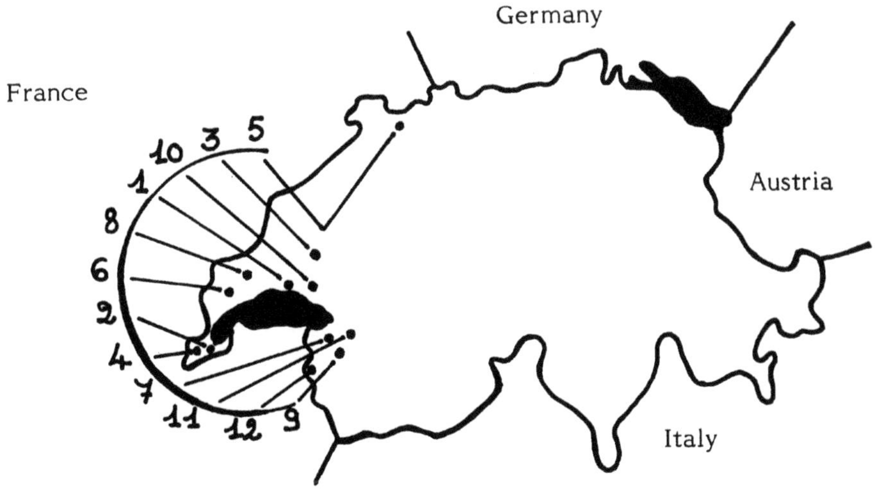


Figure 1. Map of the prospected area.

21 physico-chemical parameters have been computerised over the period to describe the different locations. Of these features, only the most relevant ones have been selected for the table below.

RESULTS

The present study investigates the zoogeographical zonation of the species of caddis flies in relation to regional conditions and seasonal changes.

**Regional conditions**

A detailed study of the 117 species in relation to biotopes has now been made. However, I feel it would be of more interest to the reader to find overleaf, in Table 2 a brief summary of the situation using some typical examples from five significant families.

Where a family represented no more than 4% of the total captured, further analysis had little significance.

Table 1. List of locations in order of altitude.

| Name of the location | Altitude in meters | Surroundings | Importance of the river width m | depth m | velocity* cm/s |
|---|---|---|---|---|---|
| 1. Lutry | 385 | Orchards | 3 | 0.10 | + |
| 2. Croix de Rozon | 470 | Meadows | river 15 m away | | |
| 3. Lucens | 480 | Meadows | 20 | 1 | + |
| 4. Sézenove | 509 | Orchards | river $1\frac{1}{2}$ km away | | |
| 5. Delémont | 510 | Orchards | river 1 km away | | |
| 6. Bois-de-Chênes | 516 | Deciduous tree forests | 1 | 0.05 | + |
| 7. Verschiez | 543 | Deciduous tree forests | river 3 km away | | |
| 8. Bière | 650 | Deciduous tree forests | 3.50 | 0.50 | +++ |
| 9. La Peufaire | 780 | Pine tree forests | 2 | 0.50 | +++ |
| 10. Les Pueys | 1'150 | Pine tree forests | 4 | 0.15 | ++ |
| 11. Taveyannaz | 1'530 | Meadows | 2 | 0.10 | + |
| 12. Bretolet | 1'923 | Meadows | river 3 km away | | |

* Velocity of surface waters measured in mid-current by deplacement of standard cork.
+ : 5-25 cm/s
++ : 26-50 cm/s
+++ : over 50 cm/s

At most of the locations, half of the trapped caddis flies belonged to the Limnephilidae which comprise a great number of European species. In three locations, samples indicated dominant families other than Limnephilidae ; these locations were situated at low altitudes either on the Plateau (Lucens 480 m) or near the Lake of Geneva's shore (Lutry 385 m and Sézenove 509 m). This would confirm Ward's observations (1980) in Colorado U.S.A., where Limnephilidae were said to be absent or rare in the lower foothills and plains. At these altitudes the majority of Trichoptera thus belonged to the Psychomyiidae (54% in Lucens) or to the Leptoceridae (47% in Lutry and 49.9% in Sézenove) clearly showing the lake's influence at the last two locations.

Rhyacophilidae were normally present except in Bretolet (1923 m) where they were totally absent ; the combination of a severe climate and the shortage of water explains the absence of this family. Four locations had 1% or less Rhyacophilidae: Croix de Rozon, Delémont, Bois-de-Chênes and Verschiez. This was probably due to the fact that running waters were either scarce or consisted of streams that dried up during mid-summer. It also indicated that Rhyacophilidae do not fly far. Rhyacophilidae, as their name indicates, are found in fast running waters with good oxygenation. These conditions are present in mountain resorts.

Hydropsychidae, unlike Rhyacophilidae, seemed to thrive in calm and shallow waters, conditions mostly found in the plains to which they are almost restricted. They were therefore absent from Taveyannaz and Bretolet, two high altitude locations. Unlike Rhyacophilidae, they seem to be good flyers as most of the locations situated relatively far from water had about 25% of their caddis fauna belonging to this family. (Croix de Rozon, Delémont, Sézenove).

Another point of interest was how the population numbers changed during the season with a shifting of the peak of abundance towards autumn due to change of climate with increased altitude. (Fig. 2).

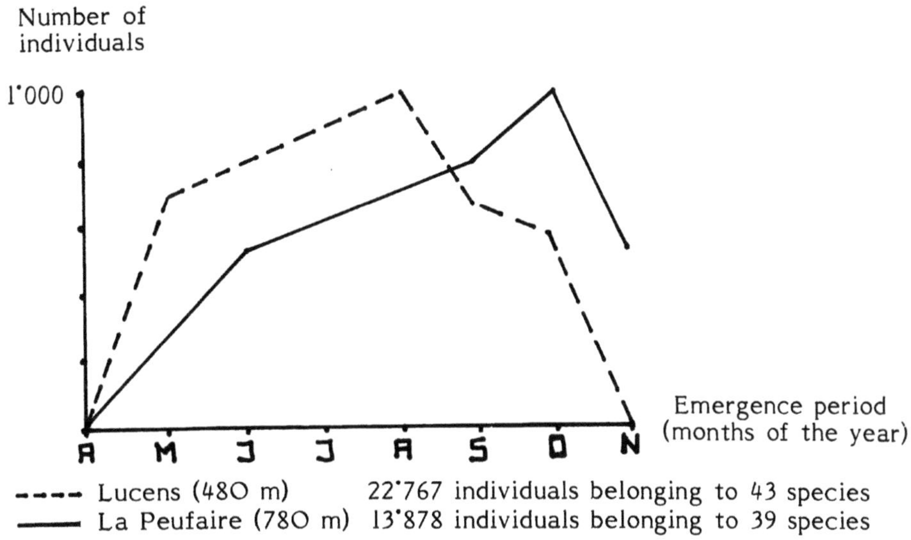


Figure 2. Shifting of abundance peak with season and altitude.

## Seasonal changes

After demonstrating the already described seasonal changes in numbers, I charted the changes in seasonal numbers for each species.

In addition to the shifting of the peak of abundance towards October for mountain locations, this study revealed a contraction of the period of flight activity as the climate became more severe with altitude. This confirmed the impression I had gained from trapping large numbers of insects. (Table 3).

Table 3. An example of the contraction phenomenon observed for some species, as shown by *Limnephilus sparsus* (Curt.).

| | IV | V | VI | VII | VIII | IX | X | XI | emergence period (months of the year) |
|---|---|---|---|---|---|---|---|---|---|
| Verschiez (543 m) | | | | | | | | | |
| Les Pueys (1.150 m) | | | | | | | | | |
| Taveyannaz (1530 m) | | | | | | | | | |

The shifting and the contraction to the flight period described above is only found in species which have a wide distribution. For those which have a well defined niche, the answer to a change of conditions is to leave or die.

Table 2. Percentage family representation for each location.

| | Lutry | Croix de Rozon | Lucens | Sézenove | Delémont | Bois-de-Chênes | Verschiez | Bière | La Peufaire | Les Pueys | Taveyannaz | Bretolet |
|---|---|---|---|---|---|---|---|---|---|---|---|---|
| Rhyacophilidae | 2 | 1 | 13 | 4 | 0,5 | 1 | 1 | 16 | 5 | 37 | 17 | |
| Glossossomatidae | | 0,7 | | | | 0,5 | | 3 | 0,07 | 1 | | 1 |
| Hydroptilidae | 4 | | 0,5 | 0,2 | | | | | | | | |
| Philopotamidae | 0,1 | | | | | | 0,1 | | 0,1 | 4 | 4 | |
| Hydropsychidae | 17 | 22,3 | 23,5 | 27 | 27 | 33 | 1 | 2 | 0,03 | 3 | | |
| Polycentropodidae | 1 | 2,5 | 0,1 | 0,5 | | 3,5 | 0,4 | 0,5 | 3,8 | 3 | 4 | 0,9 |
| Psychomyidae | 21,8 | | 54 | 0,5 | | | | 0,5 | | | 0,5 | |
| Ecnomidae | 0,7 | | 0,1 | | | | | | | | | |
| Phryganeidae | 0,3 | 0,5 | | 0,8 | 0,5 | 1 | | | | | | 0,1 |
| Limnephilidae | 5,5 | 64 | 2,5 | 17 | 68 | 56 | 97,5 | 77 | 90,5 | 47 | 73 | 98 |
| Goeridae | 0,2 | | 1,8 | | | | | | | 0,5 | | |
| Leptoceridae | 47 | 9 | 4 | 49,9 | | 0,2 | | 0,3 | | | | |
| Lepidostomatidae | 0,1 | | | | 3 | | | | 0,1 | 0,5 | | |
| Sericostomatidae | 0,1 | | | | 0,5 | 0,8 | | 0,2 | 0,4 | | | |
| Odontoceridae | 0,2 | | 0,5 | 0,1 | 0,5 | 4 | | 0,5 | | 4 | | |

## CONCLUSION

Five Trichoptera families are particularly successful in western Switzerland and have adapted to a range of biotopes. The detailed study of locations as different as possible throughout Western Switzerland provided me with some insight into the requirements and way of life of the most widespread species. Some characteristics and particular requirements are :

- Level of oxygen needed related to water flow ;
- Flying ability to ensure distribution ;
- Adaptation to change of climate : shifting of the emergence period towards autumn or contraction of peak of the flight period as the climate becomes more severe with altitude.

## REFERENCES

Crichton, M.I. 1960. A study of captures of Trichoptera in a light trap near Reading, Berkshire. **Trans. R. ent. Soc. Lond.**, 112 (12) : 319-344.

Crichton, M.I. 1971. A study of caddis flies (Trichoptera) of the family Limnephilidae, based on the Rothamsted Insect Survey, 1964-68.**J. Zool. Lond.** 163 : 533-563.

Crichton, M.I. 1974. The interpretation of light trap catches of Trichoptera from the Rothamsted Insect Survey. **Proc. first Int. Symp. on Trichoptera,** H. malicky (ed.), Junk, The Hague : 147-158.

Malicky, H. 1973. Trichoptera (Köcherfliegen). **Handb. Zool.**, 4(2) 2/29 : 1-114.

Siegenthaler, C. 1985. Un exemple de réussite chez les Trichoptères : Les Limnephilidae. **Mitt. schweiz. Ent. Ges.**, 58 : 461-466.

Ward, J.V. 1980. Altitudinal distribution and abundance of Trichoptera in a rocky mountain stream. **Proc. third Int. Symp. on Trichoptera,** G.P. Moretti (ed.), Junk, The Hague : 375-381.

# LIGHT TRAP COLLECTIONS OF TRICHOPTERA NEAR THE SOURCE OF HOPP BROOK, NEW HAVEN COUNTY, CONNECTICUT

V. A. NELSON

## SUMMARY

A single-location light trap was used to survey an area encompassing approximately five wooded acres containing a sand-bottom stream, and a swamp subject to drying in the summer. Sixty-five species of caddisflies representing fourteen families were captured from the beginning of May to the end of November during the three years of the study. The trap was set up so that the insects attracted to a UV tube flew or dropped into a collector containing a 50% solution of ethylene glycol which acts as a preservative and does not readily evaporate.

## INTRODUCTION

While studying material for preparation of "The Trichoptera of Connecticut" I decided to heed the words of C.P. Alexander, who once said that one should "learn what is in his own back yard". With this in mind, a light trap was set up using a white garage door as a backround. Insects were attracted to the light from an area across the street containing a stream in a marsh/swamp that depends on summer rainfall to maintain standing water. The stream is the source of Hopp Brook, a first order stream that flows into the Naugatuck River which ultimately empties into Long Island Sound. This arrangement proved to be a very good survey tool, and yielded a surprising diversity of caddisflies in a limited area.

## MATERIALS AND METHODS AND DESCRIPTION OF THE STUDY AREA

A Ward's light trap (Ward's Natural Science Establishment, Rochester, New York) consisting of a pail that supported an 8 watt fluorescent UV lamp surrounded by four reflecting vanes at right angles was used to attract the insects for this study. A second pail containing a 50% solution of ethylene glycol (Prestone® antifreeze) was inserted in the first and served as the collector. Insects were attracted to the light and flew or dropped into the fluid. The trap was placed in front of a white garage door that acted as an additional reflector. It was operated from one half hour before dark until approximately 11 P.M. The light trap was run from mid April to the cold part of November, and on warm days that occurred after that. Collecting averaged three times per week during the seasons over the three years of this study. The insects were removed using a sieve and the fluid reused for subsequent collections. All caddisflies were retained and stored in 80% alcohol for this study. Males were identified for recording purposes.

The source of Hopp Brook is located in a heavily wooded area, bounded on the northwest by a rock ledge approximately 50 feet high and 400 feet from the location of the light trap. The effective visual range of the light, encompassing about 5 acres, is limited by the rock ledge and the heavy vegetation consisting largely of maple and oak trees and a variety of understory plants. The stream is 250 feet from the light source.

## RESULTS AND DISCUSSION

A total of 65 species of Trichoptera representing 15 families was collected (Table 1). Twenty-seven species are represented by 1, 2 or 3 collections of one to four specimens within a time frame of less than four days. The remainder were more commonly collected over a longer period of time.

The collecting dates represent the earliest and latest that the insects were collected, and do not necessarily correspond to emergence and cessation of activity. *Dolophilodes distinctus*, for example, can be collected in Connecticut in February, but is not active enough to fly to lights (cold weather is the most likely factor). The total also represents only those species that are attracted to light ; there may also be other caddisflies in the area, although I have not collected others.

It would appear that the area surveyed supports a broadly diverse caddisfly fauna. The limited range covered by the light trap suggests that all or most of the species collected are breeding in the study area, although emergence trap studies should be conducted to verify this. Additionally, it is not known how many, if any, caddisflies are flying in (above the trees) from nearby areas.

Table 1. Seasonal occurrence of caddisflies based on light trap collections - 1982-1984.

| | First and last collections |
|---|---|
| Brachycentridae | |
| *Micrasema wataga* Ross | one collection - mid June |
| Glossosomatidae | |
| *Glossosoma nigrior* Banks | June 13 - July 18 |
| Hydropsychidae | |
| *Ceratopsyche bronta* Ross | June 13 - Sept 7 |
| *C. morosa* Hagen | June 11 - Sept 19 |
| *C. sparna* Ross | June 11 - Sept 19 |
| *C. slossonae* Banks | two collections - mid July |
| *C. ventura* Ross | June 5 - June 14 |
| *Cheumatopsyche pettiti* (Banks) | June 13 - Aug 8 |
| *Hydropsyche betteni* Ross | June 5 - Sept 7 |
| *Parapsyche apicalis* (Banks) one | one collection - mid June |
| Hydroptilidae | |
| *Hydroptila ampoda* Ross | one collection - early June |
| *Oxyethira* sp. | one collection - early August |
| Lepidostomatidae | |
| *Lepidostoma carrolli* Flint | two collections - early Sept |
| *L. griseum* (Banks) | one collection - early Sept |
| *L. serratum* Flint and Wiggins | one collection - mid Sept |
| Leptoceridae | |
| *Ceraclea transversa* (Hagen) | one collection - mid June |
| *C. maculata* (Banks) | one collection - mid July |
| *Leptocerus americanus* (Banks) | June 24 - Aug 3 |
| *Oecetis cinarescens* (Hagen) | July 4 - Sept 19 |
| *O. inconspicua* (Walker) | June 13 - Sept 11 |
| *Triaenodes tarda* Milne | two collections - mid June |
| Limnephilidae | |
| *Anabolia bimaculata* (Walker) | two collections - mid July |
| *A. consocia* (Walker) | June 24 - Sept 11 |
| *Frenesia difficilis* (Walker) | Nov 9 - Dec 14 |
| *F. missa* (Milne) | Nov 9 - Dec 14 |
| *Ironoquia lyrata* Ross | Sept 1 - Sept 19 |
| *I. parvula* (Banks) | Oct 8 - Oct 13 |
| *I. punctatissima* (Walker) | Sept 1 - Sept 21 |
| *Hydatophylax argus* (Harris) | three collections - mid June |
| *Limnephilus indivisus* Walker | June 24 - Sept 11 |
| *L. moestus* Banks | June 4 - July 15 |
| *L. sericeus* (Say) | one collection - early Oct |
| *L. submonilifer* Walker | one collection - early Sept |
| *Neophylax concinnus* McLachlan | Oct 13 - Nov 9 |
| *Platycentropus radiatus* (Say) | June 13 - July 18 |
| *Pseudostenophylax sparsus* (Banks) | two collections - mid June |
| *P. uniformis* Betten | three collections - mid June |
| *Pycnopsyche circularis* (Provancher) | one collection - mid Sept |
| *P. gentilis* (MacLachlan) | Sept 10 - Sept 19 |
| *P. guttifer* (Walker) | two collections - mid Sept |
| *P. luculenta* (Betten) | July 12 - Sept 21 |
| *P. scabripennis* (Rambur) | two collections - early Sept |

| | |
|---|---|
| Molannidae | |
| *Molanna blenda* Sibley | July 12 - Aug 7 |
| Odontoceridae | |
| *Psilotreta labida* Ross | June 5 - June 14 |
| Philopotamidae | |
| *Chimarra socia* Hagen | June 24 - July 19 |
| *Dolophilodes distinctus* (Walker) | June 11 - Sept 7 |
| *Wormaldia moesta* (Banks) | June 8 - Aug 5 |
| Phryganeidae | |
| *Agrypnia vestita* (Walker) | June 5 - Sept 10 |
| *Banksiola crotchi* Banks | June 11 - July 16 |
| *B. dossuaria* (Say) | June 4 - July 29 |
| *Phryganea sayi* Milne | July 15 - Aug 4 |
| *Ptilostomis ocellifera* (Walker) | June 4 - Aug 5 |
| *P. postica* (Walker) | June 24 - Sept 5 |
| *P. semifasciata* (Say) | one collection - early July |
| Polycentropodidae | |
| *Phylocentropus lucidus* (Hagen) | June 11 - July 18 |
| *Polycentropus cinereus* (Hagen) | one collection - mid July |
| *P. clinei* (Milne) | two collections - early Aug |
| *P. crassicornis* (Walker) | June 16 - July 16 |
| *P. interruptus* (Banks) | one collection - mid June |
| *P. pentus* Ross | one collection - early June |
| Psychomyidae | |
| *Lype diversa* (Banks) | June 13 - July 18 |
| Rhyacophilidae | |
| *Rhyacophila carolina* Banks | June 9 - July 15 |
| *R. glaberrima* Ulmer | two collections - early Sept |
| *R. vibox* Milne | two collections - mid June |
| Sericostomatidae | |
| *Agarodes grisea* Banks | June 8 - June 24 |

LeSage and Harrison (1979) used a mixture of ethylene glycol and alcohol as a collecting fluid. For the present study, the 50% solution of commercial antifreeze proved highly satisfactory. Preservation of Trichoptera and Diptera is excellent, and this method provides much better specimens than dry collecting where Lepidoptera scales get on everything. Unfortunately, moths do not fare well, nor, curiously, do most beetles which tend to disarticulate. Ethylene glycol does not readily evaporate, and the specimens can remain in the fluid for a month or more without harm.

## REFERENCE

LeSage, L. & Harrison, A.D. 1979. Improved traps and techniques for the study of emerging aquatic insects. **Entomol. News**, 90 : 65-78.

# FOREST CANOPY COLLECTION OF TRICHOPTERA

W. DOWNS, & P.A. GODWIN.

## SUMMARY

Trichoptera were saved from canopy collections of insects, obtained by spraying mixed hardwoods and pines in a second growth tree stand in lowland Connecticut, U.S.A. Trichoptera were found in 105 collections, belonging to 10 families, 22 genera and 32 species. All of the species encountered had been found in light trap captures made at ground level in nearby regions except *Triaenodes nox,* thus far encountered in Connecticut only in canopy collections.

## INTRODUCTION

In connection with studies being made on the spatial distribution of gypsy moth caterpillars by the U. S. Department of Agriculture Forest Service, Northeastern Forest Experiment Station, a program of spraying canopies of various tree species with a pyrethrum based spray was instituted. The region was in the lower Connecticut River Valley, near North Stonington, Connecticut within a few miles of Long Island Sound. The forested area is cut-over, with secondary growth of trees ranging from 20 to 60 feet in height, mixed hardwood and pine species. Individual trees were selected, height and species noted, and canopies sprayed using an extension boom. Fallen insects were collected on sheets spread beneath the tree. One of the authors (P.A.G.) took opportunity to save caddisflies from this material. Collections were begun most years in late May and continued several days each week into early July. Examination of the caddisfly records showed that May and July records added nothing of significance to the data so the following observations on collected material are limited to the June collections. Similarly, a breakdown of data to tree species yielded no significant new trends, and fragmented the data too much, so tree data are consolidated.

## RESULTS

105 collections were examined. Often there were several collections made on the same day from different trees. 10 families are represented in collections, 22 genera and 32 species. 24 of the species are identified, 8 are pending identification.

18 species were collected as single specimens, males or females, in no more than one to three collections. Several species were collected in larger numbers ; *Banksiola crotchi* showed up repeatedly, as did *Platycentropus radiatus, Oecetis inconspicua, Triaenodes injusta* males, *Triaenodes nox* males and *Triaenodes* sp. females. Thirteen collections had *Triaenodes nox* males in company with *T.* sp. females. In one instance, *Triaenodes flavescens* yielded 33 males in one collection and a sole additional male in a collection on the following day.

## DISCUSSION

The data certainly illustrate a probably random dispersal for most of the species. Whether any of the species by preference inhabit the canopy cannot be answered from data available. Unfortunately, there are no accompanying observations from captures made simultaneously in the same region at ground level. However, ground captures, in large part light trap captures have been made in nearby regions over the past decade. All of the species have been taken in such captures except *Triaenodes nox* males and *Triaenodes* sp. unidentified females.

There is evidence (Table 1) of time limited "emergences" of several species (viz. *Triaenodes flavescens*) . Such "emergences" or "hatches" have long been recognized by trout fishermen, as well as entomologists.

Earlier references to canopy collection of Trichoptera are those of Hingston (1932) and Corbet (1961). Corbet, working in East Africa, using tree towers erected for study of canopy frequenting mosquitoes in connection with yellow fever studies (Haddow *et al.* 1961) lists 5 families, 8 genera and 11 species of

**Table 1. Trichoptera Canopy Collections : June 1982, 1983, 1984, 1985 by Family, Genus, Species, Sex, # of Coll'ns & # Collected**

| Family, Genus, Species | Numbers | | | | |
|---|---|---|---|---|---|
| | Males Specimens | | Females Specimens | | Coll'n Total |
| Rhyacophilidae | | | | | |
| *Rhyacophila* sp. | (1)* | 1 | | | (1) |
| Philopotamidae | | | | | |
| *Chimarra obscura* (Wlk.) | | | (1) | 1 | (1) |
| *Dolophilodes distinctus* (Wlk.) | (9) | 14 | | | (9) |
| Psychomyiidae | | | | | |
| *Psychomyia flavida* Hagen | | | (1) | 1 | (1) |
| Polycentropodidae | | | | | |
| *Polycentropus crassicornis* Wlk. | (4) | 4 | (1) | 1 | (5) |
| *Polycentropus cinereus* Hagen | (4) | 17 | (4) | 21 | (6) |
| *Polycentropus remotus* Banks | (2) | 3 | | | (2) |
| *Polycentropus interruptus* Banks | (4) | 19 | (1) | 1 | (5) |
| *Nyctiophylax vestitus* (Hagen) | (1) | 1 | | | (1) |
| *Cernotina spicata* Ross | (1) | 1 | | | (1) |
| *Lype diversa* (Banks) (1 in May) | | | | | |
| Hydropsychidae | | | | | |
| *Hydropsyche betteni* Ross | (2) | 2 | | | (2) |
| *Cheumatopsyche* sp. | (1) | 1 | | | (1) |
| Hydroptilidae | | | | | |
| *Oxyethira serrata* Ross | | | (1) | 1 | (1) |
| Phryganeidae | | | | | |
| *Banksiola crotchi* Banks | (24) | 102 | (12) | 22 | (29) |
| *Ptilostomis* sp. | (1) | 1 | | | (1) |
| *Phryganea* sp. | (2) | 2 | | | (2) |
| Limnephilidae | | | | | |
| *Arctoecia consocia* (Wlk.) | (1) | 1 | | | (1) |
| *Limnephilus indivisus* (Wlk.) | (1) | 1 | | | (1) |
| *Limnephilus sordidus* (Hagen) | | | (2) | 2 | (2) |
| *Limnephilus* sp. | (1) | 1 | | | (1) |
| *Limnephilus* sp. | | | (1) | 1 | (1) |
| *Platycentropus radiatus* (Say) | (7) | 7 | (8) | 8 | (15) |
| *Pycnopsyche* sp. | (1) | 1 | | | (1) |
| Molannidae | | | | | |
| *Molanna blenda* (Sibley) | (1) | 2 | | | (1) |
| Leptoceridae | | | | | |
| *Ceraclea alagma* Ross | (1) | 1 | (1) | 1 | (2) |
| *Oecetis inconspicua* (Wlk.) | (20) | 34 | (14) | 17 | (28) |
| *Oecetis cinerascens* (Hagen) | (3) | 4 | (2) | 8 | (4) |
| *Triaenodes injusta* (Hagen) | (13) | 29 | (1) | 1 | (13) |
| *Triaenodes flavescens* Banks | (2) | 34 | | | (2) |
| *Triaenodes nox* Ross | (16) | 54 | | | (16) |
| *Triaenodes* sp. (all same sp.) | | | (18) | 32 | (18) |

* signifies number of collections (   ) positive for a given species

Trichoptera picked up incidental to other studies.

There are examples in groups of blood-sucking diptera of species which, although breeding at ground level, are to be found almost exclusively in tree tops. *Simulium samboni* is one such, observed in Trinidad, W.I. and *Aedeomyia squamipennis* another, widespread through tropical America and South America. Whether any Trichoptera have similar habit is not known.

ACKNOWLEDGEMENTS

The authors acknowledge encouragement from Dr. Charles Remington, Peabody Museum, Yale, of Dr. Andrew S. Main, Yale Arborvirus Research Unit, Yale Medical School, of Dr. Vernon Nelson, Southern Connecticut State College, and of Dr William E. Wallner, U.S.D.A. Forest Service, Northeastern Forest Experiment Station, all of New Haven.

REFERENCES

Hingston, R.W.G. 1932. A Naturalist in the Guiana Forest. New York, xvi + 384 p.

Haddow, A.J., Corbet, P.S. & Gillette, J.D. 1961. Entomological studies from a high tower in Mpanga Forest, Uganda : Introduction. **Trans. r. ent. Soc. London** 113 : 249-256.

Corbet, P.S. 1961. Entomological studies from a high tower in Mpanga Forest, Uganda : XII. Observations on Ephemeroptera Odonata and some other orders. **Trans. r. ent. Soc. London,** 113 : 356-361.

# PATTERNS OF GEOGRAPHIC VARIATION IN ADULT FLIGHT PHENOLOGY OF SOME NEARCTIC TRICHOPTERA

J.S. RICHARDSON

## SUMMARY

Analyses of published records of adult capture periods of some nearctic Trichoptera reveal significant linear regressions against latitude for many species. For most species first date of capture was later with increasing latitude. Furthermore, length of capture period for a given taxon decreased with increasing latitude. For autumn-emerging species these patterns are reversed, with date of first capture earlier, and capture period longer with increasing latitude. The slopes of these relationships depend on how early (or late) in the season a species emerges as adults. The linearity, absence of discontinuities, and the large amount of variance explained by latitude suggest reduced developmental synchrony within a cohort and/or a continuous trend to multiple, overlapping cohorts with decreasing latitude.

Patterns of life cycle phenology with respect to latitude and altitude are well known for many organisms (Bradshaw 1976). The most common pattern is delayed emergence with increasing latitude or altitude. This observation has also been noted for Trichoptera (Ross 1944 : 84, 173, 195). To date no-one has made a detailed examination of how much of the variance in trichopteran emergence phenology is attributable to latitude as opposed to annual or longitudinal variation, etc. Moreover, the form of such relationships has not been documented. In this paper I explore some of these patterns and speculate on what kind of hypotheses these patterns may generate.

The data for these analyses came from published records of adult collections across North America. Criteria for the inclusion of a record depended upon sample size, collection methods, locality (single site, state) and length of collection period (single year, several years integrated). I attempted to avoid species which are known to have an adult diapause, however for many species it is not known whether adult diapause exists. There were sufficient records of emergence for analysis of patterns of 16 species (from 8 families). This included anywhere from 4 to 17 records per species. I have used the terms "emergence period" and "flight period" interchangeably.

Linear regressions of first date of emergence against latitude were significantly different ($p<0,05$) from a slope of zero for 9 of the 16 species (Fig. 1A, B). In general, those species with earlier emergence periods had steeper (and significant) slopes than species emerging in mid-season. This is as would be predicted if the selective pressure was variable early (or late) season weather. The slopes of 13 out of the 16 regression lines were positive (note that latitude is the independent variable), significantly different from random (sign test, $p = 0.0085$). The three species with negative slopes were all species emerging in the latter half of the year, a pattern regularly seen in terrestrial insects (Hardwick 1971). The imposition of linear regressions on these data especially for species like*Oecetis inconspicua* and *Ceraclea tarsi-punctata* with an extensive range, may obscure biologically important patterns, but higher order regressions were not considered.

There were 7 significant ($p < 0.05$) regressions out of 16 comparisons of the length of the emergence period versus latitude. Of these 7 regressions all but *Pycnopsyche lepida* had negative slopes. Considering the slopes of all regression equations 13 of 16 had negative slopes, a higher frequency than expected at random (sign test $p = 0.0085$). The three species with positive slopes were the same 3 that had negative slopes in the latitude-first date of emergence regressions (Fig. 1B, Table 1).

There was a lot of residual variation about the latitude regressions for many species (Fig.2). The amount of variance in first date of emergence accounted for by latitude, for those species with significant regressions, ranged from 59% for *Ceraclea tarsi-punctata* to over 98% for *Lype diversa* (Table 1). There are many potential sources for this variance component, such as altitude, stream order, gradient, etc..., however the available data were insufficient for testing any of these. Where it has been studied, the duration (as well as the onset and termination) of the flight period is correlated with the sample size in adult insect collections (Hardwick 1971).

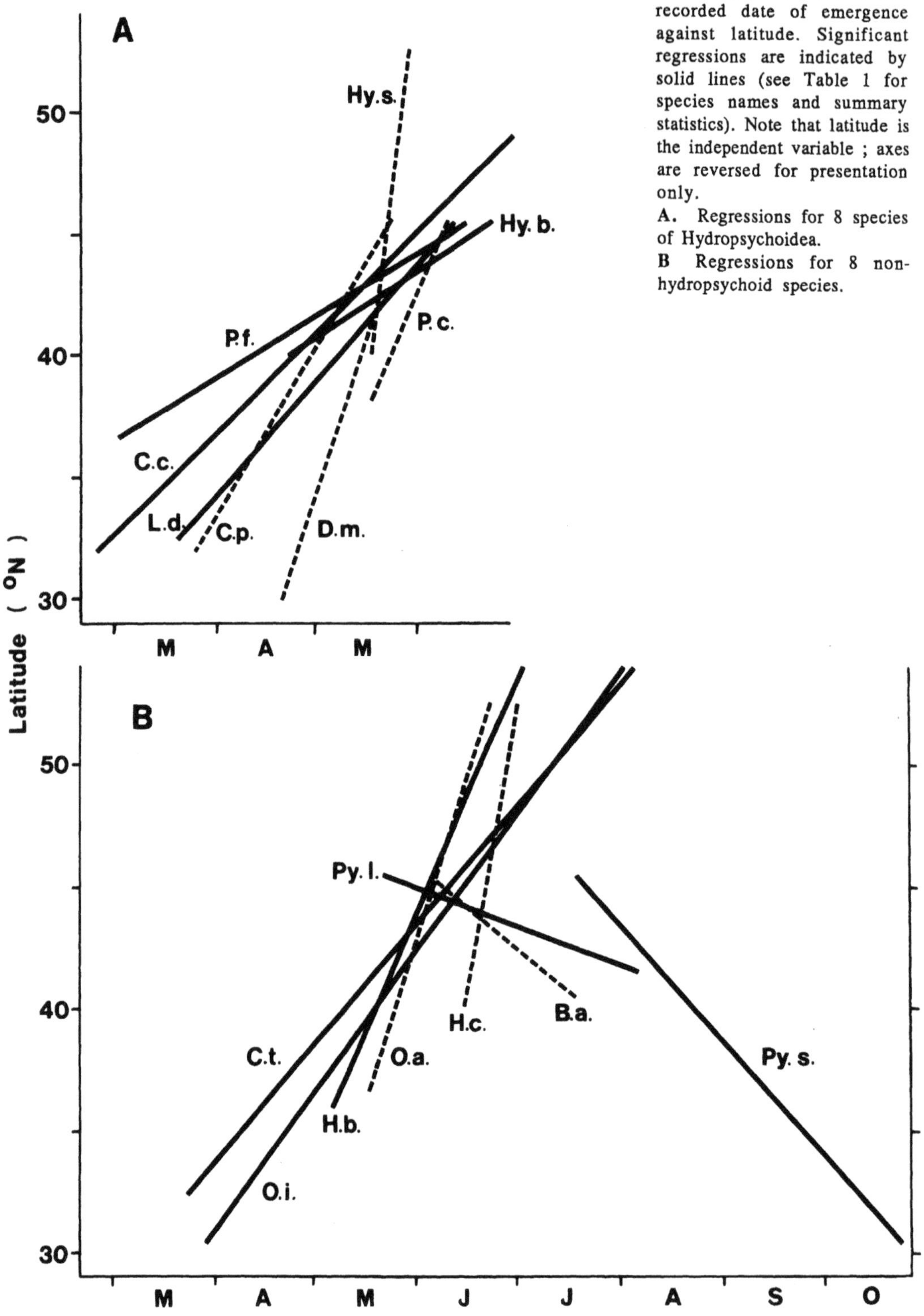


Figure 1.
Linear regression lines of first recorded date of emergence against latitude. Significant regressions are indicated by solid lines (see Table 1 for species names and summary statistics). Note that latitude is the independent variable ; axes are reversed for presentation only.
A. Regressions for 8 species of Hydropsychoidea.
B Regressions for 8 non-hydropsychoid species.

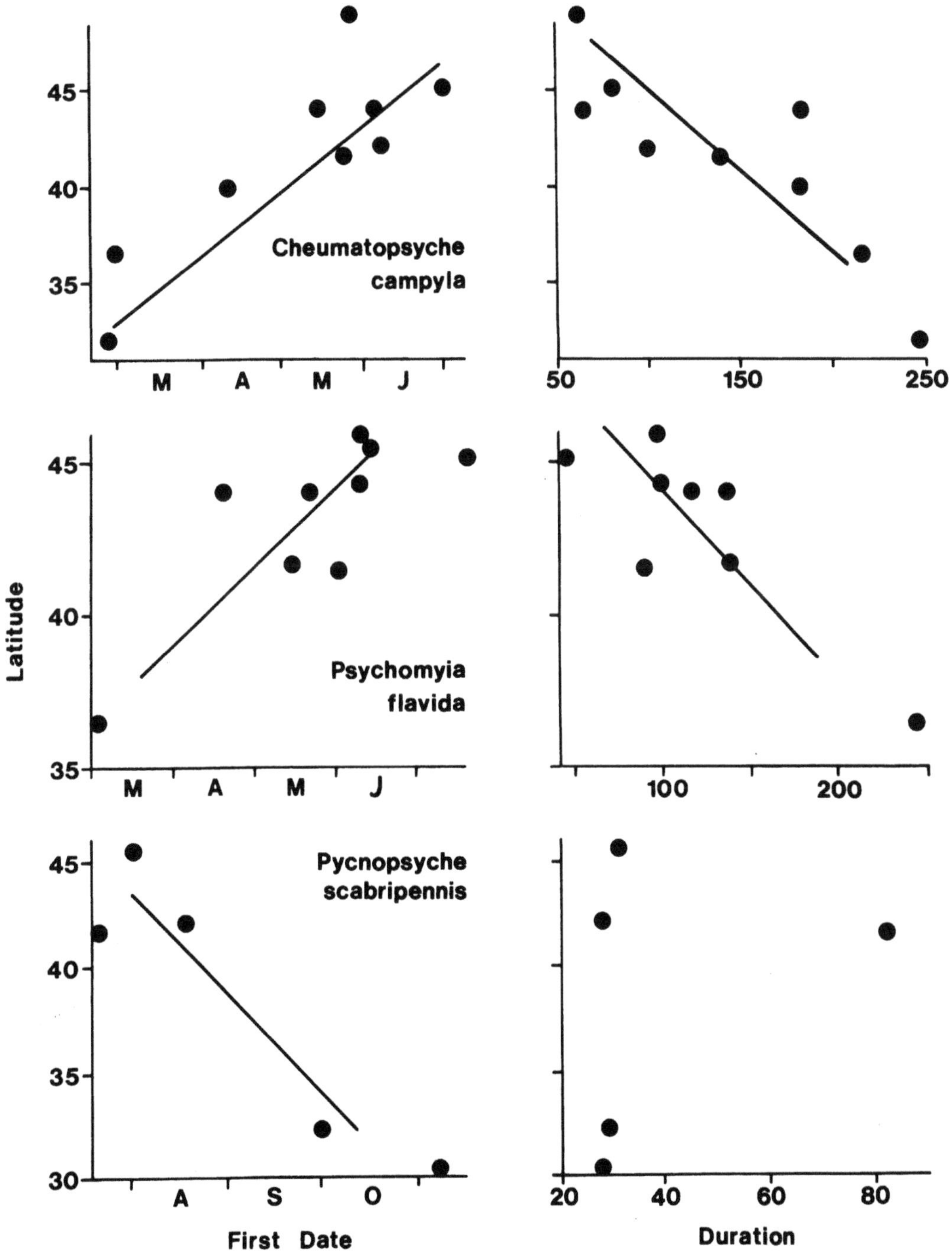


Figure 2. Examples of the relationship between first date of emergence, or duration of the emergence period, against latitude.

Data on sample sizes, as well as mean and median emergence dates would be valuable for use in any kind of comparison.

Table 1. Summary of test statistics for 16 species of nearctic Trichoptera, including regression coefficients (R) for first date of emergence versus latitude and duration of flight period versus latitude.

| CODE | TAXON | REGRESSION COEFFICIENT | | SAMPLE SIZE |
|---|---|---|---|---|
| | | FIRST DATE | DURATION | |
| C.c. | *Cheumatopsyche campyla* | 0.84 * | -0.86 * | 9 |
| C.p. | *C. pettiti* | 0.51 | -0.70 * | 9 |
| D.m. | *Diplectrona modesta* | 0.50 | -0.40 | 6 |
| Hy.b. | *Hydropsyche betteni* | 0.79 * | -0.83 * | 7 |
| Hy.s. | *H. slossonae* | 0.41 | -0.42 | 12 |
| L.d. | *Lype diversa* | 0.99 * | -0.42 | 5 |
| P.c. | *Polycentropus cinereus* | 0.46 | -0.76 * | 8 |
| P.f. | *Psychomyia flavida* | 0.80 * | -0.84 * | 9 |
| B.a. | *Brachycentrus americanus* | -0.55 | 0.48 | 7 |
| C.t. | *Ceraclea tarsi-punctata* | 0.77 * | -0.40 | 8 |
| H.b. | *Helicopsyche borealis* | 0.78 * | -0.46 | 17 |
| H.c. | *Hydroptila consimilis* | 0.46 | -0.40 | 6 |
| O.a. | *Oecetis ovara* | 0.39 | -0.41 | 10 |
| O.i. | *O. inconspicua* | 0.94 * | -0.94 * | 12 |
| Py.l. | *Pycnopsyche lepida* | -0.97 * | 0.96 * | 4 |
| Py.s. | *P. scabripennis* | -0.93 * | 0.55 | 5 |

* $p<0.05$

In several species a large amount of variance in the length of the flight period is explained by latitude, for instance *Cheumatopsyche campyla* ($R^2$= 0.74), *Psychomyia flavida* ($R^2$=0.70), and *Oecetis inconspicua* ($R^2$=0.88). There was no evidence from these results to indicate discrete "steps" in regressions of first date, or duration of emergence against latitude. Since it's unlikely that this simply represents reduced synchrony of a cohort with decreasing latitude, it suggests a change in age-structure from non-overlapping cohorts with a univoltine pattern, to overlapping cohorts and more than a single generation per year. This gradual trend indicates that simple patterns of voltinism are likely to break down, or perhaps there is some alteration of habitat. Such "cohort-splitting" is known from other aquatic insects.

Most insects in temperate latitudes use some combination of photoperiod and temperature to cue the timing of ontogenetic events (Danilevsky *et al.* 1970). Many insects use diapause to co-ordinate development with such environmental cues ; in Trichoptera diapause is known in all ontogenetic stages (Wiggins 1977). The actual details of how life cycle phenology is co-ordinated with seasonal changes is known for very few trichopteran species. Such ecophysiological responses are considered to be adaptive for each local population and therefore should be heritable traits (Tauber and Tauber 1986). We know little about the actual developmental cues or their heritability for most aquatic insects.

There are many selective agents which could affect the evolution of life cycle phenology. The significant relationships of the length of flight period decreasing with increasing latitude suggests that variability in early and late growing season weather may act as stabilizing selective agents, resulting in increased synchrony of more northerly populations. One should not lose sight of the fact that life-histories evolve in response to a suite of environmental characteristics not just temperature variation. Predator satiation has been suggested for a high degree of emergence synchrony in mayflies, but has not been implicated in trichopteran emergence patterns. Avoidance of pupation, or oviposition during periods of spates may evolve if the timing of spates is predictable between years. It will be difficult to distinguish the effects of different environmental factors because of the high degree of covariance among them, for instance spring runoff and temperature. Unfortunately we know almost nothing of the mortality rates and sources of mortality for aquatic insects. In addition, interactions of larvae with their food supply, competitors and predators may also affect rates and timing of ontogenetic events.

Since evolutionary hypotheses are not amenable to direct testing, one potential approach might be to look for patterns of emergence phenology and use analysis of variance and multiple regression techniques to factor out potential selective agents. The use of long-term records (e.g. Ross 1944) in this data set likely reduced interannual and between site variation so that single data points may integrate many sources of variation. Ideally, emergence data from distinct populations should be used if deviations from the regression are to be used to look for relationships with other variables. This approach depends on having slightly more detailed data than is available from light trapping. The use of emergence traps allows for quantification, and hopefully, replicated sampling of single populations. Other advantages of emergence traps include the sampling of a discrete population, and observation of true emergence dates (as apposed to flight dates). Sampling of discrete populations allows one to make measures of the larval habitat, density and perhaps adult biomass. All these measures are essential to comparing populations between years, study sites and larger geographical regions.

Regressions such as those I have presented here may also be used to predict timing and duration of emergence of species in parts of their range where they have not been previously studied. One may also be able to predict the northern extent of a species' range from these relationships (e.g. Richardson and Clifford in press). While these relationships are not unexpected, they do show that the timing of emergence follows simple patterns and may be predictable, especially when more data becomes available. With better defined latitudinal patterns we may be able to define the degree to which deviations from the regression are due to factors other than latitude.

## REFERENCES

Bradshaw, W.E. 1976. Geography of photoperiodic response in diapausing mosquito. **Nature (London)**, 262 : 384-386.

Danilevsky, A.S., Goryshin, N.I. & Tyshchenko, U.P. 1970. Biological rhythms in terrestrial arthropods. **Ann. Rev. Ent.**, 15: 2O1-244.

Hardwick, D.F. 1971. The "phenological date" as an indicator of the flight period of noctuid moths. **Can. Ent.**, 1O3 : 1207-1216.

Richardson, J.S. & Clifford, H.F. (in press). Phenology and ecology of some Trichoptera in a low gradient boreal stream. **J. N. Am. Benthol. Soc.**.

Ross, H.H. 1944. The caddisflies, or Trichoptera, of Illinois. **Bull. III. Nat. Hist. Survey**, 23 : 1-326.

Tauber, C.A. & Tauber, M.J., 1986. Ecophysiological responses in life-history evolution : evidence for their importance in a geographically widespread insect species complex. **Can. J. Zool.**, 64 : 875-884.

Wiggins, G.B. 1977. Larvae of the North American Caddisfly Genera (Trichoptera). **Univ. Toronto Press**, 4O1 p.

# THE COMPARISON OF LIGHT TRAP AND STICKY TRAP CATCHES OF ADULT TRICHOPTERA (LYON, FRANCE)

P. USSEGLIO-POLATERA

SUMMARY

Light trap and sticky trap catches of adult Trichoptera during five weeks in the summer of 1982 are compared. The relationship between the two samples would suggest that the light source is poorly attractive. The deformation of the aerial biocenosis due to the principle of trapping seems rather limited. Our results confirm the upstream migration of imagines. However it appears that wind has a strong effect on flight direction.

## INTRODUCTION

Over a number of years light traps have been used to study the systematics, biology and ecology of aquatic and other insects in their natural environments (Crichton 1960, Ulfstrand 1970, Svensson 1972, Andersen 1979...).

In addition, the trapped insects, particularly Trichoptera, may be regarded as indicators of the quality of the biotope and of its qualitative evolution, although this trapping method does have certain drawbacks (Resh 1976, Malicky 1978, 1980, 1981, Chantaramongkol 1983, Usseglio-Polatera 1985). The results of light trapping depend on the distribution of the available insects but the choice of the trapping method and its operating conditions are important and induce an empirical selectivity that must be estimated. Therefore, in the following experiments, a sticky trap, which is been used for a passive trapping device, and a light trap, of a type which as already been used for a systematic survey of Trichoptera in the Rhône since 1958 (Usseglio-Polatera 1985), have been operated simultaneously and the results of this comparison are given below.

## MATERIALS AND METHODS

The light trap has already been described by FONTAINE (1982, trap n°1). It uses a mercury vapour lamp (MAZDA MA 80) emitting both ultraviolet and visible light. The sticky trap is made of a rectangular transparent polyethylene sheet (70cm x 60cm) covered with a non-drying glue, which is soluble in turpentine, and hung transversely about 10 cm above the water surface. Trichoptera seemed to show no avoiding reactions.

## STUDY AREA

The traps were set up on a canal-boat moored in Lyons, on the left bank of the Rhône river, downstream from the Pont Lafayette, where the river flows between two artificial embankments. Below the Rhône-Saône junction, several kilometers downstream (near Pierre-Bénite), a reservoir has been created, causing a significant slowing down of the flow, and a permanent raising of the level of the water, within the trapping area.

Both traps were so arranged to eliminate as far as possible any interaction. The trapped insects were collected every day at the same time.

## RESULTS

### 1. Captures of Trichoptera

The results from the two traps, during five weeks in the summer of 1982, are shown in Figure 1.

Of the 26 species recorded, 13 were caught by both traps, ten in the light trap only and three by the sticky trap only.

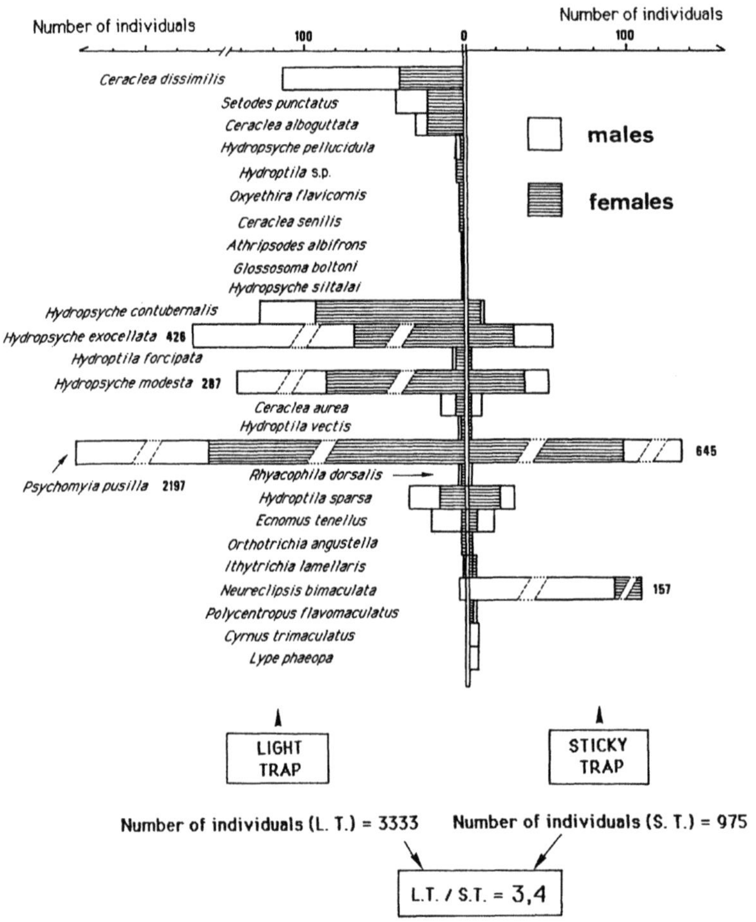


Figure 1. Comparison of catches of Trichoptera in the mercury vapour light trap (L.T.) and in the sticky trap (S.T.) from the end of July to the beginning of September, 1982.

The relationship between the two samples (light trap/sticky trap = 3.4) would suggest that the light source is poorly attractive (Robinson & Robinson 1950, Hollingsworth *et al.* 1968). The area of attraction is not known, but it seems that insects must fly very close to the trap to be influenced. On the other hand, most of the trapped insects belonged to species active daytime although they may also fly during the first hours of night (Brindle 1958, Müller 1973, Andersen 1979). The continuous exposure of the sticky sheet increases its trapping efficiency and makes up for the lack of an attractant factor (low relationship L.T./S.T.).

The proportion of Psychomyidae caught, mainly *Psychomyia pusilla*, is the same for the two sampling systems (L.T. 66.0%/S.T. 66.4%).

The adults of Hydropsychidae (L.T. 25.4%/S.T. 11.3%) and Leptoceridae (L.T. 6.1%/S.T. 0.3%) form more or less stationary swarms (Roos 1957, Crichton 1960, Benz 1975, Solem, 1978, Statzner 1978); hence they are more easily caught in a light trap. Their capture by the sticky trap depends on the position and height of the swarm of each species in relation to the position of the sticky sheet. Seven of the ten species of these two families were not caught by the sticky trap (Fig. 1).

In contrast, there were more Polycentropodidae (L.T. 0.1%/S.T.16.6%) and Hydroptilidae (L.T.

1.6%/S.T. 3.5%) with the sticky trap. Their mainly diurnal activity (Hydroptilidae : Nielsen 1948, *Cyrnus trimaculatus*, Taylor 1963, *Polycentropus flavomaculatus*; Ulfstrand 1970) and the fact that they fly swiftly close to the water surface (*Neureclipsis bimaculata*, Statzner 1978) results in their capture on the sticky sheet.

## 2. Sex-ratio

The sex proportions of the flies caught, although often divergent from 1:1, were similar for the two traps (Fig. 2). From the evidence of emergence traps it can be assumed that about the same number of males and females are present over the course of a year so that the differences in sex-ratio observed in light trap collections must be due to behavioural differences in the activity of the sexes rather than to one sex being more strongly attracted to light. Where the sticky trap catches more females, it may be because they are more often flying close to the water surface while swarming or egg-laying.

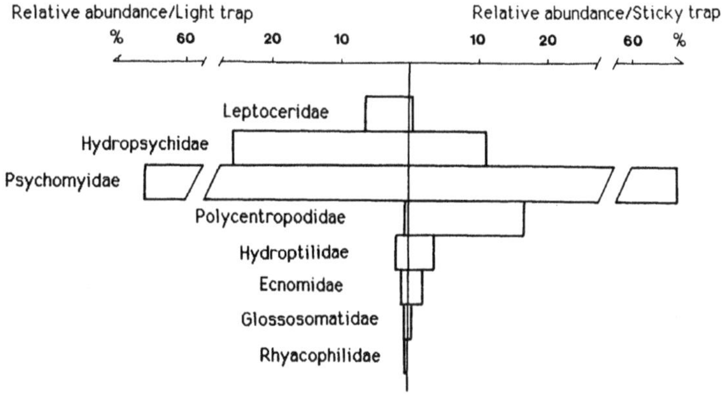


Figure 2. Sex ratio : percentage of males in the total catches of the most abundant species, from the light trap and the sticky trap.

## 3. Trapping and flight direction

Adult females fly mainly upstream; this behaviour is often considered to balance the downstream drift of larvae according to the hypothesis of a "colonization cycle" (Müller 1954).

Many trapping observations and experiments have shown that Trichoptera imagines display upstream migration (Roos 1957, Svensson 1974, Statzner 1978 Bird and Hynes 1981, Coutant 1982). In my experiments, a majority of insects (67.8%) were trapped on the downstream side of the sticky sheet so that they had been flying upstream (Fig. 3a).

In all catches, without considering wind effects, the sample on the downstream side was significantly greater than that on the upstream side ($p<0.0089$ to $p<0.0003$). This upstream migration tendency was found in males ($p<0.035$ to $p<0.0017$) as well as females ($p<0.01$ to $p<0.0005$) in the most numerous species : *Psychomyia pusilla* and *Neureclipsis bimaculata*. These results seem to confirm the upstream migration of imagines.

However, if wind direction and average speed are taken into account in every trapping period, it appears that wind has a strong effect on flight direction. Generally, the insect fly against the wind. Thus, for a moderate north wind (v<3.0m/s), 78.2% of catches were obtained on the downstream side (Fig. 3b). This ratio was slightly decreased for a stronger wind blowing in the same direction (v>3.0m/s). The upstream/downstream discrepancies were significant in every case ($p<0.025$ to $p<0.005$) except for *Neureclipsis bimaculata* females. On the other hand, the mean capture ratio on the downstream side, was only 57.3% for a east or west wind, and decreased to 40.9% for a south wind. For these two modalities, the discrepancies upstream/downstream are not significant assuming the 5% limit.

Thus, there would appear to be a true upstream migration of adults, but this may be altered by the direction of the prevailing wind.

## DISCUSSION

Light traps and sticky traps provide an incomplet sample of the species existing within an area since flight activity depends on the ethological and physiological characters of each species, and on the meteorological conditions.

Field collections based on the light trap (89% species collected during the period of simultaneous working) were larger than those based on sticky trap (62% of the corresponding species). Furthermore, the sticky trap failed to collect a number of species such as *Ceraclea dissimilis* (L.T. 114/S.T. 0), *Setodes punctatus* (L.T. 42/S.T. 0) or *Ceraclea alboguttata* (L.T. 29/S.T. 0) although dredging and artificial substrates confirmed their existence. *Neureclipsis bimaculata* (L.T. 3/S.T. 157) seemed to be the only species abundant at the site which was rare in the light trap possibly because of activity limited to daytime and weak sensitivity to the light spectrum involved.

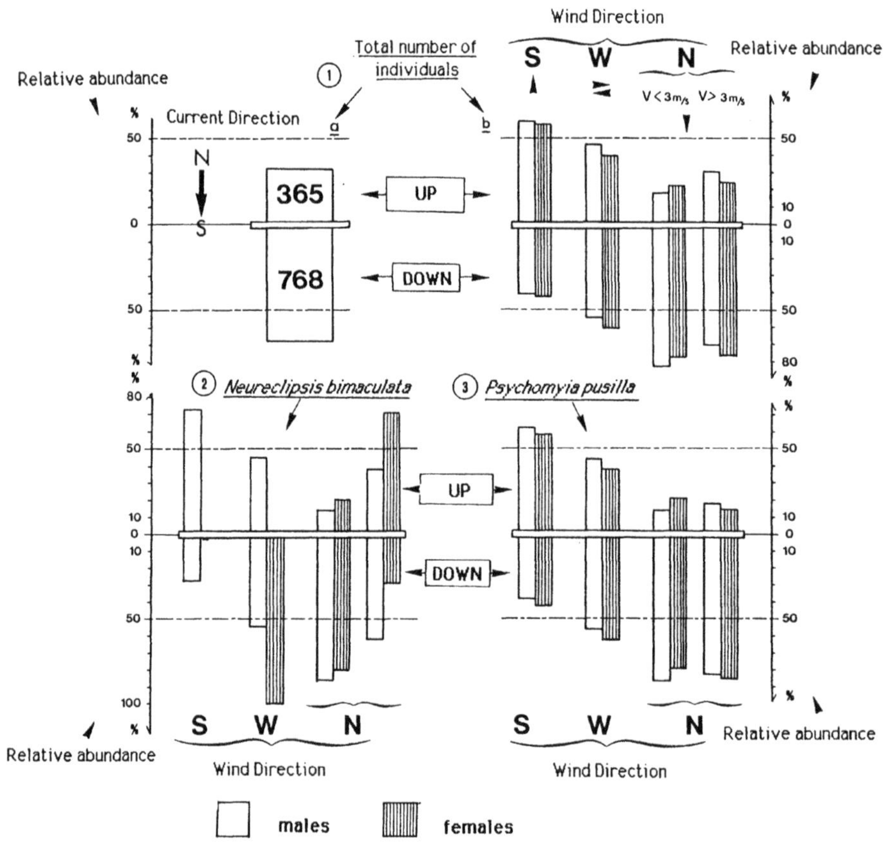


Figure 3. The percentage distribution of catches on the two sides of the sticky trap, (1a) total numbers without consideration of wind, (1b) total numbers related to wind direction, (2) *Neureclipsis bimaculata* related to wind, (3) *Psychomyia pusilla* related to wind.

Thus, the deformation of the aerial biocenosis due to the principle of trapping seems rather limited in the above described case. Bournaud *et al.* (1983) using a similar method, obtained adult ratios for six *Hydropsyche* species, very close to the ratios of corresponding 5th instar larvae collected in artificial substrates at the same time. Crichton (1960, 1965) and Svensson (1974) also showed the superiority of adult sampling with light traps in contrast to the passive suction or Malaise traps.

In conclusion, it can be said that the light trap is an efficient and valuable method of sampling, because of its convenience, its ability to sample rare or not easily accessible species as a result of the attractant feature, and also because adults are more easily identified than larvae. Thus, light trap collections can survey the qualitative evolution of an aquatic biocenosis by using the water quality indicator values of insects, especially Trichoptera.

ACKNOWLEDGEMENT

I thank Dr. M. I. Crichton for correcting the English text and for his comments on this manuscript.

REFERENCES

Andersen, T. 1979. Some caddis flies (Trichoptera) in western Norway, and their arrival pattern in light traps. **Fauna norv.** Ser. B, 26 : 12-17.

Benz, G. 1975. Über die Tanzschwärme der Köcherfliege *Hydropsyche pellucidula* Curtis (Trichoptera : Hydropsychidae). **Mitt. Schweiz. ent. Ges.**, 48 : 147-157.

Bird, G.A. & Hynes, H.B.N. 1981. Movement of immature aquatic insects in a lotic habitat. **Hydrobiologia**, 77 : 103-112.

Bournaud, M., Arens, M.F., Tachet, H. & Usseglio-Polatera, P. 1983. The problem of sampling Trichoptera in a large river. **Aquatic Insects**, 5 (3) : 167-172.

Brindle, A. 1958. Night activity of Trichoptera. **Entomologist's mon. Mag.**, 94 : 38-42.

Chantaramongkol, P. 1983. Light-trapped caddisflies (Trichoptera) as water quality indicators in large rivers : results from the Danube at Veröce, Hungary. **Aquatic Insects**, 5 (1) : 33-37.

Coutant, C.C. 1982. Evidence for upstream dispersion of adult caddisflies (Trichoptera : Hydropsychidae) in the Columbia River. **Aquatic Insects**, 4 (2) : 61-66.

Crichton, M.I. 1960. A study of captures of Trichoptera in a light trap near Reading, Berkshire. **Proc. R. ent. Soc. Lond.**, 112 (12) : 319-344.

Crichton, M.I. 1965. Observations on captures of Trichoptera in suction- and light-traps near Reading, Berkshire. **Proc. R. ent. Soc. Lond.**, 40 (7-9) : 101-108.

Fontaine, J. 1982. Le piégeage lumineux, moyen d'approche de la faune entomologique d'un grand fleuve (Ephéméroptères en particulier). **Bull. Soc. Linn. Lyon**, 51 (3) : 81-89.

Hollingsworth, J.P., Hartstack, A.W. & Lindquist, D.A. 1968. Influence of near ultraviolet output of attractant lamps on catches of insects by light-traps. **J. Econ. Entomol.**, 61 (2) : 515-521.

Malicky, H. 1978. Köcherfliegen-Lichtfallenfang am Donauüfer in Linz (Trichoptera). **Linzer biol. Beitr.**, 10 (1) : 135-140.

Malicky, H. 1980. Lichtfallenuntersuchungen über die Köcherfliegen (Insecta, Trichoptera) des Rheins. **Mainzer Naturw. Archiv.**, 18 : 71-76.

Malicky, H. 1981. Der Indikatorwert von Köcherfliegen (Trichoptera) in grossen Flüssen. **Mitt. dtsch. Ges. allg. angew. Ent.**, 3 : 135-137.

Müller, K. 1954. Investigations on the organic drift in North Swedish streams. **Ann. Rep. Inst. Freshwat. Res. Drottningholm**, 35 : 133-148.

Müller, K. 1973. Circadian rhythms of locomotor activity in aquatic organisms in the subarctic summer. **Aquilo Ser. Zool.**, 14 : 1-18.

Nielsen, A. 1948. Postembryonic development and biology of the Hydroptilidae. **Biol. Skr.**, 5 (1) : 1-200.

Resh, V.H. 1976. Life histories of coexisting species of *Ceraclea* caddisflies (Trichoptera : Leptoceridae) : the operation of independant functional units in a stream ecosystem. **Can. Ent.**, 108 : 1303-1318.

Robinson, H.S. & Robinson, P.J. 1950. The observed behaviour of Lepidoptera in flight in the vicinity of light sources. **Ent. Gaz.**, 1 : 3-20.

Roos, T. 1957. Studies on upstream migration in adult stream dwelling insects. **Ann. Rep. Inst. Freshw. Res. Drottningholm**, 38 : 167-192.

Solem, J.O. 1978. Swarming and habitat segregation in the family Leptoceridae (Trichoptera). **Norw. J. Ent.**, 25 : 145-148.

Statzner, B. 1978. The effects of flight behaviour on the larval abundance of Trichoptera in the Schierenseebrooks (North Germany). **Proc. 2nd Int. Symp. Trichoptera,** M.I. Crichton (ed.), Junk, The Hague, 121-134.

Svensson, B.W. 1972. Flight periods, ovarian maturation and mating in Trichoptera at a South Swedish stream. **Oikos,** 23 (3) : 370-383.

Svensson, B.W. 1974. Population movements of adult Trichoptera at a South Swedish stream. **Oikos,** 25 : 157-175.

Taylor, L.R. 1963. Analysis of the effect of temperature on insects in flight. **J. Anim. Ecol.,** 32 : 99-117.

Ulfstrand, S. 1970. Trichoptera from River Vindelälven in Swedish Lapland. A four-year study based mainly on the use of light-traps. **Ent. Tidskr.,** 91 (1/4) : 46-63.

Usseglio-Polatera, P. 1985. Evolution des peuplements de Trichoptères et d' Ephéméroptères du Rhône à Lyon (1959-1982). Résultats de piégeage lumineux. Dipl. Doct. Univ. Lyon I, 451 p.

# FLIGHT PATTERNS OF THREE SPECIES OF LOTIC CADDISFLIES

J.O. SOLEM and T. BONGARD

## SUMMARY

Two-sided Malaise traps were set across streams, on the banks of streams and at standing water to reveal patterns of the flight behaviour of *Rhyacophila nubila* (Zetterstedt), *Halesus radiatus* Curtis and *Potamophylax cingulatus* (Stephens). The males had similar flight patterns, a random flight, in all habitats. Females flew upstream above the running water only, while they elsewhere flew at random. Females of all species and males of *R. nubila* and *H. radiatus* flew in less numbers on the banks than above the running water. Males of *P. cingalatus* flew in significantly higher numbers on the bank than above the running water. Characteristics of the running water must be a major factor initiating the upstream flight in females.

## INTRODUCTION

The colonization cycle theory proposed by Müller (1954) states that females of aquatic insects with larvae in running water habitats fly upstream to oviposit and thus compensate for downstream drift of larvae. This theory has provoked reseachers to look at flight patterns of running water insects, and upstream flight above running water has been reported by many authors (Müller 1982, and references therein). Other data show upstream, downstream, and random flight patterns (Elliott 1967, Svensson 1974, Madson *et al.* 1973, Bird & Hynes 1981, Gullefors 1983).

The air space used by adults of aquatic insects may be divided into several habitats, e. g. above standing water, above running water, above the banks of standing water and above the banks of running water. But, our knowledge of how aquatic insects behave elsewhere than above running water is poor. The present investigation was planned to compare flight patterns in the different habitats mentioned, and in this way shed light on factors governing the flight direction.

The species studied were *Rhyacophila nubila* Zetterstedt, *Potamophylax cingulatus* (Stephens), and *Halesus radiatus* Curtis. The larvae of *R. nubila* and *P. cingulatus* are typical inhabitants of fast flowing water, and those of *H. radiatus* of more slowly flowing water.

## STUDY AREA AND METHODS

The study was conducted during 1981 to 1985 at five streams and one lake near Kongsvoll Biological Station in the Dovrefjell mountains, Central Norway (62°17' N, 09°59' E). The collecting sites were between 900 and 1350 m elevations in the subalpine and low alpine zones. Two-sided Malaise traps were placed across streams, on banks and at standing water. The traps on the banks caught insects flying parallel to the running water. The upstream and downstream catches were kept separatly.

At the lake Gavalivatn and the outflowing stream Jerosbekken the sampling sites were located within a distance of about 400 m, and sampling was done simultaneously above fast-flowing water, standing water, and on the banks. These collecting sites are open areas and only low bushes (c. 0.5 m high) of willows and birch occur ; the terrain has a gradual slope, and the climatic conditions were almost similar at these sampling sites. Every year the sampling period covered the flight season of the caddis flies in the area. The traps were emptied once a week.

Considering the differences between numbers of males/females flying above running water and on the banks, only data from traps regarded as one set up and shown in Figure 1, were used.

Chi-square tests have been used for statistical analyses.

Figure 1.
Malaise traps and their position at streams.

RESULTS

Males of *R. nubila* had no consistent flight pattern in either of the three habitats sampled (Tab. 1). At some traps more males flew upstream and at others more males flew downstream, but in general the data indicate a random flight pattern in all habitats. Females of *R. nubila* on the other hand, showed two types of flight patterns (Tab. 1) ; above the fast flowing water they had a significant upstream flight, while a random flight was performed on the banks of the stream and at standing water.

The sexes of *H. radiatus* had similar flight patterns and flew at random in all habitats (Tab. 2). However, when descriminating between ovarian maturation stages, females with fully developed eggs flew upstream above the running water (Tab. 3). No such tendencies were found in the traps on the banks, but the numbers are small.

Table 1. Number of males (M) and females (F) of *Rhyacophila nubila* caught in two-sided Malaise traps in the Dovrefjell mountains, Central Norway. Up = upstream flight, Down = downstream flight. Asterisks denote significance levels (Chi-square test), * = $p < 0.05$, ** = $p < 0.01$, *** = $p < 0.001$.

| | | Up | Down | Up | Down | Up | Down |
|---|---|---|---|---|---|---|---|
| Fast flowing water | M | 54 | 57 | 45 | 44 | 63*** | 24 |
| | F | 71*** | 33 | 73*** | 14 | 29*** | 13 |
| | M | 60 | 68 | 165*** | 108 | 134 | 113 |
| | F | 53*** | 18 | 51*** | 22 | 129*** | 21 |
| Banks of streams | M | 41* | 23 | 42* | 23 | 42 | 22 |
| | F | 7 | 7 | 9 | 7 | 7 | 5 |
| | M | 7 | 19 | 61 | 73 | 19 | 13 |
| | F | 3 | 4 | 8 | 12 | 9 | 3 |
| Standing water | M | 6 | 7 | 14 | 17 | | |
| | F | 1 | 5 | 19 | 14 | | |

Table 2. Number of males (M) and females (F) of *Halesus radiatus* caught in two-sided Malaise traps at the stream Jerosbekken in the Dovrefjell mountains, Central Norway. Up = upstream flight, Down = downstream flight. Asterisks denote significance levels (Chi-square test), * = p < 0.05, ** = p < 0.01, *** = p < 0.001

| | | Up | Down | Up | Down | Up | Down |
|---|---|---|---|---|---|---|---|
| Fast flowing | M | 109** | 70 | 73 | 91 | 119 | 106 |
| water | F | 37 | 40 | 39 | 19 | 31 | 17 |
| Banks of | M | 47 | 49 | 61*** | 25 | 28 | 24 |
| stream | F | 27 | 17 | 18 | 11 | 11 | 9 |
| | M | 88*** | 48 | 44 | 72** | | |
| | F | 14 | 18 | 11 | 13 | | |
| Standing | M | 21 | 19 | 103 | 74 | | |
| water | F | 10 | 8 | 29 | 20 | | |

Table 3. Number of females of *Halesus radiatus* in different ovarian maturation stages (OMS) flying upstream and downstream at Jerosbekken, Dovrefjell, Norway, Differentiation of OMS followed Novak & Sehnal (1963).

| Year | | 1984 | 1985 |
|---|---|---|---|
| Downstream flight, | OMS 1&2 | 4 | 0 |
| | OMS 3&4 | 5 | 10 |
| Upstream flight, | OMS 1&2 | 4 | 0 |
| | OMS 3&4 | 13 | 32 |

Table 4 . Number of males (M) and females (F) of *Potamophylax cingulatus* caught in two-sided Malaise traps in the Dovrefjell mountains, Central Norway. Up = upstream flight, Down = downstream flight. Asterisks denote significance levels (Chi-square test), * = p < 0.05,** = p < 0.01,*** = p < 0.001.

| | | Up | Down | Up | Down | Up | Down |
|---|---|---|---|---|---|---|---|
| Fast flow- | M | 115*** | 74 | 43 | 55 | 34 | 21 |
| ing water | F | 129*** | 69 | 48 | 34 | 13 | 9 |
| Banks of | M | 170 | 137 | 161 | 129 | 69 | 56 |
| streams | F | 59* | 35 | 70 | 22 | 20 | 28 |
| | M | 87 | 72 | 35 | 85 | 72 | 79 |
| | F | 35* | 17 | 12 | 11 | 11 | 15 |

Table 5. Number of individuals of *Potamophylax cingulatus* in different ovarian maturation stages (OMS) flying upstream and downstream at Jerosbekken, Dovrefjell, Norway in 1982.

| | | |
|---|---|---|
| Downstream flight | OMS 1&2 | 28 |
| | OMS 3&4 | 27 |
| Upstream flight | OMS 1&2 | 19 |
| | OMS 3&4 | 97 |

When not discriminating between the ovarian maturation stages the females of *P. cingulatus* had a random flight pattern in all three habitats (Tab. 4). But like *H. radiatus*, when discriminating between the various ovarian maturation stages *P. cingulatus* females also showed a significant upstream flight (Tab. 5). The *P. cingulatus* males had a significant upstream flight at two traps, but in spite of that the general trend in the present study is that the males had a random flight pattern.

As expected more females were flying above the running water than on the banks (Tabs. 1,2,4). The males of *R. nublia* and *H. radiatus* also flew in higher numbers above running water than on the banks. However, the males of *P. cingulatus* flew in significantly higher numbers on the banks than above the running water (Tab. 6).

Tables 6. Number of males (M) and females (F) of *Potamophylax cingulatus* caught in Malaise traps placed above running water and on the banks at the stream Jerosbekken in the Dovrefjell mountains. D = downstream flight, U = upstream flight. Statistics : Chi-square test.

| 1982 | | Bank | Water | Bank | Chisq. test |
|---|---|---|---|---|---|
| M | D | 137 | 74 | 129 | $p \leq 0.001$ bank > water |
| | U | 170 | 115 | 161 | |
| F | D | 35 | 63 | 22 | $p \leq 0.001$ water $\geq$ bank |
| | U | 59 | 129 | 70 | |
| M | D | 57 | 53 | 55 | $p \leq 0.05$ bank $\geq$ water |
| | U | 68 | 42 | 87 | |
| F | D | 28 | 34 | 17 | $p \leq 0.01$ water $\geq$ bank |
| | U | 20 | 48 | 35 | |
| 1984 | | | | | |
| M | D | 73 | 26 | 52 | $p \leq 0.001$ bank $\geq$ water |
| | U | 78 | 45 | 34 | |
| F | D | 19 | 17 | 13 | ns |
| | U | 13 | 28 | 21 | |
| 1985 | | | | | |
| M | D | 85 | 21 | 79 | $p \leq 0.001$ bank $\geq$ water |
| | U | 35 | 34 | 72 | |
| F | D | 11 | 9 | 15 | ns |
| | U | 12 | 13 | 11 | |

## DISCUSSION

The data on flight patterns above running water for females of *R. nubila, H. radiatus* and *P. cingulatus* conform to the original idea of the colonization cycle. The data also strenghten the suggestion of Roos (1957) that the instinct for upstream migration is restricted to mature females.

*R. nubila* females had a strong upstream flight at the streams investigated in the Dovrefjell mountains, and a similar pattern was reported for *R. nubila* by Roos (1957) and Gullefors (1983). Svensson (1974) reported *Rhyacophila fasciata* Hagen as a strong upstream flyer, and we predict that this is a common feature in *Rhyacophila*. A strong upstream flight for mature females of *H. radiatus* and *P. cingulatus* was also found above the water of the stream Jerosbekken. This is in contrast to the flight behaviour of *H. radiatus* and *P. cingulatus* females found by Svensson (1974). However, it has been shown in the present study that females have different flight patterns in different habitats, and it may be that the differences in flight pattern between this and Svensson's study are related to differences in stream habitats, e.g. current velocity, substrate, etc.

The seasonal and daily flight periods of *R. nubila, P. cingulatus* and *H. radiatus* overlap greatly (Solem 1985a, and unpubl. data). They therefore experienced much the same environmental conditions in the experiments. If an important environmental factor like wind should govern the flight direction, as suggsted by Ross (1957), nearly similar flight patterns should be found in all three species and in all habitats in the present study. The males of the three species in question performed nearly similar flight patterns, a random flight, in all habitats, and as such conformed to wind as a factor determining the flight direction. However, males of *R. nubila* are attracted to females by long distance female sex pheromones (Solem 1985b). The scent is spread through the air and the males will fly against the wind to locate a female, and a random flight pattern is therefore most likely when a whole flight season is considered. At present there is no evidence for female sex pheromones of *H. radiatus* and *P. cingulatus,* but if they are present their males would behave like males of *R. nubila.*

The females of the three species showed different flight patterns in the different habitats, and for them it seems unlikely that the wind should be an important factor that generally determine their flight direction in this area. Because females flew upstream above the running water only, there must be some characteristics of the running water that initiate the upstream flight. This may be current velocity or some other charateristics. Neither does wind govern the direction of the flight behaviour in swarming *Athripsodes* spp. and *Ceraclea* spp. (fam. Leptoceridae). These species fly back and forth over a particular area of water and are not affected by wind direction when the wind is below the treshold that prohibits flight (Solem 1978).

The low number of females caught in the Malaise traps on the banks compared with the catches above running water show that flight activity is low on the banks. Solem (1985a) had data on emergence and flight periods and found that the parthenogenetic species *Apatania muliebris* McLachlan had a low flight activity before the species was ready for oviposition. Such relevant data on *R. nubila, H. radiatus* and *P. cingulatus* are not available, but it is likely that they have a similar pattern as *A. muliebris.*

Considering the sites where most of the males fly, there was a clear difference between, *R. nubila* and *H. radiatus* as one group and *P. cingulatus* as an other. Like the females most males of *R. nubila* and *H. radiatus* flew above the running water, while the males of *P. cingulatus* flew in highest numbers on the banks of the stream. That males and females prefer different habitats for flight have great implications if sex ratios are estimated from Malaise trap collections, and such data must be treated with caution. However, when precise information about the site of the traps is given, sex ratios may be valuable information about habitat preference of the sexes.

## REFERENCES

Bird, G.A. & Hynes, H.B.N. 1981. Movements of adult aquatic insects near streams in Southern Ontario. **Hydrobiologica,** 77 : 65-69.

Elliot, J.M. 1967. Invertebrate drift in a Dartmoor stream. **Arch. Hydrobiol.,** 63 : 202-237.

Gullefors, B. 1983. Flight movements of caddisflies (Ins. Trichoptera) along the mouth part of a coastal stream in Northern Sweden. **Fauna Norrlandica** 3 : 1-13.

Madsen, B.L., Bengtsson, F. & Butz, I. 1973. Observations on upstream migration by imagines of some Plecoptera and Ephemeroptera. **Limnol. Oceanogr.,** 18 : 678-681.

Müller, K. 1954. Investigations on the organic drift in North Swedish streams. **Inst. Freshwater Res. Drottningholm, Rep.,** 34 : 133-148.

Müller, K. 1982. The colonization cycle of freshwater insects. **Oecologia (Berl.),** 52 : 202-207

Novak, K. & Sehnal, F. 1963. The development cycle of some species of the genus *Limnophilus* (Trichoptera). **Cas. Cs. Spol. Ent.,** 60 : 68-80.

Roos, T. 1957. Studies on Upstream Migration in Adult Stream-Dwelling Insects. **Inst. Freshwater Res. Drottningholm, Rep.,** 38 : 167-193.

Solem, J.O. 1978. Swarming and habitat segregation in the family Leptoceridae (Trichoptera).**Norw. J. Ent.,** 25 : 145-148.

Solem, J.O. 1985a. Distribution and biology of caddisflies (Trichoptera) in Dovrefjell mountains,Central Norway. **Fauna norv. Ser. B.** 62-79.

Solem, J.O. 1985b. Female sex pheromones in *Rhyacophila nubila* (Zetterstedt) and arrival patternto sticky traps (Trichoptera, Rhyacophilidae). **Fauna norv. Ser. B.,** 32 : 80-82.

Svensson, B.W. 1974. Population movements of adult Trichoptera at a South Swedish stream. **Oikos,** 25 : 157-175.

# CHANGES IN FLIGHT DIRECTION OF CADDIS FLIES WHEN MEETING CHANGES IN THE ENVIRONMENT.

B. GULLEFORS


## SUMMARY

The flight direction of adult Trichoptera along a North Swedish stream section was studied using bilateral window traps.

The frequency of upstream-flying caddisflies decreased when the velocity of the water current decreased at one side of the stream after an artificial change of the coarse bottom substrate. More females than males were trapped moving upstream. The predomination of upstream flight at one side of the stream and downstream at the other demonstrates the great importance of the position of a trap for the result of the capture.

The results support the hypothesis of the colonization cycle. It is suggested that egg bearing females are mainly moving in an upstream direction until a suitable oviposition area is found, where one of all possible directions may be optimal for the short range "egg-laying flight". In some cases this flight direction can be downstream.


## INTRODUCTION

Müller's colonization cycle (Müller 1954) has been supported for certain species of lotic caddis flies by means of trapping upstream-flying females or by direct observation (Roos 1957, Botosaneanu 1957, Nishimura 1967, Elliott 1969, 1981, Ulfstrand 1970, Olsson 1971, Göthberg 1972, Pearson & Kramer 1972, Müller 1973, 1982, Svensson 1974, Statzner 1978, Bird & Hynes 1981, Gullefors 1983).

Statzner (1978) has emphasized the importance of the flight behaviour of imagines for the larval distribution of several caddis fly species. The flight behaviour and the flight direction can be affected by environmental factors. "Channel patterns, riparian and emergent aquatic vegetations, width of the water surface , and also the flight of other species" are examples of such factors (Statzner 1978). Also lakes can represent important flight barriers (Ulfstrand 1970, Statzner 1978).

This paper describes how lotic caddis flies can change their flight direction when the bottom substratum is locally changed.

## INVESTIGATION AREA AND METHODS

The investigation was carried out at the small stream Forsån (approx. 63°00'N, 17°32'E) in the North of Sweden. The stream rises from the lake Valasjön and flows eastward through agriculture country for about 1100 m before entering the lake Forssjön.

On the trap site the width of the stream was 4 m and the water depth 0.3 m during the trapping time. pH varied from 5.8 to 6.7 during the summer of 1984. The vegetation surrounding the trap site mainly consists of alders (*Alnus incana*), bird-cherries (*Prunus padus*) and willows (*Salix* spp).

Three bilateral window traps (50 x 40 cm) were fastened on a plank 50 cm from each other. The plank with the three window traps was placed across the stream with the containers of the traps within 10 cm above the water surface (Fig.1). Flying insects hit the glass and fell into the containers filled with 50% ethylene glycol. The trap construction and the arrangement made it possible to separate upstream and downstream flying insects across the stream section.

The three trapping periods P1 - P3 are shown in Table 1. The traps were emptied at noon on the last day of each period. In period 2 boulders and cobbles were placed under the left window trap seen in the upstream direction (Fig.1). Those boulders and cobbles were barely visable above the surface. In period 3 the substrate under the left trap was restored to be the same as in period 1.

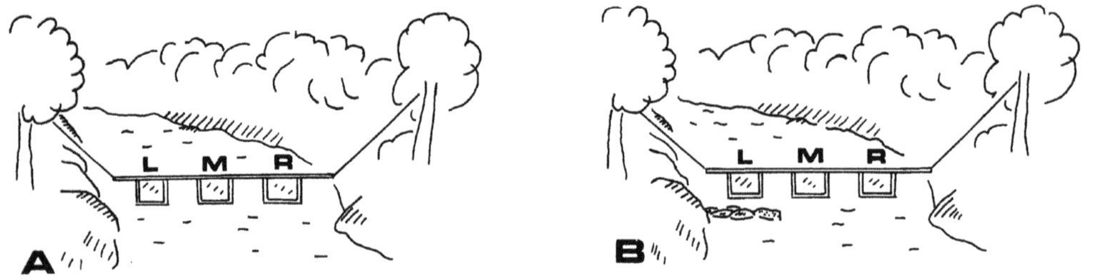


Figure 1 A. Three window traps placed across the stream 27 July - 1 August and 6 August - 8 August 1984.
Figure 1 B. From 1 August - 6 August 1984 rearrangement of some stones changed the water current velocity in the left part of the stream.

RESULTS

A total number of 6387 caddis fly specimens belonging to 25 species was captured during the three investigation periods (P1 - P3).

The manipulation of the bottom substrate caused the water current to flow more slowly in P2 in the left part of the stream. The rearrangement of the stones was also followed by lower frequencies of upstream moving adults of some species. The dominant*Wormaldia subnigra* McLachlan, which normally moves upstream (Gullefors unpubl.), decreased its total upstream frequency from 63% to 53% ($p < 0.001$) from P1 to P2 (Tab. 2). However, still more evident changes were noted for some other species. From P1 to P2, the upstream moving frequency of trapped *Rhyacophila nubila* (Zetterstedt) decreased from 70.0% to 33.8% ($p < 0.001$, N = 102), *Hydropsyche siltalai* Döhler from 67.1% to 36.6% ($p < 0.001$, N = 115), *Ceraclea alboguttata* (Hagen) from 58.7% to 39.3% (NS, N = 97) and *C. dissimilis* (Stephens) from 75% to 40% ($p < 0.001$, N = 147). In contrast to the majority of captured species, the second most common species, *Athripsodes commutatus* (Rostock) was found mostly to fly downstream in this stream section and this was still more pronounced after the substrate manipulation. Thus, the frequency of downstream moving adults of *A. commutatus* increased from 66.2% in P1 to 82.6% in P2 ($p < 0.001$) (Tab. 2). For most species the proportions of upstream and downstream numbers were similar for P1 and P3 (Tab. 2). Thus, 63.0% of *W. subnigra* were trapped when flying upstream in P1 and 62.3% in P3. The corresponding figures for *A. commutatus* were 33.8% and 32.4%, respectively.

Table 1. The trapping periods, catch per period and per cent upstream and downstream flying caddis flies.

| | | Number | | % Flying | |
|---|---|---|---|---|---|
| Period | Time | Species | Specimens | Up | Down |
| 1 | 27 July - 1 Aug. | 20 | 2326 | 57.7 | 42.3 |
| 2 | 1 Aug. - 6 Aug. | 18 | 3439 | 46.4 | 53.6 |
| 3 | 6 Aug. - 8 Aug. | 16 | 622 | 56.3 | 43.7 |
| Total | 27 July - 8 Aug. | 25 | 6387 | 51.4 | 48.6 |

As expected, the changes in flight direction applied mostly to the left part of the stream. Accordingly, *W. subnigra* was recorded for only 37.8% upstream movements in the left trap L in P2, which means that the two most common species showed a preference for downstream movements over the stream section with exposed boulders and stones. The fraction of upstream travelling *A. commutatus* in the left hand trap (L) in P2 was extremely low, 7.3 % (Tab.2).

Significantly higher upstream frequencies were recorded for females than for males of the two most common species (*W. subnigra* $p < 0.001$, *A. commutatus* $p < 0.05$). For *W. subnigra* this finding was especially evident in the trap R, i. e. the one beneath which the water current was obviously not at all affected (Tab. 3).

The lateral dispersal of most species over the stream section was very little affected by the substrate change. Species which in the first period were about equally frequent in the three traps, showed the same pattern in the second period too (*R. nubila, H. siltalai* and *W. subnigra* Tab. 2). The leptocerids *A. commutatus, C. alboguttata* and *C. dissimilis*, on the other hand, more or less avoided choosing the right part of the stream section at any period.

Table 2. The upstream flight and lateral distribution of the total catch and of the two most abundant caddisfly species.

| | |
|---|---|
| U % u<br>D<br>(S)<br>{% L} | U = Upstream flying individuals, D = Downstream flying individuals, %u = Percentage upstream flying individuals, S = Total sum of all specimens caught in the trap and {%L} = Percentage of the capture in each trap. |

| | LEFT (L) | % Up | MIDDLE (M) | % Up | RIGHT (R) | % Up | TOTAL | % UP |
|---|---|---|---|---|---|---|---|---|
| Trichoptera<br>P1 | 490<br>493<br>(983)<br>{42.3} | 49.8 | 508<br>309<br>(817)<br>{35.1} | 62.2 | 344<br>182<br>(526)<br>{22.6} | 65.4 | 2326 | 57.7 |
| P2 | 361<br>911<br>1272)<br>{37.0} | 28.4 | 641<br>559<br>(1200)<br>{34.9} | 53.4 | 592<br>375<br>(967)<br>{28.1} | 61.2 | 3439 | 46.4 |
| P3 | 101<br>126<br>(227)<br>{36.5} | 44.5 | 106<br>80<br>(186)<br>{29.9} | 57.0 | 143<br>66<br>(209)<br>{33.6} | 68.4 | 622 | 56.3 |
| W. subnigra<br>P1 | 294<br>194<br>(488)<br>{32.1} | 60.2 | 365<br>201<br>(566)<br>{37.2} | 64.5 | 298<br>168<br>(466)<br>{30.7} | 63.9 | 1520 | 63.0 |
| P2 | 317<br>522<br>(839)<br>{31.5} | 37.8 | 532<br>384<br>(916)<br>{34.4} | 58.1 | 563<br>347<br>(910)<br>{34.1} | 61.9 | 2665 | 53.0 |
| P3 | 70<br>51<br>(121)<br>{27.5} | 57.9 | 76<br>51<br>(127)<br>{28.9} | 59.8 | 128<br>64<br>(192)<br>{43.6} | 66.7 | 440 | 62.3 |
| A. commutatus<br>P1 | 98<br>246<br>(344)<br>{76.1} | 28.5 | 53<br>52<br>(105)<br>{23.2} | 50.5 | 2<br>1<br>(3)<br>{0.7} | [66.7] | 452 | 33.8 |
| P2 | 26<br>329<br>(355)<br>{70.0} | 7.3 | 58<br>90<br>(148)<br>{29.2} | 39.2 | 4<br>0<br>(4)<br>{0.8) | [100.0] | 507 | 17.4 |
| P3 | 23<br>57<br>(80)<br>{74.1} | 28.8 | 12<br>15<br>(27)<br>{25.0} | 44.4 | 0<br>1<br>(1)<br>{0.9} | [0.0] | 108 | 32.4 |

The relative lateral distribution of *A. commutatus* was unaffected by the substrate manipulation (Tab.2), but of the two latter species especially *C. dissimilis* changed the distribution pattern and greatly preferred the mid section of the stream in P2. The number was very small, however.

Table 3. The upstream flight and lateral distribution of the males and females of *Wormaldia subnigra*.

U %u — U = Upstream flying individuals, D = Downstream flying individuals, %u = Percentage upstream
D — flying individuals, S = Total sum of all specimens caught in the trap and {%L} = Percentage of
(S) — the capture in each trap.
{% L}

| | LEFT (L) | % Up | MIDDLE (M) | % Up | RIGHT (R) | % Up | TOTAL | % UP |
|---|---|---|---|---|---|---|---|---|
| W. subnigra<br>P1<br>males | 99<br>54<br>(153)<br>{32.6} | 64.7 | 92<br>55<br>(147)<br>{31.3} | 62.6 | 83<br>87<br>(170)<br>{36.1} | 48.8 | 470 | 58.3 |
| females | 195<br>140<br>(335)<br>{31.9} | 58.2 | 273<br>146<br>(419)<br>{39.9} | 65.2 | 215<br>81<br>(296)<br>{28.2} | 72.6 | 1050 | 65.0 |
| P2<br>males | 57<br>117<br>(174)<br>{30.7} | 32.7 | 86<br>74<br>(160)<br>{28.2} | 53.8 | 117<br>116<br>(233)<br>{41.1} | 50.2 | 567 | 45.9 |
| females | 260<br>405<br>(665)<br>{31.7} | 39.1 | 446<br>310<br>(756)<br>{36.0} | 59.0 | 446<br>231<br>(677)<br>{32.3} | 65.9 | 2098 | 54.9 |
| P3<br>males | 15<br>10<br>(25)<br>{24.8} | 60.0 | 13<br>10<br>(23)<br>{22.8} | 56.5 | 31<br>22<br>(53)<br>{52.4} | 58.5 | 101 | 58.4 |
| females | 55<br>41<br>(96)<br>{28.3} | 57.3 | 63<br>41<br>(104)<br>{30.7} | 60.6 | 97<br>42<br>(139)<br>{41.0} | 69.8 | 339 | 63.4 |

## DISCUSSION

The agreement in proportions of upstream and downstream flying individuals of almost all studied species before and after the substrate manipulation, confirms that the local bottom substrate in shallow streams may have a great impact on flight directions of caddis flies. The bottom substrate structure, and especially the occurrence of emerging stones and boulders, not only affects the water current velocity but probably also the local wind conditions and indirectly the flight behaviour, e.g. direction and height, of adult insects. The less pronounced frequencies of upstream movements when stones were exposed, however, may also have been influenced by the new "visual environment", referring then to the possibility of stones having constituted possible landmarks for swarming males (cf Sullivan 1980), especially leptocerids, whose activity is known to be influenced by environmental factors (Solem 1978, Statzner 1978, Solem & Resh 1981), e.g. stones, vegetation, light intensity and wind conditions.The difference in the lateral distribution over the stream between leptocerids and other species, is therefore supposed to be caused principally by species differences in flight behaviour, rather than different preadult distribution in the stream.

Differents kinds of traps have been used to demonstrate upstream migration of insects along streams and rivers. Types of traps that have been used are Malaise traps, window traps, transparent adhesive sheets, suction traps and light traps (see Müller 1974, 1982 for references). Several workers (e.g. Svensson 1974, Bird & Hynes 1981) have emphasized that the position of a trap influences the result obtained, especially when the trap does not reach across the whole stream. This study has shown the great importance of the lateral position of a trap over a stream in order to reveal the true upstream/downstream moving numbers of amphibiotic insects. Thus, at one side of the stream imagines can be flying upstream and at the other side in the opposite direction (e.g. Trichoptera total in all three periods and *W. subnigra* in P2, Tab.2).

The hypothesis of the colonization cycle applies particularly to egg bearing females. The findings of higher frequencies of upstream egg bearing females than of males of several species in this study, support the model. It seems credible that females are mainly flying upstream until they meet a suitable area for oviposition. These sites can possibly be recognized by physical characteristics of the water current as well as other factors (Badcock 1953, Statzner 1978). When an oviposition area is found one of all possible flight directions may be optimal for the subsequent short range "egg-laying flight", to use the terminology proposed by Solem (1984). In extreme cases this flight direction can be downstream.

## REFERENCES


Badcock, R.M. 1953. Observation of oviposition under water of the aerial insect *Hydropsyche angustipennis* (Curtis) (Trichoptera). **Hydrobiologia**, 5 : 222-225.

Bird, G.A. & Hynes, H.B.N. 1981. Movements of adult aquatic insects near streams in Southern Ontario. **Hydrobiologia**, 77 : 65-69.

Botosaneanu, L. 1957. Recherches sur les Trichoptères (imagos) de Roumanie. **Bull. Ent. Pologne**, 26 (25) : 383-433.

Elliott, J.M. 1969. Life history and biology of *Sericostoma personatum* Spence (Trichoptera). **Oikos**, 20 : 110-118.

Elliott, J.M. 1981. A quantitative study of the life cycle of the net-spinning caddis *Philopotamus montanus* (Trichoptera, Philopotamidae) in a Lake District stream. **J. Anim. Ecol.**, 50 : 867-883.

Göthberg, A. 1972. Gerichtete Flugbewegungen von Insekten an Fliessenden Gewässer. **Aquilo Ser. Zool.**, 13 5-7.

Gullefors, B. 1983. Flight movements of caddis flies (Ins. : Trichoptera) along the mouth part of a coastal stream in Northen Sweden. **Fauna Norrl.**, 3 : 1-22.

Müller, K. 1954. Investigations on the organic drift in north Swedish streams. **Rep. Inst. Freshw. Res. Drottningholm**, 35 : 133-148.

Müller, K. 1973. Life cycle of stream insects. **Aquilo Ser. Zool.**, 14 : 105-112.

Müller, K. 1974. Stream drift as a chronobiological phenomenon in running water ecosystems. **Ann. Rev. Ecol. Syst.**, 5 : 309-323.

Müller, K. 1982. The colonization cycle of freshwater insects. **Oecologia (Berl.)**, 52 : 202-297.

Nishimura, N. 1967. Ecological study on net-spinning caddisfly, *Stenopsyche griseipennis* McLachlan II. Upstream migration and determination of flight distance. **Mushi**, 40 : 39-46.

Olsson, T. 1971. Ljusfällefångst av Trichoptera och Plecoptera vid Rickleå 1970 : Artsammansättning, flygtider och flygriktning. **Rapp. Rickleå Fältstation**, 23 : 1-31. (Mimeogr.).

Pearson, W.D. & Kramer, R.H. 1972. Drift and production of two aquatic insects at a Mountain stream. **Ecol. Monogr.** 42 : 365-385.

Roos, T. 1957. Studies on upstream migration in adult stream-dwelling Insects.**Rep. Inst. Freshw. Res. Drottningholm**, 38 : 167-193.

Solem, J.O. 1978. Swarming and habitat segregation in the family Leptoceridae (Trichoptera). **Norw. J. Ent.**, 25 : 145-188.

Solem, J.O. 1984. Adult behaviour of north European caddisflies. **Proc. 4th Internatn. Symp. Trichoptera.** ed. J.C. Morse, Junk, The Hague, 375-382.

Solem, J.O. & Resh, V.H. 1981. Larval and pupal description, life cycle, and flight behaviour of the sponge-feeding caddisfly, *Ceraclea nigronervosa* (Retzius), in Central Norway (Trichoptera). **Ent. scand.**, 12 : 311-319.

Statzner, B. 1978. The effects of flight behaviour on the larval abundance of Trichoptera in the Shierenseebrooks (North Germany). **Proc. 2nd Internatn. Symp. Trichoptera**, ed. M.I. Crichton, Dr. W. Junk Pub., The Hague, 121-134.

Sullivan, R.T. 1980. Insect swarming and mating behaviour. **Florida. Ent.**, 64 (1) : 44-65.

Svensson, B.W. 1974. Population movements of adult Trichoptera at a south Swedish stream. **Oikos** 25 : 157-175.

Ulfstrand, S. 1970. Trichoptera from River Vindelälven in Swedish Lapland. A four-year study based mainly on the use of light traps. **Ent. Tidskr.**, 91 : 46-63.

# 8 - ECOLOGY

## 8.1. LIFE CYCLES, HABITAT AND FEEDING HABITS



## 8.2. ZONATION AND DISTRIBUTION

# THE LIFE CYCLE AND FOOD OF *APATANIA ZONELLA* (ZETT.) IN A SPRING-FED STREAM IN SW ICELAND (TRICHOPTERA : LIMNEPHILIDAE)

G. M. GÍSLASON and A. Th. SIGFÚSSON


## SUMMARY

The life cycle of *Apatania zonella* (Zett.) was studied in 1974-75 in a spring- fed stream in SW Iceland. The stream had relatively stable temperatures and chemical composition. The population had a one year life cycle ; most larval growth took place in August to September and pupation in April to September.

The larvae of *A. zonella* fed mainly on diatoms. Slower growth of larvae in winter was probably associated with lower density of diatoms.


## INTRODUCTION

Two main factors affect the life cycles of freshwater insects, food and temperature (Sweeney 1984). Spring-fed streams are relatively stable in regard to physical parameters, especially discharge, chemical composition and temperature.

*A. zonella* is one of the commonest Trichoptera in Iceland, except in the Northeast and East. It is commonly found in streams and rivers as well as on stony shores of lakes and has been found in the lower reaches of streams periodically inundated by seawater (Gíslason 1981 a, b).

This paper describes the life cycle and food of *A. zonella* in a spring-fed stream. Such studies could give information on which factors affect its biology in a stable physical environment.

## STUDY AREA

Samples were obtained from the spring-fed stream Hafnarfjardarlaekur (64° O4' N, 21°56'W), about 8 km south of Reykjavik, Iceland. The sampling site was 300 m below the source. The stream was about 4 m wide and 25-30 cm deep at the sampling site. The sampling site was divided into two areas, a fast flowing stretch where the main channel of water ran, with water velocity of about 50 cm/s, and a slow flowing stretch near the other bank, with water velocity around 25 cm/s. Sampling from the fast flowing stretch began on 29 April 1974, and in the slow flowing stretch on 29 September 1974.

The stream Hafnarfjardarlaekur is similar to other spring-fed streams in Iceland (Rist 1956). There is little or no fluctuation in water chemistry (Table 1) or water-level throughout the year. The pH was 7.7-8.6, and the specific conductance 89-95 µS/cm at 25°C. Two complete analyses, in April and September, indicate little seasonal change in the concentration of the major cations ($Na^+$ and $Ca^{++}$) and major anions ($HCO_3^-$ and $Cl^-$). This resembles other analysis from spring-fed streams in Iceland (Gíslason 1974).

TABLE 1. Chemical composition of the water in the stream Hafnafjardarlaekur

| | | meq/l | | | | | | | | µS/cm |
|---|---|---|---|---|---|---|---|---|---|---|
| Date | pH | $Ca^{++}$ | $Mg^{++}$ | $Na^+$ | $K^+$ | $HCO_3^-$ | $SO_4^{--}$ | $CL^-$ | TOTAL | (25 °C) |
| 29/4/74 | 8.5 | 0.20 | 0.07 | 0.56 | 0.01 | 0.56 | 0.07 | 0.38 | 1.01 | 89 |
| 2/6/74 | 7.7 | | | | | | | | | 90 |
| 6/7/74 | 8.4 | | | | | | | | | 93 |
| 5/9/74 | 8.6 | 0.19 | 0.06 | 0.60 | 0.01 | 0.64 | 0.07 | 0.38 | 1.09 | 95 |
| 23/9/74 | 8.5 | | | | | | | | | |
| 7/8/75 | 8.4 | | | | | | | | | 95 |

Water temperature did not fluctuate much (Fig. 1). In summer and autumn 1974 it was 3-8°C, and in winter 1974-5 -1 to + 4° C. The low value of -1°C recorded in January was presumably due to anchor ice.The stream was never covered by ice .

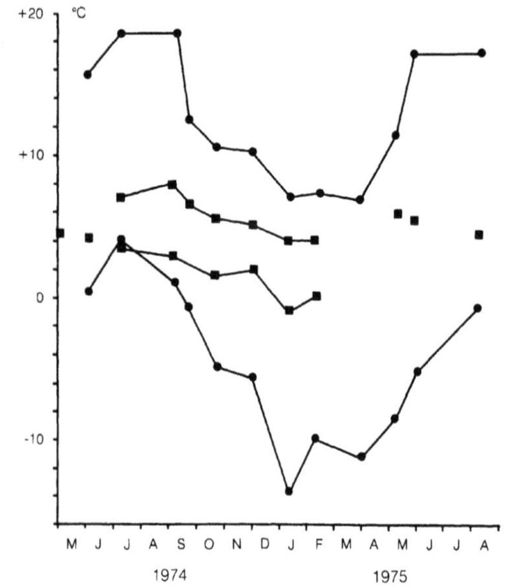


Figure 1. Maximum and minimum water temperatures in the stream Hafnarfjardarlaekur from June 1974 to February 1975 (squares connected with lines), and water temperatures on 8 and 29 May and 11 August 1975 (single squares). Maximum and minimum air temperatures in Reykjavik in 1974-75 (circles).

One other Trichoptera species, *Limnephilus griseus L.*, was found in the stream, but only in small numbers.

MATERIAL AND METHODS

Larvae, pupae and empty cases were collected at monthly intervals from stones, 7-20 cm in diameter. The stones were very porous and irregular in shape, and the larvae were usually found sheltering in cavities. All larvae were collected from each stone and examined. An attempt was made to collect aver 50 larvae in each visit. On each occasion larvae were taken from several stones. The larvae were anaesthetized with $CO_2$ and preserved in 70% alcohol immediately after collection.

Larval head-capsules were measured with an accuracy of ± 0.03 mm and the length of preserved larvae and pupae was measured with an accuracy of ± 0.1 mm under a dissecting microscope. Their weight was determined to ± 0.1 mg. Before the larvae were weighed, they were rolled gently over absorbent paper to dry off excess alcohol.

Since only 4 adults *A. zonella* were found at the stream, the flight period was estimated by the ratio of empty pupal cases against the total number of larvae and pupae.

For food analyses, 57 fourth and fifth instar larvae were studied ; 42 (including 12 empty) were from the fast flowing stretch and 15 from the slow flowing stretch. The larvae were dissected under a microscope. The foregut and midgut were removed and put in distilled water, they were opened and the contents washed out and analysed, following the method of Wotton (1977).

Samples were examined with x125 magnification under a phase contrast microscope for estimating proportional cover of diatoms and detritus. Detritus was divided into remains of vascular plants, where distinct cells could be seen, and other detritus, i.e. diatoms frustules and organic and inorganic remains where cells were not seen. For further identification of diatoms to genus, x1250 magnification was used. Two or three transects, 0.07 mm wide each, were taken through the samples and all algae, usually between 200-400 cells, were identified. Identification was based on Hustedt (1930), Foged (1974) and Barber and Hayworth (1981).

Water analyses were carried out to ± 0.01 meq/l. $Na^+$, $K^+$, $HCO_3^-$ and strong acid salts were analysed using the methods of Mackereth (1963), $Ca^{++}$ and $Mg^{++}$ were measured using a Varian Techtron AA-5 and $Cl^-$ on chloride titrator. Measurements of pH to ± 0.1 were made with a portable pH meter, EIL model 30, at the sampling site, or in the laboratory shortly after the samples were taken, usually by pouring the sample into a clean glass beaker, and stirring constanlty whilst measuring. Specific conductance (μS/cm at 25°C) was measured with a Müller conductivity meter.

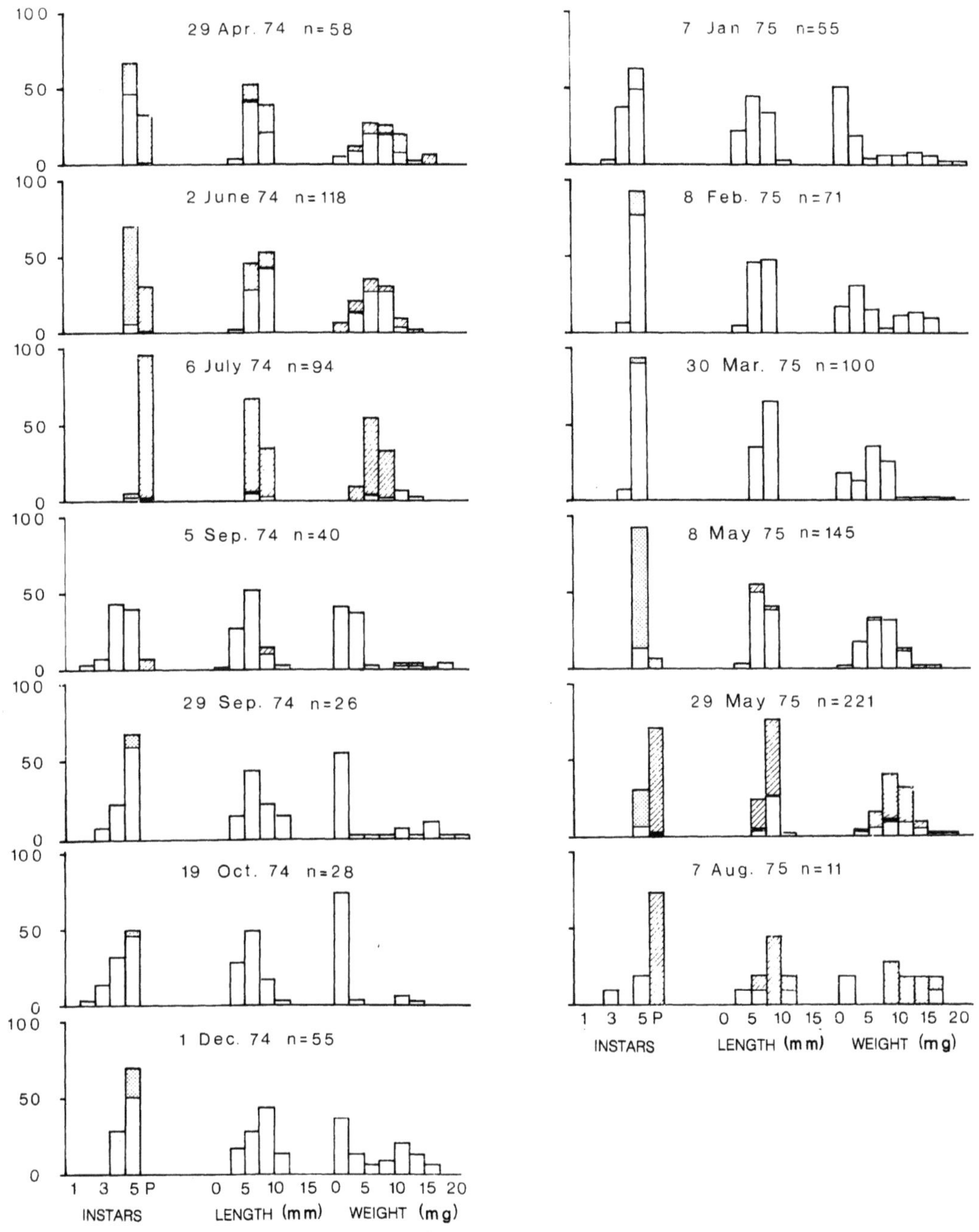


Figure 2. Life cycle of *A. zonella* in a stream Hafnarfjarlaekur. Per cent frequency distribution of instars, length and wet weight is given for each sampling date. White: larvae. Stippled: prepupae. Black: male pupae. Hatched : female pupae.

A maximum-minimum thermometer was read each the site was visited, until February 1975 ; subsequantly single readings of water temperature were taken on each visit.

## RESULTS

### Life cycle

The population of *A. zonella* in the stream Hafnarfjardarlaekur was univoltine (Fig. 2). Five larval instars were distinguished (Table 2).

Since only 4 adult females were caught near Hafnarfjardarlaekur, on 6 July 1974, all with fully mature eggs and 3 about to oviposite, an indirect method was used to indicate the flight period at Hafnarfjardarlaekur (Fig. 3). The main emergence period seems to have been in July and August. Small larvae were found in August and small and large larvae in early September. Most larvae had reached the 5th instar in late September. They pupated in April and May. The pupal stage was present until September and was the dominating stage in July. A small proportion of the larval population were found as prepupae in September, and was in an inactive stage during the winter.

| Instars | Head capsule width (mm) |
|---|---|
| 1 | 0.18-0.20 (calculated) |
| 2 | 0.25-0.30 |
| 3 | 0.35-0.48 |
| 4 | 0.52-0.70 |
| 5 | 0.72-0.98 |

Table 2 : Head-capsule width of the instars of *A. Zonella.*

Growth was rapid in late summer and autumn. The larvae reached a mean wet weight of 6-7 mg in late September and remained so until late March when the mean weight of larvae and pupae increased again, to 8-9 mg in early May and 9-10 mg in late May. In August the pupae had reached the mean weight of 11.6 mg (Fig. 2).

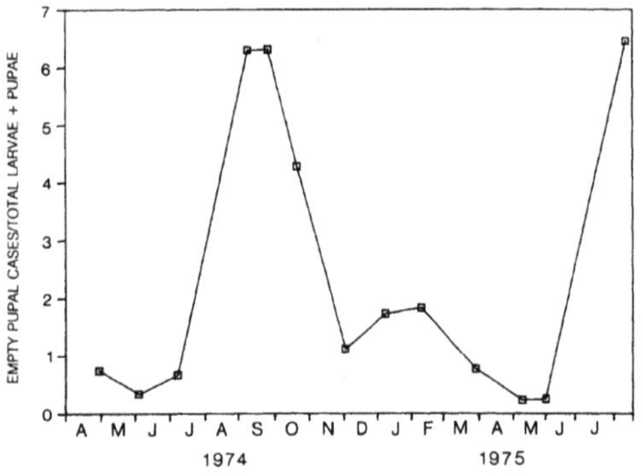


Figure 3. Flight period of *A. zonella* in the stream Hafnarfjardarlaekur, expressed as the ratio of empty pupal cases against total number of larvae and pupae.

### Food

Diatoms were the main food (67-93% by cover) of *A. zonella* in the fast flowing stretch all the year round (Fig. 4a). Detritus was 7-33% of the gut content, but remains of vascular plants were insignificant (≤1%). The proportion of detritus and plant remains was greatest during the winter, from October to late April and early May. In the slow flowing stretch, diatoms also dominated the food, except in March and April, when diatoms were in similar quantity as the detritus. The proportion of remains of vascular plants was insignificant (<1 %) (Fig. 4b).

In the fast flowing stretch of the stream 17 genera of diatoms were found. Six of these, *Cocconeis, Fragilaria, Nitzschia, Rhoicosphenia, Achnantes* and *Navicula* were most abundant (Fig. 4a). Other diatom species contributed less than 5% each of the gut content. From September to April,*Cocconeis* was most abundant and reached about 84% ot the total gut contents in April. During the summer *Cocconeis, Fragilaria,* and *Nitzschia* were the most common genera, about 10-20% each. *Rhoicosphenia* spp. were about 10-40% of the gut contents from May to December 1974, but were less than 5% from January to May 1975. Other genera reached their maxima in the spring or early summer, but were scarce during the winter.

DISCUSSION

*A. zonella* is univoltine in Iceland, with a flight period from May to August (Gíslason 1978). In the stream Hafnarfjardarlaekur, the emergence seemed to start in June (Fig. 3). Most of larval growth took place in late summer and autumn (Fig. 2). The growth was halted during the winter, but resumed in early spring (Fig. 2).

Nielsen (1943) found the last three instars and adults of *A. zonella* in a mountain lake in Jötunheim, Norway, in August, a similar instar distribution as in Iceland at the same time. However, in a study in the Dovrefjell mountains, Norway, Solem (1985) found an identical instar distribution during the summer, but inferred that the life cycle was completed in two years. His conclusion was based on the presence of 4th and 5th instar larvae in mid-July, during the main flight period. Solem (1985) concluded that these larvae were unlikely to complete their life cycle during that summer. However, most 5th instar larvae in Hafnarfjardarlaekur developed into pupae within a month in May 1975. Both in the lowlands (e.g. Hafnarfjardarlaekur) and in the highlands of Iceland (Gíslason 1977), there is a wide range in instar frequency distribution, reflecting a long emergence and flight period (Gíslason 1978).

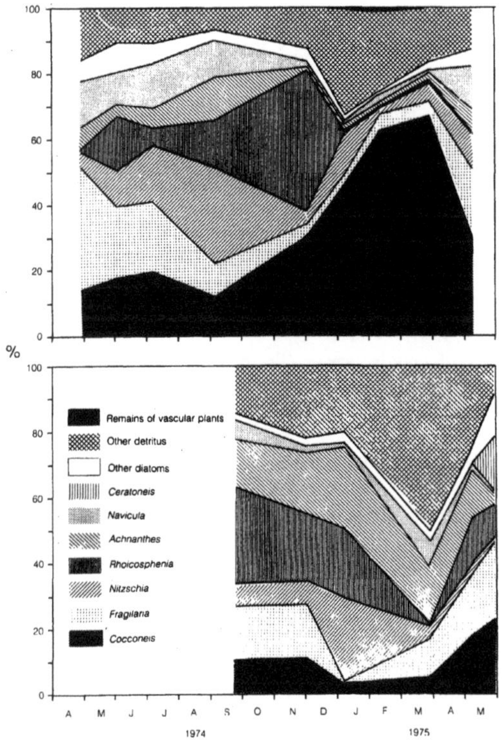


Figure 4. The food of *A. zonella* in the fast flowing stretch (a) and the slow flowing stretch (b) of Hafnarfjardarlaekur.

Life cycles of ***Apatania muliebris*** McL. also appear to differ from one region to the other. *A. muliebris* had a two year cycle in the Dovrefjell region (Solem 1985), but was univoltine in Bohemia (Klapelek 1889), in spring-fed streams in Jutland, Denmark (Nielsen 1942, 1950a, b) and in a stream in England (Elliot 1971). Also within Jutland, *A. muliebris* had different life cycle in different spring-fed streams (Nielsen 1942, 1950b, 1969). *Apatania* species are parthenogenetic (Nielsen 1950b). Of 219 adults of *A. zonella* collected in 1968 to 1975 in Iceland, only 2 (0.9%) were males, and of 494 pupae collected in 1972 to 1975, only 11 (2.2%) were males.

The blade-like toothless mandibles of *A. zonella* are specialized for scraping stone surfaces for food, mainly epiphytic algae (Mackay and Wiggins 1979). Changes in gut contents of caddis larvae reflect changes in food abundance (Hanna 1957 and Haage 1970). The gut contents of *A. zonella* presumably reflect the relative abundance of diatom species in the stream.

Nutrients, light and temperature presumably limit the growth of diatoms (Reynolds 1973). In Hafnarfjardarlaekur, water chemistry and temperature is relatively stable (Table 1, Fig 1) and light is probably the limiting factor for algal growth. The proportion of diatoms in the guts of *A. zonella* was lowest in

December in the fast flowing stretch and in March in the slow flowing stretch (Fig. 4 a, b). This is probably explained by less growth of diatoms on the stones during the period of shortest day in Iceland, rather than by increasing detritus on the bottom, which might have affected the growth of the larvae.

## ACKNOWLEDGEMENTS

Part of this study was made under the supervision of Dr. G. N. Philipson, when the first author was studying for Ph. D. at Newcastle University. Thanks are due for his encouragement and advice during the study. We should like to thank the following for various assistance during this work. Prof. A. Gardarsson, Drs A. Einarsson and V. Jóhannsson, Messrs. U. Antonsson, G. S. Jónsson, University of Iceland, J. Olafsson, Marine Research Institute in Reykjavík and J. O. Solem, University of Trondheim. Records of air temperature were obtained from the Meteorologicol Office in Reykjaví k.

## REFERENCES

Barber, H. G. & Hayworth, E. Y. 1981. A guide to the morphology of the diatom frustule. Sc. **Publs Freshwat. biol. Ass.**, N° 44, 112 p.

Elliott, J.M. 1971. The life history and biology of *Apatania muliebris* McLachlan (Trichoptera). **Ent. Gaz.**, 22 : 245-251.

Foged, N. 1974; Freshwater diatoms in Iceland. **Bibliotheca phycologica**, 15 : 1-118.

Gíslason, G. M. 1974. *Potamophylax cingulatus* (Stephens) (Limnephilidae : Trichoptera) recorded from Iceland. **Náttúrufr.**, 44 :129 - 139.

Gíslason, G.M. 1977. Aspects of the biology of Icelandic Trichoptera, with comparative studies on selected species from Northumberland, England. 412 p.

Gíslason, G.M. 1978. Flight periods and ovarian maturation in Icelandic Trichoptera. **Proc. 2nd int. Symp. Trichoptera,** M.I. Crichton (ed.), Junk, The Hague, 135-146.

Gíslason, G.M. 1981 a. Predatory exclusion of *Apatania zonella* (Zett.) by *Potamophylax cingalatus* (Steph.) (Trichoptera : Limnephilidae) in Iceland. **Proc. 3rd int. Symp. Trichoptera,** G.P. Moretti (ed.), Junk, The Hague, 93-98.

Gíslason, G.M. 1981 b. Distribution and habitat preferences of Icelandic Trichoptera. **Proc. 3rd int. Symp. Trichoptera,** G.P. Moretti (ed.), Junk, The Hague, 99-109.

Haage, P. 1970. On the feeding habits of two Baltic species of caddis larvae (Trichoptera). **Ent. scand.**, 1 : 282 -290.

Hanna, H.M. 1957. A study of the growth and feeding habits of the larvae of four species of caddisflies. **Proc. R. ent. Soc., London (A),** 32 : 139-146.

Hustedt, F. 1930. Bacillariophyta (Diatomeae). *In* Die Süsswasserflora Mitteleuropas, A. Pascher (ed.) : 1O, 466 p.

Klapelek, F. 1889. The Metamorphoses of *Apatania muliebris* McLach. : a chapter in parthenogenesis. **Ent. mon. Mag.,** 25 : 241-242.

Mackay, R. J. & Wiggins, G. B. 1979. Ecological diversity in Trichoptera. **Ann. Rev. Entomol.,** 24 : 185-208.

Mackereth, F. J. H. 1963. Some methods of water analysis for limnologists. **Sc. Publs Freshwat. biol. Ass.** , N°21, 70 p.

Nielsen, A. 1942. Uber die Entwicklung und Biologie der Trichopteren mit besonderer Berücksichtigung der Quelltrichopteren Himmerlands. **Arch. Hydrobiol., Suppl.,** 17 : 255-631.

Nielsen, A. 1943. *Apatidea auricula* Forsslund from a Norwegian mountain lake. Description of the imago and notes on the biology. **Ent. Medd.,** 23 : 18 -30.

Nielsen, A. 1950 a. On the zoogeography of springs. **Hydrobiologia,** 2 : 313-321.

Nielsen, A. 1950 b . Notes on the genus *Apatidea* MacLachlan . With descriptions of two new and possibly endemic species from the springs of Himmerland. **Ent. Medd.,** 25 : 384-404.

Nielsen, A. 1969. On the subspecies of *Apatania muliebris* McL. (Trichoptera). **Ent. Medd.,** 37 : 313-318.

Novak, K. 1960. Entwicklung und Diapause der Köcherfliegenlarven *Anabolia furcata* Br. (Trichoptera). **Cas. Cesk. Spol. ent.,** 57 : 2O7-212.

Reynolds, C. S. 1973. The seasonal periodicity of planktonic diatoms in a shallow eutrophic lake. **Freshwat. Biol.,** 3 : 89-110.

Rist, S. 1956. Icelandic freshwaters 1. Raforkumalastjori - Vatnamaelingar. Reykjavik. 127 p.

Solem, J. O. 1985. Norwegian *Apatania* Kolenati (Trichoptera : Limnephilidae) : identification of larvae and aspects of their biology in a high-altitude zone. **Ent. scand.,** 16 : 161-174.

Sweeney, B. W. 1984. Factors influencing life-history patterns of aquatic insects. *In* The ecology of aquatic insects. V. H. Resh and D. M. Rosenberg (ed.). N. Y. : 56-100.

Wotton, R. S. 1977. The size of particles ingested by moorland stream blackfly (Simuliidae). **Oikos,** 29 : 332-335.

# THE LIFE HISTORY OF *CRYPTOCHIA PILOSA* (TRICHOPTERA : LIMNEPHILIDAE) IN AN OREGON COAST RANGE WATERSHED

R.W. WISSEMAN and N.H. ANDERSON

## SUMMARY

*Cryptochia* is a small genus confined to montane areas of western North America. In the Oregon Coast Range, *C. pilosa* requires 2 years to complete its life cycle. The semi-aquatic larvae are usually found in damp organic matter 1-30 cm above the stream waterline. *C. pilosa* consumes leafy and woody debris. Case architecture appears to serve multiple functions in the dispersal of larvae and in the occupation of a unique habitat.

## INTRODUCTION

*Cryptochia* (subfamily Dicosmoecinae) is a genus of seven species confined to western North America. Denning (1975) reviewed the known distribution of all species and provided a key to the males. *C. pilosa* (Banks) and *C. furcata* Denning are widely distributed. Five other species are each known from single males collected in Oregon or northern California.

The genus *Cryptochia* was erected by Ross (1950) for *C. pilosa*, which had previously been assigned to the European genus *Parachiona*. Ross concluded that *Cryptochia* has "several characteristics which indicate an extremely primitive condition for the Limnephilidae, especially the male genitalia...". Schmid (1955) reviewed the male and female taxonomy of *C. pilosa*, and concluded that "*Cryptochia* presents a bizarre mixture of primitive and evolved characters, with multiple affinities."

*Cryptochia* larvae have a prominent carina on the head and a dense setal fringe on the pronotum (Wiggins 1977). The smooth, strongly tapered case of wood fragments is also distinctive. Based on collection notes for larvae, Wiggins (1977) postulated a semi-terrestrial existence in damp,streambank organic material.

We report here the life history of *C. pilosa* in an Oregon Coast Range watershed, augmented with observations of *C.* prob. *pilosa* from other locales in the Pacific Northwest. We believe that life history and behavioral attributes of the larval stage indicate a more specialized condition than is suggested by adult characters.

Collections of larvae, pupae, and adults were made from Flynn Creek, Lincoln Co. at approximately monthly intervals from July 1981 to November 1982 as part of a larger investigation of the life histories of Trichoptera in this forested, maritime watershed. Site descriptions are provided in Wisseman and Anderson (1984).

## GEOGRAPHICAL DISTRIBUTION

*Cryptochia* spp. are found in the montane areas of western North America. *C. pilosa* inhabits the coastal mountains from southeast Alaska to northern California. It is widely distributed throughout the Coast and Cascade Ranges, and has been collected once from the inland mountains of Idaho. *C. furcata* appears to be confined to the Rocky Mountains. The remaining species are currently known from the Sierra Nevada Moutains of northern California and the Blue Mountains of eastern Oregon.

All *Cryptochia* collections have been from forested, montane watersheds. They occur between sea level and 3000 m in altitude.

## LARVAL HABITATS

We collected more than 500 *Cryptochia* larvae from semiterrestrial habitats in the Flynn Creek watershed in two years by handpicking, floatation, and Berlese extraction from organic debris taken above the water surface. In contrast, only five larvae were found in more than 500 benthos samples.

Larvae were most often recovered from damp leaves and wood on stream margins or in debris jams 1-30 cm above the waterline. They were abundant in floating material behind debris jams after freshets.

During the autumnal pulse of leaf input (primarily *Alnus rubra*), larvae were most frequently associated with damp alder leaves on the stream bank. During periods of low leaf availablity, they were found clinging to wet wood along the stream. In the dry summer months, larvae burrowed into debris on stream margins.

The longitudinal distribution of larvae ranged from springheads to small rivers, though the population was centered in the 3rd order region of Flynn Creek (discharge range : 0.0025-3.94 $m^3$ $sec^{-1}$).

LIFE CYCLE

### Eggs

Egg masses were collected in April and May from hillside seeps in the upper watershed. They were suspended 5-20 cm above surface water on exposed plant rootlets that had water dripping from them. Masses were 6-8 mm in diameter, and consisted of clear, tacky gelatin with no surface texturing or skin. Eggs were arranged in rough rows, filling the entire matrix. Eggs averaged 166 per mass (n=5, range 132-198) and were 0.40-0.45 mm in diameter when developed embryos were visible. The incubation period lasted approximately three weeks, both for lab-reared masses and those marked in the field.

Oviposition sites were restricted to springs or seeps and maximal larval densities occurred further downstream suggesting that adult flight behavior is oriented upstream.

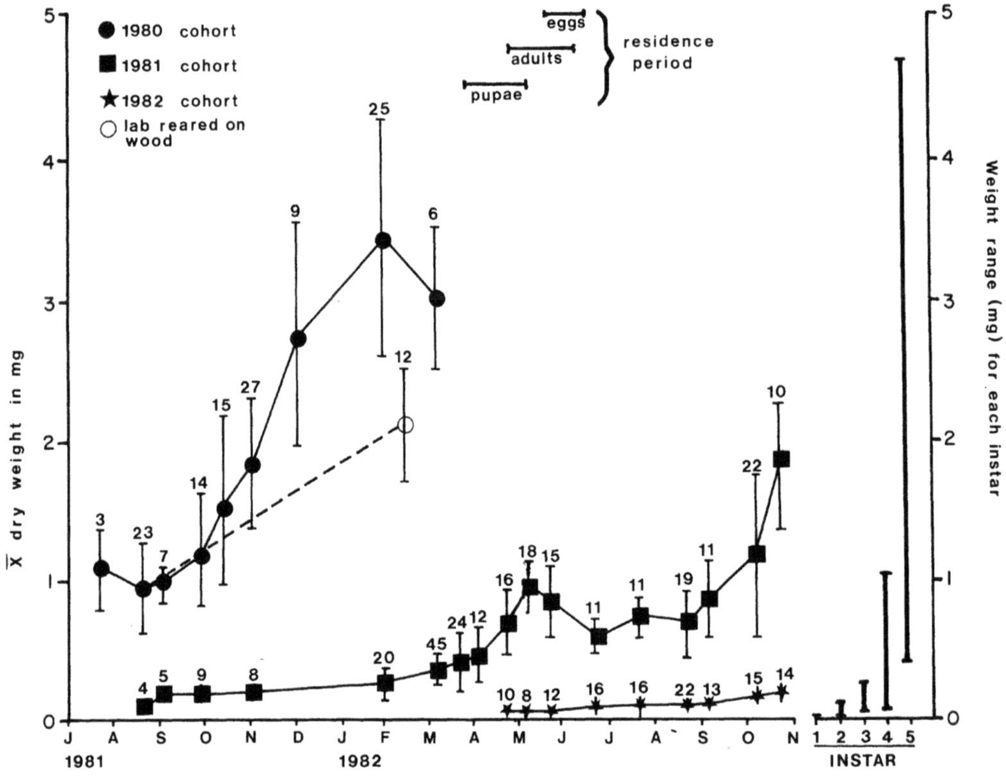


Figure 1. Life history of *Cryptochia pilosa* at Flynn Creek, OR. Mean dry weight of three larval cohorts (designated by year eggs laid) with ± 1 s.d. and number of larvae weighed provided. Growth trajectory of 1980 cohort larvae reared on wood in laboratory aquaria is included. Residence period of pupae, adults and eggs is shown. Weight range exhibited by each larval instar is appended on the right.

### Larva

Five larval instars were readily distinguished by head capsule width. Measurements were made with an ocular micrometer across the widest portion of the head when viewed dorsally. The range (mm) for each instar was : I, 0.20-0.22 (n=50) ; II, O.28-0.30 (n=15) ; III, 0.36-0.44 (n=66) ; IV, 0.56-0.66 (n=132) ; V, 0.80-1.00 (n=125).

Two distinct cohorts of larvae were present in all collection intervals indicating that *C. pilosa* requires two years for development at Flynn Creek (Fig. 1). Larval collections from other areas suggest that a biennial life cycle is the norm with populations at high elevations perhaps requiring more than two years to complete a cycle.

Growth curves for larvae collected in 1981 and 1982 indicate slow weight gain during the first year (Fig. 1). Although larvae had molted to the fourth instar in the first year, their mean weight was 0.6 mg, or about 15% of their maximum size. Rapid growth and molting to the final instar occurred during late spring or early summer followed by a mean weight loss during late summer, possibly due to a quiescent period associated with drought conditions.

The most rapid weight gain occurred during the fall and winter months of the second year. Larvae were common on the surface of leaf packs and wood debris during this time. About 80% of the total larval weight was achieved during this period. A small decrease in larval mean weight was observed from February to March, probably due to loss of mature larvae that moved to pupation sites.

Gut contents suggest that larvae feed on decaying leaf and wood substrates. Guts were often full of amorphous, fine particulate organic matter, with few recognizable wood or leaf cells. We believe larvae may feed preferentially on fungal fruiting bodies and hyphae. Larvae showed significant growth (Fig. 1) over a six month period in the lab when kept only on wood with surfaces conditioned by decay fungi. Guts of these larvae contained few recognizable wood fibers. The amorphous, fine particulate material in the guts was most likely of fungal origin.

**Pupae and adults**

Only seven pupae were found during the course of this study and were collected in April. All were wedged in damp, rotted wood on the stream margin. Pupae were not found on any substrates submerged in the stream in three years of general collecting. No adults were collected in emergence traps placed over open water (10 traps 0.25 $m^2$, run for two years).

Eight adults were caught by net or sticky traps along first-order tributaries from April 21 to June 19. Adults did not respond to blacklights placed in the watershed, nor have they been captured in light traps elsewhere in their geographic range. The flight period over the entire range, as indicated by fewer than 20 records from museum collections, is April 21 to August 10. Early season records are from maritime coastal sites and later records from higher elevations or more northerly populations.

## MORPHOLOGY AND ADAPTATION

**Dispersal**

*C. pilosa* larvae construct cases from fragments of woody material in such a manner that case plus larvae are fully buoyant. Larvae of the Flynn Creek population constructed cases from fragments of salmonberry (*Rubus spectabilis*) stems which have vascular fibers of large diameter capable of retaining air when wetted.

When dislodged into the stream from the stream bank, larvae were carried by the current on the water surface. Larvae placed manually into the current withdrew into their case and became quiescent. Upon lodging on the stream margin or in flotsam behind debris jams, they crawled out of the water onto leaves or wood.

Cases appear to provide a mechanism for dispersal of larvae to new habitats. Floatation increases the probability that larvae displaced into the drift (e.g. by spates) will come to rest against streambank habitats suitable for foraging. Whether larvae also actively enter the surface drift when they fail to find suitable food is not known.

**Terrestrial existence**

Capillary action from damp surfaces through the case appears to provide a saturated atmosphere for the larva, aiding terrestrial existence by reducing desiccation. The dense setal fringe of the pronotum and head may act in concert with the flattened head to form a "lid and gasket" during periods of moisture deficit.

Case form and surface tension forces generated between the damp case and substrate allows larvae to cling to near-vertical pieces of wood. Larvae and case can maintain this position without assistance from tarsal claws or silken ties.

**Predation**

Case and larva are very cryptic. The case material closely matches the substrates in color and texture. A flattened head and setal fringe seal the retracted larva tightly into the case. Larvae move slowly when foraging. From observations in laboratory aquaria, larvae forage at night. This may reduce risk of predation.

Carabid beetles on the stream margin and hydrophilid beetles in flotsam are suspected of being predators. Gut contents of both occasionally contained caddisfly larval sclerites. Empty cases nearly bisected by a longitudinal incision were often found in flotsam behind debris jams, possibly the remains of invertebrate attack upon larvae.

## CONCLUSIONS

We believe that the peculiar aspects of *Cryptochia* larval habitat and behavior, combined with the unique morphology of case and larva, indicate unusual features for a member of the subfamily Dicosmoecinae. Terrestrial existence probably represents a secondary invasion, necessitating the evolution of a spectrum of behavioral and morphological adaptations to cope with this environment.

## ACKNOWLEDGEMENTS

The following individuals and museums generously supplied records : G.B. Wiggins (Royal Ontario Museum), O.S. Flint (Smithsonian), J.Unzicker (Illinois Natural History Survey), F. Schmid (Canadian National Collection), D.G. Denning (Moraga CA), A.P. Nimmo (Univ. of Alberta), N.A. Erman (Univ. of California), and S.D. Smith (Central Washington State Univ.). Peter McEvoy and Mary Jo Wevers kindly reviewed the manuscript. This project was supported in part through funding from N.S.F. Grant BSR-8022190. This is Technical Paper No. 7941 of the Oregon Agricultural Experiment Station.

## REFERENCES

Denning, D.G. 1975. New species of Trichoptera from western North America. **Pan-Pacific Ent.**, 51 : 318-326.

Ross, H.H. 1950. Synoptic notes on some Nearctic Limnephilid caddisflies (Trichoptera, Limnephilidae). **Amer. Midl. Nat.**, 43 : 410-429.

Schmid, F. 1955. Contribution à l'étude des Limnophilidae (Trichoptera). **Mitt. Schweiz. Entomol. Ges.** 28 : 1-145.

Wiggins, G.B. 1977. Larvae of the North American caddisfly genera (Trichoptera). Univ. Toronto Press., 401 p.

Wisseman, R.W. & Anderson N.H. 1984. Mortality factors affecting Trichoptera eggs and pupae in an Oregon Coast Range watershed. **Proc. 4th Internat. Symp. Trichoptera**, J.C. Morse (ed.), Junk, The Hague, 455-460.

# LIFE CYCLES AND GROWTH PATTERNS OF TRICHOPTERA IN THE MATARRAÑA, A KARSTIC RIVER

L. RECASENS and M.A. PUIG

## INTRODUCTION

Some papers have shown the importance of temperature in the life cycles and growth patterns of caddis flies (Butler 1984), but there is little information on rivers with a small range of thermal variation. This is the case of the River Matarraña (Ebro basin) where the life cycles of the most abundant Trichoptera species have been studied. The possibility of a growth pattern related to the first instar weight and the number of instars and/or moults is demonstrated.

## STUDY AREA AND METHODS

The Matarraña is a tributary of the River Ebro (NE Spain) which drains the Prelitoral Catalan Mountains at 1100 m (a.s.l.) into a karst system. Four sampling sites were studied along the river axis (Fig. 1). The first station at 600 m (a.s.l.) and its caddis-fly population were the subject of this paper. The main characteristics of this station are the small thermal fluctuation (10°C to 18°C) and minimal variation of water level (Puig *et al.* in press), except for an autumnal flood, in November for the last three years, but the effect of this flood was not recorded 12 hours after.

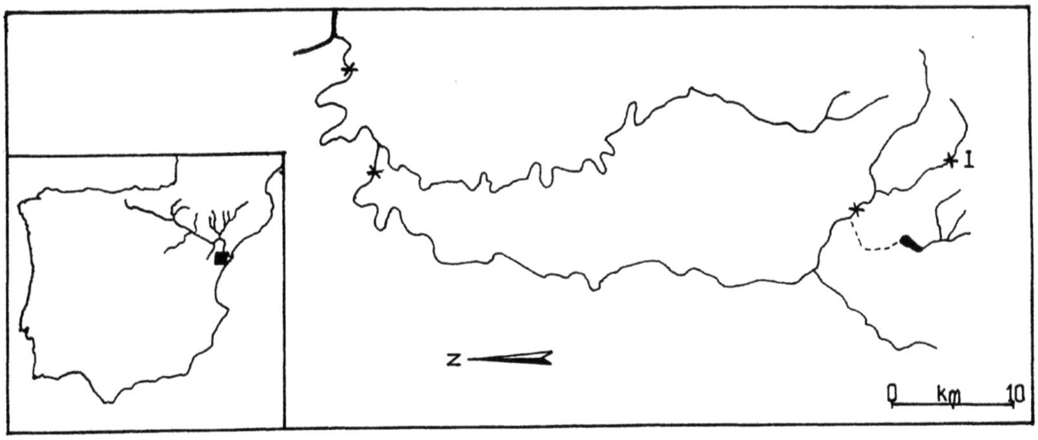


Figure 1. Location of sampling sites. The inset shows its geographical location in Spain.

Trichoptera were sampled with a modified Surber net of 250 and 100 µm pore size in the terminal part. The number of Surbers per month was 6 to 8. The sampling period was from April 1984 to March 1985. The material was preserved at -18°C and transferred to 70% alcohol for determination.

## RESULTS

Sixteen species of caddis flies were captured at the first station in the Matarraña , the six most abundant species were *Oxyethira cf. frici* Kl, *Hydroptila vectis* Curtis, *H. insubrica* Ris, *H. martini* Marshall, *Polycentropus flavomaculatus* Pictet and *Hydropsyche instabilis* Curtis. The other species captured throughout the study period , but in small numbers were*Rhyacophila munda* Mcl., *Glossosoma bifidum* Curtis, *Agapetus* *berbericus* Giudicelli & Dakki, *Philopotamus montanus* Don., *Psychomyia pusilla* Fab., *Tinodes dives* Pict., *Agraylea* sp., *Stactobia* sp., *Hydroptila forcipata* Eat.,*Mystacides azurea* L.

Three types of life cycles can be distinguished : one annual generation ; two generations per year and three generations.

*Oxyethira cf. frici* and *Hydroptila vectis* have only per year one annual generation (Fig. 2), both were captured in late autumn and winter, and the generation time was four months for each species.

Only *Hydroptila insubrica* has two clearly defined generations per year (Fig. 3). This species has a long generation from late autumn to spring and another short one, in summer. It is interesting to note the absence of this species in larval form in early autumn in contrast to the other species.

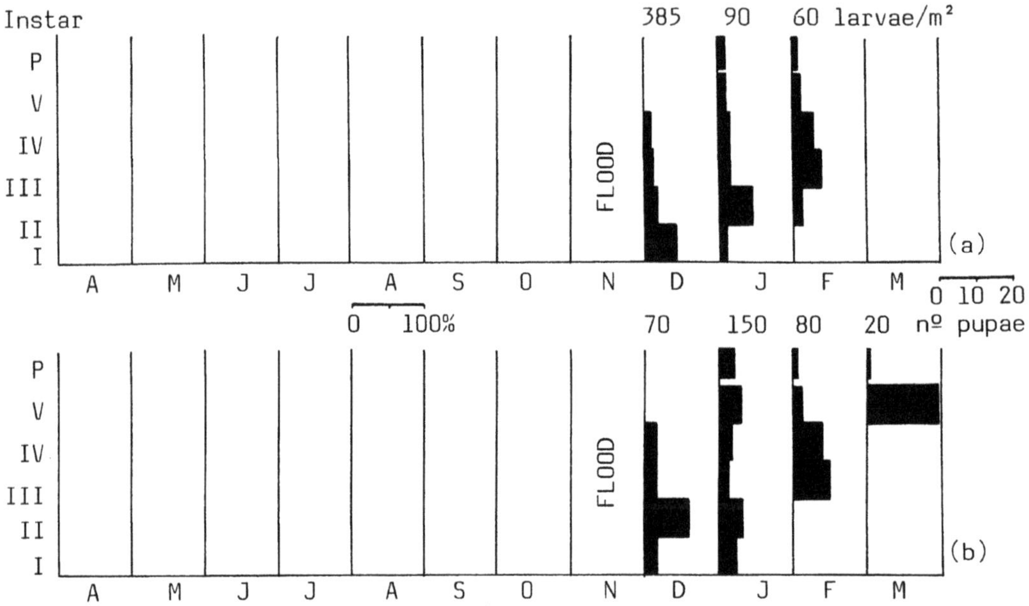


Figure 2. Instar-frequency data of a) *Hydroptila vectis* and b) *Oxyethira cf. frici*. The number of pupae/ 0.1 $m^2$ and larvae/ $m^2$ are also indicated.

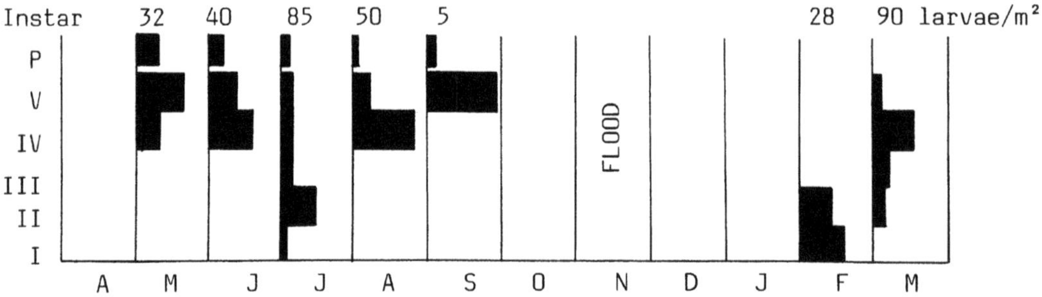


Figure 3. Instar-frequency data of *Hydroptila insubrica* in the Matarraña.

*Hydropsyche instabilis* has one winter generation, from November to June, and a second generation in the summer (Fig. 4b). The fifth instar larvae predominate for about four months and the adults emerge during three months leading to a confused situation in summer. It is possible to distinguish the development of two populations at this time. The start of these populations is separated by two months. The second population is infrequent, with a small number of specimens. Therefore and because of the long period of emergence of the winter generation we propose the existence of two cohorts in the summer and not two generations.

More than two generations per year, but not clearly three, have been recorded for *Hydroptila martini*

(Fig. 4a). This species has a summer generation and a second one that begins in September and ends in winter. Later, in the winter (March 1985) another generation has begun. The time observed for two complete generations during the study year was four months for each generation.

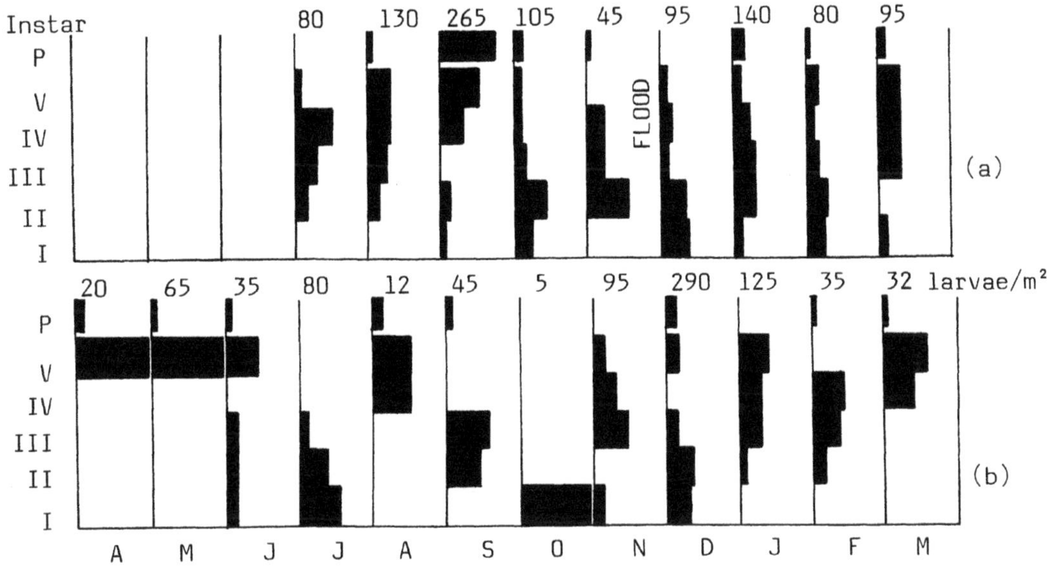


Figure 4. Instar-frequency data of a) *Hydroptila martini* and b) *Hydropsyche instabilis*. The number of pupae/ 0.1 $m^2$ and larvae/$m^2$ are also indicated.

The only species which clearly shows three annual generations is *Polycentropus flavomaculatus*. One winter-spring generation (from March to June), a summer generation (from June to September) and another in autumn (September-October to January) were observed (Fig. 5), this species has a new generation every four months.

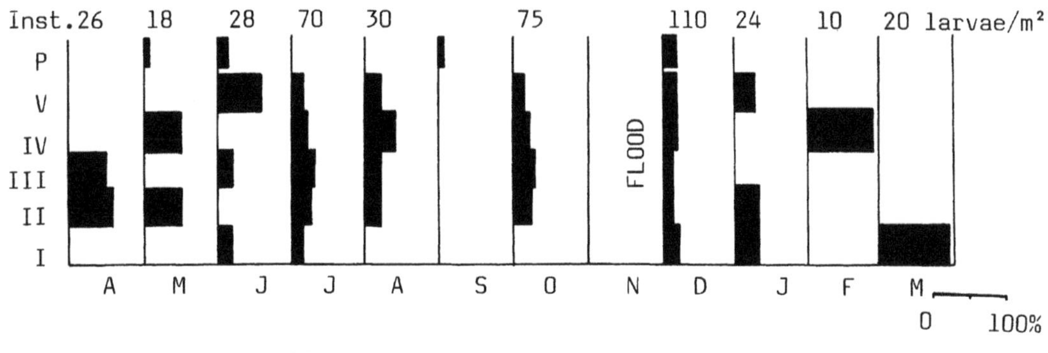


Figure 5. Instar-frequency data of *Polycentropus flavomaculatus*.

The November flood recorded in 1984 had two main effects in all species studied, the begining of a new generation and in some cases a sudden end of a generation by the disturbance of the fifth instar.

A study was made of the growths of three species :*H. instabilis* and *Hydroptila martini* (Matarraña populations) and *H. siltalai*. Two populations of this last species were studied (Vallvidrera and Avencó), each having a long generation (autumn-spring generation in each case). The partial data for *Agapetus fuscipes* Curtis were also studied.

The dry weight of each instar of each species and population was determined by drying the individuals at 100°C for 6 hours. The different replicates of each instar were weighted on a precision balance.

The weight increases observed were not linear. Usually one or two moults presented the maximum increase (Table 1). Two patterns can be observed, one with proportional increases (*H. instabilis* ) and the other with a maximum increase at the first moult and similar small increases for the others (*Agapetus fuscipes* ). Some species exhibit erratic growths. Thus, in *H. siltalai* the maximum increase may take place at different moults in different populations ; it was in the fourth winter instar in the Avencó (Recasens 1985) and in the fifth instar in the Vallvidrera.

Table 1. Mean weight, weight increase and weight increase average per instar (+ Matarraña, * Avencó, ** Vallvidrera).

| Species | | Mean weight (mg.) | Weight increase/ instar | Weight increase average/instar |
|---|---|---|---|---|
| *H. siltalai(*)* | I | 0.007 | | |
| | II | 0.028 | 4.00 | 4.41 |
| | III | 0.120 | 4.29 | 4.41 |
| | IV | 0.642 | 5.35 | 4.41 |
| | V | 2.572 | 4.01 | 4.41 |
| *H. siltalai (**)* | I | 0.017 | | |
| | II | 0.069 | 4.06 | 3.75 |
| | III | 0.218 | 3.16 | 3.75 |
| | IV | 0.710 | 3.26 | 3.75 |
| | V | 3.209 | 4.52 | 3.75 |
| *H. instabilis (+)* | I | 0.019 | | |
| | II | 0.051 | 2.68 | 3.51 |
| | III | 0.172 | 3.37 | 3.51 |
| | IV | 0.636 | 3.70 | 3.51 |
| | V | 2.727 | 4.29 | 3.51 |
| *Hydroptila martini (+)* | I | ----- | | |
| | II | 0.0073 | | |
| | III | 0.0541 | 7.44 | |
| | IV | 0.0694 | 1.28 | |
| | V | 0.1740 | 2.51 | |
| *Agapetus fuscipes (**)* | I | 0.00054 | | |
| | II | 0.00570 | 10.56 | |
| | III | 0.00964 | 1.69 | |
| | IV | 0.01265 | 1.31 | |
| | V | 0.01500 | 1.19 | |
| | VI | ------- | ---- | |
| | VII | ------- | ---- | |

The weight increases in each instar do not follow a mathematical relationship, but considering the average weight increase in Trichoptera with five instars (4 moults) this relationship approaches 4 (see Table 1). The mathematical relationship between the weight of the first instar and that of the last can be formulated as :

$Wn = 4^{(n-1)} . W_1$ or $Wn = 2^{a\,(n-1)} . W_1$ if $a = 10/n$

(n=n° of instar ; n-1= n° of moults ; Wn = weight of the last instar ; $W_1$ = first instar weight).

The results show that this hypothesis is apparently confirmed in the case of five instars (Trichoptera) and ten instars (Copepoda) (Margalef 1974 and Løvtrup *et al.* 1974). Hutchinson (1959) concluded that the corresponding body volum (weight) ratio of about 2 is an ecological requirement for co-existence of species. It is possible extend this ratio to coexistence of instars.

However, the study of species with more than five instars is necessary to complete the data and to confirm or refute the proposed relationship between weights.

ACKNOWLEDGEMENTS

We would like to thank Dr. Margalef for suggesting this study of Trichoptera growth patterns, J. Murillo for his criticism and acknowledge the help and support of the Fishgroup of the Matarraña Project. Financial support was provided by Project #CCA8411065 from the United States-Spain Joint Committee for Scientific and Technological Cooperation.

REFERENCES

Butler, J. 1984. Life histories of aquatic insects. *In* Resh J.V. & Rosenberg D.M., The Ecology of Aquatic insects. New York, Praeger Eds. : 24-55.

Hutchinson, G.E. 1959. Homage to Santa Rosalia or Why are there so many kinds of animals ? **Am. Nat.**, 93 : 145-159.

Løvtrup, S., Rahemtulla, F. & Höplund, NG. 1974. Fisher's axiom and the body size of animals. **Zool. Scv.** 3(2) : 53-58.

Margalef, R. 1974. Ecologia. Barcelona : Omega Ed., 951 p.

Puig, MA., Sostoa, A. & Lobón-Cerviá, J. in press. Geomorfología, físico-química y características biológicas del río Matarraña. **Nat. Hisp.**

Recasens, L. 1985. Cicle vital d' *Hydroppsyche siltalai* (Trichoptera : Hydropsychidae) a la riera de l'Avencó (Vallès Oriental). **Butll. Inst. Cat. Hist. Nat.**, 52 (Sec. Zool., 6) : 129-133

# REMARKS ON THE LIFE HISTORY OF *ANABOLIA FURCATA* BRAUER (TRICHOPTERA)

J. MAJECKI


## SUMMARY

On the basis of laboratory experiments and the observations in the field conditions, the differentiation of larval stages, diapause time and the period of emergence of adults were determined. Examining the eggs and larva under experimental conditions at the temperatures close to natural for a given month the period of larvae hatching was found to last very long from November till April.

Eggs laying at the time of autumn and the development of the very eggs at relatively low temperatures seem to be beneficial because of a smaller amount and lower activity of probable predators both aqueous vertebrates and invertebrates.

Thus the larval diapause not only maximizes the use of food sources but also decreases the danger of predators.


## INTRODUCTION

The interaction of temperature and the quality and quantity of food significantly affects the growth and development of water animals. Both low temperature and insufficient amounts or low quality of food can delay the emergence of adults and/or decrease their fecundity (Anderson 1977, Anderson and Cummins 1979). Owing to the accomodation abilities of the species, these negative influences can be compensated by a change of diet or an increased consumption of low caloric food (Iversen 1974, Anderson and Sedell 1979). These abilities can be enhanced by the occurrence of diapause in different developmental periods.

In the life cycle of many caddisflies species, there occurs the delay of development commonly called diapause, which is indispensable for maximal optimalization of emergence and egg laying. The development of sexually mature adults (imaginal diapause) in only a few months, or the egg diapause, constitutes a classical example of the adaptation to life in temporary waters (Novak and Sehnal 1963, Mackay and Wiggins 1979, Mackay and Smith 1980) and the strong pressure of dessication, is surely the reason for such developmental strategies. Summer larvae diapause, which is characteristic for many univoltine autumn species, seems to be a different phenomenon. In the time of this diapause, larvae stop eating, fix themselves to the bottom and remain in these conditions for a few weeks. Decreased daylength induces the end of diapause, the continued development and the emergence of sexually mature adults. This phenomenon was thoroughly analysed by Novak (1960) for *Anabolia furcata* larvae. Mackay and Wiggins (1979) stated that this waiting period is required if the new generation of larvae is to coincide with their food, autumnal leaves. On the other hand, Otto (1981) maintains that only relatively large larvae, which are already present in autumn and capable of feeding on fresh fallen leaves, can use this kind of food most effectively since the decomposition of the leaves decreases their nutritional value. In case of *A. furcata,* the most abundant emergence of sexually mature adults takes place at the end of September and the beginning of October, the largest number of eggs is surely laid at this time as well. Novak (1960) found about 12.6% of the larvae of *A. furcata* (1st stage), 46.4% (2nd), 37.5% (3rd) and 3.5% (4th) at the end of March. He reported that most of the larvae were present during winter in the 2nd and 3rd stages and the overwintering of eggs can not be excluded. Keeping in mind the time necessary for hatching as well as the low temperature of water in autumn, it seems unlikely that many larvae *A. furcata* are in the 2nd or even 1st larval stage in the autumn months. It can be supposed that this special developmental strategy, i.e. summer larval diapause, could also bo formed in *A. furcata* by other selection factors and not only by a rich source of foods in autumn.

## METHODS

*Anabolia furcata* females bred under laboratory conditions were placed in separate small glass chambers.

The egg mass was then put into tubes with water. The tubes were kept in a climatic-box first for 3 months at 10°C ; from February to April, the temperature was lowered to 5°C and again elevated to 10°C. This temperature remained unchanged until the end of the experiment, i.e. till the moulting of pupa.

After hatching the tube bottoms were covered with sand ; dry alder leaves, lettuce, filamentous algae and small portions of mixed Oligochaeta were served as food. Water in the breeding chambers was changed at least once a week, and then every 2 or 3 days with larval growth. To prevent larvae from excessive density as well as to avoid differentiation of the size of bred larvae the egg mass was put into a new tube every month. Breeding was observed every few days by means of a stereomicroscope and on the basis of exuviae presence, the period of moulting was determined. On other observations concerning the behaviour of these larvae was also noted.

Larvae that developed under laboratory conditions were compared with those living in the natural environment. Starting from the beginning of April, larvae were collected from a pool situated inside a deciduous forest nearby Lodz. Their developmental stage was determined on the basis of head capsule and in some larvae the gut content was investigated.

RESULTS

The *A. furcata* females laid 4 egg masses in total ; one mass was properly shaped and contained many eggs. It was spherical : after watering, its diameter was 0.7 cm and there were about 70 eggs. The other three egg masses were smaller, irregular in shape and the amount of eggs ranged from 10 to 14. The most important stages of egg development and larvae are presented in Figure 1. The largest number of larvae appeared at the

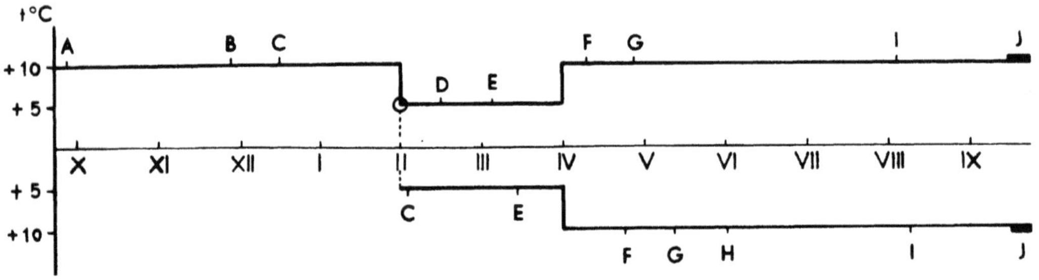


Figure 1. Development of *Anabolia furcata* larvae. A - egg laying ; B - several larvae visible inside the egg mass ; C - hatching of the first larva ; D - moulting after the 1st stage ; E - moulting after the 2nd stage ; F - moulting after the 3rd stage ; 6 - moulting after the 4th stage ; H - hatching of last larva ; I - fixation to the bottom ; J - adult emergence ; O - separation of the egg mass.

end of January i.e about 4 months after the egg masses were deposited at 10°C. The decrease of temperature inhibited neither hatching nor ecdysis. The investigated egg masses yielded 51 larvae in total until the beginning of June ; 35 larvae completed the development. 34 larvae emerged from the biggest egg mass (24 completed the development), 20 eggs were most probably not fertile and 13 larvae most likely fully shaped by the end of June did not leave the egg mass at all. Comparing the differentiation of stages of those larvae caught in the middle of April in pools with those reared under experimental conditions, some differences in the participation of particular stages can be observed (Figure 2).

In the examined pool, the larvae starting from July left the coastal zone and moved to deeper water layers and fixing themselves to the bottom. During rearing in August at 10°C, the larvae stopped food consumption and fixed to the bottom. After hatching, immature larvae fed both on cut alder leaves and periphyton growing on filamentous algae, Larvae were often observed consuming the fungi (*Saprolegnia glomerata*) that grow on egg masses. Though the larvae were not associated with mixed Oligochaeta, it is possible that they still ate them. The cannibalism of large larvae in relation to small ones was noted.

The analysis of the larvae's natural diet indicated that, except fragments of fallen leaves, the larvae of all stages consumed algae *Lyngbya* sp., *Oscillatoria* sp., *Trachylomonas* sp., *Gomphonema* sp., *Synedra* sp., *Fragilaria* sp., *Cymbella* sp., *Chlorella* sp. *Oedogonium* sp., *Scenedesmus* sp., *Pediastrum* sp., *Eudorina* sp., *Pleurotaenium* sp., *Botryococcus* sp. , fungal sporae, and pollen. The predominant food types consumed by larvae was detritus. However, in the end of April *Chlorella* sp. made up an important part of the gut content of larvae in the 2nd stage.

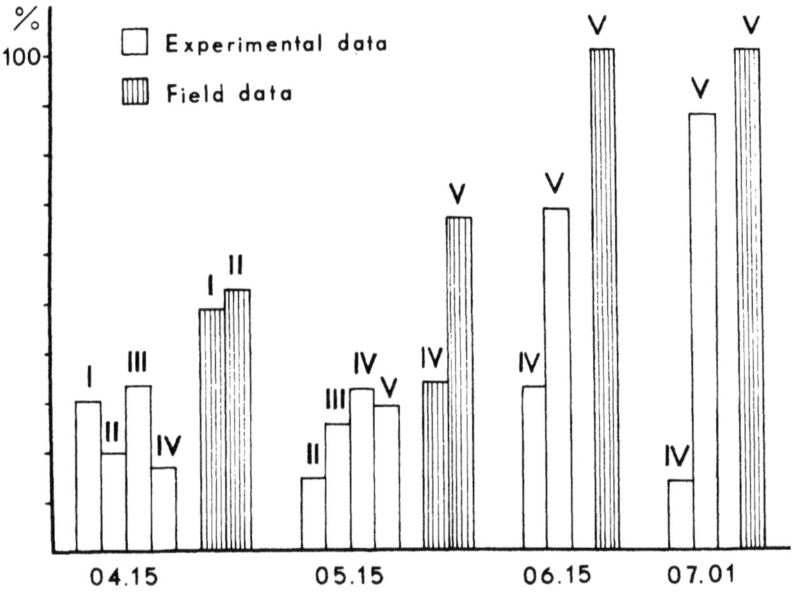


Figure 2. Differentiation of larval stages of *Anabolia furcata* under natural and experimental conditions.

Under experimental conditions the majority of larvae after hatching started building their cases from leaf fragments ; sometimes, larvae were seen burrowing into the substrate. Some larvae were delayed in building the cases up to 2 weeks. Several started their cases but failed to finish them as they entered free-living. Three larvae at the 1st stage showed no building activity for 1 month and then died.

## DISCUSSION

The experiments carried out indicate that hatching of larvae from a single egg mass takes a very long time. Hence, the differentiation of developmental stages observed in spring especially Novak 1960, is not only caused by relatively small desynchronization of the egg laying time but by the prolonged period of hatching for particular egg masses as well. According to Ward and Stanford (1982), prolonged hatching has an obvious adaptive value especially in unpredictable habitats. It should be concluded that newly-hatched larvae can be under the influence of selective factors, which act with different intensity at different periods of hatching. Taking into account the first larvae appearance and their ability to use various sources of food, as well as Otto's (1981) suggestions, it can be supposed that the pressure of aquatic predators, smaller in low temperatures and bigger in higher temperatures, is the selective factor responsible for the formation of "careful" hatching. Predation influences the morphology, physiology and behaviour of prey (Sih *et al.* 1985) and caddisflies often constitute a significant component of the fish diet (Crisp *et al.* 1978, Losos *et al.* 1980, Moretti *et al.* 1981). According to Flecker (1984), fish predation may be most significant during late summer and the intensity of this consumption should be estimated, remembering that fish digestion is far slower at low temperatures (Backiel 1971). Similarly, the activity of invertebrate predators evidently decreases at low temperatures (Thompson 1978). Larvae that leave the egg mass in winter are less exposed to the danger since the aquatic predators are not so active and ice cover makes it difficult to detect the preys.

As the development of eggs evidently depends on temperature (Jones 1976), their delayed laying decreases the possibility of acceleration in their development and hatching under the temperatures higher than the autumnal average. Even if this unpredictable increase of temperature occurred, some larvae would still emerge in winter because of long hatching.

In summation, it can be concluded that larval diapause increases synchronization of emergence and egg laying by specimens living under differential environmental conditions. Thus larval diapause enhances the flow of genes among various populations and, together with prolonged hatching, decreases the pressure of predators on at least some part of new larval generation. This very pressure, and not just the large amount of food in autumn, could be the factor that has a decisive influence on the formation of this developmental strategy.

ACNOWLEDGEMENTS

The author thanks to Dr. E. Kalinowska for identification of algae in the larval diet. This work was supported by CPBP 04.03 Research Problem Polish Academy of Science.

REFERENCES

Anderson, N.H. 1977. Continuous rearing of the Limnephilid Caddisfly, *Clistronia magnifica* (Banks). **Proc. 2nd Internatn. Symp. Trichoptera,** M.I. Crichton (ed.), Junk, The Hague : 317-329.

Anderson, N.H. & Cummins, K.W. 1979. Influences of diet on the life histories of aquatic Insects. **J. Fish. Res. Board Can.,** 36 : 335-342.

Anderson, N.H. & Sedell, J.R. 1979. Detritus processing by macroinvertebrates in stream ecosystems. **Ann. Rev. Entomol.,** 24 : 351-377.

Backiel, T. 1971. Production and food consumption of predatory fish in the Vistula River. **J. Fish. Biol.,** 3 : 369-405.

Flecker, A.S. 1974 The effects of predation and detritus on the structure of a stream insect community : a Field Test. **Oecologia,** 64 : 300-305.

Iversen, T.M. 1974. Ingestion and growth in *Sericostoma personatum* (Trichoptera) in relation to the nitrogen content of ingested leaves. **Oikos,** 25 : 273-282.

Jones, N.V. 1976. The Trichoptera of the stony shore of a lake with particular references to *Tinodes waeneri* (L.) (Psychomyiidae). **Proc. 1st Internatn. Symp. Trichoptera,** H. Malicky (ed.), Junk, The Hague : 117-130.

Losos, B., Prenaz, M. & Kubickova, J. 1980. Food and growth of fishes of the Jihlava River. **Acta Sc. Nat. Brno,** 14 : 1-46.

Mackay, R.L. & Wiggins, G.B. 1979. Ecological diversity in Trichoptera. **Ann. Rev. Entomol.,** 24 : 185-208.

Novak, K. 1960. Entwicklung und Diapause der Köcherfliegenlarven *Anabolia furcata* Br. (Trichoptera). **Cas. Cs. Spol. Entomol.,** 57 : 207-212.

Novak, K. & Sehnal, F. 1963. The development cycle of some species of the genus *Limnephilus* (Trichoptera). **Cas. Cs. Spol. Entomol.,** 60 : 68-80.

Otto, C. 1981. Food related adaptations in stream living caddisfly larvae feeding on leaves. **Oikos,** 37 : 117-122.

Sih, A., Crowley, P., Mc Peek, M., Petranka, J. & Strohmeier, K. 1985. Predation, competition, and prey communities : a review of field experiments. **Ann. Rev. Ecol. Syst.,** 16 : 269-311.

Thompson, D.J. 1978. Toward a realistic predator-prey model : the effect of temperature on the functional response and life history of larvae of the damselfly, *Ischnura elegans*. **J. Anim. Ecol.,** 47 : 7576-767.

Ward, J.V. Stanford, J.A. 1982. Thermal responses in the evolutionary ecology of aquatic Insects. **Ann. Rev. Entomol.,** 27 : 97-117.

Wiggins, G.B. Mackay, R.J. & Smith, I.M. 1980. Evolutionary and ecological strategies of animals in annual temporary pools. **Arch. Hydrobiol. Suppl.,** 58 : 97-206.

# THE EFFECT OF TEMPERATURE AND DIET ON THE LARVAL GROWTH OF *GLOSSOSOMA NIGRIOR*

M. P. OEMKE


## SUMMARY

The effects of diet and temperature were investigated on the growth of *Glossosoma nigrior* (banks) larvae. Early instars were collected from a 3rd and 1st order stream in Southern Michigan. The 3rd order stream appeared visibly rich in periphyton food resources and had warmer mid-summer water temperatures. Instars collected from this stream were larger in weight than instars from the first order stream and the respective estimated larval growth curves were significantly different, with instars from the 3rd order stream showing larger individual weights and faster growth. Experiments were conducted investigating the relative importance of diet and temperature on larval growth. Larvae collected from the first order stream were reared on their natural periphyton diet and on periphyton removed from the 3rd order stream, at temperatures of 10°, 15° and 20°C for four weeks. Results indicated that diet was not significant in effecting weight gains. Temperature, however, significantly ($P < 0.001$) increased larval growth rates, increasing mean total growth from 0.074 ± .02 mg per larva at 10°C to 0.13 ± .07 mg per larva at 20°C after 4 weeks. The direct affects of temperature on larval growth may therefore be more important than diet affects in many streams.


## INTRODUCTION

Previous word on the diatom feeding preferences of *Glossosoma nigrior*, a stream grazer (Oemke 1984) indicated a preferential selection by larvae for certain species and growth forms of diatoms. Similar diatom species were selected by larvae inhabiting two distinctly different streams ; Spring Brook, a small, first order stream with a dense riparian development which shades the stream in the summer, and Augusta Creek, a larger, third order stream which is open to sunlight in the mid-channel regions even in the summer. Both stream sections studied are located in Kalamazoo County, Michigan (USA).

While *G. nigrior* larvae were found to ingest very similar food resources in both streams, analysis of instar weights collected from the streams showed that generally larger instars were found in Augusta Creek than in Spring Brook. Examination of larval growth curves extrapolated from the dry weights of individual instars for the two streams (Fig. 1) indicated the more dramatic increases in body weight recorded for Augusta Creek larvae for successive instars.

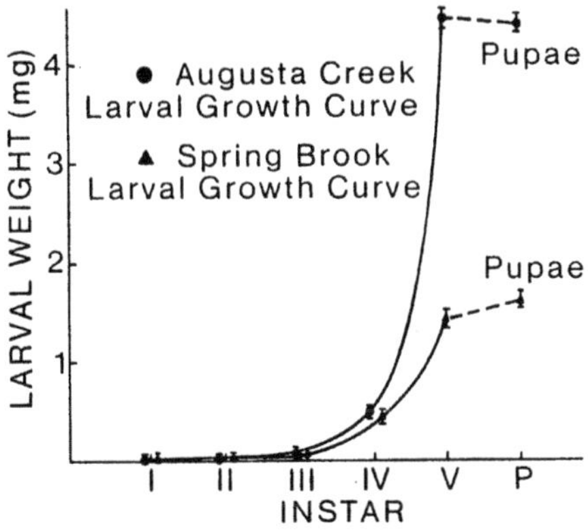


Figure 1. Growth curve based on the mean instar specific dry weights for Spring Brook and Augusta Creek (curves were fitted by eye).

These weight differences were very highly significant ($P < 0.001$) between V instars and between pupae collected from both streams. It was theorized that these larval weight differences were due to the different nutritional nature of each stream's periphyton as a result of the different algal species composition.

This paper will present the results of studies designed to elucidate the relative importance of periphyton diet and stream water temperature as contributors to larval growth rates for this stream grazer.

## MATERIAL AND METHODS

Larvae were collected on the same date from both streams together with epilithic periphyton samples to determine ash-free dry weights and dry weights of feces and the periphyton food sources. Feces were obtained by placing groups of larvae in small containers for 1 h and then filtering any fecal material. Stream periphyton was collected by carefully scraping the top of stream rocks. Samples of feces and periphyton were dried, weighed, and heated at high temperature following the procedure of Conover (1966), to determine ash-free dry weights.

Field collected first instar larvae were taken from Spring Brook in September and were used to compare growth of larvae on different periphyton food diets at temperatures of 10°, 15°, and 20°C. Thirty eight larvae were randomly selected, dried, and weighed to determine a mean original weight of .0196 ± .002 mg. larva (X ± s.e.). Groups of 30 larvae were placed in each of four, one liter jars at each of the three temperatures (360 total larvae) and filled with stream water filtered through a 0.4 micron filter. To assure an overabundance of available periphyton, small rocks with naturally occurring periphyton from Augusta Creek and Spring Brook were used to provide food sources for the larvae. Fresh filtered stream water was added daily, and each week rocks with fresh periphyton replaced the previously grazed rocks. All jars were aerated with bubbling airstones. Four replicates (jars) were used for each temperature and diet. Each week for 4 weeks several specimens were removed from each of the four jars and placed in a drying oven at 60°C for 24 h. Larvae were weighed on a Cahn, model 21 electrobalance.

A three way ANOVA was run without replication because the cell variances were close and the number of specimens per treatment combination was similar. All main effects of temperature, time, and diet were fixed treatments, fitting a model I ANOVA (Sokal and Rohlf 1969). The analytical procedure for unbalanced factorial data or the Federer-Zelen method (Gill 1978) was also used.

Table 1. Estimates of assimilation efficiencies for larvae from Spring Brook and Augusta Creek based on techniques of Conover (1966).

| STREAM | INSTAR[1] | FECES A.F.D.W./D.W. | PERIPHYTON A.F.D.W./D.W. | % ASSIMILATED |
|---|---|---|---|---|
| SPRING BROOK | V | .5143 | .7858 ± .0481 | 71.13 |
| | V | .5178 | .7858 ± .0481 | 70.73 |
| | V | .5065 | .7858 ± .0481 | 72.02 |
| | IV | .4291 | .7858 ± .0481 | 79.51 |
| | IV | .4859 | .7858 ± .0481 | 74.24 |
| | IV | .4748 | .7858 ± .0481 | 75.36 |
| AUGUSTA CREEK | IV-V | .4786 | .7852 ± .0273 | 74.89 |
| | IV-V | .5093 | .7852 ± .0273 | 71.61 |
| | IV-V | .4781 | .7852 ± .0273 | 74.94 |

[1] n = 28 V instars from Spring Brook; n = 86 IV instars from Spring Brook; n = 46 IV-V instars from Augusta Creek.

RESULTS

Comparisons of the ash-free dry weight to dry weight ratios between the food source and the feces can roughly determine the percent of the food assimilated (Conover 1966). The mean percent of food assimilated was calculated as 73.83 ± .01(X ± s.e.) for Spring Brook IV and V instars and 73.81± .001 for Augusta Creek instars (Table 1). These results are similar to the 73% assimilation efficiency found by Hargrave (1970) for *Hyalella azteca* L., when grazing on epiphytic diatoms.

The nearly identical assimilation efficiencies for larvae from both streams (74%) indicates that each streams periphyton appears to be a good nutritional source. Differences in organic matter levels between each stream's periphyton, do not appear to explain the observed weight differences between the respective instars from both streams.

Weeks three and four showed the largest increases in larval weights, particularly at the higher temperatures (Fig. 2). Mortality appeared the highest for the Spring Brook diet at the 20°C temperature with only a single larva surviving the four week test. Statistical analysis indicated that interaction between main effects was not significant and that only the time factor significantly increased the larval weights ($P < .01$). The analysis was then repeated using the error mean square to test for significance of the three main effects ; diet, temperature, and time in weeks. The differences in diets from Spring Brook and Augusta Creek were not significant in contributing to differences observed in weight gains, while temperature was the most significant main effect ($P < .001$) contributing to weight gains, followed by development time in weeks ($P < .01$) (Table 2).

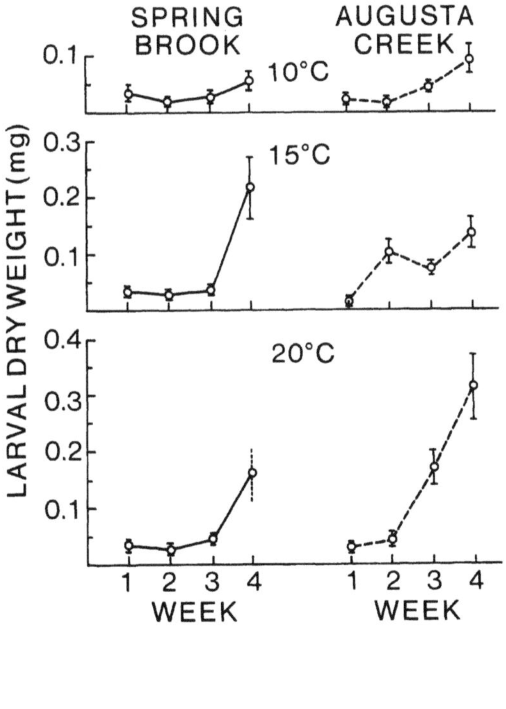


Figure 2. Comparisons of larval weights (X ± s.e.) held at three temperatures (10°, 15°, and 20°C), supplied with two different diets, and measured for four weeks.

Table 2. Analysis of variance of the weight gains of *G. nigrior* larvae supplied with Augusta Creek periphyton and Spring Brook periphyton.

| SOURCE OF VARIATION | D.F. | S.S. | M.S. | F-RATIO |
|---|---|---|---|---|
| DIET A | 1 | .0008 | .0004 | .04 |
| TEMP B | 2 | .2266 | .1133 | 11.33*** |
| WEEKS C | 3 | .1242 | .0414 | 4.14** |
| ERROR | 252 | 2.5188 | .010 | |

*** F.001 [2,252] = 10.8

**F.01 [3,252] =3.87

The greatest percentage of growth occurred at 15°C on the Spring Brook diet (Table 3). The Augusta Creek diet however, showed a constant increase in percentage growth per day as the temperature increased. The total growth was maximum at 15°C on the Spring Brook diet (.2073 mg per larva). The drop in growth rates for larvae reared at 20°C on the Spring Brook diet corresponded with the highest mortality levels at this treatment combination (90 %).

Table 3. Comparison of *G. nigrior* growth rates (mg dry wt.) on two diets at three temperatures.

| TEMP | REL GROWTH RATES[1] (mg/mg • $d^{-1}$) X ± S.E. | % TOTAL GROWTH INCREASE | GROWTH % /DAY | TOTAL GROWTH mg/larva X ± S.E. |
|---|---|---|---|---|
| | | SPRING BROOK DIET | | |
| 10° | .0518 ± .008 | 240.31 | 8.58 | .0471 ± .014 |
| 15° | .1062 ± .040 | 1,057.65 | 37.77 | .2073 ± .047 |
| 20°[1] | .0402 ± .006 | 142.35 | 6.78 | .0279 ± .002 |
| | | AUGUSTA CREEK DIET | | |
| 10° | .0662 ± .012 | 376.53 | 13.45 | .0738 ± .021 |
| 15° | .0595 ± .023 | 583.67 | 20.85 | .1144 ± .028 |
| 20° | .0894 ± .056 | 1,494.90 | 53.39 | .1364 ± .068 |

[1] the means are from sets of 4 chambers with intitial larval density the same.

DISCUSSION

Larval growth measured in the laboratory, using two periphyton diets from Augusta Creek and Spring Brook, indicated that temperature, not diet was the most significant factor contributing to variability in weight gains. Several researchers have found the interaction of temperature and food quality and quantity hard to separate (Anderson and Cummins 1979, Ward and Cummins 1978, Merritt *et al.*1982), although others consider temperature the major factor affecting growth rate, as well as final body size (Vannote and Sweeney 1980, Colbo and Porter 1981).

It may be virtually impossible to separate the contributing factors of temperature and food which may affect larval growth in stream grazers, particularly if a few or a single species of diatom, preferentially selected, might greatly influence the grazer's overall nutrition. Changes in water temperature can profoundly influence the diatom community by completely changing the major diatom species composition (Patrick *et al.*1969, Patrick 1971, Whitford and Schumacher 1963, Dillard 1971). It would be hard to separate the effects of temperature change which may be so closely linked to algal species changes in the available periphyton food sources. It is possible that the observed low growth rate for larvae feeding on the Spring Brook diet at 20°C may reflect some critical temperature threshold level for particularly sensitive diatom species (Dillard 1971) which were critical to the nutrition of the larvae. The importance of diatoms in the diet of the mayfly, *Cloeon triangulifer*, was also found to be especially evident at intermediate temperatures of 15°C and 20°C by Sweeney and Vannote (1984). A total of 55 unique diatom species were found only in Augusta Creek, while only 30 taxa were unique to the Spring Brook flora. A total of 159 diatom taxa were identified from Augusta Creek and 134 taxa from Spring Brook (Oemke 1983). The greater species diversity of the Augusta Creek diatom flora or the presence of diatom species accustomed to 20°C temperatures, which are close to summertime maximum temperatures for Augusta Creek, may have provided a species pool from which a vigorous periphyton growth could be maintained. Spring Brook periphyton, having a smaller species pool, and with species accustomed to cooler water temperatures, may thus have been adversly affected by temperatures above 15°C. The 20°C temperatures may have reduced the Spring Brook diatom populations to low levels, reducing larval growth rates, as well as survivorship. Larval survivorship at 20°C on the Augusta Creek diet was three times greater than for larvae supplied the Spring Brook diet.

Thus, while temperature proved the statistically significant factor affecting larval growth, it is very likely to be closely linked to species changes in the periphyton. The interactions between temperature changes and periphyton species changes which may affect growth of stream grazers needs to be examined further.

REFERENCES

Anderson, N.H. & Cummins, K.W. 1979. Influences of diet on the life histories of aquatic insects. **J. Fish. Res. Bd. Can.**, 36 : 335-342.

Colbo, M.H. & Porter, G. N. 1981. The interaction of rearing temperature and food supply on the life history of two species of Simuliidae (Diptera) **Can. J. Zool.**, 59 : 158-163.

Conover, R.J. 1966. Assimilation of organic matter by zooplankton. **Limnol. Oceanog.** , 11 : 338-345.

Dillard, G.E. 1971. An epilithic diatom community of a North Carolina sandhills stream. **Rev. Algol.**, 1O : 118-127.

Gill, J.L. 1978. Design and analysis of experiments. **Iowa State Univ. Press**, Iowa, USA. 409 p.

Hargrave, B.T. 1970. Utilization of benthic microflora by *Hyalella azteca* (Amphipoda). **J. Anim. Ecol.**, 39 : 427-437.

Merritt, R.W., Ross, D. H., & Larson, G.J. 1982. Influence of stream temperature and seston on the growth and production of overwintering larval black flies (Diptera : Simuliidae). **Ecology**, 63 : 1322-1331.

Oemke, M.P. 1983. Diatom utilization by the stream grazer, *Glossosoma nigrior* (Banks) (Trichoptera : Glossosomatidae) in two southern Michigan streams. Ph. D. Dissertation, Michigan State University, E. Lansing, Michigan. USA. 162 p.

Oemke, M.P. 1984. Diatom feeding preferences of *Glossosoma nigrior* (Banks) larvae, stream grazers in two southern Michigan streams. **Proc. 4th Intern. Symp. Trichoptera**, J. Morse (ed.), Junk, the Hague, : 291-299.

Patrick, R. 1971. The effects of increasing light and temperature on the structure of diatom communities. **Limnol. Oceanog.**, 16 : 4O5-421.

Patrick, R., Crum, B. & Coles, J. 1969. Temperature and manganese as determining factor in the presence of diatoms or blue-green floras in streams. **Proc. Nat. Acad. Sci.**, 64 : 672-678.

Sokal, R.R. & Rohl, F. J. 1969. Biometry. W.H. Freeman, San Francisco, USA. 776 p.

Sweeney, B.N. & Vannote, R.L. 1984. Influence of food quality and temperature on life history characteristics of the parthenogenetic mayfly, *Cloeon triangulifer*. **Freshwater biol.**, 14 : 621-630.

Vannote, R.L. & Sweeney, B.W. 1980. Geographic analysis of thermal equilibria : a conceptual model for evaluating the effect of natural and modified thermal regimes on aquatic insect communities. **Am. Nat.**, 115 : 667-695.

Ward, G.M. & Cummins, K. W. 1978. Life history and growth pattern of *Paratendipes albimanus* in a Michigan headwater stream. **Annls. Ent. Soc. Am.**, 71 : 272-284.

Whitford, L. A. & Schumacher, G.J. 1963. Communities of algae in North Carolina streams and their seasonal relations. **Hydrobiolgia**, 22 : 133-196.

# LARVAL DEVELOPMENT, SUBSTRATE PREFERENCE, AND FEEDING HABITS OF *POLYCENTROPUS VARIEGATUS* MILNE IN MODEL STREAM CHANNELS (TRICHOPTERA : POLYCENTROPODIDAE)

M.J. WEVERS and R.W. WISSEMAN

## SUMMARY

Indoor model stream channels in the Oregon Coast Range (USA) supported a comparatively high density of *Polycentropus variegatus* larvae in contrast to natural streams in the immediate vicinity.

*P. variegatus* exhibited a univoltine life cycle. Small size classes and early larval instars dominated the population throughout the fall and winter months, although all size classes and instars were represented in benthic and drift samples. Growth was rapid during May and June. Pupation, emergence, and oviposition occurred throughout the summer months.

Larval densities were greater on graval substrates than on bedrock and silt. Larvae were generalist predators. Mayflies, crustaceans, chironomids, stoneflies, *Polycentropus* larvae, and oligochaetes contributed to their diet.

## INTRODUCTION

*Polycentropus variegatus* Milne, 1936 exhibits a western montane distribution in North America, ranging from southeast Alaska to Mexico in coastal and interior mountain ranges. In Oregon, *P. variegatus* adults have appeared in low numbers in light and emergence trap samples from rivers and streams (unpubl. data) while larvae were rarely encountered in benthic samples. Emergence traps set along Oak Creek in the Oregon Coast Range for two flight seasons failed to recover *P. variegatus* adults (Anderson and Wold 1972) and larvae have rarely been recovered (unpubl. data). Little is known of the life cycle of *P. variegatus* in western Oregon, although adult records indicate a summer flight period from mid-June through mid-September.

In Scandanavian streams, *P. flavomaculatus* is thought to have a two year life cycle (Ulfstrand 1968). Although Bass *et al.* (1982) note other studies in which a one year life cycle was found for *P. flavomaculatus*, they found a fast growing summer population and an overwintering population in their recirculating stream channel. This indicates a bivoltine life cycle. Other work in Britain revealed a bimodal peak in the adult summer flight period (Bass *et al.*1982). A wide range of size classes throughout the year, overlap of body lengths of a single instar with other instars, and overlap of head capsule measurements with other instars may make the life cycle of *Polycentropus* difficult to ascertain.

*Polycentropus* appears to demonstrate a propensity for model stream conditions. Bass *et al.* (1982) report densities of *P. flavomaculatus* as high as 20,515/ $m^2$ in their recirculating stream channel. At the Oak Creek Laboratory of Biology located on Oak Creek in the Oregon Coast Range (5 mi. W. Corvallis) , *P. variegatus* is one of the dominant caddisflies in our model stream systems. We examined aspects of its life history, habitat preference, and diet in small indoor model stream channels and report our findings here.

## METHODS

Three different substrate types (gravel, bedrock, slit) were established in twelve wooden model stream troughs. Electrically driven paddlewheels provided current velocity. Water flow from two large outdoor model streams was equally distributed to the twelve streams by a water outlet rotating over a carousel flow-splitter box. Invertebrate and algal colonists were thus uniformly supplied to each trough. Stream water was exchanged at a rate of 2.5 l/min. The laboratory streams are described in greater detail in Warren *et al.* (in press).

Benthic, drift, and emergence samples were taken regularly from March 1980 through August 1984 (Warren *et al.* in press). Invertebrates were sorted, counted, and measured. Larval densities of *P. variegatus* in gravel, bedrock, and silt streams were determined from benthic data. Body length (from the anterior margin of the labrum to the base of the anal prolegs) was measured to the nearest millimeter.

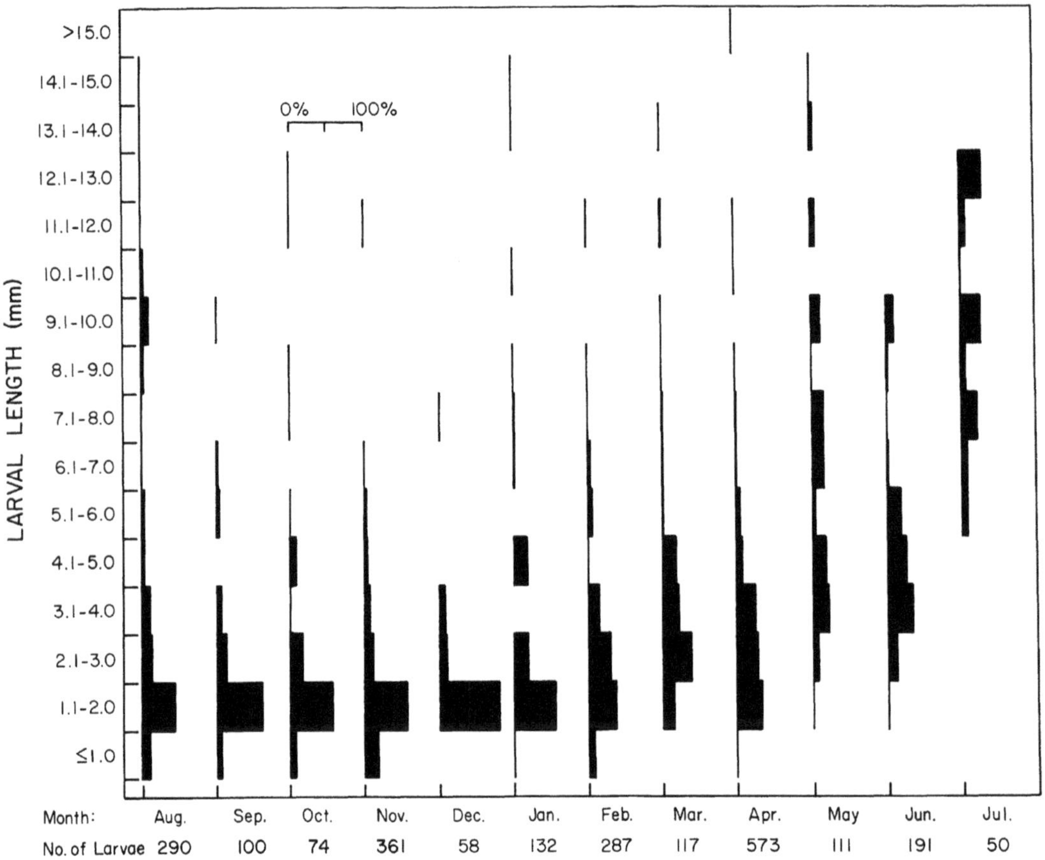


Figure 1. Length-frequency of *Polycentropus variegatus* larvae from model stream channels. Frequencies are expressed as relative percentages. Data are from benthic and drift samples taken August 1980 through August 1984.

Head capsule widths (at widest point) of *P. variegatus* larvae from benthic and drift samples were measured with the aid of 50 x magnification. The median head capsule width and range for each instar were determined.

The digestive tracts of 43 4th and 5th instar *P. variegatus* larvae from benthic samples were dissected for examination. Gut contents were teased out in 70% ethanol. Recognizable animal fragments were identified and counted.

## RESULTS

### Larval development

A wide range of size classes of *P. variegatus* was found throughout the year (Fig. 1). The small (1-3 mm) size classes predominated from late summer through early spring. The frequency of larger size classes increased during the spring and throughout the summer. Larvae as large as 18 mm in length were found.

Head capsule measurements gave clear separation of instars with no overlap (I : x = 0.24 mm, range = 0.14-0.29 mm, n=317. II : x=0.37 mm, range=0.33-0.43 mm, n=136. III : x=0.55 mm, range=0.47-0.65 mm, n=135. IV : x=0.86 mm, range=0.69-1.02 mm, n=87. V : x=1.37 mm, range=1.12-1.61 mm, n=107). First instar larvae dominated from September through February (Fig. 2). Growth was rapid during the spring and by April and May, 4th and 5th instars were predominant. Fifth instars were proportionately most abundant in July. By August, the frequency of 1st and 2nd instar larvae was similar to that of 5th instar larvae.

*P. variegatus* pupae were collected from the model streams from mid-June through August. Adults were found in emergence traps in August and early September.

These results suggest that *P. variegatus* has a univoltine life cycle with rapid larval growth in the spring and pupation and emergence during the summer. Body length is variable and does not directly correspond to instar class.

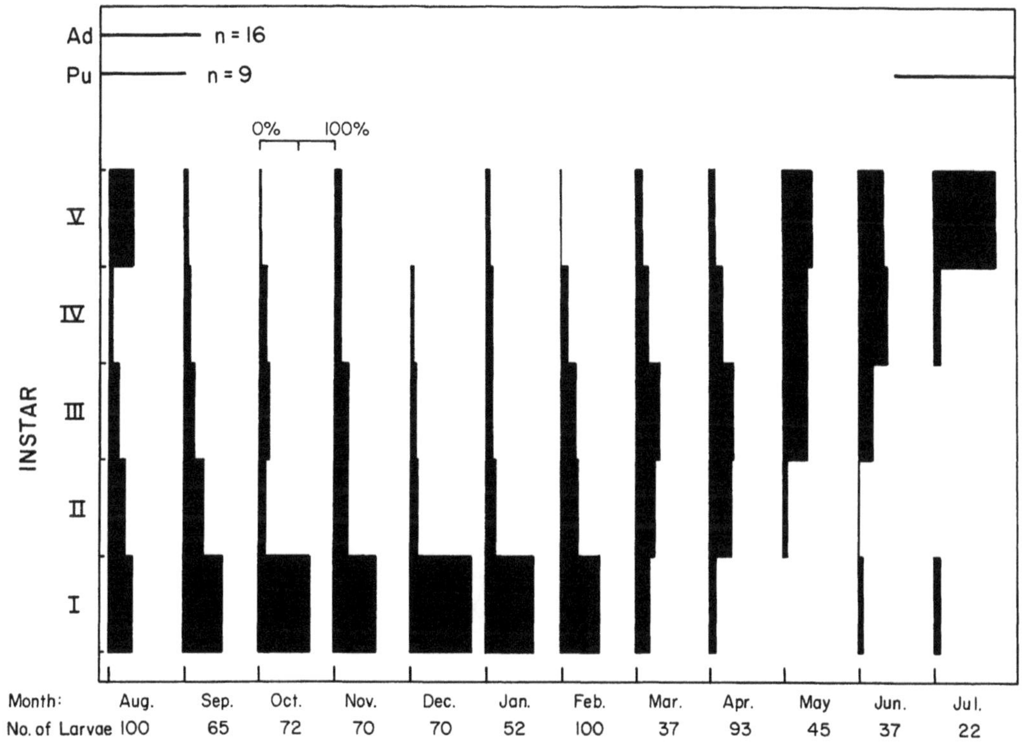


Figure 2. Instar frequency of *Polycentropus variegatus* larvae from model stream channels. Frequencies are expressed as relative percentages. Occurrence of pupae and adults are indicated. Data are from benthic, drift, and emergence samples taken September 1980 through August 1984.

## Substrate preference

The numerical density of *P. variegatus* larvae in the model streams varied seasonally (Fig. 3). Densities in all substrate types were greatest during the spring. Numbers declined into the summer as larvae pupated and emerged. During the fall months, densities rose as the number of early instar larvae increased as eggs hatched.

*P. variegatus* larvae were more abundant in gravel than in bedrock or silt substrate streams. Mean densities in gravel ranged from 1 to 96 larvae per 0.1 $m^2$. Bedrock and silt streams supported mean densities of 0 to 5 larvae per 0.1 $m^2$.

## Food of larvae

A variety of prey was found in the guts of *P. variegatus* larvae (Table 1). Mayflies and crustaceans accounted for the largest percentage of the diet, however chironomids, nemourid stoneflies, *P. variegatus* larvae and oligochaetes were also found. The frequency of prey items found in gut contents is proportional to the relative abundances of these prey in the model streams (Warren *et al.* in press). This suggests that *P. variegatus* is a generalist predator. The more abundant prey species are more likely to become ensnared by the nets of this sedentary predator.

Hildrew and Townsend (1977) found that another polycentropodid caddisfly, *Plectrocnemia conspersa,* is widely distributed and occupies all substrate types in natural streams, but is most abundant where prey density is highest. Among our model streams, those with gravel substrate had the greatest diversity and abundances of prey species (Warren *et al.* in press), which may explain our high densities of *P. variegatus* larvae there.

DISCUSSION

The univoltine life cycle of *P. variegatus* in our model streams in the Oregon Coast Range is similar to that of *P. flavomaculatus* in Europe. The wide range in larval size and instar class at all times of the year is similar to *P. flavomaculatus* and also to *P. conspersa* (Hynes 1961).

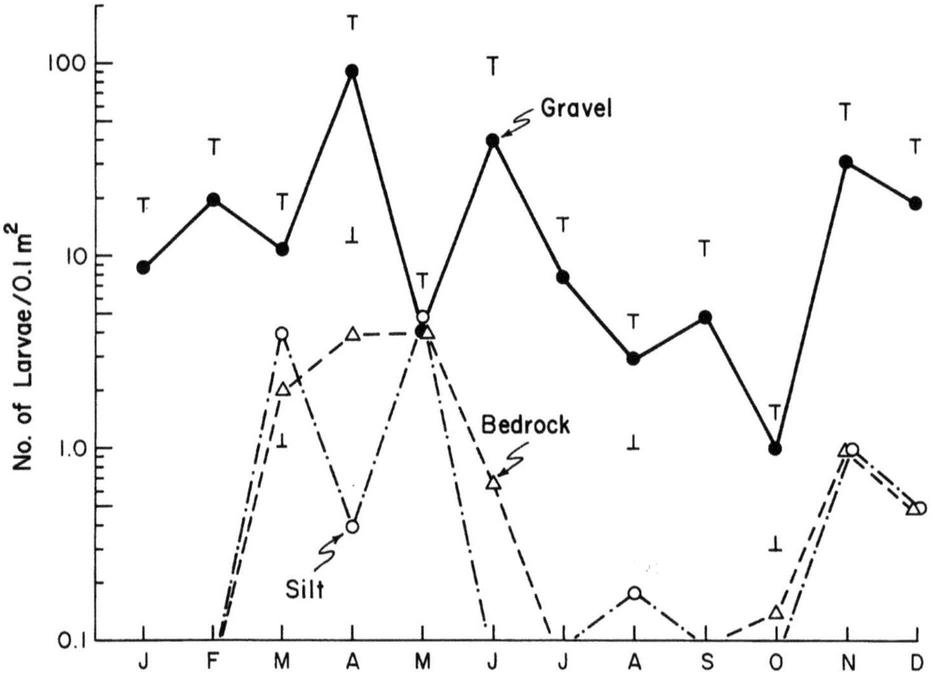


Figure 3. Seasonal variation of larval densities of *Polycentropus variegatus* in model stream channels of different substrate types (x ± 90% C.I.). Data are from benthic samples taken August 1980 through August 1984.

Table 1. Frequency and relative percentages of prey items in the guts of *Polycentropus variegatus* (n=43, 4th and 5 th instars).

| Taxa | No. | % |
|---|---|---|
| Ephemeroptera | 24 | 26 |
| Ostracoda | 23 | 25 |
| Copepoda | 17 | 18 |
| Cladocera | 11 | 12 |
| Chironomidae | 7 | 8 |
| Nemouridae | 7 | 8 |
| *Polycentropus* | 2 | 2 |
| Oligochaeta | 1 | 1 |

Hynes (1961) suggests that it may be advantageous for a carnivore to have a wide range in size and head capsule width to take advantage of a variety of prey sizes. Andrewartha and Birch (1984) discuss "spreading of risk" through phenotypic variation, such as body size or rate of larval development, and its advantages in a changing environment. For a sit-and-wait predator such as *P. variegatus*, variable size as observed in our model streams could increase the opportunities for the population to exploit available food sources.

Controlled flow conditions in model streams apparently facilitate larval net construction and maintenance. A lack of physical disturbance of retreats combined with the abundance of prey items may account for the high relative abundance of *P. variegatus* larvae in model streams as compared to its apparent scarcity in adjacent natural streams.

## REFERENCES

Anderson, N.H. & Wold, J.L. 1972. Emergence trap collections of Trichoptera from an Oregon stream. **Can. Ent.**, 104 : 189-201.

Andrewartha, H.G. & Birch, L.C. 1984. The Ecolocical Web. **Univ. Chicago Press.**, Chicago and London, 506 p.

Bass, J.A.B., Ladle, M. & Welton, J.S. 1982. Larval development and production by the net-spinning caddis, *Polycentropus flavomaculatus* (Pictet) (Trichoptera), in a recirculating stream channel. **Aquatic Insects**, 4 : 137-151.

Hildrew, A.G. & Townsend, C.R. 1977. The influence of substrate on the functional response of *Plectrocnemia conspersa* (Curtis) larvae (Trichoptera : Polycentropodidae). **Oecologia**, 31 : 21-26.

Hynes, H.B.N. 1961. The invertebrate fauna of a Welsh mountain stream. **Arch. Hydrobiol.**, 57 : 344-388.

Ulfstrand, S. 1968. Life cycles of benthic insects in Lapland streams (Ephemeroptera, Plecoptera, Trichoptera, Diptera, Simuliidae). **Oikos**, 19 : 167-190.

Warren, C.E., Liss, W.J. & Wevers, M.J. in press. Artificial streams for ecosystem toxicity studies. Final Report, Electr. Power Res. Inst. Palo Alto, California.

## ADDENDUM

Nimmo, A.P. 1986. The adult Polycentropodidae of Canada and adjacent United States. **Questiones Entomologicae**, 22 : 143-252

# A REVIEW OF BIONOMICS OF THE JAPANESE LEPIDOSTOMATIDAE (TRICHOPTERA), WITH PARTICULAR REFERENCE TO THE CHANGE OF CASE MATERIALS AND ITS ECOLOGICAL SIGNIFICANCE.

T. ITO

SUMMARY

The case making habit, habitat, annual life cycle and geographical distribution are reviewed for the 15 species of Lepidostomatidae. By the comparison of the bionomical characters, the developmental change of case materials from sand grains to plant fragments may be understood in the speeding up of larval growth and, hence, the increase of the number of descendants in a given time.

## INTRODUCTION

The change from mineral to organic cases during the larval development is known only in Lepidostomatidae (Tsuda 1936, Nielsen 1942, Ulmer 1957, Flint and Wiggins 1961, Hickin 1962, Lepneva 1971, Hansell 1972, Anderson 1976, Ito 1978-1985), although the reverse change is found in many species of Limnephilidae (Nielsen 1942, Flint 1960, Lepneva 1971, Elliott 1971, Mackay 1972, Wiggins 1977, Otto and Svensson 1980). But the ecological meaning of the change of case materials has little been discussed in Lepidostomatidae (see, Hansell 1974).

In this paper I will review the bionomics of the 15 species, 3 genera and 3 species groups, of Lepidostomatidae with particular reference to their case making habit and discuss the ecological significance of the change of case materials.

## REVIEW OF BIONOMICS

### Case making habit (Fig.1-A)

The 3 species of *Dinarthrum* occurring in Japan use sand grains and build cylindrical cases throughout the larval and pupal life (Ito 1984a). The species of *Goerodes* also make cylindrical cases with sand grains at first, but change the materials to leaf pieces, building four-sided cases, in the course of development ; the change takes place in the 5th instar in the *naraensis* group (Ito 1985c, d), in the 4th in the *japonicus* group (Ito 1978), and in the 2nd or 3rd in the *orientalis* group (Ito 1985a). Newly hatched larvae of *Neoseverinia crassicornis* make tubular cases with detritus and sand grains, and add square leaf pieces on the top of the case ; in the latter half of the 2nd instar to the first half of the 5th the cases, exclusively make of leaf pieces, are four-sided with lateral and dorsal flanges ; in the latter half of the 5th instar the cases are four-sided but without flange and mainly made of fragments of bark and twigs (Ito 1983b).

### Habitat and geographical distribution (Fig.1-B, 2)

Larvae of the Japanese Lepidostomatidae occupy a wide range of lotic habitats, from small water flows to lower reaches of streams, and also near mouths of influents at mountain lakes (Fig. 1-B). No species is confined to lentic habitats and temporary waters. In the streams the species of the *japonicus* group of *Goerodes* are found and successively replace one another from upper reaches to the lower (Ito 1983c).

The larvae of the 3 species of the *orientalis* group of *Goerodes* live in small peculiar flows, which are characterized by brown colored water and abundant undecomposed organic matter such as dead semi-aquatic plants ; the temperature of the habitats may be fairly high in summer partly because of the small water volume and partly because of the absence of plant shade (Ito 1985a).

The 7 other species are essentially confined to small and cool water flows (Ito 1983b, 1984a, 1985c, d).

All species of the *japonicus* group of *Goerodes*, *G.naraensis* and *N. crassicornis* are widely distributed, but the 8 other species are more or less limited in distribution. This difference in distribution presumably reflects the extensiveness of habitats, that is, the size of populations. Remarkable local variation of male genitalia is found in *G. naraensis* and *N. crassicornis* (Ito 1983a, 1985c), which though widely distributed, inhabit small habitats.

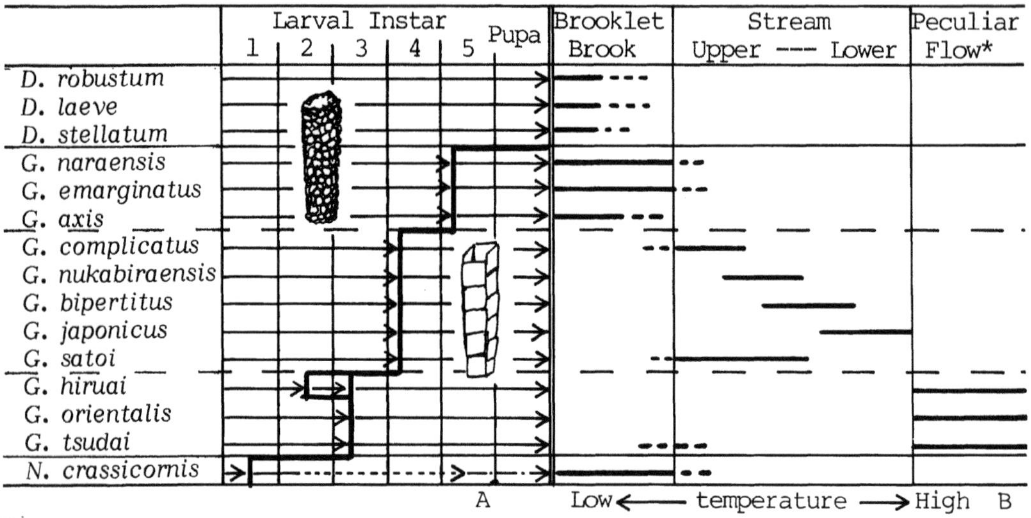


Figure 1-A. Developmental change in the materials and shape of the cases. Broken line, with flanges. Chained line, mainly made of barks and twigs.

Figure 1-B. Habitats. Water flows are classified based on width, i.e., brooklet (under 30 cm), brook (30-60 cm) and stream (over 60 cm). *, see text.

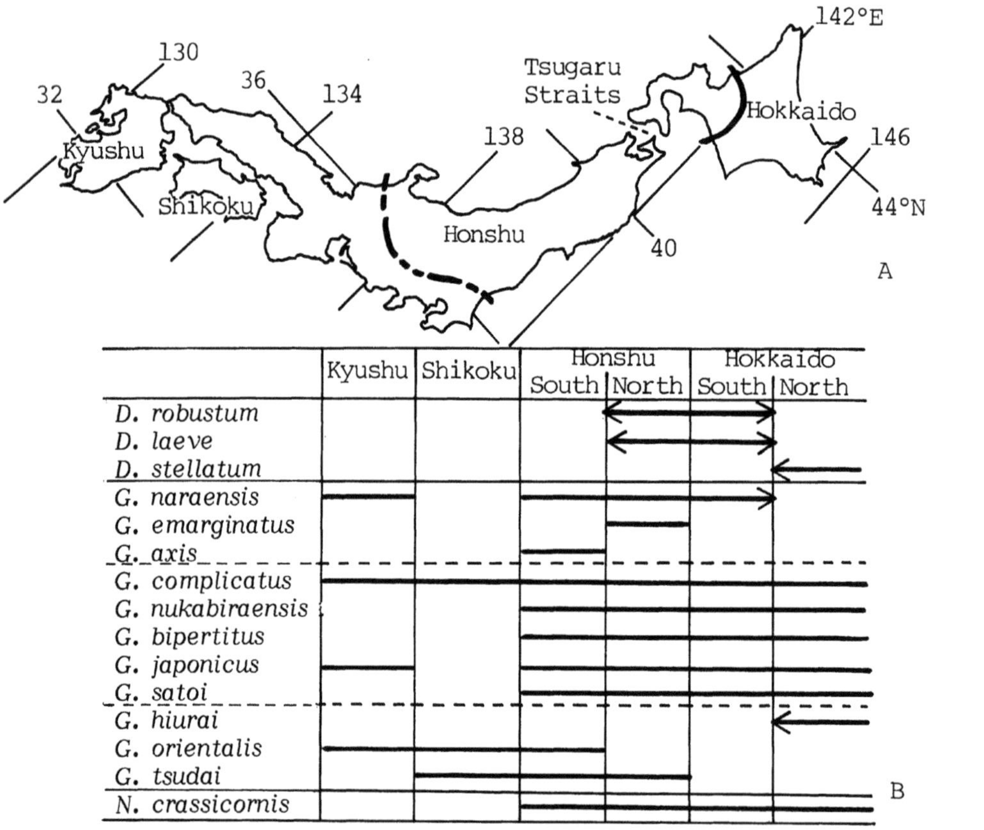


Figure 2. Geographical distribution. A, map showing boundaries. B, distribution. Arrow, established margin. Poor records in Shikoku and Kyushu probably owing insufficient surveys.

As shown in Figure 2, the lepidostomatid fauna is not uniform in Hokkaido and also in Honshu. So far as based on my insufficient observations, there is a boundary in Honshu, running south of Nikko and Kiso. In Hokkaido, the boundary of distribution is located near Mt. Shokambetsu and west of Mt. Yubari ; it coincides with neither Ishikari Lowland, which was invaded by sea a few thousands years ago, or peaks of mountains.

*D. robustum* and *D. laeve* are distributed across the Tsugaru Straits, and *G. naraensis* and *N. crassicornis* shows no local morphological difference between both sides of the Straits.

**Annual life cycle** (Fig.3)

The lepidostomatid caddisflies occurring in Hokkaido, northern Japan, need several months or one year or two years for completing one generation. The 3 species of *Dinarthrum* have a two year life cycle, emerging in summer (Ito 1984a). *G. satoi* and *G. naraensis* are univolutine and emerge in late spring (Ito 1980, 1985c). *G. complicatus, G. hiurai* and *N. crassicornis* are also univoltine, the adults appearing in summer (Ito 1980, 1983b, 1985a). *G. nukabiraensis* and probably also *G. bipertitus* and *G.japonicus* emerge three times in two years (Ito 1983c). *G. orientalis* and *G. tsudai* may have two or more generations in Honshu, central Japan (Ito 1985a).

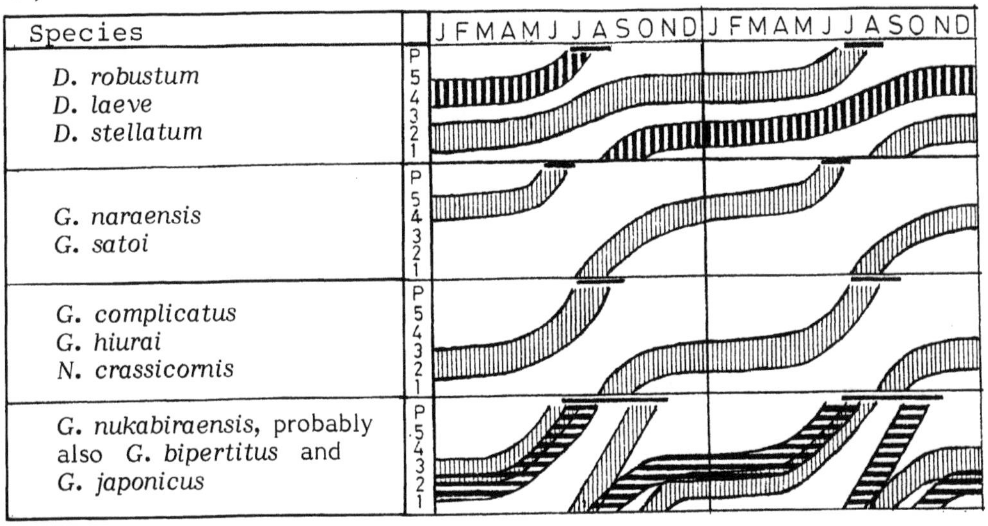


Figure 3. Schematic representation of annual life cycle in Hokkaido, northern Japan, Horizontal line, appearance of adult. 1-5, larval instars. P, pupa.

**Influence of coexistence on the bionomics of *D. laeve* and *D. robustum*** (Fig.4)

In a spring, Takino, Sapporo, where *D. laeve* occurs together with *D.robustum, D.laeve* emerge in late May to mid June, and *D.robustum* in mid August to early September. Then, the coexisting *D. laeve* emerge ca.40 days earlier than it does when occurring singly, so that the two species are completely allochronic in adult stage. In this spring, the case of *D. laeve* is fairly smooth and covered with a thin layer of silk on the exterior as well as on the interior surface, however, when *D.laeve* occurs singly, its case is rugged on the exterior surface as in other species of *Dinarthrum.*

On the other hand, the emerging period and case of *D. robustum* in this spring are not different from those in other springs occurring singly (Ito 1984a).

ECOLOGICAL SIGNIFICANCE OF CHANGE OF CASE MATERIALS

Four bionomical characters, i.e., instar of changing the case materials, habitat, number of generations in a year and wet weight of female pupa, the indicator of fecundity, of the 3 genera and 3 species groups are compared in Table 1.

In general, the growth of larva is profoundly influenced by the water temperature. By the way, *Dinarthrum, Goerodes naraensis* group and *Neoseverinia* inhabit small, cool flows where their foods, terrestrial fallen leaves, are abundant in all seasons. But the number of generations in a year and weight of female pupa are rather different among three groups, that is, (0.5, 3-10 mg), (1, 3-12 mg) and (1, 15-46 mg), respectively. As mentioned above, *Dinarthrum* constantly use sand grains, *Goerodes naraensis* group change from sand grains to leaf pieces at 5th instar, and *Neoseverinia* change at Ist instar. Then, the time required for completing one generation and/or weight of female pupa may depend not only on the water temperature but also on the case materials used.

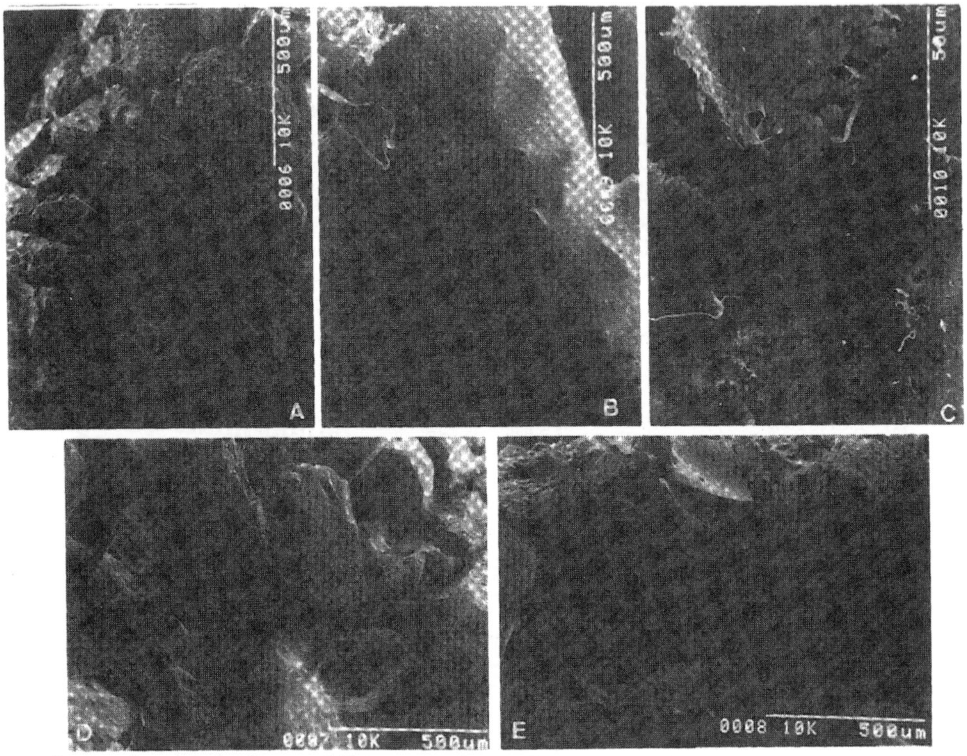


Figure 4. Exterior surface of cases of *D.laeve* (A-C) and *D. robustum* (D,E). A,D. when occurring singly. B,C,E. When the two species coexisting.

Table 1. Bionomical characters compared for the three genera and three species groups of the Japanese Lepidostomatidae.

| Genus and species group | change of case materials from sand to leaf | Habitat | Number of generations in a year | Wet weight of female pupa |
|---|---|---|---|---|
| *Dinarthrum* | not change | small, warm flow | 0.5 | 3-10 mg |
| *G. naraensis* | at 5th instar | small, cool flow | 1 | 3-12 |
| *G. japonicus* | at 4th instar | mountain stream | 1-1.5 | 4-15 |
| *G. orientalis* | at 3rd or 3nd instar | small, warm flow | 1-2 | 7-18 |
| *Neoseverinia* | at Ist instar | small, cool flow | 1 | 15-46 |

I postulate, therefore, that to make a sand case need more energy than to make a leaf case. This idea finds the support in Otto and Svensson's (1980) comparative study of the final larval size among many species of caddisflies making organic, mineral and mixed cases. The developmental change of case materials observed in the Japanese Lepidostomatidae may be understood in the speeding up of the larval growth and the increase of the number of descendants in a given time.

## REFERENCES


Anderson, N.H.1976. The distribution and biology of the Oregon Trichoptera. **Tech. Bull., Agri. Exp. St., Oregon State Univ.**, 134 : 116-124.

Elliott, J.M. 1971. Life histories and drifting of three species of Limnephilidae (Trichoptera). **Oikos**, 22 : 56-61.

Flint, O.S.Jr. 1960. Taxonomy and biology of Nearctic limnephilid larvae (Trichoptera), with special reference to species in eastern United States. **Ent. Am., n.s.**, 40 : 1-117.

Flint, O.S.Jr. & Wiggins, G.B. 1961. Records and description of North American species in the genus *Lepidostoma*, with a revision of the *vernalis* group (Trichoptera : Lepidostomatidae). **Can. Ent.** 93 : 279-297.

Hansell, M.H. 1972. Case building behaviour of the caddis fly larva *Lepidostoma hirtum*. **J. Zool.**, Lond., 167 : 179-192.

Hansell, M.H. 1974. A progress report on some approches to the study of larval house building with particular reference to *Lepidostoma hirtum*. **Proc.1st Int. Symp. Trichoptera**, H. Malicky (ed.), Junk, The Hague : 181-184.

Hickin, N.E. 1962. Trichoptera from a Kenya high altitude stream. **Proc. Roy. ent. Soc., Lond.**, (A) 37 : 129-134.

Ito, T. 1978. Morphological and ecological studies on the caddisfly genus *Dinarthrodes* in Hokkaido, Japan (Trichoptera, Lepidostomatidae).I. The larval development and the cases on four species of *Dinarthrodes*. **Kontyû**, Tokyo, 46 : 574-584.

Ito, T. 1980. Morphological and ecological studies on the caddisfly genus *Dinarthrodes* in Hokkaido, Japan (Trichoptera, Lepidostomatidae). II.Life histories of two coexisting species *D. complicatus* and *D. satoi*. **Kontyû**, Tokyo, 48 : 311-320.

Ito, T. 1983a. Morphology and bionomics of *Neoseverinia crassicornis* (Ulmer) (Trichoptera, Lepidostomatidae) I. Morphology of adult and pupa. **Kontyû**, Tokyo, 51 : 207-213.

Ito, T. 1983b. Morphology and bionomics of *Neoseverinia crassicornis*(Ulmer) (Trichoptera, Lepidostomatidae) II.Larvae, egg and bionomics. **Kontyû**, Tokyo, 51 : 322-329.

Ito, T. 1983c. Longitudinal distribution and annual life cycle of the *japonicus* group of *Goerodes* (Trichoptera, Lepidostomatidae). **Jpn. J. Limnol.**, 44 : 269-276.

Ito, T. 1984a. Three new species of *Dinarthrum* (Trichoptera, Lepidostomatidae). **Kontyû**, Tokyo, 52 : 1-20.

Ito, T. 1984b. On the genus *Goerodes* (Trichoptera, Lepidostomatidae) in Japan. **Kontyû**, Tokyo,52 : 506-515.

Ito, T. 1985a. Morphology and ecology of three species of *orientalis* group of *Goerodes* (Trichoptera, Lepidostomatidae). **Kontyû**, Tokyo, 53 : 12-24.

Ito, T. 1985b. Females, pupa and larvae of the *japonicus* group of *Goerodes* (Trichoptera, Lepidostomatidae). **Kontyû**, Tokyo, 53 : 261-269.

Ito, T. 1985c. Description, geographical variation and ecology of *Goerodes naraensis* (Tani) (Trichoptera, Lepidostomatidae). **Jpn. J. Limnol.**, 46 : 199-211.

Ito, T. 1985d. Two new species of the *naraensis* group of *Goerodes* (Trichoptera, Lepidostomatidae). **Kontyû**, Tokyo, 53 : 507-515.

Lepneva, S.G.1971. Fauna of the U.S.S.R. Trichoptera. II. Larvae and pupae of Integripalpia. Trans. 1964 Rud., Ed., Israel Prog. Sci. Trans., Jerusalem.

Mackay, R.J. 1972. Temporal patterns in life history and flight behaviour of *Pycnopsyche gentilis, P.luculenta* and *P. scabripennis* (Trichoptera : Limnephilidae). **Can. Ent.**, 104 : 1819-1835.

Nielsen, A. 1942. Über die Entwicklung und Biologie der Trichopteren mit besonderer Berücksichtgung der Quelltrichopteren Himmerlands. **Arch.Hydrobiol.,Suppl.**, 17 : 255-631.

Otto, C. & Svensson, B.S. 1980. The signigicance of case material selection for the survival of caddis larvae. **J. Animal Ecol.** , 49 : 855-865.

Tsuda, M. 1936. Untersuchungen über die Japanischen Wasserinsekten II. Lepidostomatidae (Trichoptera). **Annot. Zool. Japon.**, 15 : 400-409.

Ulmer, G. 1957. Köcherfliegen (Trichopteren) von den Sunda-Inseln. II. **Arch. Hydrobiol., Suppl.**, 21 : 408-608.

Wiggins, G.B. 1977. Larvae of the North American caddisfly genera (Trichoptera). Univ.Toronto Press.

# CADDISFLY ADAPTATIONS TO THE VARIABLE HABITATS AT THE LAND-WATER INTERFACE

N. A. ERMAN

## SUMMARY

Adaptations for life at the land-water interface may be found in all phases of the Trichoptera life cycle. Some, such as terrestrial pupation or eggs laid on land, are genetically fixed ; others, such as upstream migration and variable case construction, are flexible traits that can be used by the organism under changing conditions. Six families, out of 14 in the Sagehen Creek Basin, have members especially adapted to these habitats. Limnephilidae exhibits the greatest number of adaptations and comprises the largest group living at the land-water interface. Eighteen traits are identified as having possible adaptive advantages for life in the changeable aquatic habitats of the study area.

## INTRODUCTION AND STUDY AREA

An intensive study of caddisflies has been made in a basin on the east side of the Sierra Nevada, CA, U.S.A. since 1979. The area studied is 2700 hectares and includes Sagehen Creek from its headwaters (2256 m) to its end in a reservoir (1804 m).

Emergence traps, light traps, and hand-netting were used extensively to determine adult distribution ; and bottom sampling of larvae was done in all aquatic habitat types in the area. A detailed picture of Trichoptera distribution and habitat preference has emerged from this study.

Of the 14 families (75 species total) collected in the area, species from 6 families prefer the variable habitat at the land-water interface (Table 1). These habitats include 1) spring streams braiding through wet and dry meadows and fens that sometimes get very warm and/or dry completely in the lower reaches by late summer, 2) snow-melt streams and ponds that are filled only during April and May, and then dry up, 3) a shallow headwater lake (maximum depth, 1.5 m) that dries completely by autumn in 2 out of 3 years.

The caddisflies that live in these habitats are not subjected to fish predation but they are preyed on by invertebrates (e.g., *Yphria californica* and *Sialis* spp.) and birds. All are univoltine.

Although some of the species listed in Table 1 are found in other habitats, they are either not abundant or are not permanent there. To do detailed life history studies of individual species, I have tried to study them in their preferred habitat (that habitat in which they consistently occur in the highest numbers). It is only in such optimum habitats that some of the most unique adaptations of small organisms can be observed.

The variable habitats of this area do not include the spring sources (stable flow and constant temperatures) or the larger continuously-flowing streams. However, it should be noted that the environment at the very edge of these latter streams is variable and that a distinctive group of caddisfly larvae live there (e.g., *Lepidostoma* spp., *Cryptochia* spp., *Onocosmoecus unicolor*) rather than out in the main stream channel. They often make cases of and may feed on root materials and grass hanging along the stream bank. But they will not be further discussed in the present paper.

## RESULTS AND DISCUSSION

Eighteen behavioral traits have been identified for various species living at the land-water interface in the Sagehen Creek Basin (Table 2). Whether or not all of these traits are specific adaptations for life in these habitats is unknown. Certainly some of the traits are not limited to caddisflies in these areas (e.g., terrestrial eggs). Species are listed only with traits they are known to possess. I have made no attempt to list which of the species do not possess a certain trait because in some cases I do not know.

Terrestrial egg masses and larval hatching within the eggs masses (traits 1 and 2, Table 2) have been discussed extensively for other species (Ross 1944, Wiggins 1973). In the present study terrestrial egg masses have been found for *Lenarchus rillus*, *Limnephilus externus*, *Limnephilus peltus* and *Parthina linea*. Larval hatching within egg masses has been observed in *Lenarchus rillus* and is assumed for *Limnephilus externus*.

*Lenarchus rillus* and *Limnephilus externus* both live in the headwater lake, and are the only Trichoptera there. *L. rillus* inhabitats many other places in the basin but is abundant only in the lake while *L. externus* is found nowhere else in the basin.

Table 1. Trichoptera of variable habitats at the land-water interface, Sagehen Creek Basin, Sierra Nevada, CA, U.S.A.
(1a - spring streams, warmest reaches only ; 1b - spring streams ; 2 - Snow melt streams and ponds ; 3 - headwater lake ; 4 - intermittent stream).

| Trichoptera of variable habitats | Preferred habitat | Total Trichoptera families in basin | N° species per family |
|---|---|---|---|
| Hydroptilidae | | Brachycentridae | 3 |
| * *Hydroptila rono* Ross 1941 | 1a | Calamoceratidae | 1 |
| *Ochrotrichia trapoiza* Ross 1947 | 1a | Glossosomatidae | 3 |
| Lepidostomatidae | | Goeridae | 1 |
| * *Lepidostoma roafi* (Milne) 1936 | 1b | Hydropsychidae | 7 |
| *Lepidostoma veroda* Ross 1948 | 1b | Hydroptilidae | 3 |
| Limnephilidae | | Lepidostomatidae | 6 |
| * *Chyranda centralis* (Banks) 1900 | 1b | Limnephilidae | 30 |
| *Clostoeca disjuncta* (Banks) 1914 | 2 | Odontoceridae | 1 |
| * *Desmona bethula* Denning 1954 | 1b | Philopotamidae | 4 |
| * *Hesperophylax designatus* (Walker) 1852 | 4 | Phryganeidae | 1 |
| * *Lenarchus rillus* (Milne) 1935 | 2 & 3 | Polycentropodidae | 1 |
| * *Limnephilus externus* Hagen 1861 | 3 | Rhyacophilidae | 13 |
| * *Limnephilus peltus* Denning 1962 | 1b | Sericostomatidae | 1 |
| * *Psychoglypha mazamae* Denning 1970 | 1b | | |
| Odontoceridae | | | |
| *Parthina linea* Denning 1954 | 1a | | |
| Phryganeidae | | | |
| * *Yphria californica* (Banks) 1907 | 1b | | |
| Sericostomatidae | | | |
| * *Gumaga griseola* (McLachlan) 1871 | 1a | | |

* adults reared from larvae

*Yphria californica* was observed diving underwater and attaching its eggs to the underside of permanent vegetation, usually moss mats and sometimes *Veronica americana*. This trait may insure that at least some eggs are laid in stream sections with permanent water flow.

Traits 4, 5, and 6 (Table 2) are related. Larvae apparently withstand high maximum temperatures by burrowing beneath the surface of the substrate (silt and peat in these cases) where temperatures are cooler. In the laboratory *Gumaga griseola* larvae were able to survive totally dry conditions for several days by pulling into their cases (see also Imhoff and Harrison 1981). This finding was by accident when a rearing tank dried. When water was added again some larvae began to move within a few minutes.

With the surface area between water and land so great it seems only logical that some species would feed on living plants as do *Chyranda centralis, Desmona bethula* and *Psychoglypha mazamae*. *Veronica americana* is extensively used. *Desmona bethula* with one of the most unusual adaptations also leaves the water to feed on several plant species as reported previously (Erman 1981).

These same three species also move upstream (trait 9) prior to pupation during dry years. A detailed study of the movements of *Chyranda centralis* has been discussed elsewhere (Erman, 1986).

Some species show a staggered larval development with rates varying by as much as 6 months. For example, some larvae of *Chyranda centralis* reach the fifth instar as early as December ; others, as late as May and yet they all pupate by mid-summer. In March it is possible to find four instars of *Desmona bethula* at one

TABLE 2 : Traits of Trichoptera living at land-water interface, Sagehen Creek Basin.

**EGGS**

1
Laid on land
*Clostoeca disjuncta**
*Lenarchus rillus*
*Limnephilus externus*
*Limnephilus peltus*
*Parthina linea*

2
Larvae hatch and remain in egg mass
*Lenarchus rillus*
*Limnephilus externus*

3
Egg masses attached Underwater to permanent aquatic vegetation
*Yphria californica*

**LARVAE**

4
Withstand high maximum daily water temperatures (22-27°C)
*Gumaga griseola*
*Hydroptila rono*
*Ochrotrichia trapoiza*
*Parthina linea*

5
Burrow into substrate during day
*Gumaga griseola*
*Hydroptila rono*
*Ochrotrichia trapoiza*
*Desmona bethula*

6
Can withstand total drying of environment for at least several days
*Gumaga griseola*

7
Feed on living plants
*Chyranda centralis*
*Desmona bethula*
*Psychoglypha mazamae*

8
Feed outside stream
*Desmona bethula*

9
Move upstream prior to pupation (toward permanent spring source) during dry years
*Chyranda centralis*
*Desmona bethula*
*Psychoglypha mazamae*

10
Staggered larval development but synchronized emergence
*Chyranda centralis*
*Desmona bethula*
*Gumaga griseola*

11
Variable cases depending on habitat
*Desmona bethula*
*Limnephilus peltus*
*Psychoglypha mazamae*
*Yphria californica* (inhabits other species' cases)

12
Contruct sturdier, darker cases for pupation
*Chyranda centralis* green vegetation → bark
*Lenarchus rillus* green vegetation → bark and pine needles
*Limnephilus externus* green vegetation → bark
*Yphria californica*

13
Larval diapause
*Desmona bethula*

**PUPAE**

14
Pupate out of water
*Limnephilus peltus*

*Clostoeca disjuncta*, *Parthina linea* } ? larvae disappear prior to pupation

15
Larvae burrow and pupate in stream or pond bottom
*Chyranda centralis*
*Desmona bethula*
*Lenarchis rillus*
*Psychoglypha mazamae*
*Yphria californica*

**ADULTS**

16
Mate on vegetation close to ground. Females probably flightless.
*Parthina linea*

17
Emerge just prior to stream completely drying. Date may differ greatly from year to year depending on conditions
*Hesperophylax designatus*

18
Possible adult diapause
*Chyranda centralis*
*Limnephilus peltus*
*Lenarchus rillus*

* eggs never found but females observed flying up dry stream beds in early summer

site but emergence of adults is synchronized in the autumn. Staggered larval development may be an advantage in changeable habitats by insuring the survival of at least some larvae.

Several species make variable cases in these habitats depending on availability of materials. *Limnephilus peltus* usually constructs a tightly woven case of root pieces (Fig. 1) but occasionally uses sand grains. The same is true for *Desmona bethula* (Erman 1981). *Psychoglypha mazamae* changes its case material at least three times during development. It often uses the cases of other larvae to construct its own (Fig. 2). *Yphria californica* frequently inhabits cases of other larvae, particularly those of *Limnephilus peltus,* and adds material to these "stolen" cases as it grows.

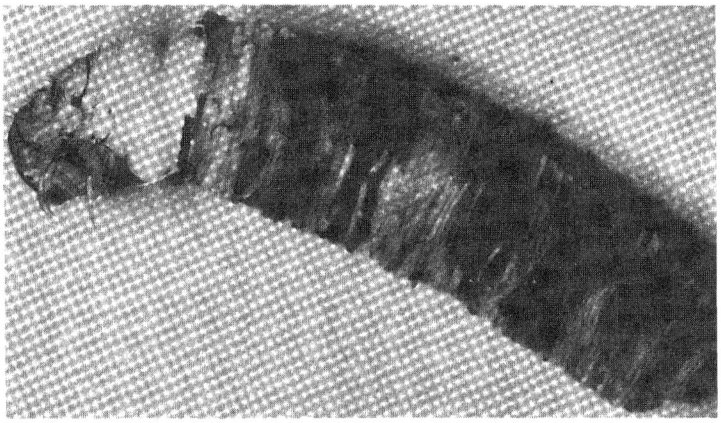

Figure 1. *Limnephilus peltus* larva constructs case of tightly woven root pieces. Leaves water to pupate in moss.

Figure 2. *Psychoglypha mazamae* with cases of other Trichoptera (*Gumaga griseola, Desmona bethula,* and *Limnephilus peltus*) and clams incorporated in its case.

Making sturdier cases prior to pupation (trait 12) is not limited to larvae in these habitats but is common here. Case color is often changed from green to brown, possibly reducing attraction to shredders.

Larval diapause, as defined by Novak (1960), has been described previously for *Desmona bethula* (Erman 1981). Pupation out of water (Flint, 1958) is known to occur in only one species in the basin, *Limnephilus peltus* (Erman, 1984a), possibly the most ubiquitous and abundant.

Several larvae burrow prior to pupation. This habit may protect them from some predators, insure that cases will not be washed away, and provide a cooler environment (see also Otto 1981). *Lenarchus rillus* burrows vertically along the water's edge and constructs a unique tubular pupal case of two chambers (Fig. 3). Adults emerge just as water recedes from the case location.

Mating behavior of *Parthina linea* is restricted to vegetation just above the water and females probably do not fly (Erman 1984b). *Hesperophylax designatus* inhabits and is abundant in one intermittent stream in the basin. Except for its capacity to emerge just prior to the disappearance of water in spring or summer (trait 17), no other traits are known about it. Whether it lays eggs in the dry stream bed or has an ovarial diapause is unknown. It does not occur in the areas inhabited by many of the other species discussed here.

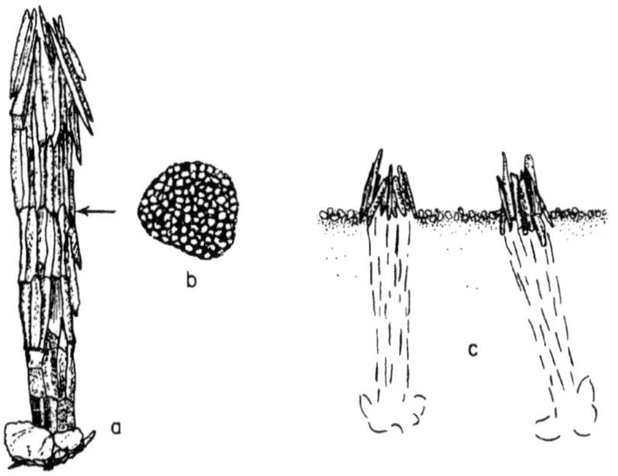


Figure 3. a)Two-chambered pupal case of *Lenarchus rillus*. b) silk grid (enlarged) separating chambers, pupa in lower chamber. c) cases buried in substrate, only top section protrudes.

*Limnephilus peltus, Chyranda centralis* and *Lenarchus rillus* are suspected of having an ovarial diapause. They emerge in early to mid-summer, are caught at light traps in late summer, and eggs and early instars are not found until autumn.

Many other investigators have described for other species some of the adaptations discussed here. It is significant that several adaptations to life in variable habitats may be present in the same species and in different developmental stages. Some of the more flexible traits such as burrowing, upstream movement, food selection and case building can vary within one species under different environmental conditions.

ACKNOWLEDGEMENTS

I would like to thank Skip Hodges for photopraphs. Donald Denning and Glenn Wiggins confirmed identities of some of the species. This work was partially funded by the University of California Water Resources Center, Project UCAL-WRC-W-645.

REFERENCES

Erman, N.A. 1981. Terrestrial feeding migration and life history of the stream-dwelling caddisfly, *Desmona bethula* (Trichoptera : Limnephilidae). **Can. J. of Zool.,** 59 : 1658-1665.

Erman, N.A. 1984a. The use of riparian systems by aquatic insects. *In* California Riparian Systems, R.E. Warner and K.M. Hendrix (eds), Univ. of California Press, Berkeley, : 177-182.

Erman, N.A. 1984b. The mating behavior of *Parthina linea* (Trichoptera : Odontoceridae), a caddisfly of springs and seeps. **Proc. 4th Internat. Symp. Trichoptera,** ed. J.C. Morse, Junk, The Hague, : 131-136.

Erman, N.A. 1986. Movements of self-marked caddisfly larvae, *Chyranda centralis* (Trichoptera : Limnephilidae) in a Sierran spring stream, California, U.S.A. **Fresh. Biol.,** 16 : 455-464

Flint, O.S. 1958. The larva and terrestrial pupa of *Ironoquia parvula* (Trichoptera : Limnephilidae). **J. New York Entomol. Soc.,** 66 : 59-62.

Imhof, J.G.A. & Harrison A.D. 1981. Survival of *Diplectrona modesta* Banks (Trichoptera : Hydropsychidae) during short periods of dessication. **Hydrobiologia,** 77 : 61-63.

Novak, K. 1960. Entwicklung und Diapause der Köcherfliegen larven *Anabolia furcata* Br. (Trichopt.).**Cas. Cs. Spol. Entomol.,** 57(3) : 207-212.

Otto, C. 1981. Food related adaptations in stream living caddisfly larvae feeding on leaves. **Oikos,** 37 : 117-122.

Ross, H.H. 1944. The caddisflies, or Trichoptera, of Illinois. **Bull. Ill. Nat. Hist. Survey,** 23(1) : 1-326.

Wiggins, G.B. 1973. A contribution to the biology of caddiflies (Trichoptera) in temporary pools. **Life Sci. Contr., Roy. Ont. Mus.,** 88.

# DISTRIBUTION OF THE *HYDROPSYCHE* LARVAE AND THE STRUCTURE OF THEIR NETS

H. TACHET, J.P. PIERROT et M. BOURNAUD

## SUMMARY

In three different types of watercourses in the neighbourhood of Lyons, there are five species of *Hydropsyche* (*H. angustipennis, H. exocellata, H. modesta, H. pellucidula* and *H. siltalai*). Th structure of the net and especially the size of meshes and the symmetry of the net vary according to the different species, but these differences can not be related to the longitudinal distribution of larvae. A hypothesis suggesting phenomena of clogging is proposed.

## INTRODUCTION

Several authors after Kaiser (1965) have shown that the mesh size of hydropsychid nets varies according to species. Wallace (1975), Wallace *et al.* (1977), Alstad (1980,1982), Fuller and Mackay (1980), Fuller *et al.* (1983) established a relation between the longitudinal distribution of different species and the structure of their net : the species of small rivers spinning nets with larger meshes than those of large rivers.

In the neighbourhood of Lyons, five species of *Hydropsyche* (*H. angustipennis, H. exocellata, H. modesta, H. pellucidula* and *H. siltalai*) occur in very different relative abundance in different types of watercourses (Fig. 1). Is this difference related to the structure of the net as suggested above ?

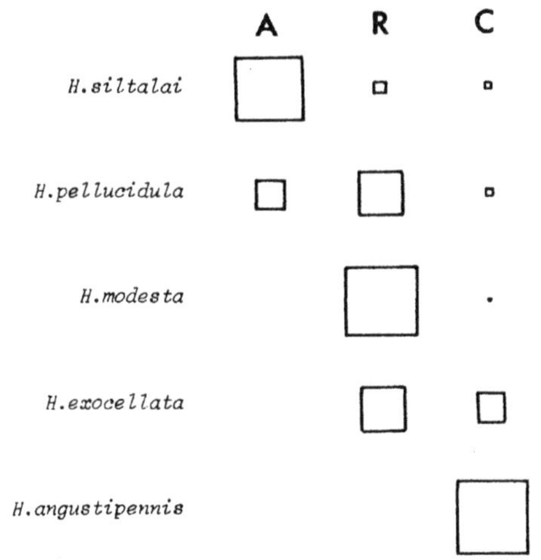


A : Ardière stream,hills of the Beaujolais,altitude 600 m.
R : Rhône river upstream Lyons,altitude 185 m.
C : Cotey stream,small tributary of the Rhône river,altitude 185 m.

Figure 1 : Distribution of the five species of *Hydropsyche* in three types of watercourse in the neighbourhood of Lyons (the squares are proportional to the abundance of larvae in each type of watercourse).

MATERIAL AND METHODS

There are two methods to study the structure of the net. In the first one (see for example Wallace 1975), the net, generally sampled in the stream, is carefully cut and then spread on a slide. Mesh size may be measured under a microscope. The second method used especially by Sattler (1958), Kaiser (1965), Décamps *et al.* (1973), Besch *et al.* (1977) implies the building of a net under laboratory conditions. The frame supporting the net may be photographed. This last method has the advantage of keeping the net undamaged. We have used an adaptation of this last method, but, instead of photographing it directly, we put the frame supporting the net under the optic system of a microfilm reader. The measures are made directly on the screen.

We tested about twenty fifth instar larvae of each species. Some nets have been built with a 'low' current velocity (v ≤ 15 cm $s^{-1}$) and others with 'high' current velocity (v ≥ 30 cm $s^{-1}$). We obtained a total number of 106 nets.

On account of the remarks of Sattler (1958), Besch *et al.* (1977) and Petersen and Petersen (1983), we measured seven parameters in the central part of the net where the distances between the threads A and B varies least (Fig. 2) :

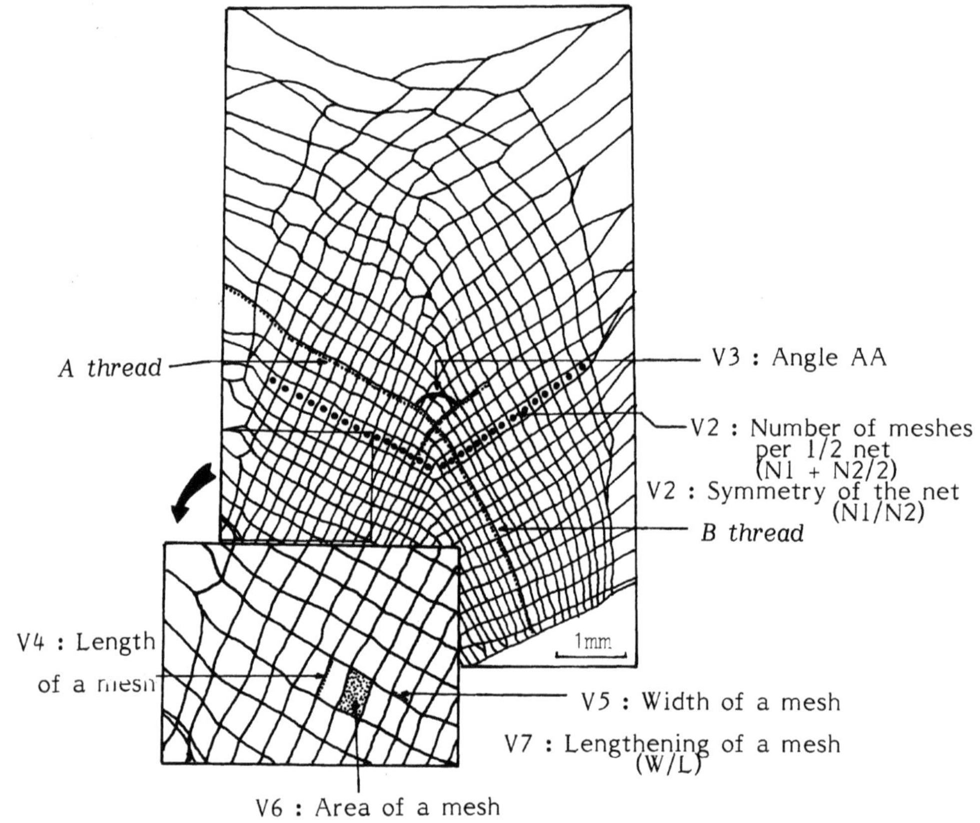


Figure 2 : variables (V1 to V7) measured on the nets of *Hydropsyche* (see text)

1. **Symmetry of the net** : along a thread A, we compute for one row, the number of meshes on the right and the left part of the net, then we calculate the ratio of the number of meshes of each half row.
2. Mean **number of meshes** for one half row.
3. **Angle AA** : angle defined by the crossing of two A threads.
4. Mean **length of a mesh** (= mean distance between two A threads).
5. Mean **width of a mesh** (= mean distance between two B threads).

6. Mean **area of a mesh** (*i.e.* width x length).
7. **Lengthening of a mesh** (ratio width/length).

RESULTS

In order to define some potential relations between these different variables, we analysed the table (106 nets x 7 variables) using a Principal Component Analysis (Daget 1976 ; Carrel *et al.* 1986).

The first two factors $F_1$ and $F_2$ explain 66% of the whole variability. When the co-ordinates of these seven variables are projected onto the circle of correlations (Fig. 3), we noticed that variables 4, 5 and 6 form a group opposed on the axis 1 to variable 2. These four variables are largely redundant. Variable 6 (area of a mesh) has the strongest correlation.

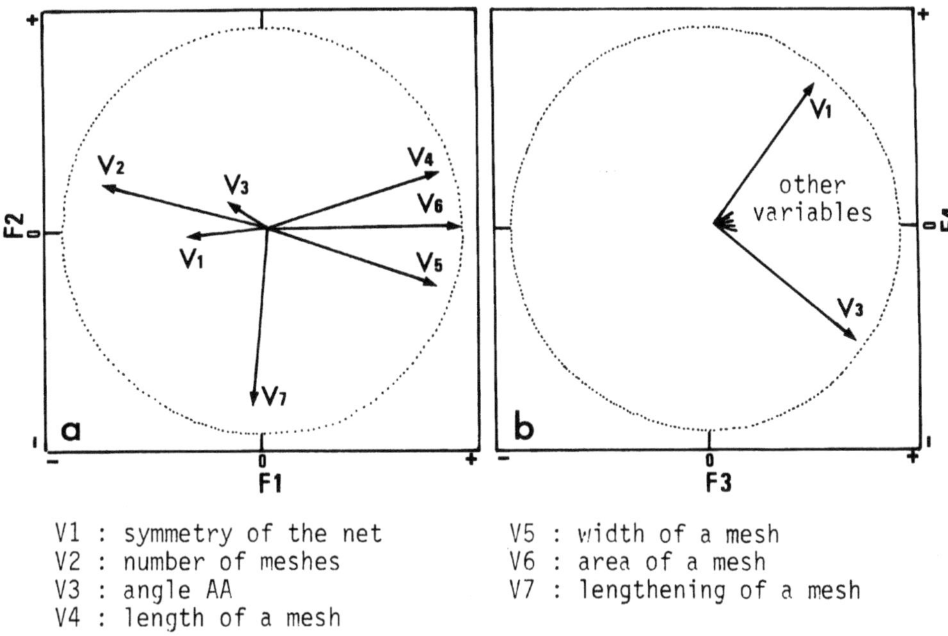


Figure 3 : Projection of the co-ordinates of the seven variables (V1 to V7) onto the circle of correlations upon the plane defined (a) by the axis $F_1F_2$ and (b) the axis $F_3F_4$ of the Principal Component Analysis.

Variable 7 (axis 2) is independent of the six others. Variables 3 (axis 3) and 1 (axis 4) (Fig.3) are interdependent and weakly dependent on the first group (2, 4, 5 and 6).

If, according to the results of this PCA, we study graphically the relation between variable 6 and the other three (*i.e.* 1,3 and 7), we noticed that the relation between variable 6 (area of a mesh) and 1 (symmetry of the net) gives the best discrimination between these five species. The relation between variables 6-3 or 6-7 divides just one species from four others.

The projection of points corresponding to each net for the different species (Fig. 4 and 5) onto the plane defined by variables 1 and 6 shows that *H. angustipennis* spins symmetrical nets with small meshes whereas *H. modesta, H. pellucidula* and *H. siltalai* spin more asymmetrical nets with larger meshes than *H. angustipennis*. The characteristics of the net of *H. exocellata* are intermediate between these of *H. angustipennis* and those of three other species.

Beyond these interspecific differences we looked for a relation between these two variables and the current velocity. For three species (Fig. 4) only we have an equal number of nets built at 'low' ($v \leq 15$ cm $s^{-1}$) and 'high' ($v \geq 30$ cm $s^{-1}$) velocities. There is a significant difference only for *H. siltalai* (Fig. 4). This species seems to build nets with larger meshes at low velocity rather than at high velocity.

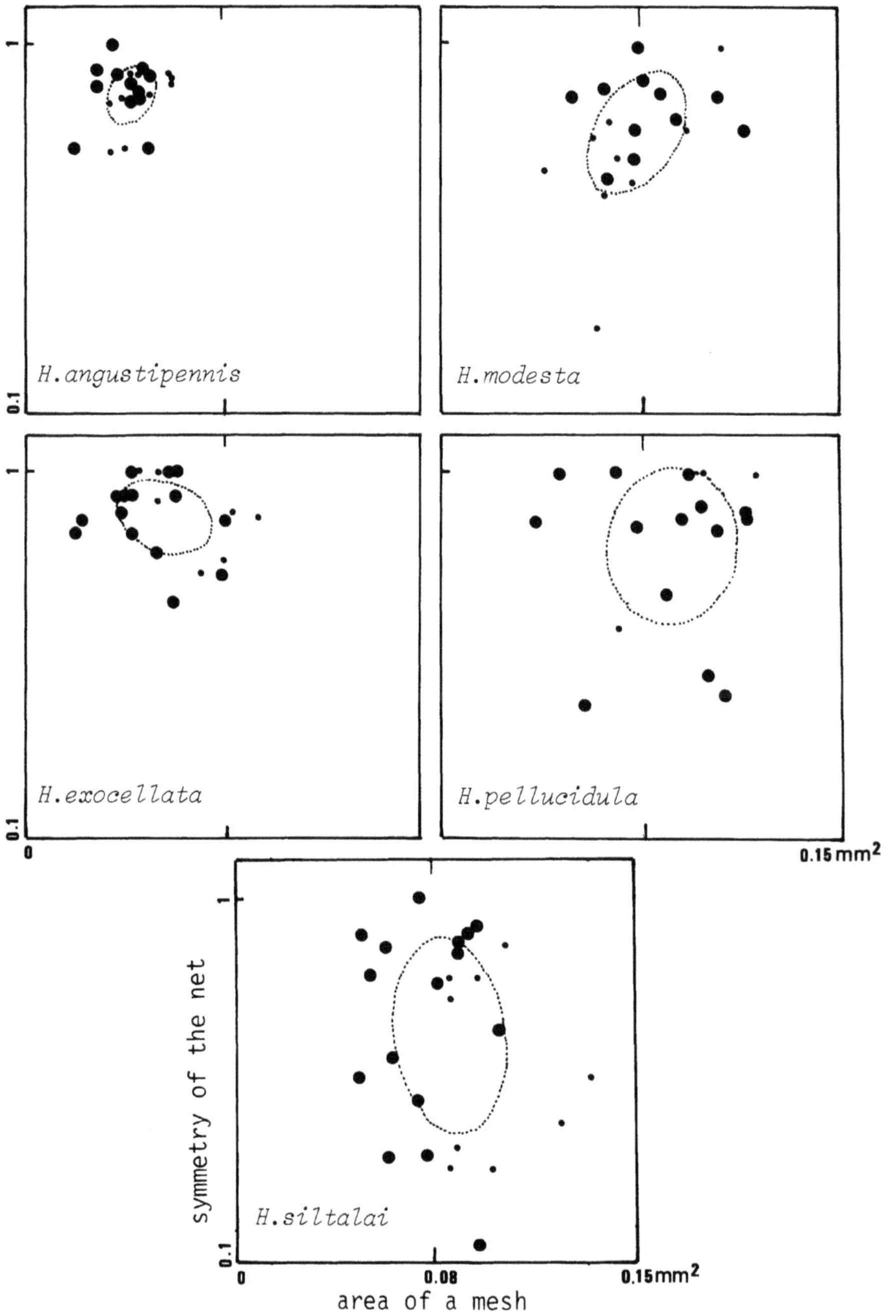


Figure 4 : Plot of values of variable 1 (symmetry of the net) versus the values of variable 6 (area of a mesh) for the five species of *Hydropsyche*. Each square corresponds to one net. The size of squares is proportional to the two current velocities (small squares : v ≤ 15 cm $s^{-1}$ and large squares : v ≥ 30 cm $s^{-1}$). The ellipse of dispersion is defined by variances and covariances of data.

DISCUSSION

The best criteria to discriminate these five species are the area of a mesh and the symmetry of the net. These differences are not related to the size of larvae (Table 1), that means they are related to the spinning behaviour of larvae and/or some characteristics of their habitat.

Table 1 : mean head-capsule length of five species of *Hydropsyche* (fifth instar) and mean size of meshes of their net

| | *H. siltalai* | *H. modesta* | *H. exocellata* | *H. angustipennis* | *H. pellucidula* |
|---|---|---|---|---|---|
| mean head-capsule length in mm | 1,59 | 1,65 | 1,72 | 1,75 | 1,92 |
| mean width and length of a mesh in µm | 229x376 | 227x365 | 175x284 | 170x259 | 215x382 |

The observed differences do not seem to support the hypothesis presented in the introduction firstly because a potamic species like *H. modesta* spins nets with meshes as large as those of a rhithronic species like *H. siltalai*, secondly because a species like *H. angustipennis* spinning a net with small meshes is always absent from the Rhône river, although its larvae are abundant in several small tributaries of the Rhône in the alluvial plain.

In any watercourse, a net of *Hydropsyche*, like a drift net, is more or less rapidly clogged by the suspended material and the larva must always brush its net with its labrum, swallowing edible elements and pulling out inedible particles. If the density of particles is too important or if the size of the meshes is too small, the net will quickly become clogged, compelling the larva to continuous (therefore energetically expensive) work of cleaning.

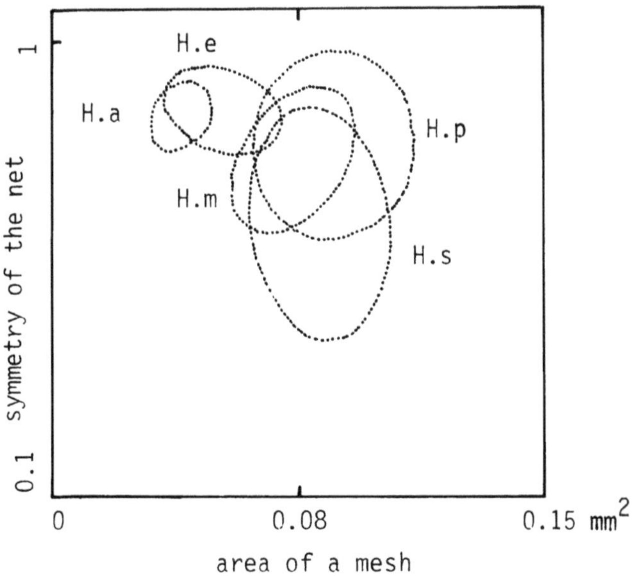


Figure 5 : Ellipse of dispersion or each species (H.a : *H. angustipennis* ; H.e : *H. exocellata* ; H.m : *H. modesta* ; H.p : *H. pellucidula* and H.s : *H; siltalai*) defined by variances and covariances of data plotted in Figure 4.

This hypothesis would explain why a species like *H. angustipennis* is never found in the Rhône river where the suspended material is abundant whereas it is present in the small tributaries of the alluvial plain of the Rhône where the waters, partially with a phreatic origin, are clearer. This hypothesis would also explain the occurrence of *H. exocellata* at one and the same time in these small tributaries and in the Rhône itself (Fig. 1), its net (Fig. 4 and 5) having intermediate characteristics.

This hypothesis emphasizing the clogging aspect rather than the feeding aspect (the *Hydropsyche* being typically an opportunist consumer) must be, of course, corroborated by other experiments.

AKNOWLEDGEMENT

We thank D. Chessel and Y. Auda for their invaluable statistical and computing advice and G. Thoiron for correcting the text.

REFERENCES

Alstad, D.N. 1980. Comparative biology of the common Utah Hydropsychidae (Trichoptera). **Amer. Midl. Nat.**, 103 : 167-174.

Alstad, D.N. 1982. Current speed and filtration rate link caddisfly phylogeny and distributional patterns on a stream gradient. **Science**, 216 : 553-554.

Besch, W.K., Schreiber, I. & Herbst, D. 1977. Der *Hydropsyche* toxisitätstest erprobt an Fenethcarb. **Schweitz. Z. Hydrol.**, 39 : 70-85.

Carrel, G., Barthelemy, D., Auda, Y. & Chessel D. 1986. Approche graphique de l'analyse en composantes principales normée : utilisation en hydrobiologie. **Acta oecologica**, 7 : 189-203.

Daget, J. 1976. Les modèles mathématiques en Ecologie. Masson (ed.), 172 pages.

Décamps, H., Besch, W.K. &Vobis, H. 1983. Influence de produits toxiques sur la construction du filet des larves d'*Hydropsyche* (Trichoptera, Hydropsychidae). **C. R. Acad. Sci., Paris**, 276 : 375-378.

Fuller, L.R., Mackay, R.J. & Hynes, H.B.N. 1983. Seston capture by *Hydropsyche betteni* nets (Trichoptera, Hydropsychidae). **Arch. Hydrobiol.**, 17 : 251-261.

Kaiser, P. 1965. Über Netzbau und Strömungsinns bei den Larven der Gattung *Hydropsyche* Pictet (Trichoptera). **Int. Revue ges. Hydrobiol.**, 50 : 191-213.

Petersen, L.B.M. & Petersen, R.C. 1983. Anomalies in hydropsychid capture nets from polluted streams. **Freshwat. Biol.**, 13 : 185-192.

Sattler, W. 1958. Beitrage zur Kenntnis von Lebenweise und Körperbau der Larve und Puppe von *Hydropsyche* Pictet (Trichoptera) mit besonderer Berüksichtigung des Netzbaues. **Z. morph. v. Okol. Tiere**, 47 : 115-192.

Wallace, J.B. 1975. Food partitioning in net-spinning Trichoptera larvae : *Hydropsyche venularis, Cheumatopsyche etrona* and *Macronema zebratum*. **Ann. Ent. Soc. Amer.**, 68 : 463-472.

Wallace, J.B., Webster, J.R. & Woodall, W.R. 1977. The role of filter feeders in flowing waters. **Arch. Hydrobiol.**, 79 : 506-532.

# SESTON QUALITY AS A FACTOR INFLUENCING TRICHOPTERAN POPULATIONS

R. C. PETERSEN, Jr

SUMMARY

The preemergence biomass of seston feeding, trichopteran guilds at 8 locations in southern Sweden was found to be correlated with the amount of animal material in the seston and not with the organic content or total seston. The number of species varied among the 8 locations and dominated by *Hydropsyche siltalai*, *H. angustipennis*, *H. saxonica*, *H. pellucidula*, *Neureclipsis bimaculata*, and *Cheumatopsyche lepida*. The quantity, organic weight, and percent organic seston was a relatively conservative property ranging from 3-11g ash free dry mass/$m^3$. Animal material in the seston consisting of rotifers, microcrustaceans and drifting insects ranged from 25 to 1899 μg animal dry mass/$m^3$ and was correlated with the differences in trichopteran biomass at the 8 locations. It is suggested that food quality as the amount of animal material in the seston is an important factor in determining the success of filter feeding trichopteran populations.

## INTRODUCTION

The filter feeding caddisflies have been reported to be non-selective or passive filter feeders (Oswood 1979). Others assume that the combination of net mesh opening and microhabitat determine particle selection (Georgian and Wallace 1981) and while it is known that filter feeders rely upon animal seston (Benke and Wallace 1980) and can change seston quality by removing animals (Ross and Wallace 1983), particle selection is not considered an important ecological or evolutionary character (Alstad 1982). Under laboratory conditions, it is known that trichopterans can and do select particles (Sattler 1958). Petersen (1985) has shown that hydropsychids prefer animal material and feed on detritus particles in response to the frequency and type of animal material in their diet.

There still remains the idea that in nature trichopteran filter feeders are essentially opportunistic and that laboratory demonstrated particle selection is not an important aspect in determining their population success. For example, they are usually lumped with the blackflies, a truely non-selective feeder, in one functional group, as filter feeders (Merritt and Cummins 1978). The assumed similarity between blackfly and hydropsychid feeding has led to the suggestion that control methods using pesticide coated particles may be equally effective for both groups of stream organism (Wiederholm 1984).

The purpose of the present paper is to explore the causal relationship between guilds of filter feeding trichopterans and the food quality of seston and to test whether laboratory observations of food selection have any ecological meaning. This is done by determining the preemergence population density of the coexisting species at 8 different locations in southern Sweden while measuring the quantity and animal content of the seston.

## PROCEDURE

### Site descriptions

Eight locations in southern Sweden were selected to represent a wide range of trophic conditions ranging from oligotrophic to eutrophic but all having populations of filter feeding caddisflies. All were similar in width, depth, substrate, riparian cover and it is assumed that differences in these factors as they affect the habitat templet are small. It is also assumed that differences in degree days between the locations are small. A combination of continuous records at Höörsan, Lerbäcken, and Verkeån-8, and spot checks at the other 5 locations gives an estimated range of 2500 to 2700 degree days per year for the 8 locations.

The 8 locations are numbered 1 to 8 with their stream name as in Table 1. Stensån is a oligotrophic pristine stream located in the province of Halland. Alkalinity tends to be low with levels averaging <0.3 meq/L. It drains a watershed of forest and light agricultural use. Stensån-1 was located approximately 100m below lake Sjöaltesjö while Stensån-2 is located 15 km downstream near Hasslöv. There are no intervening lakes or ponds along this 15 km stretch and the two stations allow comparison of a lake outlet to the downstream location. Samples for both seston and benthos were taken on 19 May.

The other 6 locations are in the province of Scania, a nutrient rich area of Sweden with alkalinities and macronutrients typical of eutrophic waters. The location on the Höjeån was in the middle of the city of Lund at Kjellbybadet. Samples were taken on 6 May. Höörsån-4 and Lerbäcken-5 are two locations on the Höörsån which drains into Ringsjön, the largest lake in Scania. Höörsån-4 is located just above the town of Höör and is mildly polluted with non-point source runoff from agriculture and residential areas. Lerbäcken is a pristine stream located in a small nature preserve and about 50 m below a small pond. Samples were taken on 13 May in Höörsån and 14 May in Lerbäcken. The Övedsån-6 location is heavily impacted from agricultural runoff and an eutrophic lake 50 m above. Based on previous studies (Petersen et al. 1984) the Övedsån location is the most eutrophic location used in this study. Samples were taken on 15 May. The Verkeån is probably the most pristine stream in Scania and is a managed trout hatchery. There are no impoundments along its length. The upper location, Verkeån-7, is mildly influenced by agicultural runoff while the downstream location has a small influence of summer bungalows and wood lot effluents. Samples were taken at both Verkeån locations on 27 May.

## Materials and methods

**Benthos,** Quantitative measurement of the filter feeding trichopterans was made using 5 Niel samples (0.05 $m^2$) taken in a transect across each stream location. Animals were eluted from the stoney substrate in the field and preserved in Kahles preservative. In the laboratory each filter feeding trichopteran was determined to species and instar, dried at 50 C and weighed.

**Seston.** The amount and type of seston was determined on the same day as the benthos was sampled. Total seston was determined by replicated 1 Liter samples on preweighed, pre-combusted Whatman GFF filters. A Clarke Bumpus plankton sampler was used to sample 10-30 $m^3$ of flowing water. The collected seston was devided into 1 mm, 500 µm and 250 µm size classes using Nytex screens. From the 1 mm and 500 µm fractions all invertebrates were removed, determined, dried at 50°C and weighed. Particles retained on the 250 µm screen were filtered into a Millipore filter, dried, cleared, determined, and their linear dimensions converted to volumes.

All correlations were performed using the Pearson product moment correlation with tests of significance determined by ANOVA. Material was transformed using the natural logarithm, ln.

## RESULTS AND DISCUSSION

### Seston at the 8 locations

Total seston dry mass (DM) ranged from 4.1 g/$m^3$ at Verkeån-7 to 13.3 g/$m^3$ at Övedsån or barely a three fold difference in the total amount of material in the water column (Table 1). The organic fraction of the seston as measured by the ash free dry mass (AFDM) ranged from 2.8 g/$m^3$ at Höörsån to 10.8 g/$m^3$ at Övedsån with about a three fold range of values. The percent organic ranged from 58% at Höörsån to 85% at Verkeån-7.

The amount of seston in a size class that can be trapped by a net and can be handled by the larva is represented by particles > 250 µm and includes drifting insects and large zooplankton. The > 250 µm boundry is applicable to late instar caddis larvae and is based on the average net dimensions of most instar IV to V larvae and from direct observations of larval feeding (Petersen 1985, 1987).

In Table 1 both the total mass of seston > 250 µm and the animal material in this size category are reported. Coarse seston ranged from 5.9 to 28.6 mg/$m^3$ and represented 0.1 to 1% of the total seston. There was a 10 fold variation in this parameter over the 8 locations. The largest variation in any seston parameter was in the amount of animal material. This varied from a low of 25 µg/$m^3$ at Verkeån-8 to a high of 1899µg/$m^3$ at Höjeån. The Verkeån does not have impoundments along its length nor does it have tributaries with impoundments. Animal material in the Verkeån seston consisted almost exclusively of drifting insects. The Höjeån receives material from a variety of settling ponds and small impoundments which feed into the river along its length. The animal seston was mostly chironomids caught up in detrital flocs, and zooplankton.

The reduction in the concentration of seston animals, commonly called drift density, that occurred from Stensån-1 to Stensån-2 is a common observation when a lake outflow location is compared to one further downstream. Pearson and Franklin (1968) report a drop of from 64.2µg/$m^3$ below a dam on the Green River to 0.1µg/$m^3$ further downstream. Bishop and Hynes (1969) report a similar reduction in the Speed River of from

17 to 1$\mu g/m^3$. The reason for the reduction is a combination of active removal of seston by filter feeders (Ross and Wallace 1982) and settling out of the material. The rate of loss will be dependent on the turbulence, discharge, gradient as well as the type of seston.

Table 1. Quantitative measurement of total seston as either ash free dry mass (AFDM) or total dry mass (DM) at the 8 locations. Coarse seston are detrital particles >250 µm in linear dimensions. Animal seston is all animal material retained on a 250 µm nylon screen.

| Location | TOTAL SESTON AFDM g/m³ | DM g/m³ | COARSE >250 µm mg/m³ | ANIMAL >250µm µg/m³ | % ANIMAL OF TOTAL | OF COARSE |
|---|---|---|---|---|---|---|
| 1. STENSÅN | 6.5 | 8.7 | 5.9 | 158.6 | 0.0024 | 2.7 |
| 2. STENSÅN | 6.0 | 8.6 | 6.3 | 38.5 | 0.0006 | 0.6 |
| 3. HÖJEÅN | 4.9 | 7.1 | 35.3 | 1899.0 | 0.0267 | 5.4 |
| 4. HÖÖRSÅN | 2.8 | 4.8 | 28.6 | 100.0 | 0.0021 | 0.3 |
| 5. LERBÄCKEN | 3.3 | 4.9 | 23.4 | 371.0 | 0.0076 | 1.6 |
| 6. ÖVEDSÅN | 10.8 | 13.3 | 10.1 | 406.8 | 0.0031 | 4.0 |
| 7. VERKEÅN | 3.5 | 4.1 | 16.9 | 36.0 | 0.0009 | 0.2 |
| 8. VERKEÅN | 4.3 | 6.4 | 13.4 | 25.0 | 0.0006 | 0.2 |

**Hydropsychid biomass and abundance.**

Below Lake Sjöaltesjön at Stensån-1 the net spinning trichoptera were represented by 5 species having a total biomass of 9652 $mg/m^2$ and numerical abundance of 2956 individuals (Table 2). The guild was dominated numerically by *C. lepida* with 1560 individuals/$m^2$, mostly instrar V and dominated in biomass by *H. pellucidula*with 5280 $mg/m^2$, mostly as pupae. Fifteen kilometers downstream there were only two species of hydropsychid remaining, *H. pellucidula* and *H. siltalai*, with an abundance of 168 individuals with biomass of 318 $mg/m^2$. Both populations were dominated by instar V with pupae being rare. The average weight of an individual within the guild decreased from 3.3 mg at Stensån-1 to 1.9 at Stensån-2. The average weight of an instar V *H.siltalai* was 11.1mg (N=30) at the lake outlet station and 8.1mg (N=10) 15 km below ($p<0.01$, Mann-Whitney U). In summary then, there was a reduction in the number of filter feeder species, the abundance, the biomass, the indivual biomass within a population and an increase in development time at the downstream location compared to the lake outlet location. There was also a decrease in the average biomass of a guild member even though the lake outlet guild was dominated by *C. lepida* a species with a much smaller final weight than any of the hydropsychids.

At Höjeån the filter feeding trichopterans occurred as a monoculture of *H. angustipennis* with a density of 1988 individuals and biomass of 14299 $mg/m^2$. Most of these individuals were late instar V or pupae and had an average weight of 7.2mg. The average weight was not significantly different from an individual *H. angustipennis* at the Stensån-1 location.

At the Höörsån location 4 filter feeding trichopterans occurred but only two with an abundance that could be quantified. *H. siltalia* dominated the guild and along with *H. pellucidula* there was a total biomass of 174 mg among 44 individuals/$m^2$. The average individual weighed 4.0 mg and were mostly instar V *H. siltalai.* Similar abundance and biomass occurred at the Lerbäcken location but the number of species in the guild increased to 6. *H. pellucidula* is replaced at Lerbäcken by *H. saxonica.* These two species tend not to occur together. Part of the reason for this is *H. saxonica* tends to prefer small heavily woodland streams while *H. pellucidula* does well in both small and larger streams. Another reason is that both emerge in early May. This similarity in their life cycle increases the possibility of competitive interference. The average individual in the guild weighed 2.9 mg at Lerbäcken which is about one third less than in Höörsån and was due to the larger number of early instar V *H. siltalai* present in the sample.

The Övedsån guild was dominated by *H. angustipennis* which made up 99% of the 4886 individuals with a biomass of 23026 $mg/m^2$.*N. bimaculata* was the second most abundant species at Överdsån having a biomass of 266 $mg/m^2$ from 406 individuals mostly as instar IV. Additional comparisons between the Övedsån and Stensån-1 *N. bimaculata* population have been reported previously (Petersen *et al.* 1985).

The two locations on the Verkeån had very low densities of single species of hydropsychids mostly in the last instar.

As the data from the Stensån and Övedsån would suggest populations of filter feeding caddisflies do well at lake outlet locations. The abundance of filter feeders at lake outlets and their downstream decrease is a common observation and has been linked to the high levels of seston in these habitats (Cushing 1963 and others). However, it is not necessarily true that lake outlets will have guilds of trichopteran filter feeders. Erman and Chouteau (1979) report that the benthic invertebrates below a peat fen were numerous but were almost exclusively blackflies not trichopterans. Peat fens may contribute detrital material to the seston and this may be readily acceptable to blackflies but this has little effect on trichoptera which require more animal material.

Table 2. Trichoptera filter feeders occurring at the eight locations with their final population biomass (mg/m$^2$). An X indicates that the taxum is present at that location but did not occur in the 5 Niel samples. A O indicates that the species does not occur at that location.

| Taxa | 1 Stens | 2 Stens | 3 Höje | 4 Höörs | 5 Lerbäcken | 6 Öveds | 7 Verke | 8 Verke |
|---|---|---|---|---|---|---|---|---|
| *Hydropsyche angustipennis* | 254. | 0. | 14299. | X | 9 | 22720 | 0. | 0. |
| *Hydropsyche pellucidula* | 5280. | 132. | 0. | 51. | 0. | 0. | 86. | X |
| *Hydropsyche siltalai* | 2600. | 186. | 0. | 124. | 190. | 30. | X | 25. |
| *Hydropsyche saxonica* | 0. | 0. | 0. | 0. | 25 | 0. | 0. | 0. |
| *Cheumatopsyche lepida* | 1380. | 0. | 0. | 0. | 0. | 0. | 0. | 0. |
| *Philopotamus montanus* | 0. | X | 0. | 0. | 0. | 0. | 0. | 0. |
| *Wormaldia subnigra* | 0. | 0. | 0. | 0.. | X | 0. | 0. | 0. |
| *Plectrocnemia conspersa* | 0. | 0. | 0. | 0. | X | 0. | 0. | 0. |
| *Polycentropus flavomaculatus* | 0. | 0. | 0. | X | X | 10. | X | X |
| *Neureclipsis bimaculata* | 138. | 0. | 0. | 0. | 0. | 266. | 0. | 0. |
| Number of Taxa | 5 | 3 | 1 | 4 | 6 | 4 | 3 | 3 |
| Total Biomass mg/m$^2$ | 9652 | 318. | 14299 | 175 | 224 | 23026 | 86 | 25 |
| Total Abundance no./m$^2$ | 2956 | 168 | 1988 | 44 | 76 | 4886 | 9 | 3 |
| Mean Weight mg/ind. | 3.3 | 1.9 | 7.2 | 4.0 | 2.9 | 4.7 | 9.6 | 8.3 |

Of the three locations presented in Table 2 with exceptionally high caddis abundance and biomass (locations 1,6,3) only 2 of these are lake outlet locations. The third, Höjeån, is a rich agricultural stream. The relationship between filter feeder biomass and the various seston parameters presented in Table 1 was investigated by correlating each seston parameter with caddis biomass in Table 2. Caddis biomass or the natural logarithm of caddis biomass was not related to total seston or ln total seston, coarse seston or ln coarse seston (P>0.72). It was related to animal seston and ln animal seston (P<0.02) (Fig. 1).

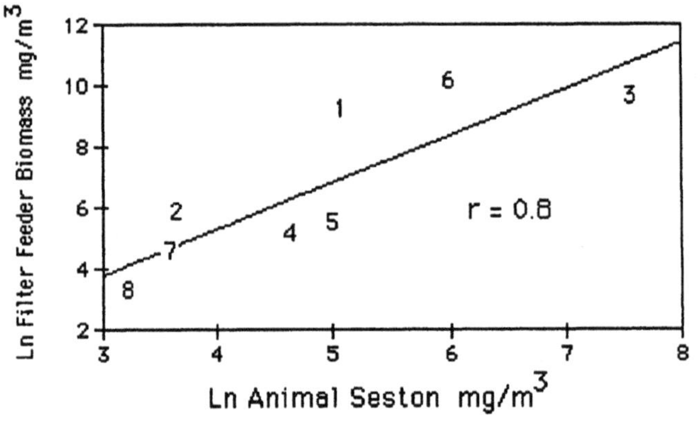


Figure 1. Correlation between natural logarithm of filter feeder biomass at the 8 study locations and the ln concentration of animal material in the seston in mg/m$^3$. Numbers refer to locations as in Table 1.

These relationships can be seen directly in Tables 1 and 2 where the total seston parameters tend not to be different between locations or at least the magnitude of the variation is much less than the differences observed among the trichopteran guilds. It is the animal material in the seston which is the variable most affecting the success of the guild at a particular location.

For deposit feeding detritivores the percent organic of detritus particles has been shown to be one index of food quality (Ward and Cummins 1979). Petersen (1985, 1987) has shown that while hydropsychids prefer animal material there is not enough of this material to match the daily consumption rate which forces that animal to rely heavily on detrital material for their diet. Therefore the organic content of the detrital material may be an important food parameter to filter feeding trichopterans as has been suggested previously (Richardson 1984). However, there was no significant correlation between percent organic and filter feeder biomass over the 8 locations. Among the locations studied the organic content did not vary and as with the values for total seston this could not match the large variability in filter feeder biomass.

From laboratory studies Petersen (1985, 1987) has shown that hydropsychid larvae select particles and for the four common Swedish species of *Hydropsyche siltalai, H. angustipennis, H. saxonica* and *H. pellucidula,* animal material is preferred above all other particles tested. The question remains whether the conditions in the natural environment allow this selection preference to take place or are food items generally so rare that the larva must take a prey not according to its preference but according to its occurrence. If a larva takes food according to its occurrence, then the biomass of the guild should be correlated to the percent of animal material in the seston. If selection is taking place in the field then guild biomass should be correlated to the concentration of seston animals. There was no correlation between ln guild biomass and ln percent animal seston. As mentioned above there was a significant correlation between ln guild biomass and the concentration of animal seston. This suggests that selection of prey items, probably by a combination of selection by the net and true behavioral selection by the larva, is occurring and affects the overall success of the filter feeding guild.

Petersen (1987) shows that for*H. siltalai* feeding on Lerbäcken seston an instar V larva must eat all the animal material, greenalgae, detritus-diatoms flocs, and detritus that lands on its net in one day to allow the larva to obtain an ingestion rate of 85% body weight per day. Only plant material which is ranked last in the selection hierarchy and which makes up about 50% of the seston can be rejected. This suggests that in Lerbäcken there is not enough high quality, animal prey, for the larva to reach a maximum growth rate and must, therefore, utilize less preferred items in its diet. These observations are supported in the present paper where the biomass of the guild at Lerbäcken is considerably smaller than at other locations where animal material in the seston is more abundant.

ACKNOWLEDGEMENTS

This research was supported by grants from the Swedish Environmental Protection Board. S. Johansson, A. Kulberg and U. Persson helped with the tedious job of sorting and enumerating the seston samples and measuring the filter feeders. L. Petersen contributed to the ideas and final preparation of the manuscript. This is a contribution from the Stream and Benthic Ecology Group, University of Lund.

REFERENCES

Alstad, D.N. 1982. Current speed and filtration rate link caddisfly phylogeny and distributional patterns on a stream gradient. **Science,** 216 : 533-534.

Bishop, J.E. & Hynes, H.B.N. 1969. Downstream drift of the invertebrate fauna in a stream ecosystem. **Arch. Hydrobiol,** 66: 56-90.

Benke, A.C. & Wallace, J.B. 1980. Trophic basis of production among net-spinning caddisflies in a southern Appalachian stream. **Ecology,** 61 : 108-118.

Cushing, C.E. 1963. Filter feeding insect distribution and planktonic feed in the Montreal River. **Trans. Am. Fish. Soc.,** 92 : 216-219.

Erman, D.C. & Chouteau, W.C. 1979. Fine particulate organic carbon output from fens and its effect on benthic macroinvertebrates. **Oikos,** 32 : 409-415.

Georgian, T.J. & Wallace, J.B. 1981. A model seston capture bynet-spinning caddisflies. **Oikos,**36 : 147-157.

Merritt, R.W. & Cummins, K.W. eds. 1978. An introduction to the aquatic insects of North America. Kendall/Hunt Pub. Dubuque. 441p.

Oswood, M.W. 1979. Abundance patterns of filter feeding caddisflies (Trichopera : Hydropsychidae) and seston in a Montana (U.S.A.) lake outlet. **Hydrobiologia,** 63 : 177-183.

Pearson, W.D., Kramer, R.H. & Franklin D.R. 1968. Macroinvertebrates in the Green River below flaming Gorge Dam, 1964-65 and 1967. **Proc. Utah Acad. Sciences, Arts and Letters,** 45 : 148-167.

Petersen, L.B.M . 1985. Food preferences in three species of Hydropsyche (Trichoptera). **Verh. inter. Verein. Limnol.,** 22 : 3270-3274.

Petersen, L.B.M. 1987. Direct observations of Hydropsyche prey selection. **Proc. 5th Int. Symp. Trichoptera,** Lyon, 21-26 July 1986, Junk.

Petersen, R.C., Wallace, J.B. & Petersen, L.B.M. 1984. Influence of velocity and food availability on catchnet dimensions of *Neureclipsis bimaculata* (Trichoptera : Polycentropodidae).**Holarctic Ecology,** 7 : 380-389.

Richardson, J.S. 19874. Effects of seston quality on the growth of a lake-outlet filter feeder.**Oikos,** 43 : 386-390.

Ross, D.H. & Wallace, J.B. 1983. Longitudinal patterns of production, food consumption, and seston utilization by net-spinning caddisflies (Trichoptera) in a southern Appalachian stream(U.S.A.) **Holarct. Ecol.** 6 : 270-284.

Ward, G.M. & Cummins, K.W. 1979. Effects of food quality on growth of a stream detritivore, *Paratendipes albimanus* (Meigen) (Diptera : Chironomidae). **Ecology,** 60 : 57-64.

Wiederholm, T. 1984. Responses of aquatic insects to environmental pollution. *In* The ecology of aquatic insects, V.H. Resh and D.M. Rosenberg (eds). Praeger, New York.

# DIRECT OBSERVATIONS OF *HYDROPHYCHE* PREY SELECTION

L. B. -M. PETERSEN

## SUMMARY

Direct observations were made on *Hydrospsyche* food selection in an attempt to find a method that augments gut analysis done by area. From these observations it was possible to rank the food types in the seston in terms of preference. By calculating volume filtered, consumption index and seston concentration, direct observations on prey selection could be used to predict gut content of filter feeding *Hydropsyche* larvae in the stream.

## INTRODUCTION

Net-spinning Trichoptera of the family Hydropsychidae are an important component of the benthic fauna in many lotic environments (Wiggins 1977), and an important processor of organic matter in the stream ecosystem (Wallace *et al.* 1977). The organic matter that is utilized has been studied through gut analysis and in the eartly trophic studies, the most common way was to record the presence or absence of a prey type in the gut to form relative frequency of occurrence (Slack 1936). Later the Millipore filter technique was developed (Mecom and Cummins 1964, Cummins 1973) and modified to fit hydropsychids (Wallace 1975). The reliability and problems of these methods have been discussed by Martin and Mackay (1982).

Recent studies have documented the competitive coexistence of congeneric net-spinning species within a stream reach (Xiang *et al.* 1984). In order to further explore how ecologically similar species are able to coexist there is a need for techniques of diet analyses which demonstrates the true prey consumption. One way to find out what a larva will eat is by direct observations of the feeding. From earlier studies it is known that *Hydropsyche* larvae are selective feeders where the larvae eat some food items, reject others, and select according to hunger (Petersen 1985). In a study where the larvae were fed detritus and *Daphnia* ,both items were eaten when the larva were hungry, while *Daphnia* were preferred as the larvae approached satiation (Petersen 1985). The feeding selectivity can be described as being inversely proportional to larval hunger state (Pastorok 1980).

In this paper the food items in the seston are ranked in order of preference by direct observations of the feeding selectivety at different satiation levels. By knowing the amount of different types of seston and the consumption index, it is possible to predict what a *Hydropsyche* larvae will eat. The method can be used alone or together with normal gut analysis to get a more exact picture of larval feeding.

## MATERIAL AND METHODS

Three species of *Hydropsyche* larvae were collected for gut analysis from Lerbäcken, a headwater stream in south Sweden. The Millipore filtration and sketching technique described by Wallace (1975) was used for the preparation and analysis of the hydropsychid foreguts. In one sample, in addition to area, the number of *Daphnia* filipods were counted and converted to number *Daphnia*. The average size, by area, of the *Daphnia* were known from seston samples, which were analyzed accordingly to Petersen (1987), making it possible to compare expected and obtained area of the *Daphnia* in the foregut.

In a laboratory experiment, three *Hydropsyche* larvae were fed 10 *Daphnia* each and killed immediately after. The foregut, which was not completely filled, was taken out and the content was put on a Millipore filter. Three additional Millipore filters received 10 whole *Daphnia* each. The area of the whole *Daphnia* were compared to the area of the *Daphnia* in the foregut to obtain a recovery rate.

Food preference was measured using fifth instar larvae of *H. saxonica*. The larvae were kept at room temperature in small recirculating aquaria (Petersen & Petersen 1984). The larvae were starved for 48 h and then fed individual particles by dropping the particle onto the net with a Pasteur pipette.

The food particles were obtained from Lerbäcken and were size fractionated (0.5-1 mm). Large particles like leaf and macroalgae were cut to obtain particles of 0.5-1 mm. The dry mass and the ash free fry mass were measured for the different food items, which were : copepods (N = 17), frozen *Diaptomus* sp.(N = 11), detritus-diatom aggregates (N = 22), plant (N = 3), detritus (N = 10), oak (N = 7), *Lemanea* sp.(N = 17), *Ulothrix* sp. A (N = 11) & B (N = 11),*Draparnaldia* sp. (N = 11), and *Batrachospermum* sp. (N = 11). N is the number of experimental runs a particular food item was used. The larvae were fed particles in a repeated a, b, c, a, b, c, ... with three to five types of food in one experimental run. The larvae could eat, attach or throw out the particle it received. This was recorded together with the length of time it took for a particle to be ingested, referred to herein as the ingestion rate. At the beginning of an experimental run all types of particles were consumed. With increasing satiation the preference of the larva would change. A food item previously eaten would be rejected. The satiation level is defined as the amount of food consumed before the larvae started to reject a particular food item. The different food items were then ranked according to the satiation levels.

To estimate the 24 h consumption index, CI, *Hydropsyche* larvae were first fed on a food item to satiation. Then the larvae were fed to satiation again at 1 h intervals for 6-8 h to get at a consumption of food per h. The wet weight of the larvae were measured and converted to dry weight. The CI was calculated as the total amount of food consumed for the 6-8 h period devided by the weight of the larvae and expressed over 24h.

## RESULTS AND DISCUSSION

When the hydropsychid larvae were fed *Daphnia* and then killed immediately, the area of the *Daphnia* in the foregut corresponded only to 70% ± 7% of the area of the whole *Daphnia* .No *Daphnia* were detected beyond the proventriculus in the posterior part of the foregut indicating that all *Daphnia* were left in the foregut. The explanation for this 30% loss is a disintegration of the *Daphnia* into minute particles which are not recognized as *Daphnia.* In the larvae from the field were the filipods of the *Daphnia* were counted and converted to number *Daphnia* of a known area, the actual measured area of the *Daphnia* in the gut were as low as 10% ± 4% of the expected. These results can be compared with those of Martin & Mackay (1982) where a large part of the *Rhyacophila* gut content was unidentifiable even if it was known what type of prey the larvae had consumed. One reason being that soft animal tissue will be mislabled as detritus. When the larvae eat the whole body as the case with *Daphnia* it is possible to identify some part of the prey to get the number of prey consumed. This is not possible when the larvae just eat the soft tissue and leave out the sclerotized parts, which are commonly the way the hydropsychid larvae consume larger prey.

Figure 1 is a ranking of different food items fed to *H. saxonica.* ., ranked as the µg food consumed before rejection of a particular item. When hungry all food items in Figure 1 are consumed. As the larvae feed , hunger decreases, and food items are rejected in an order going from the bottom of the figure to the top. This means that in an almost satiated stage only copepods will be consumed. The feeding selectivity can be described as being inversely proportional to larval hunger state.

In Figure 1 is also the ingestion rate, in µg/s, for the different food items. From the figure we can see that the time it took the larvae to eat a particle has little effect on the ranking of a food item. For example, detritus + diatom aggregates were ranked in the middle of the list but their ingestion rate, 1.7 µg/s, was the highest recorded.

The ranking of the food items in Figure 1 is a composit of several different experiments. While there was a large, up to 3 fold, variation between larvae in the amount of food that would be required to reach satiation on any particular item there was virtually no variation in the selective ranking of the items. In a few instances where a shift was observed, it was always between food items next to each other on the ranking list and was probably more a matter of the inability to obtain particles of the same quality from one day to another.

The ranking of food items in Figure 1 is based on *H. saxonica* but based on additional data it is believed to reflect the selective behaviour for *H. angustipennis* and *H. siltalai* as well. In an experiment where the three species were fed *Daphnia* and detritus all three species started to reject detritus at the same satiation level suggesting that the preference is the same for all three (Petersen 1985). Scattered experiments with *H. siltalai* have also revealed a ranking of *Daphnia* > *Ulothrix* > detritus which is in agreement with Figure 1.

Figure 1 is direct measure of seston food quality as determined by the hydropsychid larva. By quantifying the different food items from a seston sample and with the ranking list it is possible to estimate the quality and quantity of food available during different times of the year.

To calculate how much of this material will actually be consumed the consumption index has to be determined.The only estimate of consumption for a *Hydropsyche* larvae has been done by McCullough *et al.* (1979). They have two estimates, a theoretical value calculated from gut passage time for stream animals, and a measurement obtained by dual labeled food in the laboratory. The stream animals ate 142% of body weight

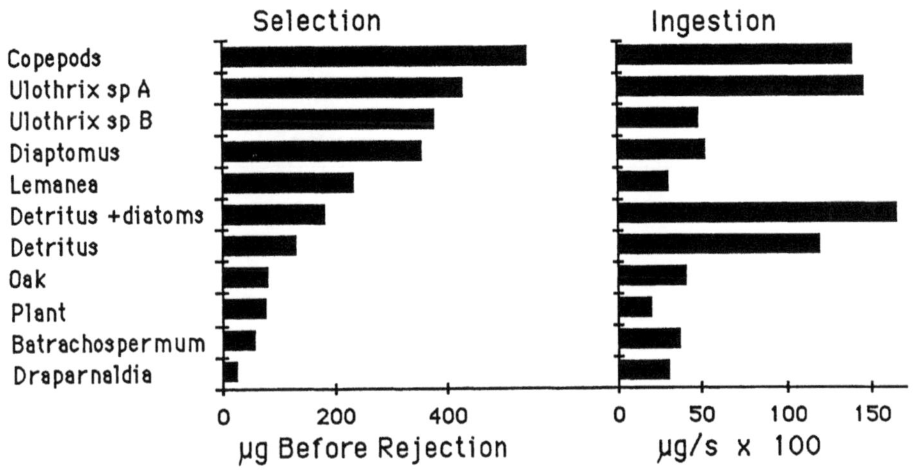


Figure 1. Selective preference of various food items by *H. saxonica* as the µg of material ingested before rejection. The ingestion rate in µg of food/s is shown for each item on the right.

per day and the laboratory animals ate 73% per day. To get additional data for the Swedish hydropsychids, larvae were fed continuously in the laboratory, with the result seen in Table 1. For *H. siltalai* the 24 h consumption index varied between 34-107% of body weight depending on food type. The food items most preferred in Figure 1, *Daphnia* and *Ulothrix*, have the lowest 24 h consumption while detritus + diatoms being of a lower rank, has a higher consumption rate. In Table 1 is the consumption rates for *H. angustipennis*. They are 65-87% lower than the rates obtained for *H. siltalai*. For *H. saxonica* consumption rates were only measured for destritus + diatoms, and they were 51% of the values for *H. siltalai*. The differences in consumption rate between the three species could possibly be explained by the weight differences. *H. siltalai* had a mean dry weight of 4.7 mg while *H. angustipennis* and *H. saxonica* had a dry weight of 6.5 respectively 7.5 mg. There was a negative correlation ($r^2 = 0.99$) between the consumption rate and body weight for the three species. The pupal weight for *H. saxonica* and *H. angustipennis* start around 7 mg, suggesting that the larvae have done most of their growth and they are close tó pupation. The lack of interest for food could also be observed more directly. For *H. siltalai* all animals fed properly (N = 23) while for *H. angustipennis* 25% (N = 16) of the larvae showed little or no interest for food and for *H. saxonica* this number was 69% (N = 16). The above results indicate a lower consumption rate prior to pupation. During the growth period of a fifth instar larva a consumption rate of 75-100% seems to be a reasonable estimate, since detritus + diatoms and detritus are the major components of the diet.

*H. siltalai* are found at medium velocities (25-45 cm/s) in the stream. The average velocity through the net has been estimated to be around 15 cm/s. The net area for a *H. siltalai* is 38 $mm^2$ which gives a filtered volume/day of 492 liter.

From the seston samples collected in the stream the amount of different prey items/liter is known. In Table 2 is the seston composition for May taken as an example. There are 4.3 particles/liter, counting particles equal to or bigger than 250 µm. These particles are available for *H. siltalai* since the larvae have a net mesh opening of 242 µm (length x width). The average larva filters 492 L/day which will yield the amount of food in the first column in Table 2. The larvae used for the gut analyses had a mean weight of 5.9 mg dry mass. If the linear relationship between weight and consumption rate discussed earlier is used, then the consumption rate will be 85%. The average larva should eat 5.0 mg per day. To obtain this amount of food the larva will need to eat all animal, zooplankton, green algae, detritus + diatoms, detritus and part of the plant material. In column 3 in Table 2 is the expected food composition in % and in column 4 is the observed % gut composition obtained by area measurement from normal gut analysis. The difference in the percent food item found by the two methods is not different (Chi square test, $P > 0.05$). The biggest difference is in the detritus being 6% more by the area method. The reason for this is plant and animal material being misidentified and put under detritus. Zooplankton has been corrected for this.

Table 1. Ingestion rate % of body weight/day by three species of *Hydrospyche* on three food types commonly found in the seston. Reported as mean and standard deviation.

| Food type | H.ang. | H.sax. | H.silt. |
|---|---|---|---|
| Detritus+diatoms | 79 (33) | 55 (30) | 107 (39) |
| Ulothrix sp. | 22 (8) | - | 34 (7) |
| Daphnia magna | 35 (9) | - | 40 (16) |

Table 2. Seston composition, gut content for *H. siltalai* and expected gut content calculated from direct feeding observations.

| Particle type | seston mg/492L | consumed mg | consumed % | by area % |
|---|---|---|---|---|
| Animal | .12 | .12 | 2 | 1 |
| Zooplankton | .07 | .07 | 1 | 1 |
| Greenalgae | .07 | .07 | 1 | 3 |
| Detritus+diatoms | 2.97 | 2.97 | 59 | 58 |
| Detritus | 1.50 | 1.50 | 30 | 36 |
| Plant | 4.77 | .27 | 5 | 1 |

It can be concluded that it is possible to rank the food items in the seston by direct observations on the food selection of *Hydropsyche* larvae. Animal material is most preferred with detritus and plant material being lowest on the rank. With estimates of consumption rate and volume filtered it is possible to estimate food composition in the gut. This method of direct observation of food preference can be used as a complement to area based gut analysis.

ACKNOWLEDGEMENTS

This research was supported by grants from the Swedish Environmental Protection Board, the Swedish Society for the Conservation of Nature and the Royal Swedish Academy of Sciences. Special thanks go to R.C. Petersen for valuable comments during experiments and manuscript writing. This is a contribution from the Stream and Benthic Ecology Group, University of Lund.

REFERENCES

Cummins, K.W. 1973. Trophic relations of aquatic insects. **Annu. Rev. Entomol.**, 18 : 183-206.

Martin, I.D. & Mackay, R.J. 1982. Interpreting the diet of *Rhyacophila* larvae (Trichoptera) from gut analyses : an evaluation of techniques. **Can. J. Zool.**, 60 : 783-789.

McCullough, D.A., Minshall, G.W. & Cushing, C.E. 1979. Bioenergetics of lotic filter-feeding insects *Simulium* spp. (Diptera) and *Hydropsyche occidentalis* (Trichoptera) and their function in controlling organic transport in streams. **Ecology**, 60 : 585-596.

Mecom, J.O. & Cummins K.W. 1964. A preliminary study of the trophic relationships of the larvae of *Brachycentrus americanus* (Banks) (Trichoptera : Brachycentridae). **Trans. Am. Microsc. Soc.**, 83 : 233-243.

Pastorok, R.A. 1980. The effects of predator hunger and food abundance on prey selection of *Chaoborus* larvae. **Limnol. Oceanogr.**, 25 : 910-921.

Petersen, L.B.M. & Petersen, R.C. 1984. Effect of kraft pulp mill effluent and 4,5,6 Trichloroguaiacol on the net spinning behaviour of *Hydropsyche angustipennis* (Hydropsychidae : Trichoptera). **Ecol. Bull.**, 36 : 68-74.

Petersen, L.B.M. 1985. Food preferences in three species of *Hydropsyche* (Trichoptera). **Verh. internat. Verein. limnol.**, 22 : 3270-3274

Petersen, R.C. 1987 (in press). Seston quality as a factor influencing trichopteran populations. **Proc. 5th Int. Symp. Trichoptera**, Lyon, 21-26 July 1986, Junk.

Slack, H.D. 1936. The food of caddis fly (Trichoptera) larvae. **J. Anim. Ecol.**, 5 : 105-115.

Wallace, J.B. 1975. The larval retreat and food of *Arctopsyche* ; with phylogenetic notes on feeding adaptations in Hydropsychidae larvae (Trichoptera). **Ann. Entomol. Soc. Am.**, 68 : 167-173.

Wallace, J.B. & Meritt, R.W. 1980. Filter-feeding ecology of aquatic insects. **Ann. Rev. Entomol.**, 25 : 103-132.

Wallace, J.B., Webster, J.R. & Woodall, W.R. 1977. The role of filter feedings in flowing waters. **Arch. Hydrobiol.**, 79 : 506-532.

Wiggins, G.B. 1977. Larvae of the North American caddisfly genera (Trichoptera). Univ. Toronto Press, Toronto and Buffalo.

Xiang, J., von Schröder, P. & Schwoerbel, J. 1984. Phänologie und Nahrung der Larven von *Hydropsyche angustipennis* und *H. siltalai* (Trichoptera, Hydropsychidae) in einem Seeabfluss. **Arch. Hydrobiol. Suppl.**, 66 (2) : 255-292.

# A NEW ORDINATION TECHNIQUE FOR ECOLOGICAL PURPOSES APPLIED TO CADDIS LARVAE IN DITCHES

L.W.G. HIGLER, R. TORENBEEK and P.F.M. VERDONSCHOT

SUMMARY

A new ordination technique has been developed, combining characteristics of the ordination programme DECORANA and the classical regression techniques. In this technique (CANOCO) a graphic representation is given of combined species clusters and abiotic parameters. The method is applied to 30 species of Trichoptera in 323 samples from peaty ditches.

The resulting classification could not be explained by the factors sampling method, season or locality. As chemical and physical factors in the ditches are more or less the same, the observed distribution must be due to factors related to the different plant species or subhabitats. CANACO is able to indicate those factors in a very homogeneous set of data. It turnes out that characteristics of the substrate like softness, tangled or open, horizontal or vertical structure and the like are discriminating.

## INTRODUCTION

During a four-year period (1980-1983) investigations have been made in wet ditches in a peat polder in order to characterize the invertebrate fauna of different subhabitats. At the Clemson meeting a provisional report has been given on the presence of caddis larvae in these ditches (Higler 1984). Samples containing caddis larvae (a total number of 323) have been selected and the caddis combination of these samples has been treated.

The purpose of this article is to reveal the preference (if existing) of caddis species for certain subhabitats with the help of a newly developed ordination technique : CANOCO.

## MATERIAL AND METHODS

### Sampling

The fauna has been sampled in various ways and seasons, so that differences in the final results can originate from different sources : 1. several methods have been used such as a handnet, artificial substrate, the isolation of a water column (including plants and bottom) over a fixed surface and removal of fixed volumina of the bank including roots of grasses and waterplants
2. the samples have been taken in all seasons and
3. in a couple of ditches (locations)
4. and finally from a variety of subhabitats, hereafter referred to as substrates. These last comprise the bottom, the banks just under the water surface, filamentous algae, several plant species and artificial substrates consisting of horizontally placed double gauzes. The samples have been taken from fixed volumina or surfaces and are mutually comparable.

### Data processing

The processing of the data is based upon a matrix of 30 caddis species and 323 samples. The analysis has been performed with 1. methods, 2. seasons, 3. locations and 4. substrates. Of these only the substrates proved to explain a substantial part of the variation, so further processing was restricted to the above mentioned substrate types such as bottom, *Typha* etc. Classification of samples was made by flexible clustering, using the program FLEXCLUS (Tongeren 1982).

The data were ordinated by detrended canonical correspondance analysis (DCCA), using the program CANOCO (Ter Braak 1985). DCCA is an extension of DCA (DECORANA ; Hill 1979), because the ordination axes are chosen in the light of the environmental variables. DCCA leads to an ordination diagram in which species (dots) and variables (arrows) are represented. The arrow of an environmental variable points into the direction of the steepest increase of weighted averages of the species for that variable and the rate of change in that direction is equal to the length of the arrow. This means that the relationship of a set of species (or one species) with one or more important environmental variables is visualised by its perpendicular projection on the environmental arrow or the imaginary extension in both directions of the arrow. Within the diagram the species and arrows can be seen as relative projections upon each other. Species-environment correlations of the axes indicate the degree of explanation , through the variables, of the variation within the data set.

## RESULTS

Thirty species have been collected, among which the Polycentropodidae and the Leptoceridae are the most important. The three most frequently occurring species belong to the Polycentropodidae, followed by three species of Leptoceridae. Among these last are the two most abundant ones. Hydroptilidae is the third important family with a relatively high frequency (Table 1). The family Limnephilidae is scarsely represented in the ditches. *Anabolia nervosa* is the most frequent and *Limnephilus lunatus* the most abundant limnephilid. Both species are very common in lowland streams and the shores of lakes (just like *Molanna angustata*). Phryganeids are also rather scarce. A total number of thirty species, however, is very high for stagnant waters in The Netherlands (Higler 1980).

Table 1. Collected caddis species with frequency and abundance data.

| SPECIES | ABBREVATION | FREQUENCY | ABUNDANCE |
|---|---|---|---|
| *Agraylea multipunctata* | AGRAMULT | 41 | 9 |
| *Agraylea sexmaculata* | AGRASEXM | 6 | 3 |
| *Orthotrichia costalis* | ORCHCOST | 3 | 3 |
| *Oxyethira flavicornis* | OXYEFLAV | 25 | 7 |
| *Tricholeiochiton fagesii* | TRLEFAGE | 15 | 3 |
| *Athripsodes aterrimus* | ATHRATER | 53 | 32 |
| *Ceraclea fulva* | CERAFULV | 5 | 1 |
| *Ceraclea senilis* | CERASENI | 4 | 1 |
| *Leptocerus tineiformis* | LECETINE | 1 | 1 |
| *Mystacides longicornis* | MYSTLONG | 62 | 15 |
| *Oecetis furva* | OECEFURV | 15 | 3 |
| *Oecetis lacustris* | OECELACU | 7 | 2 |
| *Oecetis ochracea* | OECEOCHR | 2 | 1 |
| *Triaenodes bicolor* | TRIABICO | 62 | 31 |
| *Cyrnus crenaticornis* | CYRNCREN | 81 | 12 |
| *Cyrnus flavidus* | CYRNFLAV | 85 | 10 |
| *Holocentropus dubius* | HOLODUBI | 37 | 6 |
| *Holocentropus picicornis* | HOLOPICI | 75 | 13 |
| *Ecnomus tenellus* | ECNOTENE | 5 | 2 |
| *Phryganea grandis* | PHRYGRAN | 2 | 1 |
| *Phryganea bipunctata* | PHRYBIPU | 14 | 2 |
| *Agrypnia pagetana* | AGRYPAGE | 7 | 2 |
| *Molanna angustata* | MOLAANGU | 13 | 7 |
| *Anabolia nervosa* | ANABNERV | 15 | 3 |
| *Limnephilus affinis* | LILUAFFI | 1 | 2 |
| *Limnephilus flavicornis* | LILUFLAV | 2 | 1 |
| *Limnephilus lunatus* | LILULUNA | 9 | 14 |
| *Limnephilus marmoratus* | LILUMARM | 11 | 6 |
| *Limnephilus politus* | LILUPOLI | 2 | 2 |
| *Limnephilus rhombicus* | LILURHOM | 1 | 1 |

Figure 1 shows the result of the data processing with the DCCA technique CANOCO in combination with the visualization of the species clusters as derived from the FLEXCLUS classification. These clusters are shown as elliptic planes and they are characterized shortly in table 2. The arrangement of the planes corresponds to the main arrows, which means that there is a high correlation with these variables.

The application of CANOCO thus shows a direct relationship between species and variables, where the original axes are used as a mere orientation for the projection of the species and the clusters. The real centre of the figure is the centroid and instead of the ordination axes the arrows have an explanatory function. The correlation of the arrows (substrate variables) with the ordination axes 1 and 2 is given in table 3.

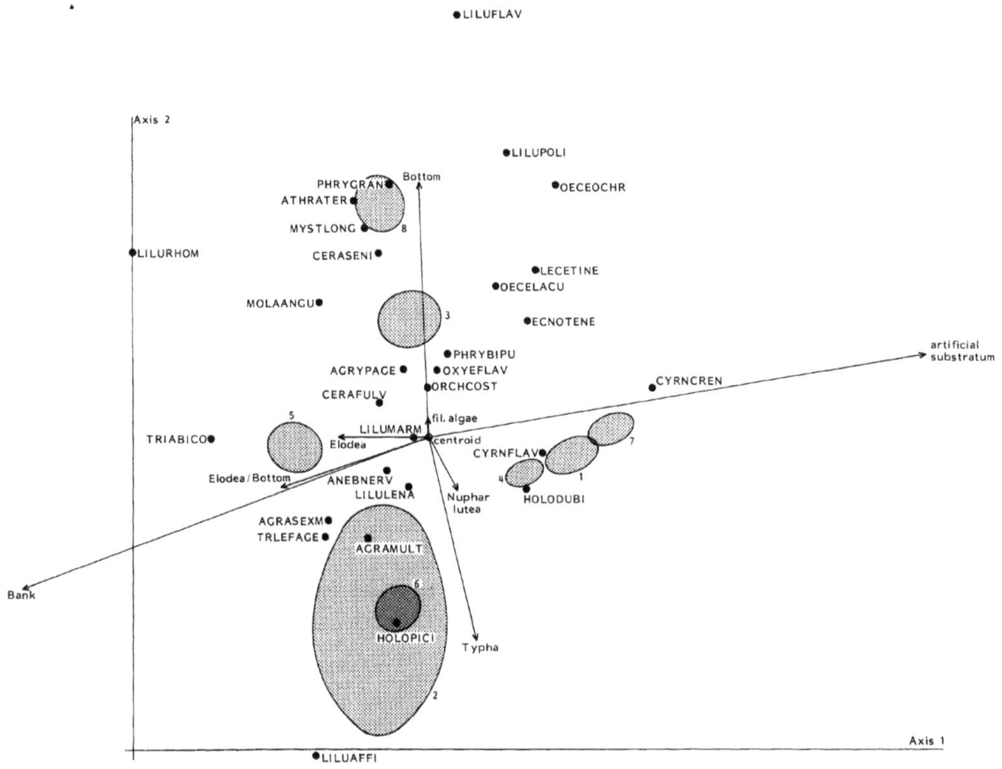


Figure 1. Ordination (CANOCO) diagram for the axes 1 and 2 with caddis larvae (dots), environmental variables (arrows) and the 90% reliability ellipses of the mean of the species scores per cluster. Codes are explained in table 1.

## DISCUSSION

The main objective of this article is to demonstrate how a new ordination technique (CANOCO) can be used for the interpretation of observed regularities in the distribution of caddis larvae in wet ditches. It has turned out that the resulting classification cannot be explained by the factors sampling method, season or locality (different ditches is one polder system). As dimensions and chemical composition of the water at the sampling stations are rather similar, the observed distribution must be due to factors related to the subhabitats or substrates.

The substrate types have been described as *Typha*, Bottom, Bank etc. with an idea that the eye of the caddis fly and the eye of the scientist registrate the same important features. If it is stated that a relationship exists between a caddis species and a certain aquatic plant species, this does not give any explanation. It only gives an indication of a certain subhabitat preference and the next step must be the investigation of food relationships, of hiding place requirements etc. A lot of autecological work must be done to understand more of the preferences of species at various instars and life stages.

Having selected substrate types as discriminating characteristics, figure 1 shall be interpreted in this light.There are four main arrows forming a new system of axes more or less parallel to the original one. The arrows of the plant species, with the exception of *Typha*, are short and they represent the overall situation in the ditches. This corresponds with the position of many of the more common species near the centroid. It is also in accordance with the difficulty to find significantly different aggregations of caddis larvae. The existing differences in the material are explained by the longer arrows of which some characteristics have been mentioned in table 4.

Table 2. Significance of the species clusters derived from FLEXCLUS.

| CLUSTER | ASSOCIATED WITH | REMARKS ABOUT SPECIES |
|---|---|---|
| 1,4,7 | Artificial substrate | |
| 1 | | Few species; predominantly Cyrnus spp. in low numbers; Mystacides and Athripsodes are absent |
| 4 | | Rich in all Polycentropodidae and Agraylea multipunctata and Oxyethira; Mystacides and Athripsodes present |
| 7 | | Like cluster 4, but both Cyrnus spp. in very high numbers |
| 2,6 | Typha spp. | |
| 2 | | Very poor in species. Only Holocentropus picicornis, Agraylea multipunctata and Anabolia nervosa present |
| 6 | | Most species present; Holocentropus picicornis is dominant |
| 5 | Banks | Rich cluster; high frequency of Oecetis furva, Orthotrichia costalis and Tricholeiochiton fagesii. Agraylea multipunctata, Oxyethira flavicornis and Triaenodes bicolor are most abundant in 5 |
| 3,8 | Bottom | |
| 3 | | Rich in species; average composition |
| 8 | | Like cluster 3, but the following spp. have a high frequency only in cluster 8: Oecetis lacustris, Ceraclea fulva and C. senilis, Phryganea bipunctata. Limnephilus flavicornis is only found in this cluster |

Table 3. Correlations of substrate with ordination axes 1 and 2.

| SUBSTRATE | AXIS 1 | AXIS 2 |
|---|---|---|
| Elodea | -0.081 | 0.083 |
| Elodea/Bottom | -0.266 | -0.146 |
| Typha | 0.080 | -0.263 |
| Filamentous algae | -0.025 | 0.008 |
| Nuphar lutea | 0.024 | -0.086 |
| Bottom | -0.052 | 0.348 |
| Bank | -0.554 | 0.120 |
| Artificial substrate | 0.708 | -0.168 |

Table 4. Some characteristics of the main arrows in fig. 1.

| SUBSTRATE | DIRECTION | POSITION | STRUCTURE | CONSISTENCY |
|---|---|---|---|---|
| Artif. substrate | horizontal | in free water | half open | solid + filamentous algae |
| Bank | vertical | protected | tangled | firm |
| Bottom | horizontal | protected | less tangled | firm/soft + filam. algae |
| Typha | vertical | in free water | open | solid |

The description of these characteristics is rather primitive as a result of the incapacity of pointing out the relevant ones for the caddis larvae. It is clear that an unprotected position in free water is very dangerous because of predators. Moreover many species do not swim. It seems less attractive to climb the bare stems of *Typha* than to creep around through loosely arranged peat and plant fragments on the bottom. A disadvantage is that oxygen conditions can be there very poor.

By combining the characteristics themselves, combinations of alternatives in the quadrants are more likely than along the arrows. The upper left quadrant contains the firm material of bottom and bank with many hiding places. Near the Bank arrow vertical structures are discriminating and this accounts also for the *Elodea* and *Elodea* /Bottom arrows. In contrast the right down quadrant is characterized by solid structures in free water with relatively few hiding places. It is the quadrant of the Polycentropodidae. Especially *Cyrnus* species prefer water with greater dimensions. Here also the directions horizontal and vertical play a role. In the upper right quadrant horizontal structures with filamentous algae can be found with in addition the nearness of the open water. The alternative is the lower left quadrant with vertical structures. Shelter places in the banks are manyfold and there are many young stages to be found between the roots of grasses and herbs. *Typha* stems offer little protection. It can be concluded that the DCCA technique enables to discern those variations in the homogeneous material that do exist and that it points out that structural characteristics of the subhabitat must be described in another way than hitherto has been practised.

## REFERENCES


Higler, L.W.G. 1980. The presence of caddis flies in The Netherlands and their role in the aquatic system. **Hydrobiol. Bull.**, 14 : 209-212.

Higler, L.W.G. 1984. Caddis larvae in ditches. **Fourth Internat. Symp. Trichopt.**, J. C. Morse (ed), Junk, The Hague : 175-178.

Hill, M.O. 1979. DECORANA - A Fortran program for detrended correspondence analysis and reciprocal averaging. Ecology and systematics, Cornell University, Ithaca, New York.

Ter Braak, C.J.F. 1985. CANOCO - A Fortran program for canonical correspondence analysis and detrended correspondence analysis. IWIS-TNO, Wageningen, The Netherlands.

Tongeren, O. van. 1982. FLEXCLUS, an interactive flexible clusterprogram. Nijmegen

# LES PEUPLEMENTS LARVAIRES DE TRICHOPTERES DES ANCIENS LITS FLUVIAUX DANS TROIS SECTEURS DE LA PLAINE ALLUVIALE DU HAUT-RHONE FRANCAIS

C. ROUX et E. CASTELLA

## SUMMARY

Larval Trichoptera populations were sampled during spring and summer in 13 former channels of the upper Rhône River and in 7 former channels of its tributary, the Ain River, upstream from Lyon (France). From 242 hand-net samples, which provided 66 Trichoptera taxa, we compared former channels of various morphological pattern, hydrological regime and successional stage in three sectors of the alluvial plain. The former channels can be set into 3 main groups according to their Trichoptera fauna : 1) the stagnant water bodies, weakly influenced by floods characterized by lenitic species (e.g. *Leptocerus tineiformis*); 2) an intermediate group is made up of the former channels (mostly connected with the river downstream) with varying flow conditions and local ground water inputs ; 3) channels highly influenced by floods, with strong ground water inputs and mostly lotic conditions where small stream species (e.g. *Plectrocnemia conspersa*) or species from the main channel (e.g. *Hydropsyche*) can be found. The fauna of group 2) consists mainly of Limnephilidae and of elements from the two other groups. Eight species of Limnephilidae can be set along the ordination of the former channels depicted by the analysis. This overall extensive approach complements related research, such as light trapping of adult Trichoptera and faunal investigations in the main channel of the river.

## INTRODUCTION

La plaine alluviale des grands fleuves est souvent caractérisée par la présence d'écosystèmes très diversifiés, depuis l'axe fluvial et ses annexes lotiques, jusqu'aux milieux semi aquatiques et terrestres qui résultent de l'abandon par le fleuve de certaines portions de son cours et de leur atterrissement progressif (Amoros *et al.* sous-presse). Entre ces deux termes extrêmes de la succession, divers types d'écosystèmes aquatiques, les bras-morts fluviaux, conservent des liens plus ou moins directs avec le cours actif et peuvent donc demeurer soumis à sa dynamique.

Le présent travail concerne les écosystèmes aquatiques abandonnés par le Rhône dans deux secteurs de son cours, et par l'Ain peu avant sa confluence avec le Rhône en amont de Lyon (Fig. 1). Des études antérieures (Castella *et al.* 1984 , Richardot-Coulet *et al.* 1983) ont montré que certains groupes d'invertébrés aquatiques (Mollusques, Odonates, Coléoptères, Trichoptères) peuvent être utilisés pour caractériser des types de bras-morts fluviaux ou comme témoins des processus évolutifs qui s'y déroulent.

Nous nous proposons de comparer, à partir de leurs peuplements en larves de Trichoptères, des écosystèmes qui diffèrent notamment par leur âge, leur mode d'évolution et leur fonctionnement hydrologique. Nous recherchons comment ces peuplements peuvent être le reflet de la diversité des anciens lits fluviaux. Pour cette comparaison, nous avons considéré chaque ancien lit comme une unité et nous n'avons pas cherché à décrire leur hétérogénéité interne. Cette étude n'est pas un recensement faunistique exhaustif mais la recherche du type d'information que peut apporter la méthode d'échantillonnage des peuplements de Trichoptères utilisée ici.

## SITES ET METHODE

20 anciens lits ont été choisis dans trois secteurs de la plaine alluviale du Haut-Rhône français (Fig. 1) ; le secteur de Jons-Villette d'Anthon A (6 anciens lits, J1 à J6) et le secteur de Brégnier-Cordon C (7 anciens lits, B1 à B7), deux secteurs où le cours actif actuel du Rhône est largement endigué et subit les impacts des ouvrages hydro-électriques ; enfin le secteur du bas-cours de l'Ain B (7 anciens lits, A1 à A7) caractérisé par des crues puissantes et soudaines et une dynamique géomorphologique très active. Les successions écologiques s'y déroulent à l'échelle de la décennie, alors que sur le Rhône, fixé par les endiguements, elles ont plutôt lieu à l'échelle du siècle.

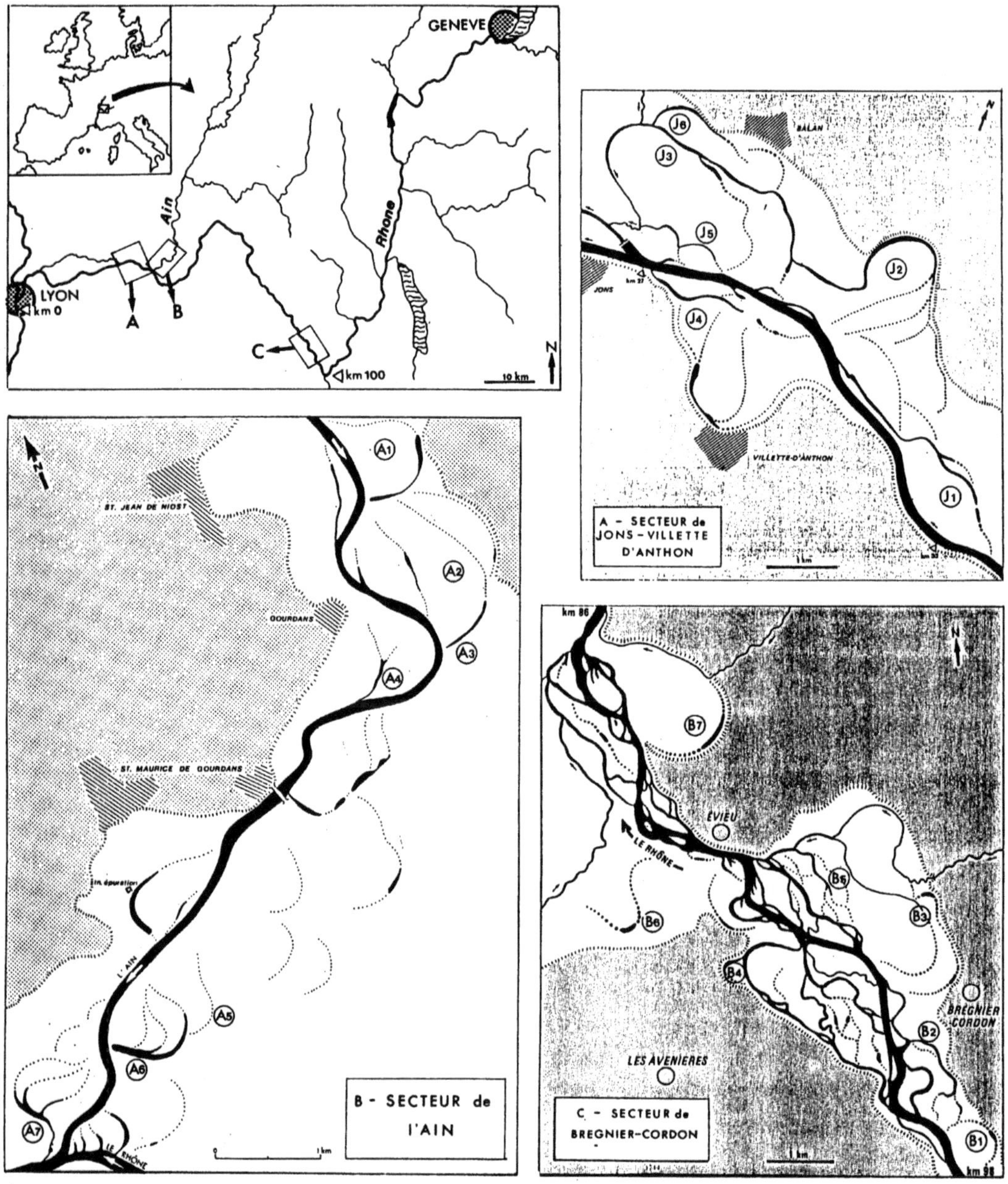


Figure 1. Le Haut-Rhône français et les trois secteurs d'étude.
*The french Upper Rhône and the 3 areas studied.*

L'échantillonnage des larves de Trichoptères a été effectué à l'épuisette (maille 500µ) qui constitue une bonne technique pour des biotopes très riches en macrophytes et relativement peu profonds (jusqu'à 1,5-2 m). Un prélèvement est constitué par la réunion de trois échantillons élémentaires effectués en transect, perpendiculairement à l'axe de l'ancien lit. Les données ont été acquises en 1981, 83 et 85, au printemps (avril-mai) et en été (juin-juillet). Les analyses préliminaires ont montré que les différences entre saisons étaient négligeables par rapport aux différences stationnelles, nous avons donc regroupé les données obtenues aux deux saisons pour chaque ancien lit.

Les effectifs sont exprimés en 4 classes d'abondance ; 1 = 1 individu, 2 = 2 à 10 ind., 3 = 11 à 100 ind., 4 = plus de 100 ind. Considérant ici chaque ancien lit comme une unité, nous n'avons conservé que la moyenne de tous les relevés effectués sur un même lit. Ainsi, partant d'un tableau initial de 242 relevés et 66 taxons, nous avons en définitive analysé le tableau 20 anciens lits x 66 taxons dans lequel est attribuée à chaque taxon la moyenne de ses notes d'abondance pour chaque ancien lit.

L'analyse statistique utilisée est "l'analyse générale" de Lebart *et al.* (1977, p.9 à 22) du tableau X comportant n lignes et p colonnes relevés, de terme général $x_{ij} = P_{i/j} - 1/P$ où $P_{ij}$ est la proportion de l'abondance totale du taxon i, rencontrée dans le relevé j. La quantité $x_{ij}$ caractérise alors l'écart entre l'espèce i et l'espèce "ubiquiste" uniformément présente dans tous les relevés (Chessel, com. pers).

## RESULTATS ET DISCUSSION

Le positionnement des 20 anciens lits sur les trois premiers axes factoriels issus de l'analyse permet de comparer ces écosystèmes d'après leur peuplement. Le premier axe factoriel f1 distingue trois groupes d'anciens lits (Fig. 2) ; -pour f1 < 0: B1, B3, B5, B6, B7 ; - pour f1 = 0 : A1, A3, A5, A6, B2, B4, J2, J3, J4, J6 ; - pour f1 > 0 : A2, A4, A7, J1, J5; le second axe f2 subdivise ce dernier groupe (Fig. 2), pour f2 < O: A7, J1; pour f2 >0 : A2, A4, J5.

Le troisième axe f3, non représenté, décrit l'originalité de J2 et J3.

Les trois premiers facteurs de l'analyse permettent donc d'établir une typologie en 5 groupes des anciens lits étudiés ;

- groupe 1 : B1, B3, B5, B6, B7
- groupe 2A : A1, A3, A5, A6, B2, B4, J4, J6
- groupe 2B : J2, J3
- groupe 3A : A2, A4, J5
- groupe 3B : A7, J1

La figure 2 propose une légende faunistique partielle de cette typologie en visualisant sous forme de carrés l'abondance de 16 espèces caractéristiques. Le fond de carte commun aux 16 cartes de répartition est le plan factoriel f1 x f2 des 20 anciens lits (logiciel graphique de Auda 1983). La légende faunistique complète est fournie par le tableau 1 des données initiales, qui a été réorganisé pour faire apparaître la typologie des anciens lits.

On remarque tout d'abord que l'analyse ne regroupe pas les anciens lits prospectés selon un découpage par secteurs géographiques, ni autour des deux cours principaux : le Rhône et l'Ain.

Le **groupe 1** correspond à des milieux d'origine variable (méandres ou chenaux d'anastomose), plans d'eau stagnante plus ou moins profonde (jusqu'à 2m.), riches en végétation au moins littorale, en voie d'atterrissement, complètement séparés du chenal vif, parfois alimentés par des arrivées d'eau souterraines (B1) ; s'y adjoint la "Lône du Mortier" (B5), qui bien qu'en liaison avec un affluent du fleuve, présente des caractères d'atterrissement et une tendance à l'état lénitique. Dans le tableau réorganisé 1, le peuplement I explique le regroupement de ces milieux. Il s'agit d'espèces inféodées à des zones riches en végétation aquatique en eau peu profonde, stagnante ou faiblement courante, les espèces les plus caractéristiques sont *Leptocerus tineiformis* et les deux espèces d'*Holocentropus*. Ces espèces peuvent apparaître dans les peuplements d'autres bras-morts, elles sont alors localisées dans les zones amont, stagnantes et en voie d'eutrophisation. Higler (1977) a montré que les espèces d'*Holocentropus* colonisent la végétation littorale en faible profondeur ; de même plusieurs auteurs ont signalé *Leptocerus tineiformis* comme une espèce d'eau stagnante, de réservoir ou d'étangs couverts de végétation (Lepneva 1970-71, Wallace 1981, Moretti 1983). Dans l'analyse des piègeages lumineux d'adultes de Trichoptères du Haut-Rhône français, Rojas-Camousseight (1985) a mis en évidence la liaison de l'espèce *Oecetis furva* avec les milieux anciens plus ou moins fermés ou en voie d'eutrophisation.

A ce premier groupe, s'oppose un type de bras-mort (**groupe 3**) en liaison plus ou moins directe avec le fleuve par l'aval (A4, A2, J5, J1 et A7), soumis à l'influence des crues, donc à des remaniements périodiques et/ou alimentés par des eaux phréatiques abondantes, qui donnent souvent à la partie moyenne de ces bras des allures de petits ruisseaux de plaine, sur le tableau 1 les groupements d'espèces IV, V, VI, caractérisent ces anciens lits. Le groupe IV composé de *Chaetopteryx villosa, Sericostoma personatum* et *galeatum, Plectrocnemia conspersa, Polycentropus irroratus, Hydroptila martini* peuple l'ensemble des bras-morts du groupe 3. Ces espèces sont généralement abondantes et constituent "une faune classique de petite rivière".

Les peuplement V et VI sont composés de 16 espèces à petit nombre d'individus, localisées uniquement dans J1 et A7. Ce sont des espèces de rivières rapides et ont, pour la plupart été retrouvées en quantités importantes dans le cours principal du Rhône ou de l'Ain dont elles constituent les hôtes habituels soit dans le

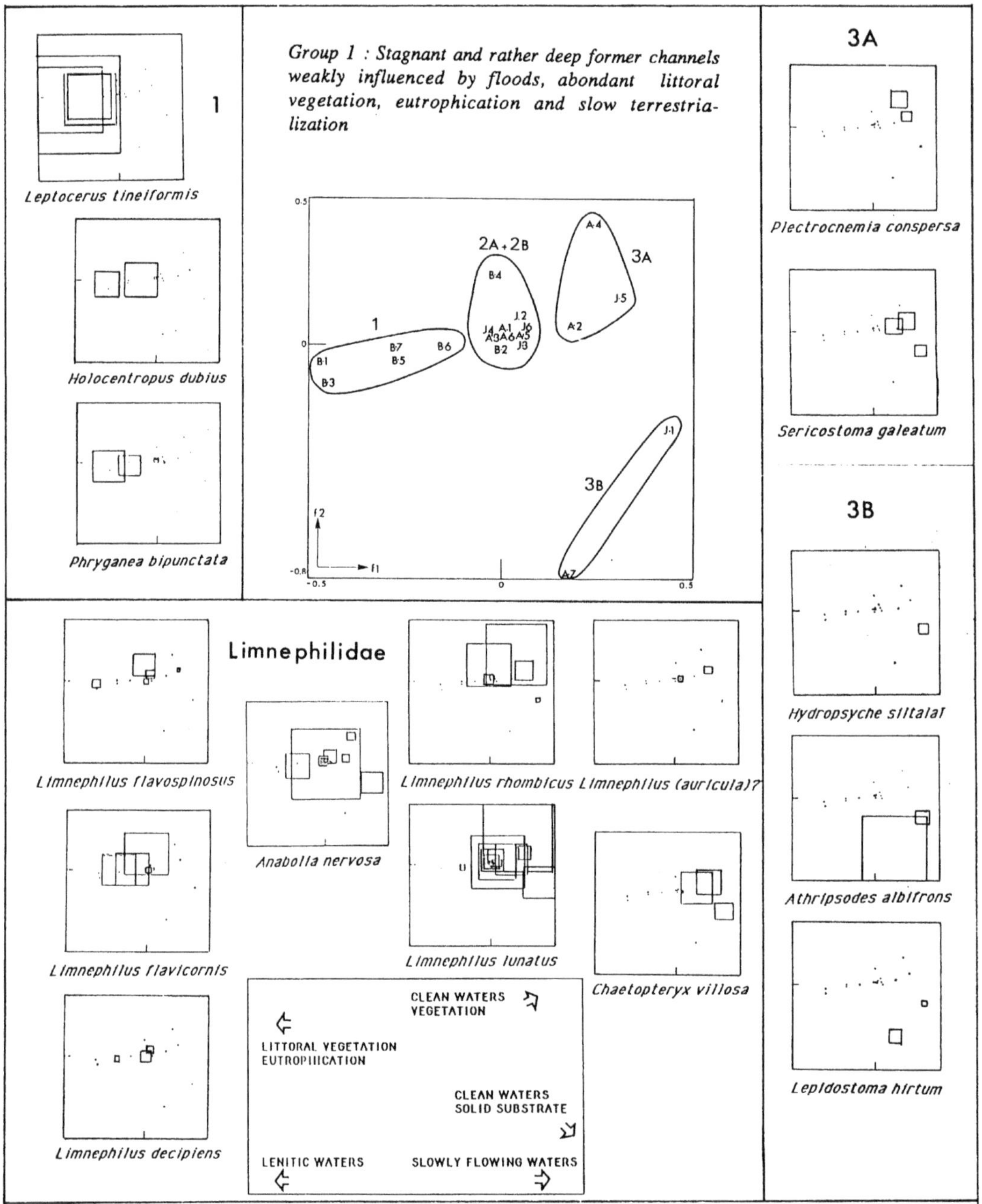


Figure 2. Plan factoriel f1 x f2 des 20 anciens chenaux et répartition de quelques espèces caractéristiques (voir le texte et le tableau 1). Logiciel graphique d'Auda (1983).
*f1 x f2 factorial map of the 20 former channels and the distribution of some distinctive species (see tabl. 1).*
*Group 2 : rather long and diversified former channels, mostly connected with the active channel. Local ground water inputs. 2B (J2 and J3) : peaty sediments and dystrophic conditions.*
*Group 3 : Channels with flowing water and local important ground water inputs. Strong influence of floods.*

Tableau 1 - Tableau des données réorganisées (voir le texte).

*66 Trichoptera species from 20 former channels of the Rhône and the Ain rivers.*

*I : lenitic fauna, associated with aquatic vegetation, shallow water of the littoral zones and eutrophication.*

*II : "ubiquitous species"*

*III : Taxa distinctive for J2 and J3.*

*IV : Fauna usually found in brooks and small streams*

*V and VI : Rheophilous potamic species coming from the main channel or its banks 1: present ; 2: a few ; 3: abundant ; 4: very abundant species.*

| | | B1 | B3 | B7 | B5 | B8 | B4 | J4 | A3 | A1 | B2 | A6 | A5 | J6 | J2 | J3 | A2 | A4 | J5 | J1 | A7 |
|---|---|---|---|---|---|---|---|---|---|---|---|---|---|---|---|---|---|---|---|---|---|
| Leptocerus tineiformis | | 4 | 4 | 4 | 4 | | | | | | | | | | | | | | | | |
| Holocentropus dubius | | 3 | | | | 4 | | 1 | | | | | | | | | | | | | |
| Phryganea bipunctata | I | 4 | | | 3 | | | 2 | | | | | | | | | | | | | |
| Oecetis furva | | 2 | | 2 | | | | | | | | 1 | | | 2 | | | | | | |
| Agraylea sexmaculata | | | 3 | | | | | | | 1 | | | | | 1 | | | | | | |
| Holocentropus picicornis | | | | 2 | | | | 1 | | | | 1 | | | | | | | | | |
| Limnephilus flavospinosus | | 2 | | | | | 3 | | | | 2 | 1 | | | 1 | | | | 1 | | |
| Limnephilus flavicornis | | | | 4 | | 4 | 4 | | | 1 | 2 | | | | | 1 | | | | | |
| Agrypnia varia | | | | 2 | | | | | 2 | | | | | | | | | 2 | | | |
| Limnephilus decipiens | | | | | 2 | | | | | 1 | 2 | | | | 2 | | | | | | |
| Athripsodes aterrimus | II | 4 | | | 4 | | 3 | 3 | 4 | 3 | | 4 | | | 2 | 2 | 4 | 4 | 3 | 2 | 2 |
| Anabolia nervosa | | | | | 3 | | | 2 | 2 | | 4 | 1 | | | 2 | 1 | 2 | 2 | | 3 | |
| Limnephilus gr. lunatus | | | | | 2 | | | 1 | 2 | 3 | 2 | 2 | 4 | 4 | 3 | 2 | 4 | 4 | 2 | 3 | |
| Limnephilus rhombicus | | | | | | | 4 | 2 | | 2 | | | | | | | | 4 | 3 | 2 | |
| Glyphotaelius pellucidus | | | | | | | 2 | 1 | | 1 | | 1 | | | | | | | | | |
| Oxyethira costalis | | | | | | | | 1 | | | | | | | | | | | | | |
| Cyrnus other sp. | | | | | | | | 2 | | | | | | | | | | | | | |
| Cyrnus trimaculatus | | | | | | | | 2 | | | | | | | 1 | | | | | | |
| Erotesis baltica | | | | | | | | 1 | | | | | | | 1 | | | | | | |
| Mystacides azurea | | | | | | | | 1 | | | | | | | 2 | 1 | | | | 1 | |
| Lype phaeopa | | | | | | | | 2 | | | | | | 2 | 1 | 2 | | | 2 | | |
| Hydroptila pulchricornis | | | | | | | | | | 1 | | | | | | | | | | | |
| Oxyethira other sp. | | | | | | | | | | 1 | | | | | | | | | | | |
| Cyrnus crenaticornis | | | | | | | | | | 1 | | | | | | | | | | | |
| Beraea pullata | | | | | | | | | | 1 | | | | 2 | | | | | | | |
| Triaenodes bicolor | | | | | | | | | | 1 | | 2 | | | 1 | | | 2 | | | |
| Limnephilus (binotatus) ? | | | | | | | | | | 2 | | 1 | | | | | | | | | |
| Cyrnus flavidus | | | | | | | | | | | | 1 | | | 2 | | | 2 | | | |
| Agrypnia pagetana | | | | | | | | | | | | | | | 2 | | | 2 | | | |
| Beraeodes minutus | | | | | | | | | | | | | | | 2 | | | | | | |
| Oecetis testacea | | | | | | | | | | | | | | | 2 | | | | | | |
| Ecnomus tenellus | | | | | | | | | | | | | | | 2 | | | | | | |
| Neureclipsis bimaculata | | | | | | | | | | | | | | | 1 | | | | | | |
| Mystacides longicornis | III | | | | | | | | | | | | | | 1 | | | | | 1 | |
| Oxyethira tristella | | | | | | | | | | | | | | | 2 | 3 | | | | | |
| Hydroptila sparsa | | | | | | | | | | | | | | | 2 | 2 | | | 1 | 2 | |
| Athripsodes cinereus | | | | | 1 | | | | | | | | | | 1 | 2 | | | | 1 | |
| Hydroptila other sp. | | | | | | | | | | | | | | | | 2 | | | | | |
| Hydropsyche angustipennis | | | | | | | | | | | | | | | | 3 | | | | | |
| Ithytrichia lamellaris | | | | | | | | | | | | | | | | 2 | | | | | |
| Phryganea grandis | | | | | | | | | | | | | | | | | 2 | | | | |
| Halesus gr. radiatus | | | | | 1 | | | | | | | | | | | 2 | | | 1 | 1 | |
| Limnephilus (auricula) ? | | | | | | | | | | | | | | 2 | | | | | 2 | 1 | |
| Hydroptila martini | | | | | | | | | | | | | | | 1 | | | | 2 | 2 | |
| Chaetopteryx villosa | | | | | | | | | | | | | | | | 1 | 4 | | 3 | 3 | |
| Sericostoma galeatum | | | | | | | | | | | | | | | | | 3 | 3 | 2 | 1 | |
| Plectrocnemia conspersa | | | | | | | | | | | | | | | | | | 3 | 2 | 1 | |
| Sericostoma personatum | IV | | | | | | | | | | | | | | | | | | 2 | | |
| Polycentropus irroratus | | | | | | | | | | | | | | | | | | | 2 | 1 | |
| Polycentropus flavomaculatus | | | | | | | | | | | | 1 | | | | | | | | 2 | |
| Hydropsyche siltalai | | | | | | | | | | | | | | | | | | | | 2 | |
| Synagapetus dubitans | | | | | | | | | | | | | | | | | | | | 2 | |
| Silo nigricornis | | | | | | | | | | | | | | | | | | | | 2 | |
| Psychomyia pusilla | | | | | | | | | | | | | | | | | | | | 2 | |
| Metalype fragilis | | | | | | | | | | | | | | | | | | | | 1 | |
| Hydropsyche modesta | V | | | | | | | | | | | | | | | | | | | 1 | |
| Cheumatopsyche lepida | | | | | | | | | | | | | | | | | | | | 1 | |
| Adicella reducta | | | | | | | | | | | | | | | | | | | | 1 | |
| Leptocerus interruptus | | | | | | | | | | | | | | | | | | | | 1 | |
| Ceraclea sp. | | | | | | | | | | | | | | | | | | | | 1 | |
| Rhyacophila fasciata | | | | | | | | | | | | | | | | | | | | 1 | |
| Rhyacophila gr. dorsalis | | | | | | | | | | | | | | | | | | | | 1 | |
| Lepidostoma hirtum | | | | | | | | | | | | | | | | | | | | 2 | 2 |
| Athripsodes albifrons | VI | | | | | | | | | | | | | | | | | | | 2 | 4 |
| Setodes argentipunctellus | | | | | | | | | | | | | | | | | | | | 1 | 2 |
| Oecetis lacustris | | | | | | | | 1 | | | | 1 | | | 1 | | | | | | 2 |
| Nombre de relevés / Number of samples | | 4 | 2 | 3 | 12 | 2 | 3 | 13 | 10 | 14 | 8 | 32 | 3 | 7 | 23 | 14 | 4 | 8 | 12 | 65 | 5 |

chenal, soit sur les rives ; ce sont par exemple : *Hydropsyche modesta, Polycentropus flavomaculatus, Psychomyia pusilla, Cheumatopsyche lepida,* des genres comme *Adicella* et *Ceraclea* et enfin *Lepidostoma hirtum, Athripsodes albifrons* et *Setodes argentipunctellus. Silo nigricornis* très localisé marque une abondante arrivée phréatique (Tachet et Bournaud 1981, Usseglio-Polatera1985, Gaschignard-Fossati 1986).

Cet ensemble montre l'influence directe du cours principal sur les anciens lits situés à proximité des axes fluviaux et constamment soumis aux crues. Si chaque espèce prise isolément parait peu significative du fait de la présence d'un très petit nombre d'individus par relevé, l'addition de toutes ces espèces apporte une information significative sur le fonctionnement de ces anciens lits et tout particulièrement sur leurs relations avec le cours actif du fleuve. La plupart des espèces citées sont accidentelles dans de tels milieux alors que dans le fleuve ou dans des conditions favorables, elles peuvent être représentées par des populations localisées de plusieurs centaines d'individus (par ex.: *Psychomyia pusilla, Cheumatopsyche lepida* ou *Hydropsyche modesta.* La faune du "Méant" J1 illustre ce phénomène. Elle est composée dans sa totalité de deux peuplements distincts : une faune de courant lent (IV) bien établie et stable, à laquelle on peut adjoindre des "ubiquistes" tels que *Anabolia nervosa* ou des Limnephilidae du groupe *lunatus*, et une faune "importée" pauvre en individus, riche en espèce (V).

Situé à la confluence de l'Ain et du Rhône, Puits Novet (A7) est soumis à des fluctuations hydrologiques intenses par suite d'arrivées de flux brutaux et apériodiques lors des crues de l'Ain entrainant un balayage du milieu, auquel se superposent des processus d'atterrissement et des phases d'étiage (Castella *et al.* sous-presse). De telles conditions pourraient expliquer sa pauvreté en larves de Trichoptère (VI) ; trois des quatre espèces présentes pourraient avoir été amenées par les crues à partir du cours principal.

Entre ces deux groupements extrêmes, I d'une part , IV V VI d'autre part, le peuplement II apparait composé d'ubiquistes, présents dans presque tous les anciens lits étudiés, très liés à la végétation des eaux lentes ou très lentes, comme par exemple *Athripsodes atterimus* et la majorité des larves de Limnephilidae.

Les larves de Limnephilidae sont bien représentées dans les milieux prospectés. Nous avons regroupé sous le terme de " *Limnephilus* groupe *lunatus*" trois espèces : *L. lunatus, L. helveticus* et *L. borealis,* dont la différenciation est difficile d'une part à cause de la variabilité de coloration de la tête de la larve de *L. lunatus* et d'autre part du fait de la ressemblance entre *L. helveticus* et *L. borealis.* Néanmoins les trois espèces ont été trouvées à l'état adulte ou de nymphes âgées, tant sur le terrain au cours des campagnes de prélèvement, que dans les pièges lumineux. Elles ont toujours été récoltées ensemble dans des proportions variables ; toutefois *L. helveticus* semble dominer nettement dans le Méant (J1). Moretti et Spinelli Batta (1984) lui attribuent une valeur écologique significative comme espèce limnocrène, liée aux étendues d'eau lotiques ou lénitiques alimentées par des fissures et riches en végétation. La figure 2 montre que l'analyse ordonne les Limnephilidae selon un gradient eau stagnante/atterrissement (*L. flavospinosus, L. flavicornis, L. decipiens*) vers des eaux claires renouvelées, plus ou moins courantes, riches en végétaux (*L. rhombicus*) et/ou sur fond solide tels que sable, cailloux, graviers (*Chaetopteryx villosa*). Une étude plus fine montre que, sauf dans J1, *L.* g. *lunatus et Anabolia nervosa* sont rarement ensemble tout au long d'un même ancien lit. Si tous les deux recherchent les zones ensoleillées, *A. nervosa* semble plus tolérant à l'eau stagnante et au substrat plus chargé en matière organique .

**Le groupe central 2** (2A et 2B) correspond à des anciens lits parfois longs et diversifiés (J2, J3, J4) généralement reliés au cours actif du fleuve soit directement par l'aval (J4, B2, A6, A1, J3) soit par l'intermédiaire d'un exutoire (B4, J2, J6, A3). L'alimentation phréatique peut y être très importante.

L'ensemble III (Tableau 1) caractérise J2 et J3, long complexe de la rive droite du Rhône formé de deux anciens méandres alimentés principalement par la nappe phréatique, peu influencé par le cours principal du fleuve et qui présente une profonde originalité. Celle-ci apparait sur le troisième facteur de l'analyse et s'explique par la présence d'espèces telles que *Cyrnus flavidus* (liée à la végétation flottante ou immergée (Nymphéacées), *Hydroptila sparsa.* D'autres espèces moins abondantes, semblent liées à un milieu légèrement courant, à végétation sur fond solide, et des milieux peu perturbés (*Mystacides azurea*) (Lepneva1970-71, Wallace 1981). C'est surtout la présence de *Hydropsyche angustipennis* qui caractérise J3. Cette hydropsyche très abondante surtout en septembre dans la végétation (*Nuphar, Potemogeton*) ne se trouve que dans les milieux avoisinant le Rhône, jamais dans l'axe principal, et seulement de façon "accidentelle" dans un canal vif (Gaschignard et Berly 1986).

En dehors des taxons qui participent directement à la définition des ensembles de bras-morts, quelques espèces ont une valeur écologique significative ; par exemple : l'espèce *Beraea pullata* est liée à des résurgences d'eau souterraine à l'amont de J6 et A1, sous couvert forestier, en zone semi-terrestre, en concordance avec les observations de Wiberg-Larsen (1979).

## CONCLUSIONS

L'étude réalisée ici sur 20 anciens lits du Rhône et de l'Ain n'est pas un inventaire exhaustif des peuplements larvaires de Trichoptères. En effet, la technique de prélèvement et l'échantillonnage réalisé à deux saisons seulement sont sélectifs vis à vis d'espèces d'eau profonde ou plus abondantes en septembre telles que plusieurs espèces de *Cyrnus* trouvées en abondance dans d'autres études plus détaillées (Richardot-Coulet *et al.* 1983). Au niveau du traitement des données nous nous sommes placés délibérément à un niveau de perception global qui n'apporte pas d'informations sur la répartition fine des larves à l'intérieur d'un ancien lit. Cependant, grâce à la multiplication de petits échantillons, cette étude apporte des précisions ou des confirmations sur la signification écologique de certains taxons (les Limnephilidae par ex.) et complète les données des pièges lumineux pour lesquels la localisation spatiale des différentes espèces est parfois difficile (Rojas-Camousseight1985). Il pourra être mis au point une méthodologie rendant ces deux types d'études plus parfaitement complémentaires. Par contre, le niveau de perception global apporte des informations sur la capacité potentielle des Trichoptères comme descripteurs écologiques dans le sens défini par Bournaud et Amoros (1984) et Amoros *et al.* (sous-presse) vis à vis de certains aspects du fonctionnement des écosystèmes abandonnés par les fleuves. Comme l'a montré l'analyse des différents groupements d'espèces, les peuplements de Trichoptères intègrent l'effet de plusieurs facteurs intervenant dans la dynamique évolutive des anciens lits étudiés, tels que les facteurs "distance au cours principal", "effets des crues", "morphologie de l'ancien chenal". Ce "pouvoir descripteur" des larves de Trichoptères demande à être confirmé par des études plus approfondies qui sont en cours et devraient permettre de caractériser plus précisément les différents types de fonctionnement et d'évolution des anciens chenaux fluviaux.

## REMERCIEMENTS

Ce travail a été réalisé dans le cadre du Programme Interdisciplinaire de Recherche sur l'Environnement (PIREN du CNRS) "Recherches méthodologiques appliquées à la gestion écologique des ressources en eaux des systèmes fluviaux", et soutenu par la Commission des Communautés Européennes (Contrat EN - 693-F/5D).

## REFERENCES

Amoros, C., Bravard, JP., Pautou, G., Roux, A.L. & Reygrobellet, J.L. Methodological research applied to the ecological management of water resources of fluvial systems. **Regulated rivers,** (sous presse).

Auda, Y. 1983. Rôle des méthodes graphiques en analyse des données : application au dépouillement des enquêtes écologiques. Thèse doctorat de 3ème cycle Lyon I.

Bournaud, M. & Amoros, C. 1984. Des indicateurs biologiques aux descripteurs de fonctionnement : quelques exemples dans un système fluvial. **Bull. Ecol.,** 15 : 57-66.

Castella, E., Richardot-coulet, M., Roux, C. & Richoux, P. 1984. Macroinvertebrates as "describers" of morphological and hydrological types of aquatic ecosystems abandoned by the Rhône river. **Hydrobiologia,** 119 : 219-225.

Castella, E., Richoux,P., Richardot-coulet, M. & Roux, C. Un diagnostic écologique de trois anciens méandres basé sur l'utilisation de descripteurs faunistiques. **Doc. Cartogr. Ecol.,** 29 (sous presse).

Gaschignard-Fossati, O. 1986. Répartition spatiale des macroinvertébrés benthiques d'un bras-mort du Rhône ; rôle des crues et dynamique saisonnière. Thèse de doctorat, Univ. Lyon I 200 pp.

Gaschignard, O. & Berly, A. Impact of large discharge fluctuations on the macroinvertebrate populations downstream of a dam. **3rd intern. Symposium on regulated streams.** Edmonton 1985. (sous-presse).

Higler, L.W.G. 1977. Macrofauna-cenoses on Stratiotes plants in dutch broads. **Res. Inst. Nature Manag.,** 11 : 1-86.

Lebart, L., Morineau, A. & Tabard, N. 1977. Techniques de la description statistique. Methodes et logiciels pour l'analyse des grands tableaux. Bordas, Paris, 351 pp.

Lepneva, S.G. 1970-71. Fauna of USSR Vol I et II Trichoptera. Israel Program for Scientific Translations. Jerusalem. 560 pp.

Moretti, G.P. 1983. Guide per il riconoscimento delle specie animali delle acque interne italiane19 : Tricotteri. CNDR, Verona, 1983. 155pp.

Moretti, G.P. & Spinelli Batta, G. 1984. Biology, zoogeography, ecology and classification of the aquatic stages of *Limnephilus helveticus* Schmid. **Proc. 4th Int. Sympos. on Trichoptera.,** J.C. Morse (ed.), Junk The Hague, 257-260,

Richardot-Coulet, M., Richoux, P. & Roux, C. 1983. Structure et fonctionnement des écosystèmes du Haut-Rhône français 29. Structure des peuplements de macroinvertébrés benthiques d'un ancien méandre. **Arch. Hydrobiol.,** 96 : 363-383.

Rojas-Camousseight, F. 1985. Etudes préliminaires sur l'utilisation des Trichoptères adultes comme descripteurs écologiques. Thèse de doctorat. Univ. Lyon. I, 215 pp.

Tachet, H. & Bournaud, M 1981. Cycles biologiques des Hydropsychidae et d'un Polycentropodidae (Trichoptera) dans le Rhône en amont de Lyon. **Proc. 3rd Int.Symp. on Trichoptera,** G.P. Moretti ed., Junk. The Hague, 347-364.

Usseglio-Polatera, P. 1985. Evolution des peuplements de Trichoptères et d'Ephéméroptères du Rhône à Lyon (1959-1982). Résultats de piégeages lumineux. Thèse de Doctorat Lyon I, vol; I et II, 248 et 203 pp.

Wallace, I.D. 1981. A key to larvae of the family Leptoceridae (Trichoptera) in Great Britain and Ireland. **Freshwater biol.,** 11 : 273-297.

Wiberg-Larsen, P. 1979. Revised key to larvae of Beraeidae in NW Europe (Trichoptera). **Ent.Scand.,** 10 : 112-118.

# PRINCIPAL COMPONENTS ANALYSIS OF TEMPORAL AND SPATIAL VARIATION IN TRICHOPTERAN FAUNAS OF NORTHEN PORTUGAL

M.C. MOLLES, Jr. and L.S. W. TERRA

## SUMMARY

A total of 20, year-round light trap collections, made at four sites on three water courses in northern Portugal, were examined using principal components analysis (PCA). Two sites were at middle elevations, one at Relva (720m) on the rio Beça (sampled two years) and one at Torno (605 m) on the Ribeiro do Ramalhoso (sampled four years). The remaining sites were at Vila do Conde (5m) on the Rio Ave, one on the riverbank, and one 50m away near some fish rearing ponds. These two sites were sampled for seven consecutive years. When all 20 annual collections were included in a PCA, each river was represented as an isolated, tight cluster of points, each point representing an annual collection. At this scale, PCA did not resolve differences between the two sites at Vila do Conde. When the other sites were excluded, however, PCA produced distinctive clusters of the riverside and upland collections. The only collection falling outside these clusters was the riverside collection made in 1983, the year a cooking oil factory began operating 200m upstream. These results lend support to the use of light traps for studies of local distributions of Trichoptera by suggesting that, in northern Portugal at least, spatial variation exceeds annual variation in the relative species composition of collections, even at fine spatial scales.

## INTRODUCTION

Williams (1939, 1940) initiated the application of rigorous statistical analysis to light trap collections. Crichton (1960) reviewed the early literature concerning light trapping of Trichoptera and applied Williams' (1937, 1939, 1940, 1954) quantitative methods. This fundamental approach of thorough collecting combined with rigorous quantitative analysis has been followed in more recent work on Trichoptera (Svensson 1972, 1974, Crichton *et al.*1978, Crichton and Fisher 1981, Novak 1981, Malichy 1981, McElravy *et al.* 1981).

The purpose of the present study was to demonstrate the application of principal components analysis (PCA) to the study of trichopteran faunas through an examination of relative annual and spatial variation in the composition of light trap collections made in northen Portugal. At the heart of this study was the following question : what degree of spatial resolution is possible when using light traps for the study of trichopteran distributions ? This question is of interest for two reasons : 1) many Trichoptera are strong fliers and are capable of dispersing long distances after emergence (Crichton *et al.* 1978, Sheldon 1984), and 2) annual variation in catches of Trichoptera at light traps may be substantial . For example, despite equal sampling effort, Crichton (1960) reported a 35% decrease in total catches of Trichoptera at Millbarn Pond between 1954 and 1955 and differences in abundances of individual species that were even greater. Both dispersal and annual variation in abundance would reduce detectable spatial differences in trichopteran faunas.

## METHODS

Adult Trichoptera were collected using a light trap (Terra 1978) at four sites near three water courses in northen Portugal. One site was located at Relva (720m) on the bank of Rio Beça , here, a medium-sized, low-gradient mountain stream. This site was sampled 22 times in 1981 and 32 times in 1982. A second site was located at Torno (605m) approximately 10m from the Ribeiro do Ramalhoso, here, a small, high-gradient mountain stream. Collections were made at this site from 1979 to 1982 (33 to 37 per year). The two remaining sites were both at Vila do Conde (5m) on the Rio Ave, here, a medium-sized, low-gradient, coastal river. One of these sites was on the riverbank and the other, an upland site, was approximately 5Om away near some fish-rearing ponds. This arrangement was designed following Crichton (1960). Light trap collections were made at these two sites every three days between March 1976 and March 1983 (approximately 1700 collections).

For each site, all Trichoptera collected within a year were identified, usually to species, and pooled to form an annual collection in which each species was represented as a proportion of the total number. These proportions were transformed (Arcsine $\sqrt{\text{proportion}}$) to correct for bimodality (Zar 1974).

The pooled annual collections were compared with principal components analysis (PCA) (Pielou 1977, Rotenberry &Wiens 1981) and a statistical package for microcomputers (Wilkinson 1984) . As applied, PCA provided a comparison of the relative species composition of collections. A PCA eases faunal comparisons by reducing the number of variables that the biologist must consider by combining species that have distributions which are similar. Species are combined into a smaller number of "factors" in such a way that the resulting factors are themselves uncorrelated with each other. The factors are then ordered in relation to the amount of variation in the data accounted for by each. Usually only the first few factors are statistically significant and provide most of the explanatory power. Collections are assigned loadings for each factor and may be plotted on factor axes to demonstrate relative similarities among collections. The relation of each species included in an analysis to a factor is expressed as a factor score, which may be used to assess the contribution of individual species to similarities and differences between collections.

Principal components analysis has some advantages for comparisons of faunas over other methods such as cluster analysis. With PCA it is : 1) unnecessary to set up arbitrary criteria for defining faunal classes or community types, and 2) unnecessary to assume that faunal classes are hierarchically related or that they even exist (Pielou 1977).

RESULTS

The 20 annual collections analyzed contained a total of over 45,000 individuals belonging to 109 species of Trichoptera. The collections at Relva contained 1491 individuals and 61 species in 1981 and 2647 individuals and 58 species in 1982. The four annual collections at Torno contained 1213 to 1887 individuals belonging to 54 to 64 species.Variation in numbers of species and individuals was considerably greater at Vila do Conde, despite uniform yearly collecting effort. At the riverside site, number of individuals varied from 1372 to 4117 and number of species from 15 to 25. Numbers and species were particularly reduced at the riverside site in 1983, the year a cooking oil factory began discharging wastes approximately 200m upstream. By comparison, variability at the upland site was somewhat reduced both for individuals, 1392 to 2728, and for species, 20 to 24. Of the 36 species at Vila do Conde, 28 were captured at both sites. Three rare species were captured at the riverside only and five, also rare, were captured at the upland site only. Therefore differences between these two sites may be mainly attributed to differences in the relative abundance of species, not to differences in species composition.

The PCA which included all collections from all four sites produced three statistically significant factors which accounted for 61.3%, 18.3%, and 9.6% of the variation respectively among collections, for a total of 89.2%. The collections at Vila do Conde were the only ones to load significantly on factor 1. The Torno collections were the only ones to load significantly on factor 2 and the Relva collections were the only ones to load significantly on factor 3. Thus, PCA indicated that the trichopteran faunas of Vila do Conde, Torno, and Relva are well-differentiated. In this comparison, spatial variations (differences between rivers) far exceeded temporal variation (differences among annual collections at each site). This result was expected since the Rio Ave, Rio Beça, and Ribeiro do Ramalhoso are substantially different physically (gradient, size, etc.) at the sites sampled. When the data were viewed at this scale, PCA did not resolve differences between the two sites at Vila do Conde.

Scores for species for factor 1 indicate that the Vila do Conde collections differed from those at Torno and Relva mainly as a consequence of higher numbers of *Hydropsyche lobata, Tinodes assimilis, T. waerneri, Ecnomus deceptor, Ceraclea sobradieli,* and *Mystacides azurea.* Species scores for factor 2 indicate that the taxa more abundant in the Torno collections were *Rhyacophila adjuncta, R. lusitanica, Philopotamus montanus perversus, Hydropsyche siltalai, Limnephilus marmoratus, Allogamus laureatus, Sericostoma vittatum,* and *Sericostoma* females. Species scores for factor 3 indicate that the Relva collections were mainly distinguished from the others by higher numbers of *Oxyethira archaica, O. frici,* polycentropodid females, *Psychomyia pusilla,* and *Athripsodes braueri.*

The PCA which included only the Vila do Conde collections produced three significant factors which accounted respectively for 34.7%, 35.9%, and 21.1% of the variation among collections. The first two of these factors, accounting for 70.6% of the variation, resulted from differences between sites. Upland collections loaded highest on factor 1 while, with the exception of 1983, the riverside collections loaded highest on factor 2 (Fig. 1).

The riverside collection of 1983 scored highly on both factors 1 and 2 and as a consequence its position falls outside the clusters indicated on Figure 1. Interestingly, 1983 was the year a cooking oil factory began operating approximately 200m upstream from the riverbank site.

The relative influence of individual species on the differences between the riverbank and upland collections may be shown by plotting their scores on factors 1 an 2. When this is done (Fig. 2) it is apparent that seven of the 36 species were most important for distinguishing between these closely approximated sites.

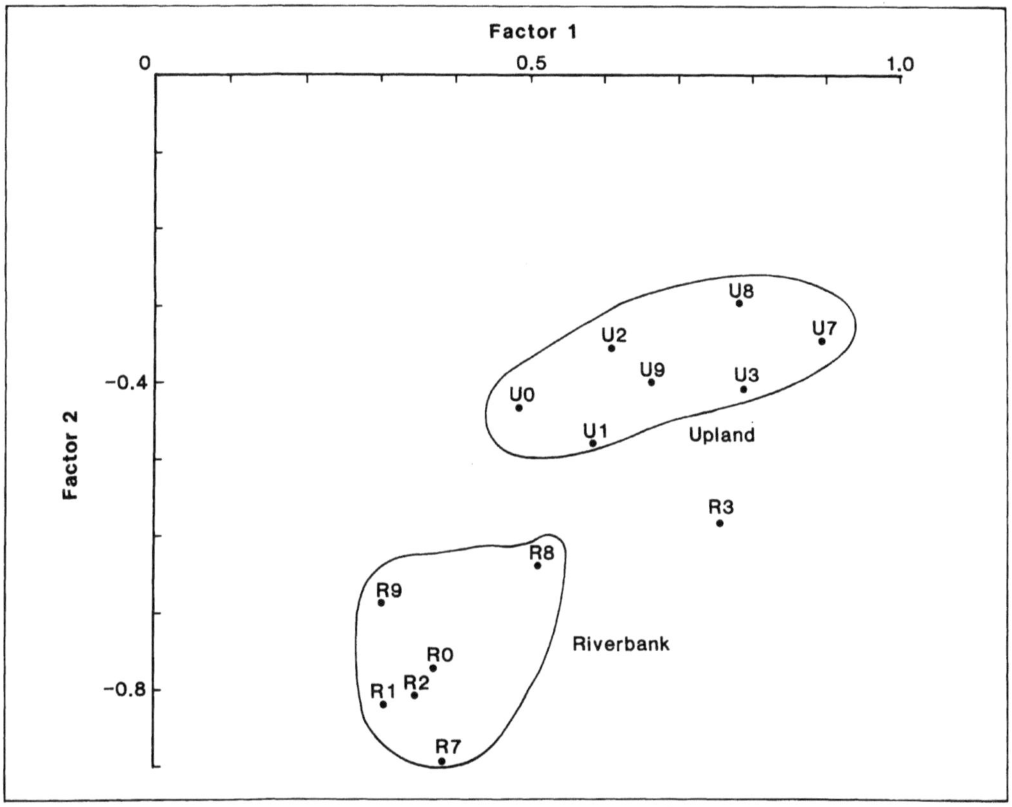


Figure 1. Loadings on factors 1 and 2 of annual collections at the upland (U) and riverbank (R) sites at Vila do Conde, 1977 to 1983.

*Tinodes assimilis* and *T. waerneri* scored highly on factor 1 and were most commonly collected at the upland site. The nearby ponds may support substantial populations of these species. *Ecnomus deceptor, Polycentropus telifer, Mystacides azurea,* and *Ceraclea sobradieli,* which had high scores for factor 2, were more commonly collected at the riverbank site. *Hydropsyche lobata* was useful for distinguishing between sites because of a difference in the pattern of captures for males and females. Females were more frequently collected at the riverbank site while males were more frequent in the upland collections. This difference between the sexes of *H. lobata* is consistent with the findings of Svensson (1972) and may be true of other species. In our analyses, however, only individuals on *H. lobata* were separated according to sex. The tight cluster of scores for the remaining 29 species (Fig. 2) indicated that each was captured with approximately equal frequency at the upland and riverbank sites.

## DISCUSSION

The results of this study suggest that, in northen Portugal at least, spatial variation exceeds annual variation in the relative species composition of light trap collections of Trichoptera. These results lend further support to the use of light traps for the study of local distributions of Trichoptera and for detecting the effects of pollution. Clearly, however, further work is necessary on the response of individual species to light, their distance and direction of dispersal following emergence, etc.

As noted above, and as pointed out by Pielou (1977), principal components anlysis has some advantages over other methods for making faunal comparisons. As presented, this study points toward another advantage.

In addition to providing an overall quantitative assessment of the relative similarities of faunas, PCA may also be used to indicate which taxa contribute most to differences between faunas. Therefore, using this method eventually leads back to the distributions of individual species, too often lost in a mathematical shuffle with many quantitative approaches.

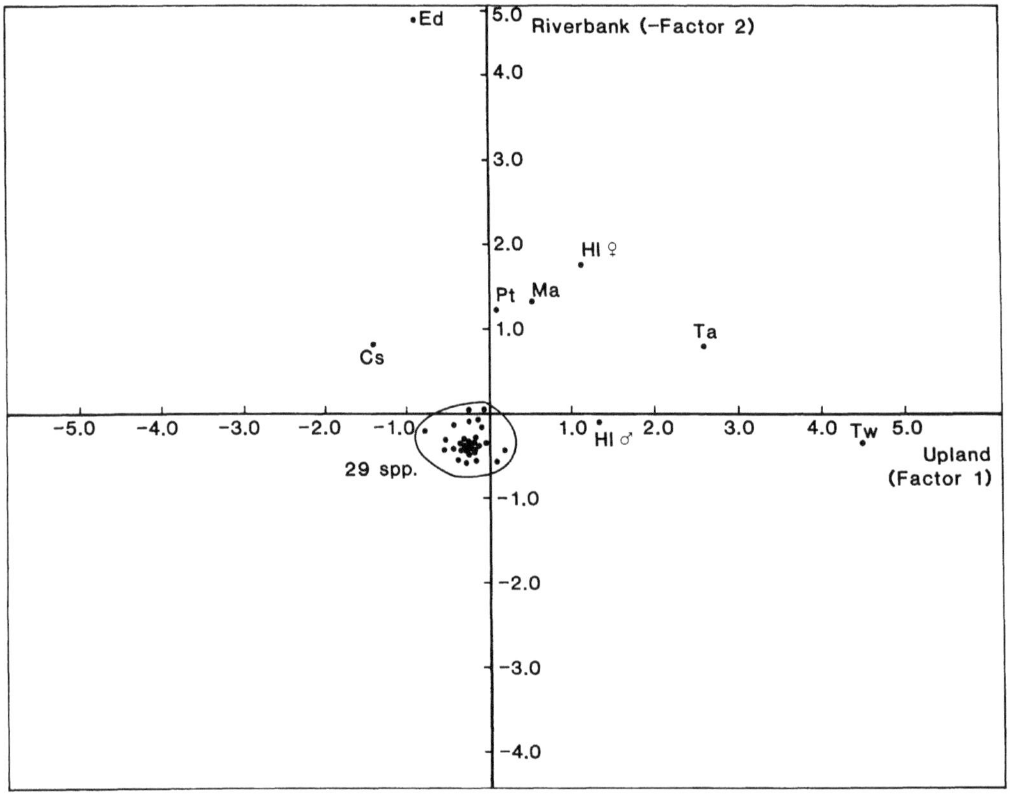


Figure 2. Scores of 36 species collected at Vila do Conde on factor 1 (dominated by upland collections) and factor 2 (dominated by riverbank collections, shown as factor 2 to ease interpretation). Symbols indicate factor scores for species more abundant at the upland or riverbank sites : *Hydropsyche lobata,* Hl, *Polycentropus telifer,* Pt, *Tinodes assimilis,* Ta, *T. waerneri,* Tw, *Ecnomus deceptor,* Ed, *Ceraclea sobradieli,* Cs, *Mystacides azurea,* Ma.

There are many ways of making faunal comparisons. The basis for all is the knowledge and logic employed by the biologist. As pointed out by Williams over thirty years ago (1954) relative to the application of mathematics to faunal studies "Mathematics is only a condensed form of logic ..." We would only elaborate by saying that mathematics also provides for standardization of methods and interpretation.

REFERENCES

Crichton, M.,I. 1960. A study of captures of Trichoptera in a light trap near Reading, Berkshire. **Trans. R. Ent. Soc. Lond., 112:319-344.**

Crichton, M.I. & Fisher, D. 1981. Further observations on limnephilid life histories, based on the Rothamsted Insect Survey. **Proc. 3rd Internat. Symp. Trichoptera,** G.P. Moretti (ed.), Junk, The Hague : 47-55.

Crichton, M.I., Fisher, D. & Woiwod, I.P. 1978. Life histories and distribution of British Trichoptera, excluding Limnephilidae and Hydroptilidae, based on the Rothamsted Insect Survey. **Holarctic Ecology,** 1 : 31-45.

Malicky, H. 1981. The phenology of dispersal of several caddisfly (Trichoptera) species in the island of Crete. **Proc. 3rd Internat. Symp. Trichoptera,** G.P. Moretti (ed.), Junk, The Hague : 157-163.

McElravy, E.P., Resh, V.H., Wolda, H. & Flint, O.S., Jr. 1981. Diversity of adult Trichoptera in a "non-seasonal" tropical environment. **Proc. 3rd Internat. Symp. Trichoptera,** G.P. Moretti (ed.), Junk, The Hague : 149-156.

Novak, K. 1981. Trichoptera distribution pattern differences found by sweeping, beating and light traps at three southern Bohemian sites. **Proc. 3rd Internat. Symp. Trichoptera,** G.P. Moretti (ed.), Junk, The Hague : 281-284.

Pielou, E.C. 1977. Mathematical Ecology. New York: John Wiley and Sons.

Rotenberry, J.T. & Wiens, J.A. 1981. A synthetic approach to principal component analysis of bird/habitat relationships. Capen D.E. (Ed.) : USDA Forest Service General Technical Report RM-87, Rocky Moutain Forest and Range Experiment Station, Fort Collins: 197-208.

Sheldon, A.L. 1984. Colonization dynamics of aquatic insects. *In* The ecology of aquatic insects, V.H. Resh & D.M. Rosenberg (eds), New York, Praeger: 401-429.

Solem, J.O. 1978. Species diversity of Trichoptera communities. **Proc 2nd Internat. Symp. Trichoptera,** M.I. Crichton (ed.), Junk, The Hague : 231-239.

Svensson, B.W. 1972. Flight periods, ovarian maturation, and mating in Trichoptera at a South Swedish stream. **Oikos,** 23:370-383.

Svensson, B.W. 1974. Population movements of adult Trichoptera at a South Swedish stream. **Oikos,** 25:157-175.

Terra, L.S.W. 1978. First results of Trichoptera collecting with light traps at Vila do Conde (Portugal). **Proc 2nd Internat. Symp. Triuchoptera,** M.I. Crichton (ed.), Junk, The Hague 75-82.

Wilkinson, L. 1984. SYSTAT. Illinois : SYSTAT Inc.

Williams, C.B. 1937. The use of logarithms in the interpretation of certain entomological problems. **Ann. Appl. Biol.,** 24: 404-414.

Williams, C.B. 1939. An analysis of four years captures of insects in a light trap. Part I. General survey ; sex proportion ; phenology and time flight. **Trans. R. Ent. Soc. Lond.,** 89:79-132

Williams, C.B. 1940. An analysis of four years captures of insects in a light trap. Part. II. The effect of weather conditions on insect activity, and the estimation and forecasting of changes in insect populations. **Trans. R. Ent. Soc. Lond.,** 90:227-306.

Williams, C.B. 1954. The statistical outlook in relation to ecology. **J. Ecol.,** 42:1-13.

Zar, J.H. 1974. Biostatistical analysis. NewJersey, Prentice-Hall, Inc.

# THE TEMPORAL AND SPATIAL DISTRIBUTION OF ADULT TRICHOPTERA ON THE UPPER SURFACES OF THE FLOATING LEAVES OF NYMPHAEIDS. A STUDY USING ADHESIVE TRAPS.

G. VAN DER VELDE and P.J.M. BERGERS

## SUMMARY

Using adhesive traps, Trichoptera were caught on the upper surfaces of floating leaves of nymphaeids during the growth season of 1977 in a Dutch pond. Trichoptera form a quantitatively very important component of the insect fauna there (35.2% of the total catch). The species caught were not evenly distributed, but there was a spatial and temporal differentiation in occurrence. There was a clear preference for certain areas within the pond studied as well as for certain subzones within the nymphaeid stands. *Cyrnus flavidus* and *C. renaticornis* showed a distinct temporal separation. In most species there are indications for two generations per season. Males outnumber the females greatly for most species.

## INTRODUCTION

Nymphaeids produce mainly floating leaves and possess flowers floating on or elevated above the water surface and rooting systems in the sediment. This macrophyte growth-form often dominates in shallow fresh water to a depth of 3 metres.

Nymphaeid floating leaves can be regarded as terrestrial islands in an aquatic environment. The air-adapted terrestrial and semi-aquatic macrofauna, including insects as well as spiders, uses these 'islands" for various purposes (Brock & Van der Velde 1983, Van der Velde & Brock 1980, Van der Velde *et al.* 1985).

The fauna on the upper surfaces of floating leaves was studied in 1977 in order to gain some insight in the temporal and spatial occurrence, species composition and diversity at the various sites within a nymphaeid system during one growth season. The present paper discusses the results of the catches of adult Trichoptera.

## STUDY AREA

All observations were made in pond F of the Oude Waal, an oxbow lake situated in the forelands of the River Waal near Nijmegen (The Netherlands) (see Van der Velde *et al.* 1985). The bottom of this eutrophic, alkaline pond is covered by a detritus layer, with the exception of the southern part where the bottom consists mainly of sand.

The vegetation of pond F consists mainly of *Nuphar lutea* (L.) Sm., and some beds of *Nymphaea alba* L. The broad helophyte zone at the north of the pond of *Typha angustifolia* L., *Rumex hydrolapathum* Huds., and *Scirpus lacustris* L. ; the narrow helophyte zone at the southern side shows the same plant species including *Mentha aquatica* L.

## MATERIALS AND METHODS

Each week, from June 16th to September 29th 1977, six undamaged floating leaves (five of *Nuphar* and one of *Nymphaea*) of the same size were gathered and covered with insect-lime (Tangletrap or Stikem), which has no odour and consequently does not attract or repel the insects.

These leaves (adhesive traps), were attached to six different floating leaves. The catch was protected from predation by birds by means of wire netting (meshes 3 cm), supported by cork floats. The wire netting was attached by means of a string to the petiole of the supporting leaf. After 24 hours, these leaves were taken to the laboratory. The fauna was washed from the leaves with refined petrol and then fixed in 70% alcohol.

Five *Nuphar lutea* leaves (1-3 and 4-5) were laid out in transects from the helophyte zone to the open water. Transect 1-3, laid out in the northern part of the pond, was 12 m wide ; transect 4-5, in the southern part of the pond, was 3 m wide. Leaf nr.6, a *Nymphaea* leaf, was laid in an isolated stand of *Nymphaea alba* surrounded by open water, at a distance of 6 m from the western bank (see Van der Velde *et al.* 1985).

The development of floating leaves was followed in plots of one square metre.

## RESULTS

A total of 21 Trichoptera species were caught. The number of specimens (8158) was 35.2% of the total insect catch ; the percentage of Trichoptera varied from 0% (29 September) to 58.5% (5 August). The genus *Cyrnus* alone made up 83.7% of the total Trichoptera catch (Table 1). Most species showed a clear predominance of males ; only in the case of *Orthotrichia costalis* and *Oxyethira flavicornis* did females dominate (Table 1).

Figure 1 shows that the total numbers of Trichoptera followed the development of the floating leaves. The maximum number was caught, however, somewhat before the maximum number of floating leaves was reached. Four peaks in the total numbers during the season can be recognized, mainly caused by changes in the numbers of the most dominant species (*Cyrnus flavidus*, *C. crenaticornis* and *Agraylea sexmaculata*).

Table 1. The distribution over the sampling sites (1-6) of adult Trichoptera caught by means of adhesive traps on the floating leaves in the Oude Waal during 1977. Ditto for the females. For each species sex ratios (% males) are indicated. The diversity was calculated from the total seasonal catches.

| species | sampling site (total) | | | | | | | sampling site (females) | | | | | | | |
|---|---|---|---|---|---|---|---|---|---|---|---|---|---|---|---|
| | 1 | 2 | 3 | 4 | 5 | 6 | total | 1 | 2 | 3 | 4 | 5 | 6 | total | % males |
| *Cyrnus flavidus* MacMachlan, 1864 | 891 | 708 | 810 | 625 | 887 | 408 | 4329 | 4 | 14 | 9 | 4 | 8 | 1 | 40 | 99 |
| *Cyrnus crenaticornis* (Kolenati, 1859) | 434 | 528 | 336 | 436 | 611 | 174 | 2519 | 5 | 19 | 21 | 6 | 20 | 8 | 79 | 97 |
| *Agraylea sexmaculata* Curtis, 1834 | 209 | 26 | 66 | 10 | 76 | 34 | 421 | 4 | - | 3 | - | - | 1 | 8 | 98 |
| *Oecetis lacustris* (Pictet, 1834) | 29 | 33 | 48 | 102 | 71 | 33 | 316 | 2 | 2 | 2 | 4 | 3 | 6 | 19 | 94 |
| *Molanna angustata* Curtis, 1834 | 1 | 9 | 7 | 70 | 49 | 5 | 141 | - | - | - | 2 | 1 | - | 3 | 98 |
| *Orthotrichia costalis* (Curtis, 1834) | 8 | 6 | 32 | 17 | 27 | 10 | 100 | 3 | 6 | 29 | 14 | 19 | 7 | 78 | 22 |
| *Ecnomus tenellus* (Rambur, 1842) | 6 | 6 | 7 | 27 | 23 | 25 | 94 | - | - | 2 | 1 | 3 | - | 6 | 94 |
| *Mystacides longicornis* (L., 1758) | 4 | 3 | 6 | 28 | 25 | 7 | 73 | 3 | 1 | 2 | 6 | 1 | - | 13 | 82 |
| *Oxyethira flavicornis* (Pictet, 1834) | 2 | 7 | 17 | 7 | 10 | 8 | 51 | 2 | 7 | 15 | 5 | 10 | 6 | 45 | 12 |
| *Triaenodes bicolor* (Curtis, 1834) | 9 | 5 | 1 | 9 | 1 | 3 | 28 | - | - | - | - | - | - | 0 | 100 |
| *Agraylea multipunctata* Curtis, 1834 | 7 | 1 | 2 | 1 | 3 | 9 | 23 | 1 | - | - | - | 1 | 1 | 3 | 87 |
| *Oecetis furva* (Rambur, 1842) | 3 | 2 | 3 | 4 | 3 | 7 | 22 | - | - | - | - | - | - | 0 | 100 |
| *Ceraclea fulva* (Rambur, 1842) | - | - | 10 | - | 3 | - | 13 | - | - | - | - | - | - | 0 | 100 |
| *Ceraclea senilis* (Burmeister, 1839) | - | 1 | - | 4 | 4 | - | 9 | - | - | - | - | - | - | 0 | 100 |
| *Holocentropus stagnalis* (Albarda, 1874) | 2 | - | - | 4 | - | 1 | 7 | - | - | - | - | - | - | 0 | 100 |
| *Tricholeiochiton fagesii* (Guinard, 1879) | 4 | - | - | - | - | - | 4 | - | - | - | - | - | - | 0 | 100 |
| *Holocentropus picicornis* (Stephens, 1836) | 2 | 1 | - | - | - | - | 3 | - | - | - | - | - | - | 0 | 100 |
| *Holocentropus dubius* (Rambur, 1842) | 1 | - | - | - | - | 1 | 2 | - | - | - | - | - | 1 | 1 | 50 |
| *Athripsodes aterrimus* (Stephens, 1836) | - | - | - | - | 1 | - | 1 | - | - | - | - | 1 | - | 1 | 0 |
| *Mystacides nigra* (L., 1758) | - | - | - | 1 | - | - | 1 | - | - | - | 1 | - | - | 1 | 0 |
| *Molanna albicans* (Zetterstedt, 1840) | - | - | - | - | 1 | - | 1 | - | - | - | - | - | - | 0 | 100 |
| Total numbers | 1612 | 1336 | 1345 | 1345 | 1795 | 729 | 8158 | 24 | 49 | 83 | 43 | 67 | 31 | 297 | 96 |
| species | 16 | 14 | 13 | 15 | 16 | 14 | 21 | 8 | 6 | 8 | 9 | 10 | 8 | 13 | |
| Shannon-Weaver (H) | 1.11 | 1.01 | 1.21 | 1.43 | 1.28 | 1.41 | 0.57 | 0.87 | 0.64 | 0.72 | 0.84 | 0.77 | 0.77 | 0.82 | |
| Evenness (e) | 0.42 | 0.38 | 0.45 | 0.50 | 0.45 | 0.51 | 0.43 | 0.96 | 0.82 | 0.80 | 0.88 | 0.77 | 0.85 | 0.74 | |

The number of species increased up to the end of July and then remained more or less constant until the beginning of September, after which it fell again. Both the Shannon-Weaver index and the Evenness showed a dip in July and then an increase towards the end of the season (Fig. 1). The catches on successive sampling dates were highly similar (Fig. 2). There is also a great deal of similarity between the catches in June and those around the end of August.

Many species showed two clear peaks in numbers during the season. There is a quantitative succession of the various species on the floating leaves during the season. *Cyrnus flavidus* and *C. crenaticornis* were both present throughout the whole study period and showed two clear peaks in numbers. Maximum numbers were not reached simultaneously but alternatingly (Fig. 3).

The number of species showed small differences between the sampling sites. The Shannon-Weaver index and the Evenness showed similar patterns over the sampling sites (Table 1).

The composition of the catches from the *Nuphar* sites appeared to be more similar than that from the isolated *Nymphaea* site (Fig. 2). This site also showed low numbers for most species (Table 1).

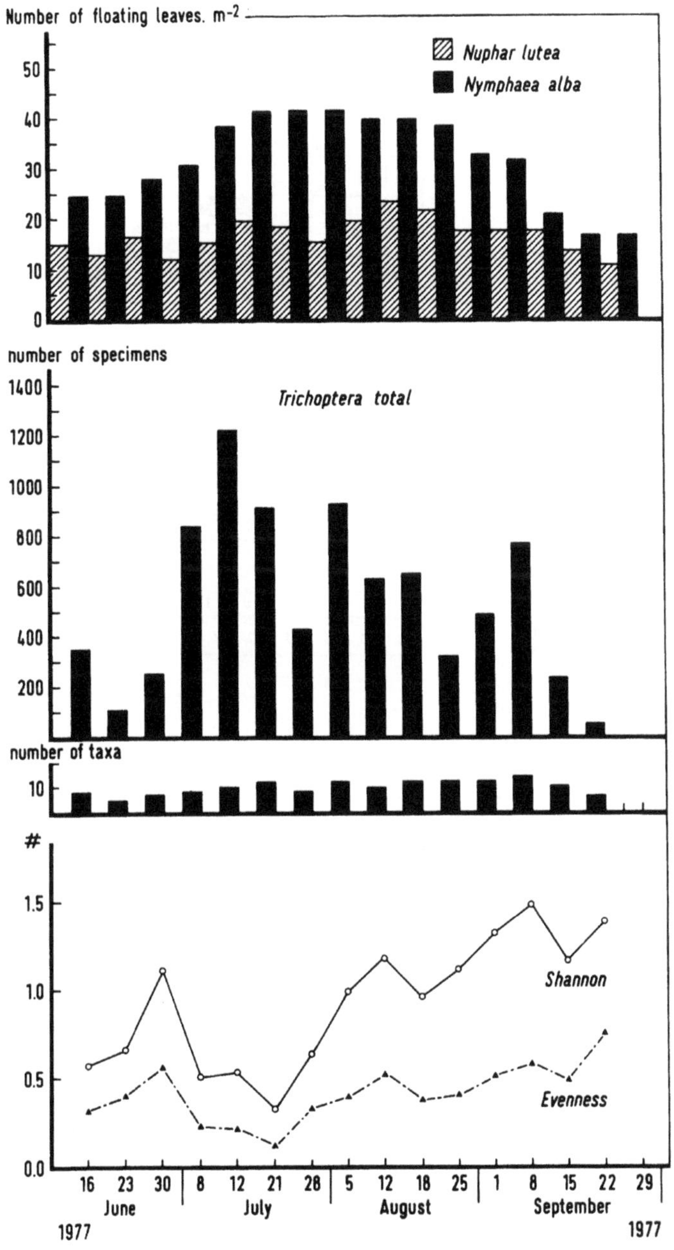


Figure 1. The numbers of floating leaves in a one square metre plot (above), total numbers of specimens and numbers of species of adult Trichoptera caught by means of adhesive traps in the Oude Waal during 1977 (middle) and the trichopteran diversity during the season (below).

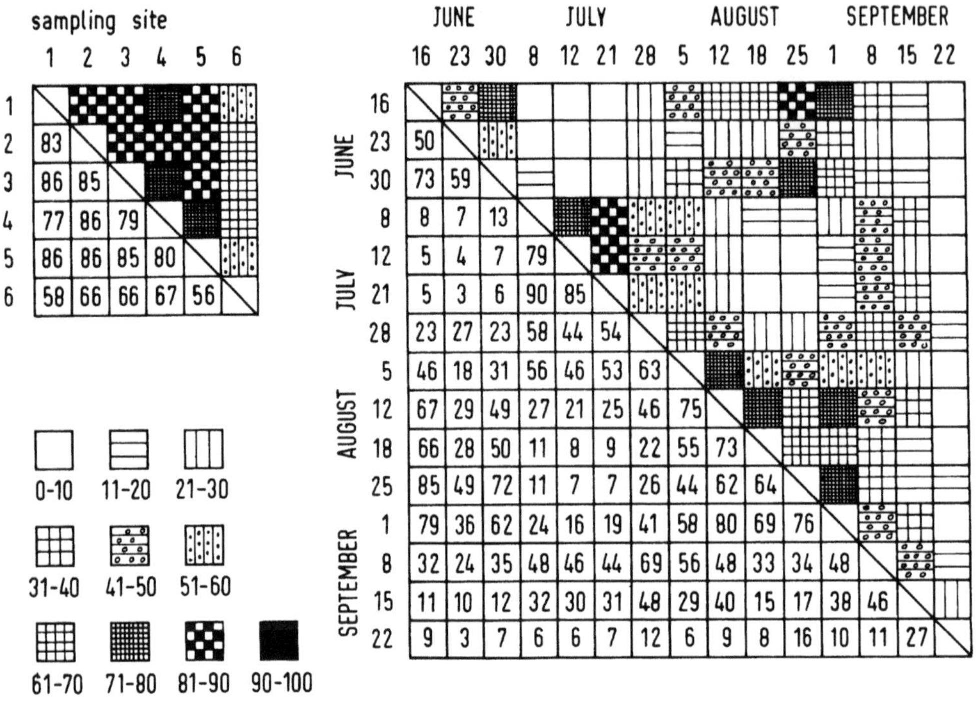


Figure 2. The similarity between the total catches of sampling sites and of dates for the adult Trichoptera caught in 1977 (for the similarity formula see Van der Velde *et al.*, 1985).

The species can be grouped according to their quantitative occurrence in certain subzones of the nymphaeid vegetation (Table 1). I.*Oecetis lacustris, Molanna angustata, Mystacides longicornis, Ceraclea senilis* and possibly *Mystacides nigra, Molanna albicans* and *Athripsodes aterrimus* were most numerous on the southern sampling sites (sites nrs. 4, 5). II.*Triaenodes bicolor, Tricholeiochiton fagesii, Holocentropus stagnalis, H. picicornis* and possibly *H. dubius* were more characteristic for the sites bordering the helophyte zones (sites nrs. 1, 4). III.*Orthotrichia costalis, Oxyethira flavicornis, Ceraclea fulva* and possibly *Agraylea sexmaculata* were more abundant at the sites bordering the open water (sites nrs. 3, 5). However, the total number of the last-mentioned species was most numerous at sampling site nr. 1 due to a mass occurrence there on one date (September 15th). IV.*Cyrnus flavidus, C. crenaticornis, Oecetis furva* and *Agraylea multipunctata* showed no clear pattern with respect to the distribution over the *Nuphar* sampling sites. V.*Ecnomus tenellus* was scarcely caught at the northern sampling sites (1-3) in contrast to the southern and western sampling sites (4-6).

## DISCUSSION AND CONCLUSIONS

Adhesive traps such as were used in the present study appeared to be very useful in studying the quantitative distribution of adult caddis-flies over the nymphaeid stands within one water body as well as over the nymphaeid subzones.

Peaks in occurrence indicated that most species have two generations during the season. This explains the two periods with high similarity as shown in Figure 2.

The occurrence of *Cyrnus flavidus* and *C. crenaticornis* showed a clear temporal separation. However, for most species a spatial separation seems to be far more important. The key factors with respect to the spatial distribution seem to be bottom composition (group I), structure of the vegetation (group II) and suitability as oviposition site (especially group III). Adults of species of which the larvae construct their cases from sand grains (group I) were caught mainly in the southern part of the pond, where there is a sandy bottom. The helophyte zone is a vertical element in a flat system and is therefore used by many species as swarming site.

The catches of females of the most numerous species reflected the distribution of their larvae on the lower surfaces of the floating leaves (Van der Velde not yet published), so that it may be suggested that females of these species use the leaves for oviposition, especially those bordering the open water. The high numbers of males in the catches suggest a swarming behaviour over the nymphaeid stands with occasional landings. Possible uses of the floating leaves for the adult caddis-flies include those of oviposition sites, resting and possibly mating sites.

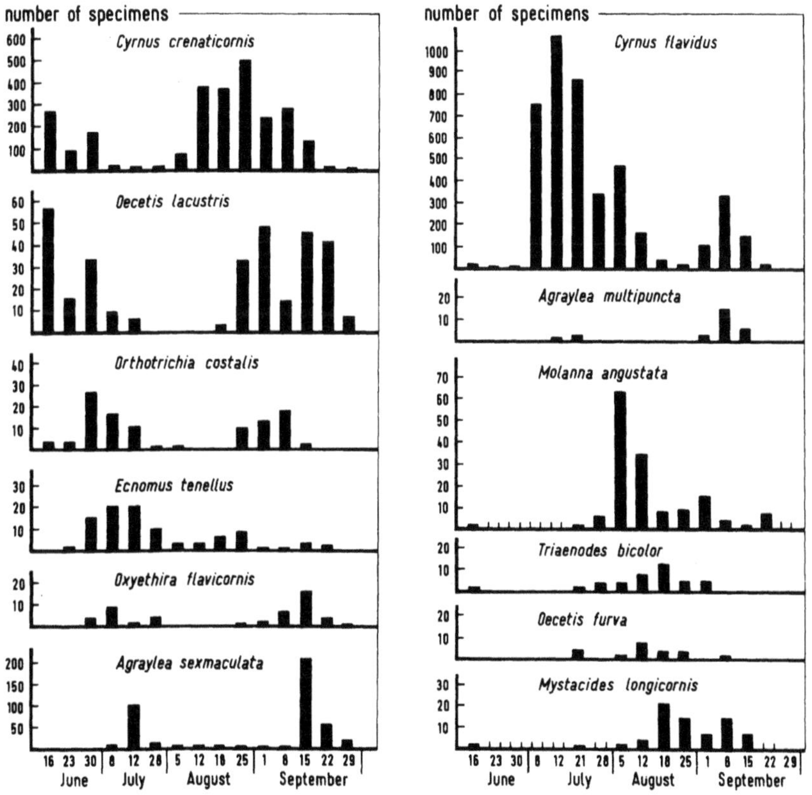


Figure 3. The numbers of specimens of various species of adult Trichoptera caught by means of adhesive traps in the Oude Waal in 1977. Only species with more than 20 specimens are represented in the figure. Catches of other species were as follows :
*T. fagesii* 30-VI (1 specimen), 12-VII (1), 5-VIII (1), 8-IX (1) ; *H. stagnalis* 8-VII (5), 12-VII (2) ; *H. dubius* 8-VII (1), 12-VII (1) ; *A. aterrimus* 21-VII (1) ; *M. albicans* 21-VII (1) ; *H. picicornis* 21-VII (1), 28-VII (1), 5-VIII (1) ; *M. nigra* 5-VIII (1) ; *C. senilis* 18-VIII (2) , 25-VIII (3), 1-IX (4) ; *C. fulva* 25-VIII (3), 8-IX (4), 15-IX (1), 22-IX (5).

As adult Trichoptera appeared to be very numerous on the upper surfaces of floating leaves, their role in the food web of this nymphaeid compartment must be important especially as prey for other animals such as predatory insects.

ACKNOWLEDGEMENTS

The authors thank Mr M. Heine, Mr P.M.P.M. Peeters and Mr L.A. Van der Heijden for assistance in the field and Prof. Dr. C. den Hartog for critically reading the manuscript.

REFERENCES

Brock, T.C.M. & Van der Velde, G. 1983. An autecological study on *Hydromyza livens* (Fabricius) (Diptera, Scatomyzidae), a fly associated with nymphaeid vegetation dominated by *Nuphar*. **Tijdschr. Ent.**, 126 : 59-90.

Van der Velde, G. & Brock, T.C.M. 1980. The life history and habits of *Notiphila brunnipes* Robineau-Desvoidy (Diptera, Ephydridae), an autecological study on a fly associated with nymphaeid vegetations. **Tijdschr. Ent.**, 123 : 105-127.

Van der Velde, G., Meuffels, H.J.G., Heine, M. & Peeters, P.M.P.M. 1985. Dolichopodidae (Diptera) of a nymphaeid-dominated system in The Netherlands : species composition, diversity, spatial and temporal distribution. **Aquatic Insects**, 8 : 189-207.

# TRICHOPTERES DE CORSE. BIOGEOGRAPHIE, ECOLOGIE ET DISTRIBUTION DES ESPECES DANS LES COURS D'EAU

J. GIUDICELLI et A. ORSINI

SUMMARY

The caddisfly fauna of Corsica includes up to 62 species distributed among 34 genera. The number of known species is low in comparison to several european countries, but some 31 of them are endemic (50%). This fauna clearly appears paleartic, dominated by elements of west-mediterranean distribution ; it also includes a small tyrrheno-maghrebine element.

Patterns of caddisfly distribution and zonation were studied in a large watershed of central Corsica. Informations are given about habitat preferences of the most characteristic components of this fauna (*Leptodrusus, Thremma, Silonella, Beraeodina*).

La première publication sur les Trichoptères de Corse remonte à 1864 ; sur les cinq espèces décrites alors comme nouvelles par Hagen, quatre sont encore valables. On doit à Mosely l'essentiel de nos connaissances sur cette faune. Au cours de deux séjours dans l'île, en 1928 et 1931, il a récolté un abondant matériel d'où il a décrit 22 espèces nouvelles. A partir de ses propres récoltes et des citations d'autres auteurs (Ris, Ulmer, Esben-Petersen), Mosely (1932) a établi un inventaire comprenant 66 espèces sûres.

Depuis on n'a pas découvert beaucoup de nouveautés faunistiques. En 1972 Tobias décrivit *Hydropsyche fumata*, précédemment mentionnée en Corse sous le nom de *H. instabilis*. Nous avons décrit deux espèces nouvelles : *Rhyacophila tarda* Giudicelli et *Hydropsyche cyrnotica* Botosaneanu & Giudicelli.

Actuellement nous arrêtons l'inventaire à 61 espèces et une sous-espèce réparties en 34 genres ; nous avons tenu compte, d'une part, des synomymies et, d'autre part, nous avons éliminé des espèces figurant dans la liste de Mosely mais dont la présence en Corse nous semble douteuse (*Silo piceus, Plectrocnemia conspersa, Wormaldia triangulifera*).

## BIOGEOGRAPHIE

Le peuplement trichoptérologique de Corse présente deux particularités.

### Nombre réduit d'espèces par rapport aux autres régions naturelles d'Europe.

L'effectif du groupe dépasse 200 espèces dans les Pyrénées et 300 dans les Alpes. En Italie Moretti et Cianficconi (1981) ont dénombré 331 espèces ; au Portugal l'inventaire de Terra (1981) comprend 141 espèces. En Corse, les lacunes les plus importantes concernent les Limnéphilides (seulement 11 espèces connues) ; on note aussi la faible diversité intragénérique : 5 espèces de *Rhyacophila* contre 20 des les Pyrénées et 30 en Italie, 3 espèces d'*Hydropsyche* contre 7 au Portugal et 6 en Italie et au Maroc.

Nous pouvons expliquer la pauvreté taxonomique de la faune corse. Au Quaternaire l'île s'est trouvée à l'écart des déplacements de faune qui ont accompagné les oscillations climatiques lors des grandes glaciations. Par contre, dans les différentes régions d'Europe continentale ces brassages de faune ont amené un enrichissement des peuplements.

### Nombre élevé d'endémiques ou d'espèces à aréal restreint.

La moitié des espèces de Trichoptères sont des endémiques au sens large.

- 9 endémiques corses :

  *Rhyacophila tarda, Philopotamus corsicanus, Hydropsyche cyrnotica, H. fumata, Diplectrona meridionalis, Polycentropus corsicus, Micrasema cinereum, Sericostoma clypeatum, Ernodes nigroauratus.*

- 17 endémiques corso-sardes :

  *Rhyacophila pallida, R. trifasciata, Agapetus cyrnensis, Hydroptila cortensis, H. bifurcata, Plectrocnemia geniculata corsicana, Polycentropus divergens, P. mortoni, Tinodes agaricinus, T. cortensis, Allogamus corsicus, Silonella aurata, Thremma sardoum, Athripsodes genei, Beraea aureomarginata, Beraeodina palpalis, Helicopsyche revelieri.*

- 3 endémiques tyrrhéniennes :
  *Diplectrona magna, Micrasema togatum, Silo rufescens.*
- 2 endémiques de Corse-Sardaigne-Baléares :
  *Agapetus quadratus, Leptodrusus budtzi.*

*Beraeodina*, genre monotypique, est le seul genre endémique corso-sarde de Trichoptère. *Leptodrusus*, autre genre monotypique, longtemps considéré comme cantonné à la Corse-Sardaigne, a été récemment trouvé aux Baléares ainsi que *Agapetus quadratus*, ancienne endémique corse (Malicky 1983).

Presque toutes ces espèces sont des endémiques anciennes car la Corse a constitué un refuge où se sont maintenues des lignées survivantes de la faune autochtone tertiaire ("Korsica ist für die Fliesswasserfauna eine Refugium für sonst nirgends in Europa mehr aufzufindende tertiare Formen" Illies 1966).

En Corse, l'endémisme, par les effets conjugués de l'isolement insulaire et de l'isolement montagnard, relève surtout de la conservation de lignées anciennes.

Le fond du peuplement est paléarctique et européen. Cependant, la majeure partie des Trichoptères de Corse est constituée d'éléments autochtones représentant le reste du peuplement de l'ancienne Tyrrhénide, vaste continent qui, au début du Tertiaire, recouvrait tout le bassin de la Méditerranée occidentale y compris l'Afrique du Nord.

Citons les Trichoptères les plus caractéristiques de cette faune tertiaire :

- le genre *Silonella* qui, en dehors de la Corse-Sardaigne, se trouve en Algérie (*S. brevipalpis*) et dans le sud de l'Espagne (De Jalón, com. verb.).
- *Hydroptila bifurcata* qui constitue avec l'espèce algérienne, *H. serrata*, une paire d'espèces-soeurs.
- *Philopotamus corsicanus* directement apparentée à une endémique marocaine, *P. ketama*.
- *Plectrocnemia geniculata corsicana* qui ne se retrouve qu'au Maroc.

Ces espèces à distribution tyrrhéno-maghrébine sont certainement de vieilles lignées, compte tenu de l'ancienneté des relations de la Corse avec l'Afrique du Nord.

Certaines endémiques se rattachent à des lignées ouest-méditerranéennes :

*Agapetus cyrnensis* à *A. segovicus* (endémique ibérique), *Diplectrona magna* à *D. moralesi* (endémique ibérique), *Allogamus corsicus* à *A. despaxi* (endémique pyrénéenne), *Athripsodes genei* à *A. tavaresi* (endémique ibérique), *Micrasema cenereum* à *M. servatum* (endémique ibérique et pyrénéenne)...

Les trois endémiques corses de *Rhyacophila* (*R. pallida, R. tarda, R. trifasciata*) forment un groupe d'espèces affines avec des endémiques italiennes (*R. rougemonti, R. italica*) et une endémique pyrénéenne (*R. rupta*).

On peut associer ces lignées et les deux espèces dont la répartition est limitée à la Corse-Sardaigne et Baléares à la présence d'un massif ibéro-pyrénéo-corso-sarde pendant presque tout le Tertiaire.

## DISTRIBUTION ECOLOGIQUE ET HABITAT DES ESPECES.

Les Trichoptères de Corse n'étaient connus qu'au plan taxonomique ; on ne savait rien de leur écologie.

Nous avons étudié le peuplement trichoptérologique d'un grand réseau hydrographique de la Corse centrale comprenant deux grandes rivières, le Tavignano et la Restonica.

Les prélèvements ont été réalisés dans 36 stations :

- 11 stations se répartissent sur le cours d'eau principal, entre 50 et 1960 m d'altitude ; 3 se situent sur le Tavignano dans son secteur de piedmont et de plaine, entre 50 et 300 m d'altitude ; 7 sur la Restonica, entre 400 et 1720 m ; une sur un torrent de haute altitude, entre 1800 et 1960 m.
- 15 stations concernent des affluents entre 300 et 1600 m.
- 10 stations sont des émissaires de sources entre 350 et 1750 m.

En raison de la nature granitique du substrat, dans l'ensemble des stations, les eaux sont très claires et peu minéralisées (conductivité : 15 à 50 µS) ; dans deux affluents et dans une source de basse altitude la conductivité est comprise entre 200 et 400 µS du fait de la présence d'affleurements calcaires.

Le peuplement du réseau hydrographique comprend 44 espèces de Trichoptères réparties en 28 genres ; 18 genres sont représentés ici par une seule espèce.

La richesse spécifique stationnelle est fonction de l'altitude. Dans le cours d'eau axial le nombre d'espèces est compris entre 16 et 20 jusqu'à 600 m, entre 14 et 17 de 700 à 1200 m, entre 6 et 12 de 1300 à 2000 m. La richesse est assez constante dans les sources, entre 9 et 12 espèces. Le peuplement le plus diversifié (20 à 22 espèces) a été relevé dans deux affluents de basse altitude avec des eaux bien minéralisées. La richesse spécifique élevée dans la partie inférieure du réseau est certainement en rapport avec l'absence de fortes pollutions et d'aménagements importants. La situation inverse (richesse spécifique élevée du cours supérieur et moyen, réduction du nombre d'espèces dans le cours inférieur) est souvent la règle dans les cours d'eau européens.

Neuf espèces sont présentes dans la moitié au moins des stations : *Hydrospsyche fumata* (27 stations), *Rhyacophila tristis* (24), *Leptodrusus budtzi* (23), *Plectrocnemia geniculata corsicana*(21), *Wormaldia variegata* (20), *Agapetus cyrnensis* (19), *Rhyacophila trifasciata* (19), *Sericostoma clypeatum* (18), *Thremma sardoum* (18). Leur fréquence élevée s'explique par leur présence à la fois dans les petits affluents et les sources et dans une portion du cours d'eau principal. Ainsi, *R. tristis, R. trifasciata, W. variegata* apparaissent dans le cours axial au dessus de 1200 m d'altitude ; elles vivent aussi, quelle que soit l'altitude, dans la plupart des affluents d'ordre 1 à 3 et les émissaires de sources. De ce fait, ces Trichoptères occupent une large zone altitudinale (de 300 à 2000 m) mais ont un habitat assez restreint : le crénal et les secteurs initiaux du rhithral.

Les espèces congénères se succèdent en altitude le long du cours d'eau principal ou occupent des zones écologiques différentes (fig. 1). Ces séquences sont souvent les homologues de celles observées dans les réseaux européens.

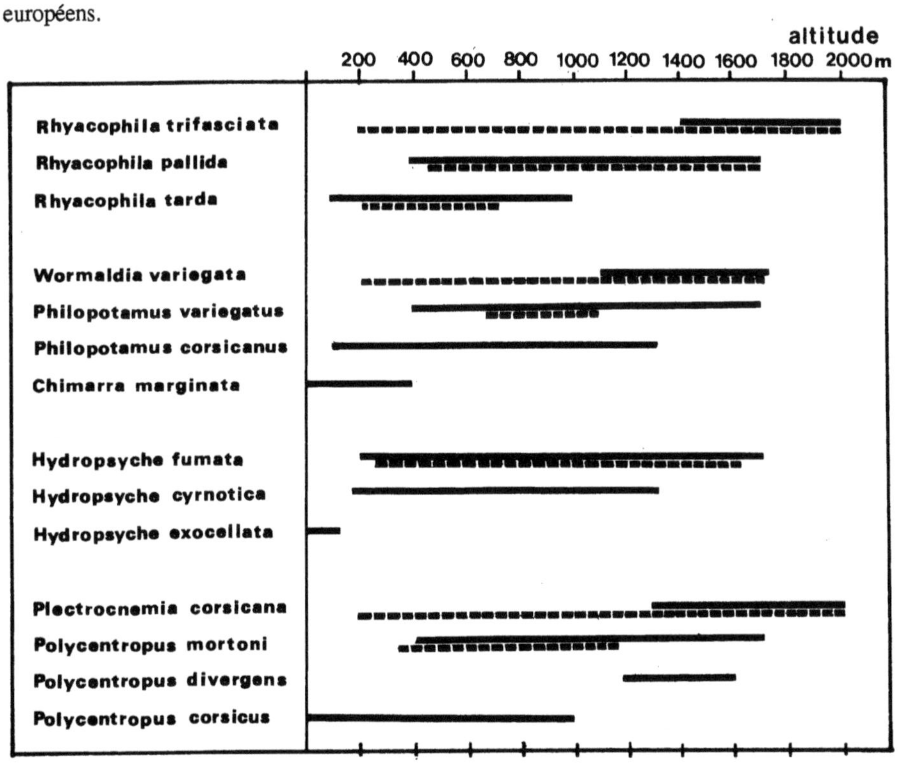


Figure 1. Répartition altitudinale des espèces de *Rhyacophila*, de Philopotamides, d'*Hydropsyche*, de Polycentropodides. (en traits pleins : répartition dans le cours d'eau principal ; en traits tirets : répartition dans les affluents et dans les sources).

*Rhyacophila trifasciata* est présente dans tous les habitats lotiques au dessus de 1400 m d'altitude ; en aval elle vit uniquement dans les sources et les affluents de l'épirhithral. *R. pallida* vit dans le cours d'eau principal entre 400 et 1700 mètres et dans les grands affluents de moyenne altitude. *R. tarda* se cantonne dans le cours principal en aval de 1000 m. Des trois espèces d'*Hydropsyche* une, *H. exocellata,* se situe dans la zone terminale du Tavignano. Les deux autres occupent un large secteur altitudinal avec une importante zone de chevauchement ; cependant leur répartition écologique est différente. *H. cyrnotica,* espèce du groupe de *pellucidula,* se trouve dans la partie inférieure et moyenne du cours principal et dans les grands affluents, entre 100 et 1300 m ; *H. fumata,* espèce du groupe d'*instabilis,* monte plus haut en altitude et a été récoltée dans tous les affluents et dans les sources. Ainsi, les *Hydropsyche* de Corse occupent les mêmes positions relatives et réalisent la même séquence que leurs homologues dans les réseaux européens.

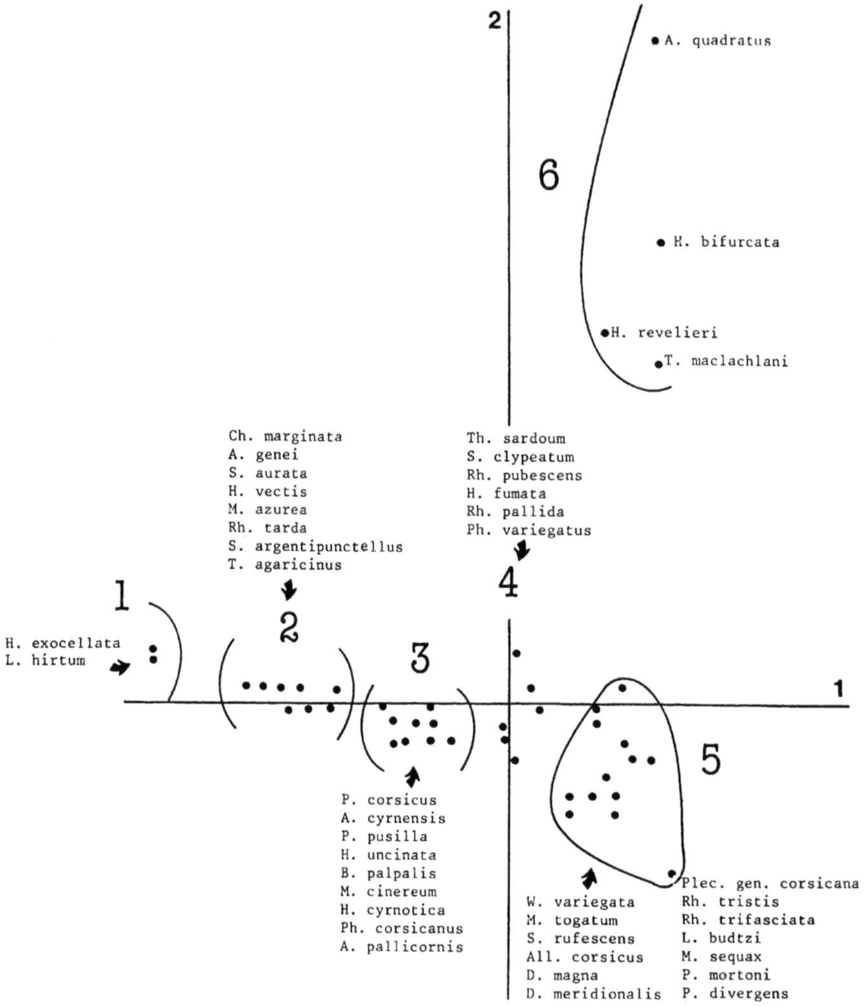


Figure 2. Carte factorielle des espèces. Mise en évidence de 6 biocénotypes.

Les quatre espèces de Philopotamides et les quatre espèces de Polycentropodides réalisent aussi deux successions altitudinales et écologiques bien nettes.

Nous apportons des précisions sur l'habitat des espèces les plus caractéristiques et originales de la faune trichoptérologique corse.

*Leptodrusus budtzi* est le Limnéphilide dominant dans les cours d'eau. Bien que sténotherme comme tous les *Drusinae*, il présente une vaste répartition altitudinale (de 300 à 2000 m) et une large amplitude d'habitat. Dans le cours d'eau axial l'espèce apparait au dessus de 1000 m. Les larves sont fréquentes dans les ruisseaux froids et dans les émissaires de source à fond pierreux au dessus de 300 m d'altitude ; elles vivent dans des courants faibles à modérés ($v \leq 50$ cm/s) et se nourrissent d'algues épilithiques et de débris végétaux grossiers.

*Beraeodina palpalis* est fréquente aux altitudes moyennes, entre 300 et 700 m, et ne dépasse pas la côte 1000 m ; elle est assez rare dans le potamal. Les larves vivent en eau calme ($v \leq 40$ cm/s) ; elles sont pétricoles et brouteuses d'algues.

*Silonella aurata* est caractéristique du potamon ; on la trouve dans le bas Tavignano avec une densité de 50 à 130 individus par $m^2$. Les larves sont pétricoles et vivent dans les zones de courants faibles ou moyens.

*Thremma sardoum* est un Trichoptère d'altitude moyenne, rare dans le potamal et ne dépassant pas 1200 m d'altitude. Il réalise son optimum écologique dans les émissaires de sources et dans les petits affluents ; ce caractère crénophile se retrouve chez les autres espèces de *Thremma* d'Europe. Les larves pétricoles broutent des diatomées.

## ELEMENTS DE BIOTYPOLOGIE

La recherche des relations stations-espèces par une méthode d'analyse multivariée (analyse factorielle des correspondances) fait apparaître la zonation des Trichoptères dans le réseau.Sur le plan factoriel 1-2 (Fig.2) on distingue six biocénotypes.

**Biocénotype 1** (station inférieure du Tavignano, alt. 50 m) : *Hydropsyche exocellata, Lepidostoma hirtum* ; il correspond au metapotamon.
**Biocénotype 2** (deux stations du bas Tavignano, alt. 100-300 m) : *Rhyacophila tarda, Chimarra marginata, Tinodes agaricinus, Hydroptila vectis, Setodes argentipunctellus, Mystacides azurea, Athripsodes genei, Silonella aurata ;* il correspond à l'épipotamon.
**Biocénotype 3** (trois stations inférieures de la Restonica, un affluent, alt. 400-800 m) : *Agapetus cyrnensis, Hydroptila uncinata, Allotrichia pallicornis, Polycentropus corsicus, Psychomyia pusilla, Hydropsyche cyrnotica, Philopotamus corsicanus Beraeodina palpalis, Micrasema cinereum* ; il correspond à l'hyporhithron.
**Biocénotype 4** (une station de la Restonica, alt. 1000 m ; quatre petits affluents, alt. 300-650 m) : *Rhyacophila pallida, R. pubescens, Philopotamus variegatus, Hydropsyche fumata, Sericostoma clypeatum, Thremma sardoum* : ce groupement ne correspond pas à une zone écologique mais fait la transition entre, d'une part l'hyporhithron et d'autre part, les secteurs supérieurs du rhithron et le crenon.
**Biocénotype 5** (stations du cours supérieur de la Restonica, affluents de haute altitude, sources ; alt. 1100-2000 m) : *Rhyacophila trifasciata, R. tristis, Diplectrona magna, D. meridionalis, Polycentropus mortoni, P. divergens, Plectrocnemia gen.corsicana, Wormaldia variegata, Allogamus corsicus, Micropterna sequax, Leptodrusus budtzi, Micrasema togatum, Silo rufescens.* Il correspond aux secteurs initiaux du rhithron (épi et métarhithron) et au crenon. *D. meridionalis, M. togatum, S. rufescens* sont plus caractéristiques du crenon que de l'épirhithron ; dans les sources d'altitudes moyennes *T. sardoum* et *H. fumata* s'ajoutent à ces espèces crénophiles.
**Biocénotype 6** (une grande source, alt. 350 m) : *Agapetus quatratus, Hydroptila bifurcata, Tinodes maclachlani, Helicopsyche revelieri.*

## REFERENCES


ILLIES, J. 1966. Die Verbreitung der Süsswasserfauna Europas. **Verh, internat. Ver. Limnol.**, 16 : 287-296.

Malicky, H. 1983. Atlas of European Trichoptera. The Haque, Junk Publ.

Moretti, G.P. & Cianficconi, F. 1981. First list of Italian Trichoptera. **Proc. 3rd Int. Symp. on Trichoptera**, ed. G.P. Moretti : 199-211, Junk, The Hague.

Mosely, M.E. 1932. Corsican Trichoptera and Neuroptera. **Eos,** 8 : 165-184.

Terra, L.S.W. 1981. Lista faunistica de Trichopteros de Portugal (Insecta, Trichoptera) **Bolm. Soc. port. Entomol.,** 12 : 1-42.

# CONTRIBUTION TO LARVAL TAXONOMY, ECOLOGY AND DISTRIBUTION OF THE CENTRAL EUROPEAN SPECIES OF THE GENUS *PHILOPOTAMUS* (TRICHOPTERA : PHILOPOTAMIDAE)

T. PITSCH

## SUMMARY

Characters are provided for the identification of larvae of the three *Philopotamus* species occurring in Central Europe : *P. ludificatus, P. montanus* and *P. variegatus*. Notes are given on their habitat and distribution.

## INTRODUCTION

In Central Europe the genus *Philopotamus* is represented by three species : *P. ludificatus* McLachlan, *P. montanus* (Donovan) and *P. variegatus* (Scopoli) (Botosaneanu and Malicky 1978, Tobias and Tobias 1981). Their larvae form an important element in the benthos fauna of most unpolluted, fast flowing streams. The adult insects can be identified using the figures given by Malicky (1983). There are descriptions of the larval stage of all three species in the literature including keys for identification (Sedlak 1985, Szczesny 1978, Ulmer 1909). But trying to identify large series of specimens from Germany, I met great difficulties and the separation of the species was unsatisfactory. Therefore the present study aimed to solve the taxonomic problems in *Philopotamus* larvae and to find characters for reliable identification of the species.

## METHODS

The three species under consideration are widely distributed and common in Central Europe. They occupy similar habitats. Quite often, two or even all three species occur together at the same locality. Thus identification of larvae by relating them to adults caught at the same station is doubtful.

The following description of differences between *P. montanus* and *P. ludificatus* is mainly based on the examination of about 200 larval exuviae from pupal cocoons. To obtain exuviae, prepupae and pupae were removed from the substratum, kept in their pupal case, and put individually into small vials with a few mm of water from the station. In this condition they were reared at about 12 to 14°C until the imago emerged. In addition ; *Philopotamus* larvae and adults were collected at each sampling station. Altogether about 500 larvae of *P. ludificatus* and 300 larvae of *P. montanus* were collected. I did not succeed in rearing *P. variegatus*. Details given for this species are derived from 5 larvae identified as *P. variegatus* due to the good accordance with the description given by Szczesny (1978). This proceeding is justified, as these larvae showed clear differences from those of *P. ludificatus* and *P. montanus* and there are no other species known from Central Europe that could be confused with it.

## TAXONOMIC RESULTS

### Separating *Philopotamus* larvae from other caddises

Larvae of the genus *Philopotamus* share the general characters of the family *Philopotamidae* : caseless, campodeiform, only pronotum sclerotised, no abdominal gills, labrum not sclerotised with the anterior border ciliated. *Philopotamus* can be distinguished from other Philopotamid genera by the following combination of characters :

- anterior edge of frontoclypeus slightly assymetrically indented
- black strip on the lateral margin of the pronotum continous with the black strip on the posterior margin
- anterior face of procoxa with two black bristles, one of them may be situated on a small, rounded protuberance.

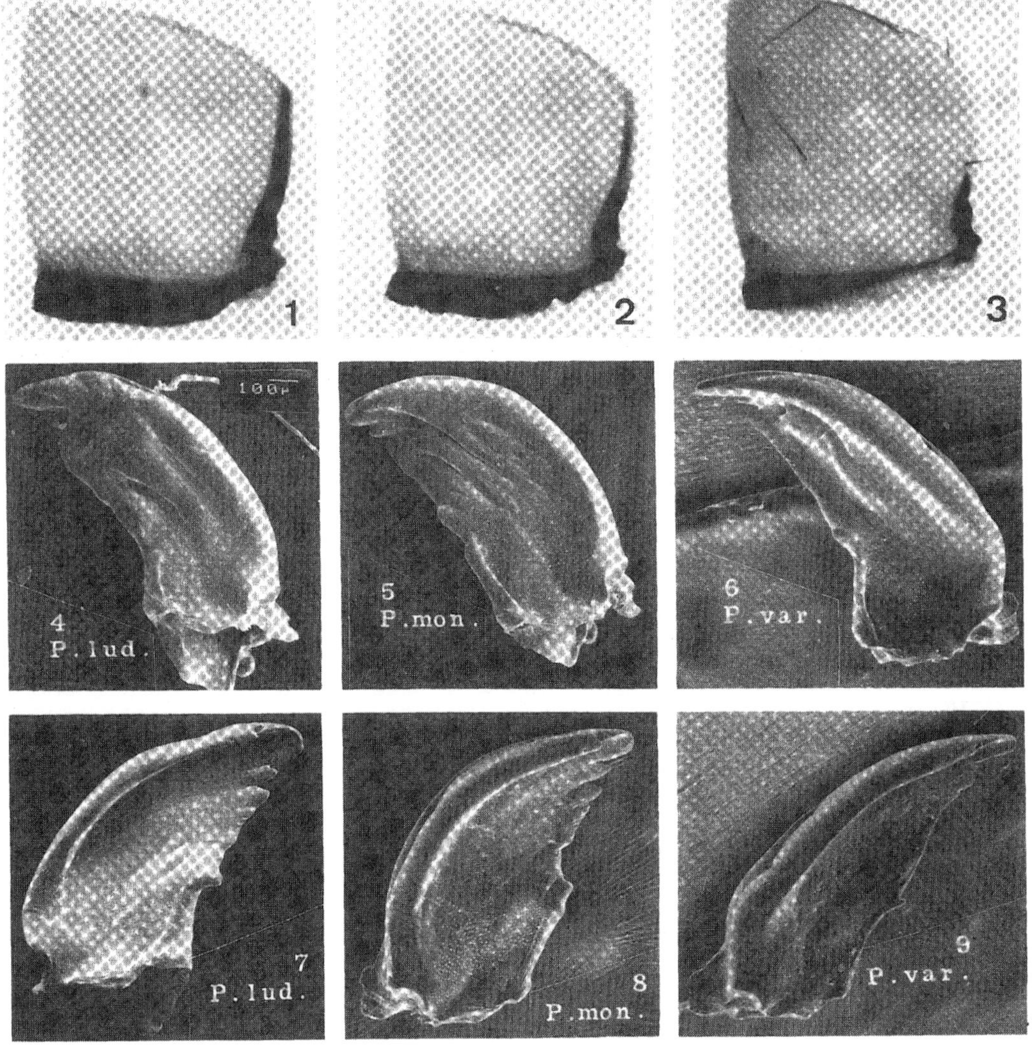


Figure 1-9. *Philopotamus*, 5th instar larvae. Figs. 1-3 : right pronotum of different individuals of *P. ludificatus*. Figs. 4-6 : right mandible, dorsal view. Figs 7-9 : left mandible, dorsal view. The scale on Fig. 4 refers to Figs. 4-9.

## Diagnostic characters

The following descriptions refer to the 5th instar larvae, if not indicated otherwise. The 5th instar is recognized by head width of usually more than about 1.3 mm.

**1. Black bands on pronotum.** The extension of the lateral strip on the pronotum is the main or only character used in the identification keys for separating the *Philopotamus* species (Sedlak 1985, Szczesny 1978, Ulmer 1909). In my samples a total continuum between the descriptions of the three species was observed, thus safe separation was impossible.

Examination of larval exuviae from pupal cocoons revealed that the average mature larvae show distinctive differences, but variation even within a population was larger than differences between species

(Figs. 1-3). Many individuals of *P. montanus* showed a black lateral strip of about two thirds of the pronotum length or even more. This coloration was rarely found in *P. ludificatus*. In all specimens of *P. variegatus* the lateral strip did not reach the base of the "x-hair" (Szczesny 1978) and in all but one specimen the posterior strip was discontinous, but many specimens of *P. ludificatus* and *P. montanus*, too, showed very short lateral bands (Fig. 3). In freshly moulded animals especially, the coloration is still darkening, the lateral band often weakly developed and the posterior band frequently discontinous. Therefore these specimens may be easily taken for *P. variegatus*.

**2. Anterior edge of pronotum.** As also stated by Szczesny (1978), the angle at the anterior edge of the pronotum is more acute in *P. ludificatus* and *P. montanus* than in *P. variegatus*. This feature is sometimes difficult to recognize, especially in specimens that have moulded only recently.

**3. Mandibles.** Ulmer (1903) presented figures showing differences between the mandibles of *P. ludificatus* and *P. montanus*, but in his key to Trichoptera larvae (Ulmer 1909), separation of the two species is based only on differences of the coloration of the pronotum. Recent works, too, do not mention the mandibles (Sedlak 1985, Szczesny 1978). In particular, the right mandible proved to show the best characters for separating *P. ludificatus* and *P. montanus*. In all *ludificatus* specimens I have observed, there was a broad indentation in the basal half of the inner margin. This feature was never seen in *P. montanus*. In addition, there are only two deep grooves in the dorsal face in *P. ludificatus* (Fig. 4), but 4 to 5 more shallow ones in *P. montanus* (Fig. 5). These features can already be observed clearly in the 4th instar and probably in even earlier stages.

**4. Lateral view of the head.** Most specimens of *P. variegatus* I examined were much larger than those of the other two species. The head of *P. variegatus* is dorsoventrally compressed, the head capsule showing a slight longitudinal edge at the level of the eye, running for the anterior two thirds of the head, whereas the parietalia of *P. ludificatus* and *P. montanus* are constantly rounded (Figs. 10-12).

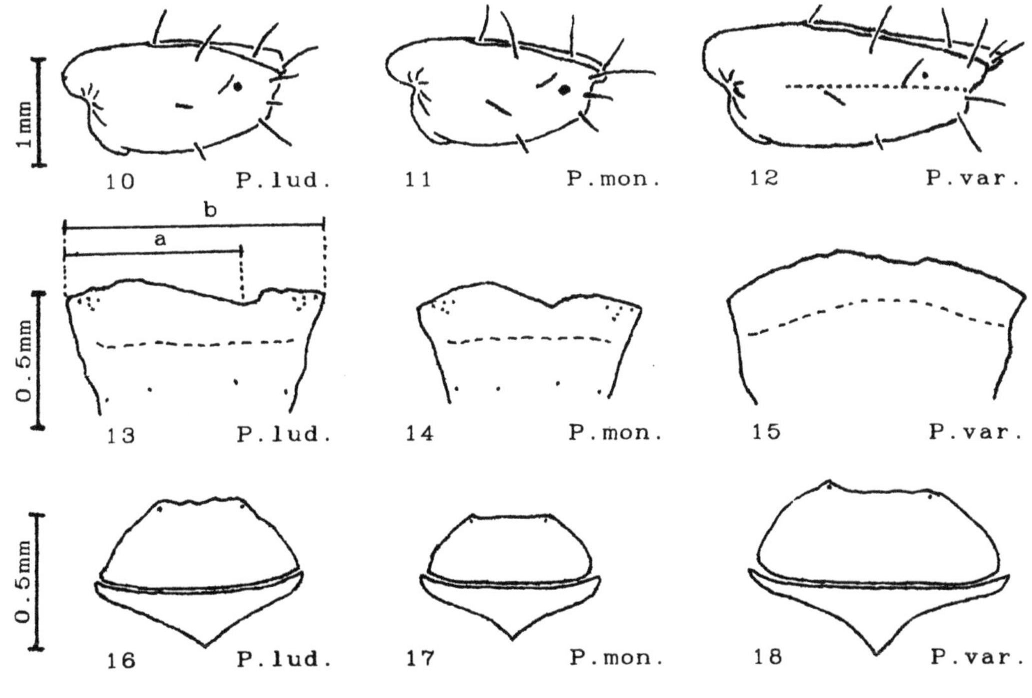


Figure 10-18. *Philopotamus*, 5th instar larvae. Figs. 10-12 : head, lateral view. Figs. 13-15 : anterior margin of frontoclypeus. Figs. 16-18 : mentum and submentum.

**5. Frontoclypeus.** In accordance with the description by Szczesny (1978) I found that the indentation of the anterior margin of the clypeus is not as deep in *P. variegatus* as it is in the other species (Fig. 15), but *P. ludificatus* and *P. montanus*, too, differ in the shape of the anterior margin, the assymetrical indentation being more laterally situated in *ludificatus* and more medially in *montanus* (Figs. 13-14). The ratio a : b, where a = distance between the left anterior edge and the point of deepest indentation and b = distance between the anterior edges, i.e. width of frontoclypeus, was between 0.55 and 0.61 in *P. montanus* (75 exuviae from 11 sites measured) and between 0.61 and 0.69 in *P. ludificatus* (118 exuviae from 24 sites).

**6. Mentum.** There are differences in the shape of the mentum between *P. ludificatus* and *P. montanus* : in *P. montanus* the sclerite is wider than in *P. ludificatus* (Figs. 16-17).

ECOLOGY AND DISTRIBUTION

The maps show the localities where samples for the present study were collected (Figs. 19-21).

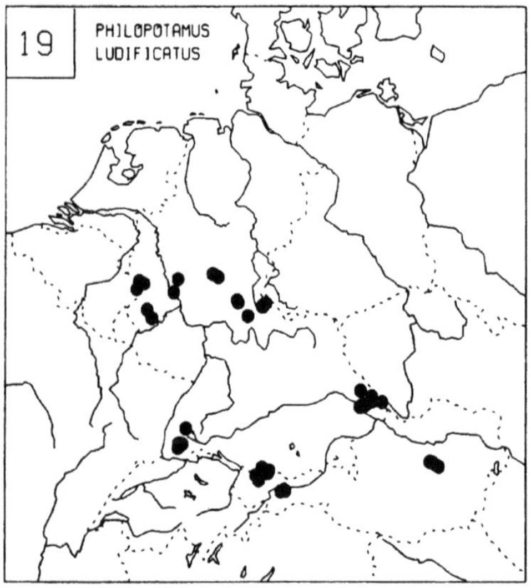


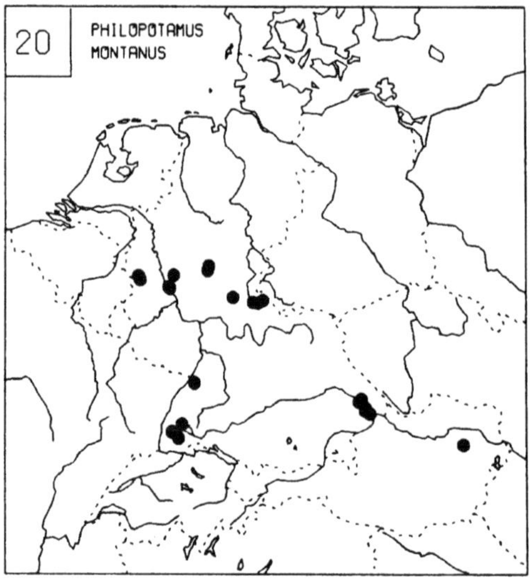


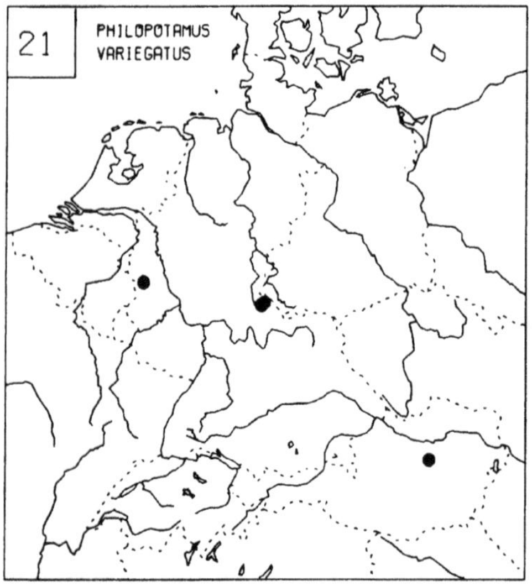


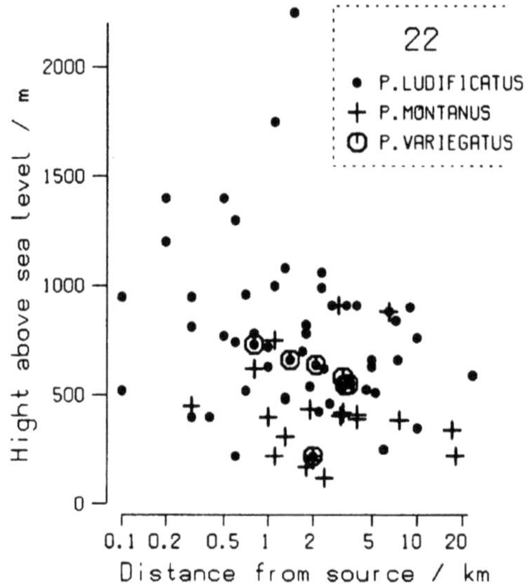


Figures 19-21. Localities where *Philopotamus* species were found.
Figure 22 : occurrence of the *Philopotamus* species in relation to altitude and distance from source.

All three species are widely distributed in mountainous areas in Central Europe. I did not find *P. montanus* in the Alps which is in accordance with Vaillant (1974, 1976). Only *P. montanus* reaches up north to Denmark and Scandinavia (Wiberg-Larsen and Andersen, in press). I do not know reliable records of any *Philopotamus* species from the North German Lowlands.

The habitat of all three species is the rhithral, mainly the epirhithral. From my findings and those of other authors (Burkhardt 1983, Szczesny 1978) *P. montanus* usually inhabits warmer streams at lower elevations, whereas the other two have a wider ecological range (see Fig. 22). However, in the Mediterranean *P. montanus* can also be found at high altitudes up to 2300 m (Décamps 1968, Vaillant 1974).

In several streams, I found *P. montanus* downstream of *P. ludificatus* populations, a situation which has been recorded in other studies as well (Burkhardt 1983, Dittmar 1955). *P. variegatus* is also noted by other authors (Burkhardt 1983, Szczesny 1978, Vaillant 1976) to be the rarest species, as in this study. I always found it together with *P. ludificatus,* the latter usually being more abundant.

All three species appear to be quite sensitive to organic pollution : they require clean, fast running water. As to the pH-value, *Philopotamus* seems to be rather tolerant. All three species were found not only in limestone areas, but also with other lithologies.

I have observed all three species flying in bright sunlight, but I have also found them in light trap samples in relatively small quantities. Malicky (in press) mentioned that the catchability of *P. ludificatus* by light traps is small and Thomas (*in* Lehmann 1970) found that light does not attract *Philopotamus.*

## ACKNOWLEDGEMENT

I thank B. J. Wilkinson for correcting the English text.

## REFERENCES

Botosaneanu, L. & Malicky, H. 1978. Trichoptera. *In* : Illies : Limnofauna Europaea, 2nd Ed., Stuttgart New York Amsterdam : 333-359.

Burkhardt, R. 1983. Untersuchungen zur Ökologie und Phänologie der Trichoptera-Arten des Vogelsberges mit besonderer Berücksichtigung ihrer Einnischung und Bedeutung als Indikatoren für den Zustand der Gewässer. Diss. Univ. Giessen.

Décamps, H. 1968. Vicariances écologiques chez les Trichoptères des Pyrénées. **Annls. Limnol.,** 4 : 1-50.

Dittmar, H. 1955. Ein Sauerlandbach. Untersuchungen an einem Wiesen-Mittelgebirgsbach. **Arch. Hydrobiol.,** 50 : 305-552.

Lehmann, U. 1970. Stromaufwärts gerichteter Flug von *Philopotamus montanus* (Trichoptera). Oecologia (Berl.), 4 : 163-175.

Malicky, H. 1983. Atlas of European Trichoptera. Junk : The Hague, Boston, London.

Malicky, H. (in press). Anflugdistanz und Fallenfangbarkeit von Köcherfliegen (Trichoptera) bei Lichtfallen.

Sedlak, E. 1985. Bestimmungsschlüssel für mitteleuropäische Köcherfliegenlarven (Insecta, Trichoptera). Wasser und Abwasser Wien, 29 : 1-146.

Szczesny, B 1978. Larvae of the genus *Philopotamus* Stephens 1829 (Insecta : Trichoptera) in Poland. **Acta Hydrobiol.,** 20 : 55-61.

Tobias, W. & Tobias, D. 1981. Trichoptera Germanica. Bestimmungstafeln für die deutschen Köcherfliegen. Teil I : Imagines. Cour. Forsch. -Inst. Senckenberg, 49 : 1-672.

Ulmer, G. 1903. Zur Trichopterenfauna von Thüringen und Harz. **Allg. Z. Ent.,** 8 : 341-350, 398-406.

Ulmer, G. 1909. Trichoptera. In : Brauer : Die Süsswasserfauna Deutschlands. Heft 5 und 6. G. Fischer : Jena.

Vaillant. F. 1974. Quelques Trichoptères *Philopotamidae* de France et d'Algerie. **Ann. Soc.ent. Fr.** (N.S.), 10 : 969-985.

Vaillant, F. 1976. Some *Philopotamidae* from France. **Proc. 1st Int. Symp. Trichoptera,** H. Malicky (ed.),Junk, The Hague : 25-31.

Wiberg-Larsen, P. & Andersen, T. In press. Revised check-list of the North-west European Trichoptera. **Ent. scand.**

# EMERGENCE AND DISTRIBUTION OF HYDROPTILIDAE IN THE LITTORAL AND OUTLET BIOCOENOSES OF LAKE KONNEVESI (CENTRAL FINLAND)

P. BAGGE


## SUMMARY

The emergence and spatial occurrence of hydroptilids in the littoral zone of L. Konnevesi (180 $km^2$) and its outlet (Siikakoski MQ 43 $m^3/s$) were studied by means of floating emergence traps (model week) and with slit traps in May-September 1983. A total of 15 species and 2281 specimens of hydroptilids were obtained. *Hydroptila tineoides* and *Oxyethira flavicornis* (which together comprised c. 80% of the material) were the dominant species in the lake, while *Hydroptila cornuta* was the most abundant hydroptilid in the slit trap material from the rapids. The total emergence of hydroptilids from the shallow stony littoral of the lake was 1276 ind/$m^2$ ant that in the outlet 691 ind/$m^2$. On shallow sandy bottoms, the corresponding densities were only 49 and 130 ind/$m^2$. The most abundant species were univoltine in the area. The maximum emergence of most species was observed in July, which coincides with the temperature maximum in the littoral.


## INTRODUCTION

In 1983, after the foundation of the Konnevesi Research Station, we started an intensive study on the distribution and emergence of aquatic insects in the littoral of L. Konnevesi and its outlet (Siikakoski). The material was sampled by means of different traps at10 localities in the lake and at 5 localities in the rapids just below the lake. It is therefore possible to compare the occurrence of the different species in the main habitats, which comprise the sandy and stony littoral of the lake, its outlet and the stony bottoms of the Siikakoski rapids. This paper deals with the seasonal and spatial occurrence of the hydroptilids whose ecology is poorly known in Finland as elsewhere.

## SAMPLING METHODS AND THE STUDY AREA

Two kinds of traps were used during May 24 - September 28 (Fig. 1). The floating emergence traps, each covering 0.36 $m^2$, were provided with a container of ethylene glycol and were emptied once a week. The slit traps (originally constructed by Kuusela and Pulkkinen (1978) for catching stoneflies), also containing a trough with ethylene glycol, were placed parallel to the stream and emptied every tenth day. The location of the traps in the SW corner of this large lake is shown in Figure 2.

The stony zone of the lake is relatively narrow and characterized by small or medium sized stones without macrophytes. In local areas, on sandy bottoms, *Lobelia dortmanna* and *Isoetes lacustris* form relatively dense meadows. In the stream bed, there are locally dense stands of *Myriophyllum alternifolium* and bryophytes. Helophytes, such as *Phragmites australis* are practically absent from the area, which may have an effect on the occurrence of some caddisflies.

In Finnish conditions, where the load of humic substances into natural waters is often very high, Lake Konnevesi is considered to be an oligotrophic clear water lake with the Secchi disk value of 5-6 m and colour value of less than 30 Pt units/1 (Table 1). In the study area the concentrations of nitrogen and phosphorus are relatively low, but both upstream and downstream of the study area the phosphorus concentrations are somewhat greater due to loading from large fish farms.

Table 1. Mean water quality in the Siikakoski rapids during the summer 1981 (Heinonen 1984).

| | | | |
|---|---|---|---|
| MQ, $m^3/s$ | 100* | pH | 7.2 |
| Temperature | 15 | Colour value | 29 |
| $O_2$-conc. mg/1 | 9.2 | Total nitrogen, µg/1 | 355 |
| Conductivity, mS/m | 4.3 | Total phosphorus, µg/1 | 9 |

*) Long term MQ = 43.5 $m^3/s$

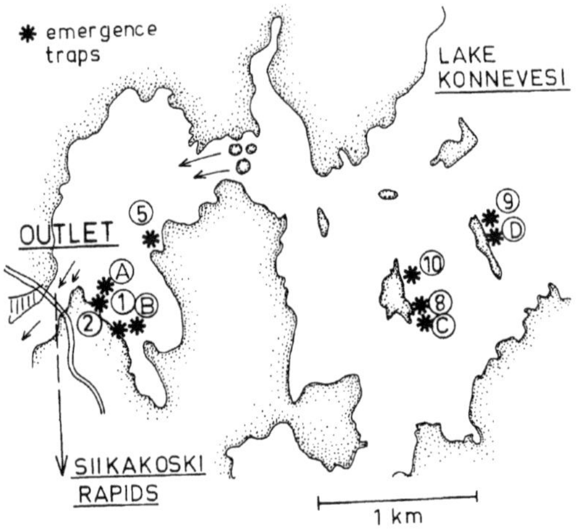


Figure 1. Sampling gears

Figure 2. The study area and location of the traps.

## RESULTS

### Seasonal occurrence of Hydroptilidae and other Trichoptera

A total of 79 species and c. 23 300 adult caddisflies were caught with traps. The corresponding numbers of hydroptilids were 15 and 2281, comprising 9.8% of the total number of caddisflies. The maximum emergence of Trichoptera was observed in the end of May and in the beginning of June when brachycentrids, especially *Brachycentrus subnubilus* was taken in large numbers. The emergence of hydroptilids was most active in July, when the temperature of the surface water was highest (17-21°C). A rapid decrease in the number of emerging hydroptilids was observed in the first half of August (Table 2). *Hydroptila cornuta* and *H. forcipata* emerged first and *H. annulata* and *H. pulchricornis* were late fliers. In the rapids of Siikakoski, emerging adults of some hydroptilids (e.g. *Hydroptila cornuta, H. lotensis* and *H. tineoides*) were observed c. two weeks earlier than in the littoral of the lake though the temperature of the surface water was about the same in both habitats. The most numerous species, *H. tineoides* and *Oxyethira flavicornis*, had relatively long emerging periods with different maxima. The results show that at least the most numerous species have only one generation per summer in Lake Konnevesi and its outlet. Females predominated in the trap material of most species (except in the population of *H. cornuta*, where the sex ratio was 1 : 1). The excess of females (3 : 1) in the populations may be true, but there is also a possibility that at least some of the females had entered the traps from outside.

Table 2 : Seasonal emergence of Hydroptilidae in 1983

| | PERIODS | | | | | | | |
|---|---|---|---|---|---|---|---|---|
| | JUNE | | JULY | | AUGUST | | SEPT. | |
| | 8.- 22. | 22.6.- 6.7. | 6.- 21. | 21.7.- 4.8. | 4.- 18. | 18.- 30. | 1.- 14. | 14.- 28. |
| *Agraylea cognatella* | - | - | 3 | 1 | 2 | - | - | - |
| *A. sexmaculata* | - | - | - | 1 | 1 | - | - | - |
| *Hydroptila angulata* | - | - | - | 37 | 19 | 1 | - | - |
| *H. comuta* | 1 | 34 | 85 | 37 | 2 | 2 | - | - |
| *H. forcipata* | 1 | - | - | - | - | - | - | - |
| *H. lotensis* | - | 1 | 2 | 30 | 8 | 4 | 1 | - |
| *H. pulchricomis* | - | - | - | 47 | 7 | 1 | - | - |
| *H. simulans* | - | - | - | 2 | - | - | - | - |
| *H. tineoides* | 2 | 326 | 499 | 100 | 4 | - | - | - |
| *Hydroptila (unidentif.)* | - | 2 | - | - | 1 | - | - | - |
| *Ithytrichia lamellaris* | - | 2 | 2 | - | - | - | - | - |
| *Orthotrichia tragetti* | - | - | - | 1 | - | - | - | - |
| *Oxyethira distinctella* | - | 5 | 40 | 14 | 1 | - | 1 | - |
| *O. flavicomis* | 1 | 63 | 238 | 575 | 53 | 6 | 4 | 1 |
| *O. frici* | | 2 | 3 | - | 2 | - | - | - |
| *O. sagittifera* | 1 | - | 1 | - | - | 1 | - | - |
| Total | 6 | 435 | 873 | 845 | 100 | 15 | 6 | 1 |

**Spatial occurrence of Hydroptilidae and other Trichoptera**

The total number of trapped caddisflies (Table 3) was highest on shallow sandy bottoms of the outlet area, where thousands of females of *Brachycentrus subnubilus* were caught in the beginning of June.

Table 3. Total emergence of Trichoptera and Hydroptilidae and the mean density of Hydroptilidae (ind./m$^2$).

| | Total n of Trichoptera | Total n of Hydroptilidae | Mean density ind./m$^2$ |
|---|---|---|---|
| Siikakoski, shore | 3252 | 22 | - |
| Siikakoski, middle | 764 | 113 | - |
| Outlet, stony zone | 2423 | 514 | 691 |
| Outlet, sandy bottom | 12638 | 94 | 130 |
| Lake, sheltered stony zone | 269 | 124 | (343) |
| Lake, exposed stony zone | 3002 | 1379 | 1276 |
| Lake, sandy bottom | 954 | 35 | 49 |
| Total | 23302 | 2281 | |

The total density of emerging hydroptilids was highest in the exposed stony belt of lake ranging from c. 350 to 3000 ind./m$^2$ (Table 3). Lowest densities (c. 50 ind./m$^2$) were observed on sandy bottoms.

The dominant species in the littoral zone of the lake were *Hydroptila tineoides* and *Oxyethira flavicornis* which reached maximum densities in the exposed stony belt. *H. pulchricornis* was most abundant at the sheltered locality (Table 4).

*H. tineoides* and *O. flavicornis* together made up more than 80% of all the hydroptilids in the littoral habitats of the lake. In other larger lakes of the Finnish Lake District, where the hydroptilid fauna has been studied more intensively, *O. flavicornis* clearly dominated and comprised c. 95% of all the hydroptilid larvae in L. Leppävesi (Valovirta 1966) and c. 93% of adults sampled by means of emergence traps in the littoral of L. Pääjärvi (Paasivirta 1975).

Table 4. Total emergence (ind./trap or ind./$m^2$) of the most common Hydroptilidae in different habitats (stony/sandy bottom).

| | | | LAKE | |
|---|---|---|---|---|
| | SIIKAKOSKI ind./trap | OUTLET ind./$m^2$ | SHELTERED ind./$m^2$ | EXPOSED ind./$m^2$ |
| *Hydroptila angulata* | 0 | 8/11 | 3 | 36/3 |
| *H. cornuta* | 24 | 38/1 | 3 | 11/0 |
| *H. lotensis* | 1 | 48/7 | 0 | 3/0 |
| *H. pulchricornis* | 0 | 28/1 | 72 | 9/0 |
| *H. tineoides* | 1 | 207/21 | 201 | 600/34 |
| *Oxyethira distinctella* | 0 | 57/18 | 15 | 1/0 |
| *O. flavicornis* | 1 | 274/71 | 78 | 580/20 |

The hydroptilid community of the outlet area was dominated by *Oxyethira flavicornis*, *O. distinctella* and *Hydroptila tineoides*.

In the rapids of Siikakoski, *Hydroptila cornuta* dominated in the traps (c. 87% of all hydroptilids). This result differs from that observed in the riffles of the river Lestijoki (Western Finland), where *Hydroptila tineoides* and *Ithytrichia lamellaris* comprised c. 78% of the larval material collected from stones (Kuusela 1979).

CONCLUSIONS

The hydroptilid fauna of Lake Konnevesi and its outlet is relatively rich in species harbouring 15 of the 27 species recorded in Finland. The great diversity of the hydroptilid fauna of the area is explained by the availability of several biotopes, the good quality of water and good growing conditions of periphytic algae, which are the main requirements of most hydroptilids (Marshall 1979). The greatest diversity of hydroptilids (13 species) was observed in the outlet area of the lake, where both limnetic and lotic species occur together. Most of the species observed in the study area occur commonly throughout the country, but there are only few records of such species as *Agraylea sexmaculata* and *Orthotrichia tragetti* in Finland and of *Agraylea cognatella* in southern Finland.

Compared with some other Finnish water bodies, where the emergence of caddisflies has been studied during the icefree period, the densities of hydroptilids are unexpectedly high in the littoral zone of L. Konnevesi (Table 5). Especially high densities were observed in the exposed stony zone of the lake, where hydroptilids made up c. 46% of all the collected caddisflies. In the outlet area of L. Konnevesi the densities of hydroptilids were lower though the diversity of species was high. Thus, the outlet has no positive effect on the abundance of the most common species.

Compared with smaller water bodies, the surface water of large lakes such as L. Konnevesi is relatively cool in spring and even in June. Thus it could be expected that the flight period of caddisflies in the lake would begin somewhat later and last longer than in smaller water bodies. In order to test this hypothesis, the emerging periods of hydroptilids from the study area are compared with data collected by Nybom (1960) from different Finnish waters (Fig. 3). Since this older material contains also data from large lakes, the differences are not so great. However, for the *Agraylea* species, *H. angulata* and *H. pulchricornis* the flight period is later than is usual in Finland. In addition, the flight periods of *Hydroptila cornuta, H. lotensis, Oxyethiera distinctella* and *O. flavicornis* are unexpectedly long in the area. In the case of *H. cornuta* and *H. lotensis* the long flight period is partly explained by the fact that their adults emerged c. two weeks earlier in the rapids of Siikakoski than in the littoral of the lake.

Table 5. Mean density (ind./$m^2$) of Hydroptilidae in some Finnish lakes and in the brackish water.

| | stony bottom | sandy bottom |
|---|---|---|
| 1 L.K-vesi, exposed | 1276 | 49 |
| 1 L.K-vesi, sheltered | 343 | - |
| 1 L.K-vesi, outlet | 691 | 130 |
| 2 L. Pääjärvi | 200 | 100 |
| 3 Gulf of Finland | - | 120 |

1 = this study
2 = Paasivirta 1975
3 = Bagge *et al.* 1980

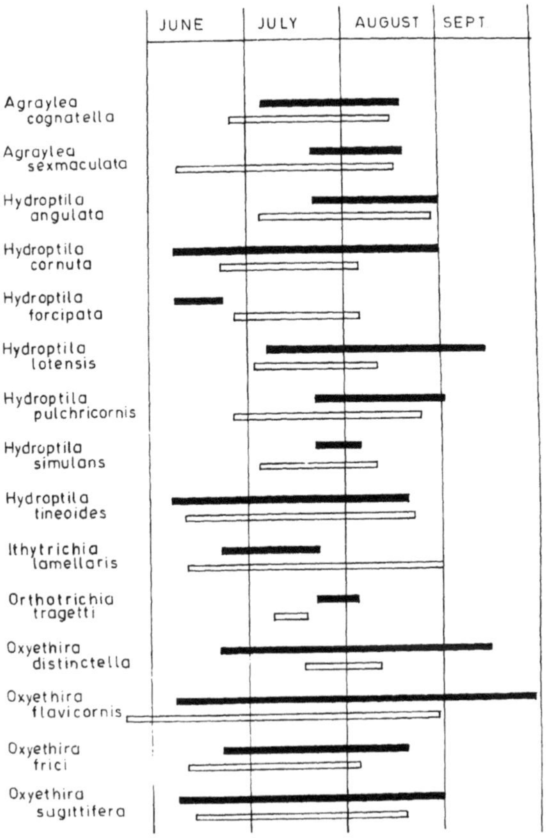


Figure 3. Flight periods of hydroptilids in the study area (▬) and in different waters of Finland (▭).

## REFERENCES


Bagge P., Ilus E. & Paasivirta, L. 1980. Emergence of insects (esp. Diptera, Chironomidae) at different depths in the archipelago of Lovisa (Gulf of Finland) in 1971. **Ann. Ent. Fenn.**, 46 (4) : 89-100.

Kuusela, K. 1979. Early summer ecology and community structure of the macrozoobenthos on stones in the Jäväjänkoski rapids on the river Lestijoki, Finland. **Acta Universitatis Ouluensis, Ser. A.**, 87 (6) : 1-123.

Kuusela, K. & Pulkkinen, H. 1978. A simple trap for collecting newly emerged stoneflies (Plecoptera). **Oikos**, 31 : 323-325.

Marshall, J. 1979. The review of the genera of the Hydroptilidae (Trichoptera). **Bull. Br. Mus. Nat. Hist. (Ent.)**, 39 (3) : 135-239.

Paasivirta, L. 1975. Insect emergence and output of incorporated energy and nutrients from the oligotrophic lake Pääjärvi, southern Finland. **Ann. Zool. Fenn.**, 12 : 126-140.

Valovirta, M. 1966. Ecological differences between caddis larvae (Trichoptera) found in lake Leppävesi (Northern Tavastia). Mimeogr. pro gradu study Dept. of Zoology, Univ. Helsinki. 85 p. (in Finnish).

# TRICHOPTERA COLONIZATION IN THE IRRIGATED MEADOWS OF NORCIA (UMBRIA, ITALY)

F. CIANFICCONI and G.P. MORETTI

SUMMARY

The repartition biozones of 20 Trichoptera species are indicated for the "Marcite" or irrigated meadows of Norcia, where only 2 of these (*Sericostoma italicum, Odontocerum albicorne*) are common to the entire water system. Five species are endemic to the Italian fauna (*Rhyacophila foliacea, R. rougemonti, Polycentropus malickyi, P. mortoni, S. italicum*). Two species were noted for the first time in Italy (*Adicella cremisa, Hydropsyche* sp.n. ?), and the former, the morphological characteristics of the larval head and case are illustrated.

In the south-east of Umbria in central Italy, at the foot of the Sibillini Mountains at an altitude of 570-590 m above sea-level, are to be found the permanently irrigated meadows called "Marcite di Norcia" (Map of Italy, F. 132 III NE - III NW). Unique in Italy, apart from those better known in Lombardy (Moretti 1957), these meadows apparently date back to the VI century according to the agricultural techniques used by the Benedictines. The "Marcite" cover an area of about 100 ha situated on the residue sediment of an ancient lake bed. The slightly sloping land allows water to flow gradually over the surface of this area. The meadows are squared off into "cortinelle" by so-called irrigation and collection ditches or "canals" which constitute the means by which the meadows are flooded (Fig. 1).

The waters which supply the "Marcite" derive from the numerous springs to be found in the Norcia area. In part, these springs constitute the source of the River Sordo, into which the larger part of the waters from the "Marcite" flow, in part they are channelled into pools (Fig. 2) and irrigation canals where flow is regulated by man-made wooden mini-locks (Fig. 3). In summer, the farmers take turns with the irrigation while in winter this is continuous thus preventing the ground from freezing. The fields are therefore always green allowing for from six to ten forage harvests per year. In autumn the last hay harvest is left to "rot" (meaning of the verb "marcire" in Italian) on the meadows in order to enrich them with organic matter, hence the Italian name "Marcite" (Fig. 4).

Besides their importance in agriculture, the "Marcite" are extremely interesting from the hydrobiological point of view for the wealth and variety of the biotopes peculiar to them, i.e., springs, pools, small falls, hygropetric walls, spring-created brooks, "fontanili" or limnocrenic springs, irrigation and collection canals and the River Sordo itself (Fig. 5).

The irrigation canals run between rows of the willow (*Salix alba*) and Lombardy poplars (*Populus nigra italica*) or are hidden by the leaves of *Petasites*. The hydrophytes *Callitriche palustris, Potamogeton crispus, Polygonum amphibium, Ranunculus trichophyllus, Nasturtium officinale* form rich clusters. The limnocrenic springs are carpeted with floating *Lemna gibba* and *L. minor* (Orsomando 1981).

Research carried out at intervals during the seasons over the five-year period 1980-85 resulted in a few physical-chemical parameters illustrated in Figure 6. Water temperature was maintained at values of 10-13° C in all biotopes, except in the case of the River Sordo where it reached 15° C. Oxygen was invariably below saturation value except for the small falls (108%), while pH varied from neutrality to slight alkalinity in the limnocrenic springs (7.8). Total hardness and $=SO_4$ were higher in the hygropetrics (43° F, 210 mg/1, respectively) and in the limnocrenic springs (39° F, 211 mg/1). Organic matter, $\equiv PO_4$, $-NO_2$ showed high values in both the collection canals and the river, as did $-NO_3$ in the pools and limnocrenic springs (4.4 mg/1), while traces of $NH_3$ were present everywhere. Conductivity varied from 487 (springs) to 622 (limnocrenic springs) µS at 18° C. The waters were clear, turbidity reaching a maximum of 10-28 JTU.

Unfortunately several of the canals receive waste discharge from both an abattoir and a hospital which has altered the chemical properties of the water, thus seriously damaging the Trichoptera fauna, a situation which it is hoped will be improved by the installation of a depurator plant which will become operative in the near future.

Figure 1-4. The "Marcite di Norcia". 1, "Cortinelle" ; 2, Pool with small falls and hygropetric wall ; 3, Collection canal with mini-locks ; 4, Final autumn harvest left to "rot" on the meadows.

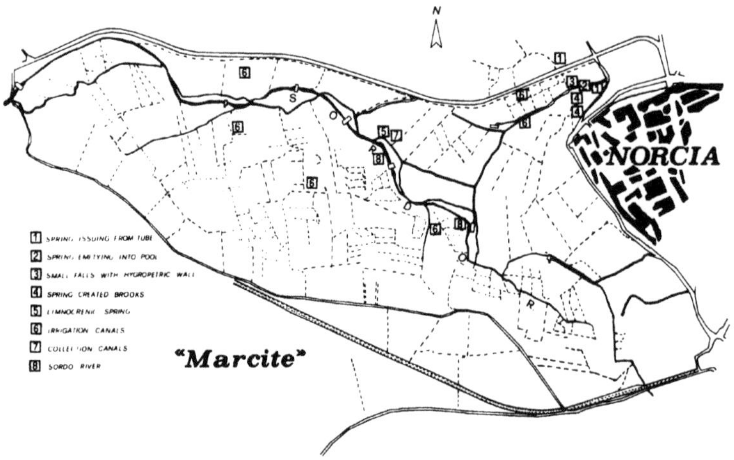


Figure 5. Map of the "Marcite" and micro-environments examined.

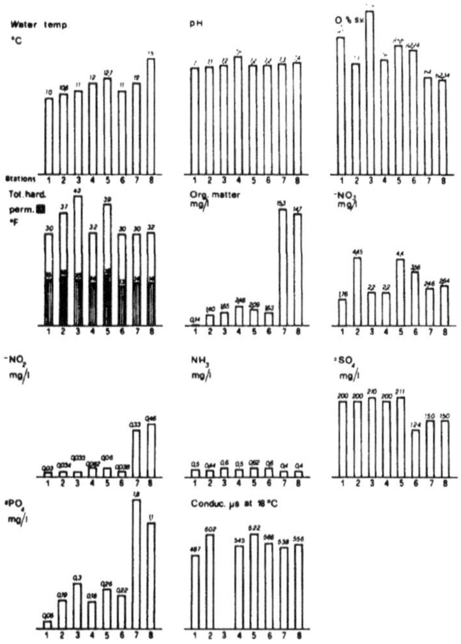

Figure 6. Average physical-chemical values of "Marcite" waters in 1980-85.

The species of Trichoptera found in the various aquatic biotopes are listed in systematic order in Table 1.These total 20, divided into 13 families. Five are peculiar to the Italian fauna : 3 (*R. foliacea, P. malickyi, S. italicum*) mainly to the central Apennines, 1 (*R. rougemonti*) to the southern central Apennines and Sicily, for which the "Marcite di Norcia" represent to date, the northern most limit to its distribution. *P. mortoni* has a wide distribution in the central southern Apennines, Sicily, Sardinia and C

Two species were noted for the first time in Italy : *Adicella cremisa* (Malicky 1972) (Fig. 7) and *Hydropsyche* sp.n. ? which is a subject of research in progress.

An overall ecological evaluation has shown a wealth of subdivision of crenophile and rheophile specimens, with a host of specializations going from eucrenal (*R. rougemonti, P. malickyi*) and limnocrenal (*L. lunatus*) through a complex succession of hygropetrics (*T. maclachlani, C. irrorata, B. maurus*) and hypocrenal (*R. foliacea, A. cremisa*) down to epirhythral (*A. nimbulus, P. mortoni*). None of the species seem to have established themselves permanently in the "cortinelle" or in the collection canals with the earthy grass-grown beds.

Only two species have a wide distribution which includes the irrigation canals where the canal beds are covered in pebbles, sand or stones. The species are *S. italicum* and *O. albicorne* which have found optimum conditions in these clear waters with their constant temperature and wealth of trophic vegetable resources.

*S. italicum* is the most abundant species representing 38% of the total specimens collected, followed by *O. albicorne* and *R. rougemonti* (Fig. 8). Each species has a different biological cycle from spring emergence (*A. cremisa, Hydropsyche* sp.n. ?), summer emergence (*P. conspersa, C. irrorata, B. maurus*), spring-summer emergence (*S. italicum, O. albicorne*), to autumn-winter (*C. gessneri*), and winter emergence (*R. rougemonti, W. occipitalis*). In nearly all the species the sex ratio is in favour of the males.

Table 1. Systematic list of the Trichoptera species found in the micro-environments examined.

| Species \ stations | 1 | 2 | 3 | 4 | 5 | 6 | 7 | 8 |
|---|---|---|---|---|---|---|---|---|
| RHYACOPHILIDAE | | | | | | | | |
| 1- *Rhyacophila foliacea* Moretti | | | | 0 | | | | 0 |
| 2- *Rhyacophila rougemonti* McL. | 0 | 0 | | | | | | |
| GLOSSOSOMATIDAE | | | | | | | | |
| 3- *Agapetus nimbulus* McL. | | | | | | | | 0 |
| HYDROPTILIDAE | | | | | | | | |
| 4- *Hydroptila martini* Marshall | | 0 | | | | | | |
| PHILOPOTAMIDAE | | | | | | | | |
| 5- *Wormaldia occipitalis* Pictet | | 0 | 0 | 0 | | | | 0 |
| HYDROPSYCHIDAE | | | | | | | | |
| 6- *Hydropsyche pellucidula* Curt. | | | | 0 | | | | |
| 7- *Hydropsyche* sp.n. ? | | | | 0 | | | | 0 |
| POLYCENTROPODIDAE | | | | | | | | |
| 8- *Plectrocnemia conspersa* Curt. | | 0 | | 0 | | 0 | | |
| 9- *Polycentropus malickyi* Moretti | | 0 | | | | | | |
| 10-*Polycentropus mortoni* Mosely | | | | | | | | 0 |
| PSYCHOMYIDAE | | | | | | | | |
| 11-*Lype reducta* Hagen | | | 0 | 0 | | | | |
| 12-*Tinodes maclachlani* Kimmins | | | 0 | | | | | |
| LIMNEPHILIDAE | | | | | | | | |
| 13-*Limnephilus lunatus* Curt. | | | | | 0 | | | 0 |
| 14-*Chaetopteryx gessneri* McL. | | 0 | 0 | | | | | |
| LEPIDOSTOMATIDAE | | | | | | | | |
| 15-*Lepidostoma hirtum* Fabr. | | | | 0 | | | | |
| 16-*Crunoecia irrorata* Curt. | | | 0 | | | | | |
| LEPTOCERIDAE | | | | | | | | |
| 17-*Adicella cremisa* Mal. | | | | 0 | | | | |
| SERICOSTOMATIDAE | | | | | | | | |
| 18-*Sericostoma italicum* Moretti | 0 | 0 | 0 | 0 | | 0 | | 0 |
| BERAEIDAE | | | | | | | | |
| 19-*Beraea maurus* Curt. | | | 0 | | | | | |
| ODONTOCERIDAE | | | | | | | | |
| 20-*Odontocerum albicorne* Scop. | 0 | 0 | | 0 | | 0 | | 0 |

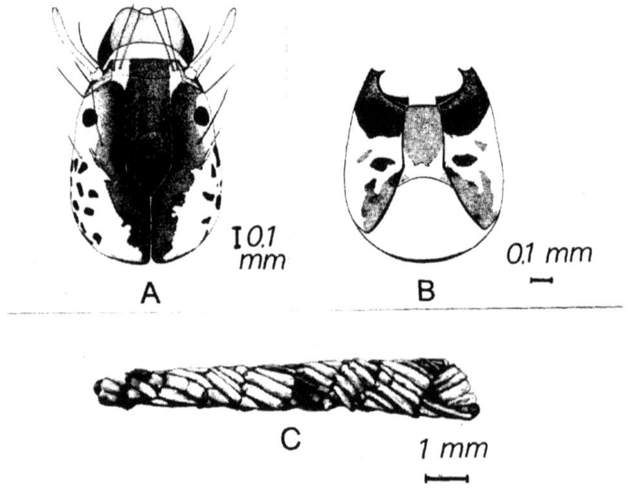


Figure 7. *Adicella cremisa*. A, larval head, dorsal ; B, larval head, ventral ; C, straight, cone-shaped larval case built of cuttings from willow roots and rhizomes of *Epilobium parviflorum* in a spiral arrangement. The pupal case is that of the larva, but shortened and anchored to submerged willow roots.

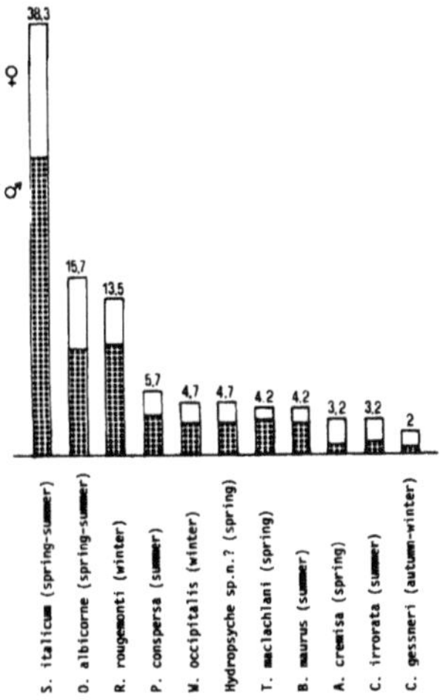


Figure 8. Number of specimens (%) for 12 Trichoptera species found in 1980-85. Sex ratio. Emergence periods.

## REFERENCES


Botosaneanu, L. & Malicky, H. 1978. Trichoptera. In : Illies, Limnofauna Europaea 333-359. Fischer, Stuttgart.

Botosaneanu, L. & Novak, K. 1965. Les espèces européennes du genre *Adicella* Mcl. (Trichoptera). **Acta ent. bohemoslov.**, 62 : 468-479.

Malicky, H. 1972. Weitere neue Arten und Fundorte von westpaläarktischen Köcherfliegen (Trichoptera), vor allem aus dem oestlichen Mediterrangebiet. **Mitt. ent. gesell. Basel**, 22 (2/3) : 39/65.

Malicky, H. 1983. Atlas of European Trichoptera. **Series Entomol.** 24, Junk, The Hague.

Moretti, G.P. 1957. I tricotteri delle Marcite. **Boll. Zool. Agr. Bach. Milano**, 22 : 189-215.

Moretti, G.P. & Cianficconi, F. 1981. First list of Italian Trichoptera. **Proc. 3rd Int. Symp. Trichoptera**, G. P. Moretti (ed.), Junk, The Hague, 201-213

Orsomando, E. 1981. Le Marcite di Norcia. **Umbria Economica, Spoleto**, 2 (3) : 63-71.

# EFFECTS OF AN ARTIFICIALLY SILTED STREAM BOTTOM ON SPECIES COMPOSITION AND BIOMASS OF TRICHOPTERA IN BREITENBACH

R.H. WAGNER

## SUMMARY

By putting a weir immediately below an emergence trap, current velocity was reduced, and thus a layer of fine sand was deposited. There was a significant decrease in specimen numbers for most of the dominant taxa, *Agapetus fuscipes*,*Tinodes rostocki*, *Sericostoma personatum*, and *Rhyacophila fasciata*. No effect was observed for *Apatania fimbriata* and *Chaetoptery villosa*.Trichoptera biomass was reduced for about 1/3 to 2/3. The amount of specimens in the sandy trap showed a heavy influence of the mode of emergence in between different species. Drift distances during emergence vary from about 50 cm to 10 m, depending on current velocity and on the species.

## INTRODUCTION

Discharge and current velocity are major determinants of the patchy distribution of vertically and horizontally variable structured sediments. Current velocity is dependent on features of the drainage area : its size, the amount of rainfall, the geology of the drainage area, the incline of the stream-bed, the topography and land-use (forest, meadows, fields).

Current velocity and its ability to move particles influence many events in the stream. Subtratum conditions have been used to define longitudinal subdivision on running waters with good fit to the distribution patterns of grazers and filter feeders.

Longitudinal zonation is clearly evident, but the pattern of sediment structure in streams is much more complicated. Patterns of current velocities in all streams are complex, ranging from high in the middle of the stream channel to reversal currents near the banks. This also leads to a transverse zonation with respect to current direction with erosion zones in the middle of the stream channel and sedimentation zones near the streambanks.

The combination of wide current amplitude and stream bottom provides a range of microhabitats, which are colonized only by well adapted species.

The interaction between current velocity, stream bottom, and biocoenoses therefore is of fundamental interest in running water ecology. A recent literature overview was provided by Tolkamp (1980). Because laboratory experiments with selected larvae and selected particle size were virtually useless (larval behaviour in the laboratory differed markedly from behaviour in the field, e.g. substratum choice was often governed by the choice of food, personal observation), a field experiment combined with the collection of adult merolimnic insects using emergence traps was planned.

## MATERIAL AND METHODS

The study was carried out in the Breitenbach, F.R. of Germany, a small stream flowing through a sandstone area. Under normal flow conditions (20 1/sec) the spaces in the sediment are filled by fine sandy material, together with fine detritus, so that water exchange is hindered or even prevented. This leads to almost or totally anoxic zones in the sediment, at depths sometimes less than 1-2 cm. Most aquatic insects, particularly caddis larvae, are restricted to the well-aerated uppermost sediment layer.

By placing a weir immediately below the more downstream emergence trap, current velocity was reduced from 40 - 80 cm/sec to 20 - 40 cm/sec in 1981.A fine uniform sand layer was thus deposited. Its maximum grain size was 2.5 mm, but almost 98% of the sediment weight was of less than 1 mm diameter grain size.

Insects were collected daily from 2 traps. They were preserved in alcohol, identified to species, dried and weighed.

Statistical comparison from catches of both sites was carried out using a sign-test for defined time intervals (decades, months, emergence curves can be used, example see figure 1).

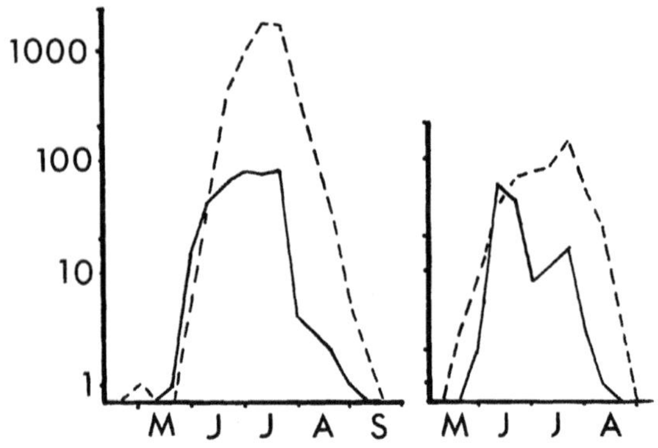


Figure 1. Emergence curves of *Agapetus fuscipes* 1983 and 1984 over a sandy stretch (full line) and over an untreated stretch (broken line) in Breitenbach.

RESULTS

In a long term study since 1969 more than 50 species of Trichoptera have been found in the Breitenbach. In the middle course *Agapetus fuscipes* Curtis, *Tinodes rostocki* McLachlan, *Apatania fimbriata* (Pictet), *Chaetopteryx villosa* (Fabricius), *Sericostoma personatum* (Kirby & Spence), and *Rhyacophila fasciata* Hagen are the dominant taxa, as far as biomass is concerned. *A. fuscipes, T. rostocki,* and *A. fimbriata* are chiefly grazers, *C. villosa* and *S. personatum* are predominantly shredders, while *R. fasciata* is a predacious insect. For specimen numbers see table 1, for biomass table 2.

Table 1. Specimen numbers of dominant Trichoptera caught in emergence traps over a sandy stretch (S) and over an undisturbed reference stretch (R) 1981 - 1984.

| | | 1981 | 1982 | 1983 | 1984 |
|---|---|---|---|---|---|
| *Agapetus fuscipes* | S | 166 | 251 | 354 | 143 |
| | R | 1597 | 2313 | 4832 | 491 |
| *Apatania fimbriata* | S | 703 | 535 | 513 | 898 |
| | R | 463 | 644 | 621 | 718 |
| *Tinodes rostocki* | S | 112 | 832 | 889 | 254 |
| | R | 346 | 648 | 2339 | 1081 |
| *Chaetopteryx villosa* | S | 303 | 121 | 266 | 629 |
| | R | 289 | 263 | 266 | 821 |
| *Sericostoma personatum* | S | 42 | 28 | 26 | 87 |
| | R | 63 | 66 | 38 | 111 |
| *Rhyacophila fasciata* | S | 31 | 27 | 10 | 11 |
| | R | 139 | 82 | 32 | 37 |

Larvae of the glossosomatid *Agapetus fuscipes* prefer to live on stones and gravel more or less exposed to the current. For a few weeks in the spring, current velocity can hinder the movement of larvae onto the surface of the stones. The majority of the population is then found in the dead water zones behind the stones, or at their sides. Larvae of *A. fuscipes* are unable to remain on unstable sediments such as sand. It was therefore not surprising, that specimen numbers and biomass decreased significantly ($p < 0.05$) over the sandy stretch.

Larvae of *Tinodes rostocki* are found in more extreme situations, because they prefer to build their galleries on the upper surface of gravel and stones. Larvae cannot change their position against the current, but they seem to be well sheltered. Larvae without a gallery transported to the laboratory were unable to build another one. Comparatively many specimens have been found in the trap over the sandy stretch. This is mainly dependent on the isolated position of the pupae in the middle of the stream. Emerging pupae have to cross areas of high current speed and normal drift distances are therefore exceeded. In the sandy stretch current velocity is lowered and numbers of emerging insects are raised, because they can here easely reach the bank. Specimen numbers and biomass decreased ($p < 0.05$) over the sandy stretch.

*Apatania fimbriata* is typical of small Central European streams. It has a univoltine life cycle, although a few percent of the population overwinter twice. Larvae are often found in the same habitats as *T. rostocki* larvae. They seem to undergo winter dormancy ; aggregations of up to 100 larvae may be found in dead water zones and underneath larger stones. More specimens of this species have been caught over the sandy stretch of the Breitenback. This is surprising but it resembles the *T. rostocki* situation. The pupae of *A. fimbriata* are also attached to larger stones in the middle of the stream and they have similar problems in reaching the bank as does *T. rostocki*. Differences in biomass and specimen numbers are not significant.

The major shredder among the Breitenback caddis flies is *Chaetopteryx villosa*. Its life history is well known and its temperature dependence has been demonstrated elsewhere (Wagner, in press). *C. villosa* is the latest caddis fly emerging from September until December. In the Breitenbach it has a univoltine life cycle but it may be semivoltine at extreme elevations in the Alps or in Scandinavia. Larval development occurs during spring but mainly in summer. It is extremely desynchronized and during April and May all five larval instars coexist in the brook. There was no evident preference of emerging specimens for one of the study areas.

The larvae of *Sericostoma personatum* are known to have a diurnal rhythm moving from the stream bottom (night) into the sediment (day). Its life cycle is not well known. Life cycle data from various authors record 1, 2, 3, or more years' life cycles. Laboratory experiments with larvae of this species show that the life cycle period is strongly temperature dependent. At constant 14°C, the life cycle is completed in 1 year., while at 7°C it will last about 3 years (experiments not yet finished). Female place the egg masses, structured like these of nearctic Sericostomatidae (Wood *et al.* 1982), underneath stones in the main currrent. A few egg masses develop immediately and rapidly (about 35 days at 12°C) while many others seem to undergo a diapause and hatch at the beginning of the next spring. This is possibly the reason for the contradictory life cycle results of different scientists. During the life cycle *S. personatum* has 7 larval instars. The results of the emergence diagrams, show that *S. personatum* avoids the sandy stretch ($p < 0.5$). There are various reasons : 1. larvae prefer places with sand under stones and pupal case are fixed on stones, but stones were not available in the sandy stretch of the Breitenbach. 2. in the sandy stretch larvae cannot live in the sediment, because it is anoxic a few mm below the surface. Most of the adults must therefore have entered the trap via drift.

The life cycle of *Rhyacophila fasciata* is difficult to interpret and while many scientists have tried to elucidate it, most results were contradictory. It is possibly of high plasticity with specimens completing their life cycle within 1, 1,5 or 2 years. Larvae have only been found in stony reaches of the Breitenbach and pupae are always attached underneath large stones. Therefore all specimens in the trap over the sandy area must have entered it via drift during emergence. The reduction of specimen numbers and biomass over the sandy stretch is significant ($p < 0.05$).

## INFLUENCE OF DRIFT ON EMERGING TRICHOPTERA

With the help of two traps and sand-avoiding species it is possible to estimate roughly the drift distances of emerging caddis flies. For example, no pupae of *Rhyacophila fasciata* occurred in the sandy stretch, because the necessary substratum was not available. Specimens occurring in the trap must therefore have entered it during emergence. If the drift distance is considered approximately constant for a single species, the percentage of specimens in the sandy trap, in relation to those in the unmodified trap, gives an estimate of drift distance. Percentage of specimens in the sandy stretch gives the drift distance, compared to the length of a trap (6 m). For results see table 3.

These drift distances will vary from time to time, because they are also dependent on the current velocity. Faster currents will result in greater drift distances, because emerging pupae have to cross areas of high current speed. They will also be influenced by discharge rates. The comparatively low value for *Agapetus fuscipes* is due to the attachment of pupal cases in reaches near the bank where current velocities are low. The maximum value for *Tinodes rostocki* is affected by the very isolated position of the pupae in the middle of the stream.

Table 2. Biomass of Trichoptera over sandy (S) and undisturbed reference stretch (R) at Breitenbach 1981-1984 (mg/yr).

| | | 1981 | 1982 | 1983 | 1984 |
|---|---|---|---|---|---|
| *Agapetus fuscipes* | S | 83.0 | 125.5 | 177.0 | 62.5 |
| | R | 798.5 | 1156.5 | 2416.0 | 236.2 |
| *Apatania fimbriata* | S | 724.1 | 551.1 | 528.4 | 781.3 |
| | R | 476.9 | 663.3 | 639.6 | 652.3 |
| *Tinodes rostocki* | S | 88.5 | 657.3 | 702.3 | 131.5 |
| | R | 273.3 | 511.9 | 1847.8 | 552.8 |
| *Chaetopteryx villosa* | S | 1327.1 | 530.0 | 1165.1 | 2681.2 |
| | R | 1265.8 | 1151.9 | 1165.1 | 3682.0 |
| *Sericostoma personatum* | S | 202.4 | 135.0 | 125.3 | 372.3 |
| | R | 303.7 | 318.1 | 183.2 | 486.2 |
| *Rhyacophila fasciata* | S | 131.4 | 114.5 | 42.4 | 28.5 |
| | R | 589.4 | 347.4 | 135.7 | 123.3 |
| others | S | 611.4 | 467.6 | 383.9 | 746.9 |
| | R | 857.7 | 1213.0 | 865.7 | 1119.0 |
| Total biomass | S | 3167.9 | 2581.0 | 3124.4 | 4804.2 |
| | R | 4565.3 | 5362.1 | 7253.1 | 6851.8 |

Table 3 : Assumed drift distance for emerging caddis flies, minima and maxima (cm).

| | min | max |
|---|---|---|
| *Agapetus fuscipes* | 45 | 150 |
| *Apatania fimbriata* | 495 | 910 |
| *Tinodes rostocki* | 145 | 775 |
| *Chaetopteryx villosa* | 275 | 630 |
| *Sericostoma personatum* | 250 | 475 |
| *Rhyacophila fasciata* | 133 | 200 |

ACKNOWLEDGEMENT

Dr. E. Cox, Schlitz, is sincerely thanked for linguistic help.

REFERENCES

Tolkamp, H.H. 1980. Organism-substrate relationships in lowland streams. Centre for Agricultural Publishing and Documentation, Wageningen, 211 p.

Wagner, R. (in print). Egg development and life cycle of *Chaetopteryx villosa* (Fabr.) (Insecta, Trichoptera). **Holarct. Ecol.**

Wood, J.R., Resh, V.H. & Mc Ewan, E.H. 1982. Egg masses of Nearctic Sericostomatid caddisfly genera (Trichoptera). **Ann. entomol. Soc. Am.**, 75 : 430-434.

# STUDIES ON *PLECTROCNEMIA CONSPERSA* (CURTIS) IN COPPER CONTAMINATED STREAMS IN SOUTH WEST ENGLAND

S.T. DARLINGTON, A.M. GOWER and L. EBDON

## SUMMARY

The life cycle of *Plectrocnemia conspersa* (Curtis) took one year in both a copper contaminated stream and in a control stream. Larvae were more aggregated and occurred at higher densities in the contaminated stream. In the larval populations of both streams there was an exponential decrease in copper concentration with increasing body weight, with the highest concentrations (up to 4812 $\mu gg^{-1}$) in the youngest larvae analysed (instar II). Copper concentrations in all the larval instars analysed exhibited seasonality. The results are discussed in relation to other observations for *P. conspersa*, and it is suggested that its importance as an indicator of metal enriched conditions is not confined to S.W.England, nor specifically to copper.

## INTRODUCTION

South West England is heavily mineralised especially in the areas of the Land's End and Bodmin Moor granites and their metamorphic aureoles. This area of Cornwall has a long history of mining activity which has resulted in elevated metal levels in the water and sediments of mine-drainage streams (Thornton, 1983). However, apart from Brown (1976) who investigated the tolerance of *Asellus meridianus* to lead and copper in two Cornish rivers, there have been few detailed investigations on the tolerant organisms associated with metal contaminated streams in S.W.England. A prominent feature of the species-poor communities in such waters, especially those affected by copper, is the relative abundance of the net-spinning caddis *Plectrocnemia conspersa* (Curtis). Although the life cycle of this species has been studied by Tachet (1967) and Hildrew & Townsend (1982) no previous study has been undertaken on its life cycle and population characteristics in conditions of such high metal concentrations.

As freshwater invertebrates can accumulate metals they may have a role as in situ biomonitors of metal pollution, and previously most workers have pooled a number of individuals of the selected species to allow analysis by flame atomic absorption spectroscopy. Thus little or no information is available on metal levels in an individual, or on the variation associated with a particular mean concentration.

This paper presents the results of a quantitative study into the life-cycle of *P. conspersa* in a copper contaminated stream and in a control stream, and information is also given on the copper concentrations in the larvae. Finally the prominence of this species in metal rich waters in S.W. England is considered in relation to selected records of its occurrence elsewhere in streams with elevated metal concentrations.

## STUDY AREAS AND METHODS

A number of streams in S.W. England have been sampled during the course of this investigation where *P. conspersa* and *P. geniculata* are evident in aquatic communities. The main site was on the edge of the Bodmin Moor granite (Fig.1) where a number of streams are affected by high copper and zinc concentrations from past mining activity.

Darley Brook is one such stream which originates as groundwater flow from the adit of an old copper mine (Nat. Grid Ref. SX 265723, altitude approximately 240 m). In the upper reaches (site 1) the water is crystal clear with virtually all of the copper (mean concentration 0.85 $mgl^{-1}$) in 'soluble' form. There is a luxuriant growth of the alga *Microspora* and the leafy liverwort *Jungermannia atrovirens*, and the macro-invertebrate community is reduced to chironomid larvae, *P. conspersa*, and an *Agabus* complex of larvae and adults. *P. conspersa* is the most conspicuous invertebrate because of its numbers, the relatively large size of the final instar larvae and the nets which are widespread and prominent in both riffles and pools. After flowing approximately 1 km an uncontaminated stream (site 2) supporting a diverse biotic community and with a mean copper concentration of < 0.04 $mgl^{-1}$ enters the Brook, which flows another 2 km before its confluence with the River Lynher. Sites 1 and 2 are comparable riffle sections.

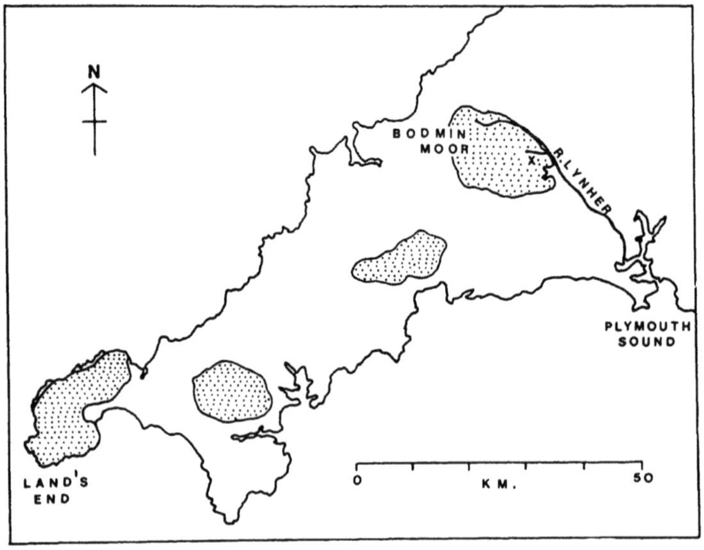


Figure 1. Location of the Darley Brook (x) and River Lynher in Cornwall, also showing the main granite areas.

The life cycle and population characteristics of *P. conspersa* were determined from quantitative samples obtained with a cylinder sampler (area 0.075 $m^2$) incorporating a removable net (mesh pore size 400 μm), and ten samples were collected at random from each site every month from December 1983-November 1985. After extraction, larvae of *P. conspersa* were starved for 48 hours at 10°C, and separated into instars (I-V) from head width measurements. They were then freeze-dried, weighed, totally digested in nitric acid and analysed by graphite furnace atomic absorption spectroscopy , as described in Darlington *et al.* (1986).

## RESULTS

Estimates of the mean number of larvae in 1000 $cm^2$ (based on sample counts) and frequency estimates of pharate pupae, pupae and emerging adults at sites 1 and 2 are shown in Figure 2. Instar I larvae were found in small numbers in the adit channel but not at either of the riffle sampling sites and are therefore omitted here. Egg masses were obtained from June to August. Changes in larval density in each year at both sites show a steady progression in growth culminating in a single main period of pupation and a peak in adult emergence in July and August. Recruitment into both populations is probably partly due to delayed egg hatching. No instar I and II larvae were found during June and July. It is therefore suggested that *P. conspersa* is univoltine in these streams. Larvae occurred at significantly higher densities in the contaminated stream.

The relationship between the arithmetic mean ($\overline{X}$) and the variance ($S^2$) for all the larval instars combined at either site follows a power law (Taylor 1961) :

$$S^2 = a\overline{X}^b$$

where a and b are constants. The constant b is an index of aggregation and was found to be 1.65 and 1.31 for sites 1 and 2 respectively, indicating that the larvae were more aggregated in the contaminated stream than in the control stream.

The copper concentrations in 73 larvae collected from sites 1 and 2 in September 1984 are summarised in Table 1. Each larval instar of *P. conspersa* from site 1 has a very much higher copper concentration than the same instar at site 2. Within each site there was a significant ($p < 0.001$) difference in copper concentrations between the instars, with the lowest concentration in instar V larvae and the highest in instar II.

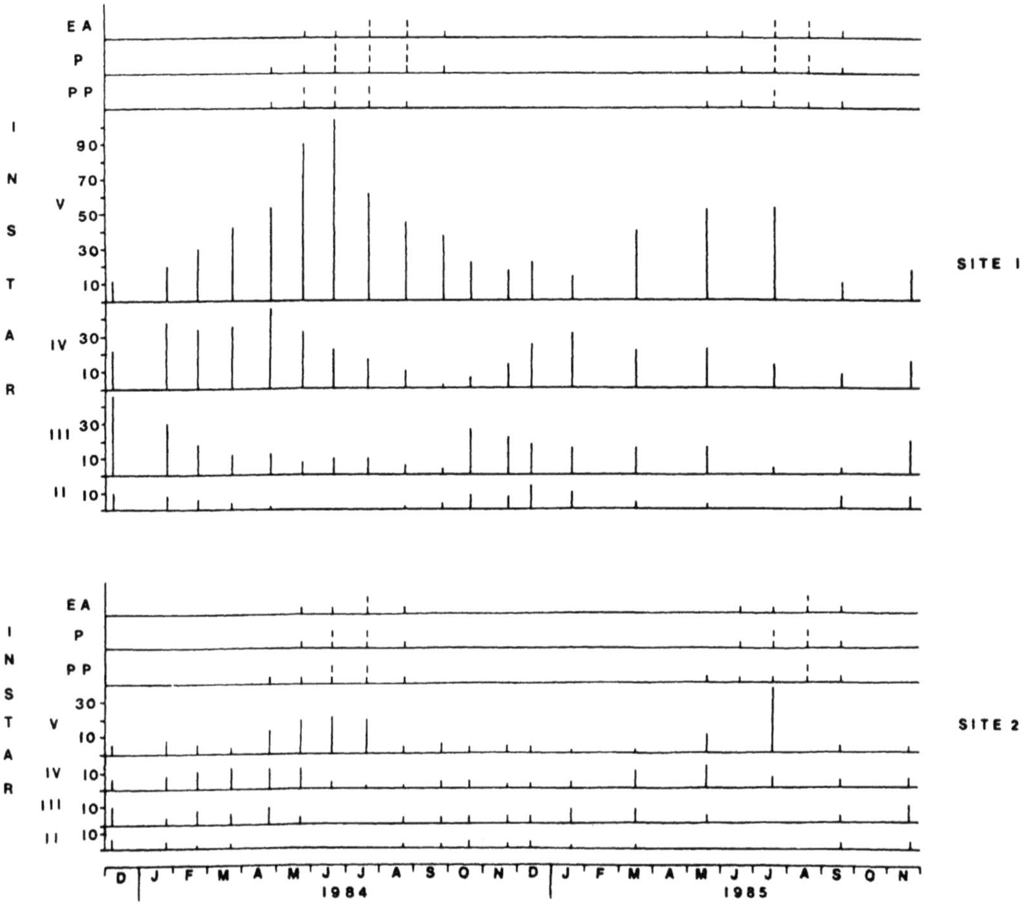


Figure 2. Mean number of larvae (instars II-V) in 1000 cm$^2$ at sites 1 and 2 from December 1983-November 1985, and frequency estimates of pharate pupae (PP), pupae (P) and emerging adults (EA) ; I rare, I common, I abundant.

Table 1. Concentration of copper ($\mu gg^{-1}$) in larval instars II to V of *P. conspersa* at sites 1 and 2, September 1984, showing range, mean, standard error (S.E.) and number (n) of larvae which were individually analysed.

| INSTAR | SITE 1 | | | | SITE 2 | | | |
|---|---|---|---|---|---|---|---|---|
| | n | MEAN | S.E. | RANGE | n | MEAN | S.E. | RANGE |
| 11 | 8 | 3767 | 278 | 2653-4812 | 5 | 813 | 37 | 705-907 |
| 111 | 10 | 1806 | 171 | 1187-2666 | 10 | 344 | 18 | 258-420 |
| 1V | 10 | 740 | 104 | 400-1237 | 10 | 136 | 9 | 97-177 |
| V | 10 | 258 | 33 | 122-424 | 10 | 70 | 8 | 38-113 |

When the copper concentration for each of the larvae is plotted against dry body weight for these samples, as in Figure 3, an exponential decrease in concentration with increasing weight is observed at both sites.

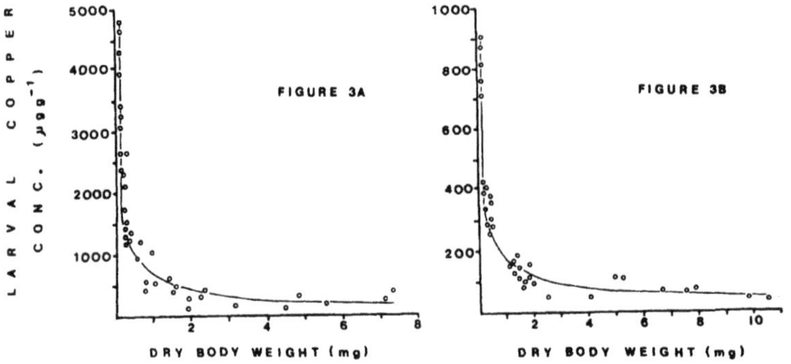


Figure 3. Relationship between dry body weight (mg) and copper concentration ($\mu gg^{-1}$ dry weight) in individual larvae (instars II-V) collected in September 1984 at (A) site 1 and (B) site 2.

The copper concentration of instar V larvae was further investigated by analysing 10 individuals in alternate months from May 1984 to July 1985. The results, presented in Figure 4, show a clear seasonal variation at both sites, with maxima of 373 and 99 $\mu gg^{-1}$ in July 1984 and minima of 132 and 27 $\mu gg^{-1}$ in January 1985 at sites 1 and 2 respectively. This seasonality has also been demonstrated in the other larval instars (II-IV) at both sites.

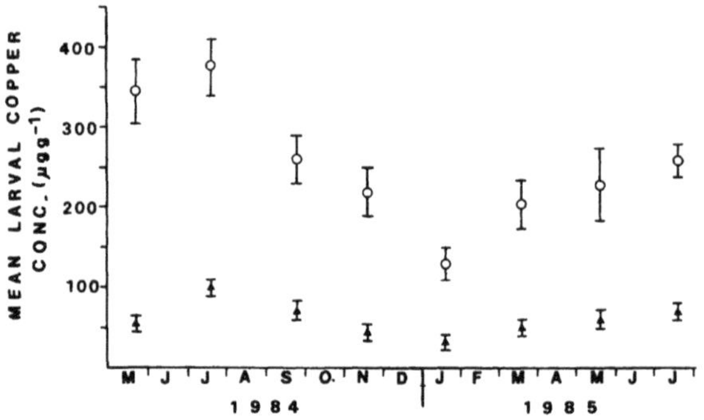


Figure 4. Mean concentration of copper ($\mu gg^{-1}$) of ten individually analysed instar V larvae collected in alternate months from May 1984-July 1985 at sites 1 = O and 2 = ▲ (vertical bars show standard errors).

## DISCUSSION

*P. conspersa* is widely distributed in S.W. England, and it is known to be a characteristic species of small acid and cool streams such as those on the moors of Devon and Cornwall. However it is clear that *P. conspersa* also occurs at lower altitudes and under a wider range of conditions.

In clean streams *P. conspersa* and *P. geniculata* may be found as part of a diverse invertebrate community, and a striking feature of copper contaminated streams in S.W. England is the prominence of *P. conspersa* . In several of the streams with copper concentrations ranging from 0.1-0.4 $mg.l^{-1}$ *P. conspersa* is relatively abundant but within a more diverse community than in the Darley Brook, and in the headwaters of the Darley Brook with very high copper concentrations (as much as 1.0 $mg.l^{-1}$ near the adit) *P. conspersa* is at its most abundant. Indeed final instar larvae attained densites of over 100 $m^{-2}$ at site 1 in June 1984. Significantly lower densities were recorded at site 2, with figures similar to those obtained by Elliott (1981) in an unpolluted stream for another net-spinning caddis *Philopotamus montanus* which also had a similar b value (index of clumping) of 1.27.

Now it is interesting to consider the reported occurrence of *P. conspersa* in other metal rich waters. Armitage (1980) found that in streams in the Northern Pennines with high zinc levels *P. conspersa* was present in a reduced community. In their study of the effects of acidification in West Wales, Stoner *et al.* (1984) recorded the occurrence of *P. conspersa* in the impoverished fauna of streams with a low pH and elevated concentrations of aluminium . The detailed investigations of Hildrew and Townsend (1976, 1982) showed *P. conspersa* to be the important predator in an iron rich stream with a low faunal diversity. More locally Brown (1977) recorded *P. conspersa* as one of only a few species in the copper rich waters of the River Hayle, Cornwall, and some correlation was found between the concentration of copper in the water and in 'free-living' Trichoptera.

The current investigation has shown that larvae of *P. conspersa* can accumulate copper from the water, with levels measurable even in larvae from the control stream. The exponential decrease in larval copper concentration with increasing body weight suggests some form of surface to volume phenomenon. A similar concentration-size relationship has been demonstrated for several metals in other insects (Smock1983). The summer maximum in the copper concentration of instar V larvae could be due to more prey being consumed and an increase in the copper concentration of the prey itself ; this is currently being investigated.

Thus it is clear that *P. conspersa* is metal tolerant and has potential as an indicator of metal enriched conditions. Metal analysis is perhaps best carried out on the instar V larvae as they are present throughout the year, are of a reasonable size and show the least variation in metal concentration . However larvae must be collected at the same time from different streams if metal concentrations are to be compared.

## REFERENCES


Armitage, P.D. 1980. The effects of mine drainage and organic enrichment on benthos in the River Nent system, Northern Pennines.**Hydrobiologia**,74: 119-128.

Brown, B.E. 1976. Observations on the tolerance of the isopod *Asellus meridianus* Rac. to copper and lead. **Wat. Res.**, 10 : 555-559.

Brown, B.E. 1977. Effects of mine drainage on the River Hayle, Cornwall (A) Factors affecting concentrations of copper, zinc and iron in water, sediments and dominant invertebrate fauna. **Hydrobiologia**, 52 : 221-233.

Darlington, S.T., Gower A.M. & Ebdon, L. 1986. The measurement of copper in individual aquatic insect larvae. **Environ. Technol. Letters**, 7 : 141-146.

Elliott, J.M. 1981. A quantitative study of the life cycle of the net-spinning caddis *Philopotamus montanus* (Trichoptera : Philopotamidae) in a Lake District stream. **J.Anim. Ecol.**, 50 : 867-883.

Hildrew, A.G. & Townsend, C.R. 1976. The distribution of two predators and their prey in an iron rich stream. **J. Anim. Ecol.**, 45 : 41-57.

Hildrew, A.G. & Townsend, C.R. 1982. Predators and prey in a patchy environment : a freshwater study. **J. Anim. Ecol.**, 51 : 797-815.

Smock, L.A. 1983. Relationships between metal concentrations and organism size in aquatic invertebrates. **Freshwat. Biol.**, 13 : 313-321.

Stoner, J.H., Gee, A.S. & Wade, K.R. 1984. The effects of acidification on the ecology of streams in the Upper Tywi Catchment in West Wales. **Environ. Pollut.** Ser A., 35 : 125-157.

Tachet, H. 1967. Quelques aspects du cycle biologique de *Plectrocnemia conspersa* (Curtis) (Trichoptera : Polycentropodidae). **Annls. Limnol.**, 3 : 177-184.

Taylor, L.R. 1961. Aggregation, variance and the mean. **Nature**, 189 : 732-735.

Thornton, I. 1983. Applied environmental geochemistry. Academic Press, London.

# THE DISTRIBUTION OF TRICHOPTERA IN DUTCH SOFT WATERS DIFFERING IN pH.

R.S.E.W. LEUVEN, J.A.M. VANHEMELRIJK and G. VAN DER VELDE

## SUMMARY

The presence and abundance of caddis larvae were examined in lentic soft waters in The Netherlands. The selected waters reflected a large variation in pH (3.5-9.5). Trichoptera accounted for 1.6-7.3% and 1.4-12.7% of the total biomass of macroinvertebrates in the water compartment and sediment, respectively. The yearly average biomass of caddis larvae varied between 2-49 mg. $m^{-2}$. Highest standing stocks were recorded in acid waters. In total, 31 taxa were encountered.*Cyrnus flavidus, Ecnomus tenellus* and *Triaenodes bicolor* showed high presence percentages and often dominated the assemblages of caddis larvae. Although 27 taxa were recorded in acid waters, the lowest pH of occurrence varied among different species. Only 17 taxa (55%) appeared to be extremely acid tolerant and were observed in waters with pH values below 4.0. The sampled systems showed remarkable differences in the structure of trichopteran communities. With hierarchical classification four major groups of waters and three species groups could be distinguished. The species richness of Trichoptera assemblages was strongly correlated with the pH of the sampling sites. Multivariate analysis of the obtained data has illustrated that apart from pH also the ionic strength, trophic level (C, P, N), aluminium content, depth and area of the waters may determine the distribution of Trichoptera.

## INTRODUCTION

All over the world oligotrophic soft waters are seriously threatened by human activities. Particularly the lack of buffering capacity and poorness of nutrients make them vulnerable to environmental perturbations. Recently it has been shown that most of the lentic soft waters in The Netherlands have been acidified and/or eutrophicated to some extent (Leuven *et al.* 1986a). These processes are associated with a variety of chemical and biological changes.

In spite of a growing interest in environmental pollution, relatively little is known about the impact of water acidification on Trichoptera. Almost all studies concerning the effects of low pH on Trichoptera (Sutcliffe and Carrick 1973 ; Wiederholm and Eriksson 1977 ; Matthias 1983 ; Otto and Svensson 1983 ; Zitschke *et al.* 1983) have been conducted in lotic waters or systems which show zoogeographical and geomorphological differences with susceptible waters in The Netherlands. Historical data on the presence and abundance of caddis larvae in Dutch soft waters are scarce and only available for a few systems or species. Therefore comparative biogeographic studies are required in order to determine possible effects of acidification on the trichopteran fauna.

In the present paper the distribution of caddis larvae in Dutch soft waters is described. The sampled waters reflect different degrees of acidification. Direct gradient and multivariate analyses of available data are applied in order to study whether pH or other environmental factors determine the distribution of Trichoptera in lentic soft waters.

## MATERIALS AND METHODS

Twice, during spring and summer of 1983, the trichopteran assemblages were examined in 35 bodies of water (sites 1-35). Most sampling sites (moorland pools and small lakes) are situated in the southern and eastern parts of The Netherlands. A few waters (dune pools) are located on the Wadden island of Terschelling. Furthermore, during 1984 the abundance of caddis larvae was regularly determined in four moorland pools (site 1, 15, 17 and 36). All study areas are dominated by mineral sandy soils which are relatively poor in lime and nutrients.

At each location water samples were taken. The pH-measurements were carried out with a Metrohm model E 488 pH-meter and a model EA 152 combined electrode. Alkalinity was estimated by titration of 100 ml of water with 0.001 M HCl down to pH 4.2. Electrical conductivity (EC) and temperature were measured with a portable conductivity meter (YSI-SCT model 33). The measured conductivity was always corrected for pH and temperature (18°C). Maximum water depth was recorded during habitat surveys, while the lake area was calculated from topographic maps (1:25000), using a planometer. The geographical position and selected limnological data are presented in Table 1. All other environmental data are given by Leuven *et al.* (1986b).

Table 1. Geographical position and selected limnological data of the sampling sites.

| Sampling sites | | Lat. | Long. | Depth m | Area ha | pH | Alk. meq/l | EC µS/cm |
|---|---|---|---|---|---|---|---|---|
| 1 | Galgeven | 51°33' | 5°09' | 3.1 | 12.0 | 3.5 | 0.000 | 38 |
| 2 | Ven near Schayk | 51°45' | 5°35' | 1.1 | 0.4 | 3.7 | 0.001 | 117 |
| 3 | Ganzenven | 51°35' | 5°16' | 0.7 | 2.8 | 3.8 | 0.000 | 35 |
| 4 | Bergven | 52°26' | 7°01' | 1.1 | 5.0 | 3.8 | 0.000 | 65 |
| 5 | Voorste Choorven | 51°34' | 5°12' | 1.2 | 5.4 | 3.8 | 0.005 | 91 |
| 6 | Rouwkuilen | 51°30' | 5°55' | 1.0 | 2.4 | 3.9 | 0.000 | 45 |
| 7 | Witven | 51°34' | 5°12' | 1.4 | 2.3 | 3.9 | 0.000 | 132 |
| 8 | Leikeven | 51°37' | 5°02' | 0.9 | 1.9 | 3.9 | 0.000 | 77 |
| 9 | Plakkeven | 51°38' | 5°03' | 0.6 | 1.1 | 3.9 | 0.019 | 35 |
| 10 | Peetersven | 51°22' | 5°28' | 1.0 | 1.4 | 3.9 | 0.000 | 19 |
| 11 | Diaconieven | 51°34' | 5°11' | 1.6 | 1.7 | 4.0 | 0.000 | 33 |
| 12 | Ganzenpoel | 52°55' | 6°18' | 1.5 | 4.6 | 4.1 | 0.010 | 50 |
| 13 | Lobeliabaai | 51°24' | 5°38' | 0.5 | 5.9 | 4.1 | 0.020 | 74 |
| 14 | Kleine Schaapsloop | 51°21' | 4°29' | 0.7 | 1.4 | 4.1 | 0.006 | 24 |
| 15 | Staalbergven | 51°34' | 5°13' | 1.5 | 5.9 | 4.3 | 0.000 | 125 |
| 16 | Gerritsfles | 52°09' | 5°49' | 1.8 | 6.8 | 4.3 | 0.000 | 117 |
| 17 | Groot Huisven | 51°35' | 5°16' | 1.5 | 4.4 | 4.3 | 0.045 | 77 |
| 18 | Dunepool Bakkeveen | 53°05' | 6°15' | 0.6 | 0.1 | 4.4 | 0.065 | 74 |
| 19 | Van Hunenplak | 53°23' | 5°14' | 0.8 | 0.9 | 4.7 | 0.050 | 198 |
| 20 | Galgenven | 51°22' | 5°29' | 1.9 | 3.0 | 5.2 | 0.040 | 87 |
| 21 | Ditch Winkelsven | 51°33' | 5°16' | 0.7 | 0.1 | 5.2 | 0.082 | 108 |
| 22 | Grote Moost | 51°18' | 5°52' | 1.0 | 15.0 | 5.4 | 0.084 | 101 |
| 23 | Zwarte Water | 51°22' | 4°56' | 0.9 | 4.0 | 5.4 | 0.081 | 167 |
| 24 | Van Esschenven | 51°34' | 5°12' | 2.0 | 5.2 | 5.8 | 0.082 | 477 |
| 25 | Kriekeput-2 | 51°16' | 4°06' | 1.8 | 1.0 | 6.3 | 0.237 | 242 |
| 26 | Broekse Wiel | 51°44' | 5°46' | 4.0 | 0.5 | 6.8 | 0.553 | 205 |
| 27 | Kriekeput-1 | 51°16' | 4°06' | 2.3 | 0.4 | 6.8 | 0.598 | 402 |
| 28 | Beuven | 51°24' | 5°39' | 0.7 | 62.5 | 6.9 | 0.621 | 198 |
| 29 | Oriltjesplak | 53°23' | 5°13' | 0.9 | 3.6 | 6.9 | 1.169 | 259 |
| 30 | Pool near Eibergen | 52°08' | 6°40' | 2.0 | 2.8 | 7.0 | 0.704 | 618 |
| 31 | Kroonpolder | 53°23' | 5°11' | 1.1 | 13.6 | 7.6 | 3.294 | 552 |
| 32 | Nieuwkuykse Wiel | 51°41' | 5°11' | 8.8 | 2.8 | 7.6 | 0.634 | 152 |
| 33 | Banen | 51°16' | 5°48' | 1.0 | 8.5 | 7.7 | 0.575 | 212 |
| 34 | Bankven | 51°30' | 5°04' | 0.4 | 10.0 | 9.4 | 0.834 | 242 |
| 35 | Badhuiskuil | 53°24' | 5°15' | 1.0 | 1.7 | 9.5 | 0.338 | 262 |
| 36 | Grootmeer | 51°26' | 5°19' | 1.0 | 14.0 | 8.1 | 1.270 | 175 |

The caddis larvae occurring in the water compartment were sampled with dip-nets (mesh size 0.5 mm) and those occurring in the bottom with core samplers. The sampling sites were distributed throughout the waters in order to cover the range of major habitats and to obtain a reasonably complete list of taxa. At the laboratory all caddis larvae were sorted after washing the samples over a 0.5 mm sieve. The larvae were identified as far as possible. The nomenclature used here follows that of Geijskes and Higler (1980).

Numerical treatments were carried out with presence-absence data of Trichoptera. For clustering the average linkage technique was used. Species-centered Principal Component Analysis (PCA), regression analysis, Spearman rank correlation test, and Kruskal-Wallis and Wilcoxon tests were performed with Statistical Analysis System (SAS 1982).

## RESULTS

Dutch soft waters are characterized by distinctive geomorphological and chemical properties. Selected limnological data of the sampling sites are presented in Table 1. Generally these waters are small, shallow, stagnant and fully mixed. Nearly all systems lack surface water connections and show periodically fluctuating water levels. The sampled waters reflect a large variation in pH, alkalinity and ionic content.

The relative abundance of trichopteran taxa are given in Table 2. In this table the taxa are arranged in such a way that the apparent pH-tolerance of caddis larvae is emphasized. A total of 31 taxa was observed in

Dutch soft waters. *Ecnomus tenellus, Cyrnus flavidus* and *Triaenodes bicolor* exhibit high presence percentages (> 25%). The larvae of *Agraylea multipunctata, Cyrnus flavidus, Dasystegia varia, D. obsoleta, Ecnomus tenellus, Glyphotaelius pellucidus, Holocentropus dubius, H. stagnalis, H. picicornis, Limnephilus subcentralis, L. vittatus, Oligotricha striata* and *Triaenodes bicolor* may be dominant species (RA > 25 %) in acid waters (pH < 5). Slightly acid and alkaline bodies of water (pH > 5) can be dominated by one or two of the following species : *Agrypnia pagetana, Cyrnus flavidus, C. insolutus, Ecnomus tenellus, Holocentropus stagnalis, Limnephilus flavicornis, L. lunatus, L. subcentralis, Oecetis lacustris* and *Triaenodes bicolor*. Many taxa (55%) tolerated pH values below 4.0. In the most acid system *Dasystegia obsoleta, Ecnomus tenellus* and *Cyrnus flavidus* were still recorded. Twelve taxa were only encountered in acid waters.

Table 2. Species richness, relative abundance (RA) and presence of caddis larvae among different sampling sites. RA : percentage of total number of caddis larvae collected in the water compartment.

| | Sites | Presence (%) | | | | |
|---|---|---|---|---|---|---|
| | 11111111112222222222333333<br>12345678901234567890123456789012345 | Total n=35 | A n=11 | BI n=6 | BII n=5 | C n=10 |
| 1 Dasystegia obsoleta | | 11 | 27 | 17 | - | - |
| 2 Ecnomus tenellus | | 26* | 45 | 33 | 40 | - |
| 3 Cyrnus flavidus | | 34 | 64 | 17 | 80 | - |
| 4 Dasystegia varia | | 17 | 45 | 17 | - | - |
| 5 Oligotricha striata | | 6 | 18 | - | - | - |
| 6 Phryganea spec. | | 14 | 36 | - | 20 | - |
| 7 Holocentropus dubius | | 6 | 18 | - | - | - |
| 8 Trichostegia minor | | 9 | - | 17 | 20 | 10 |
| 9 Oecetis lacustris | | 23 | - | 67 | 60 | 10 |
| 10 Limnephilus marmoratus | | 31 | 36 | 33 | 80 | 10 |
| 11 Holocentropus picicornis | | 17 | 9 | 50 | - | - |
| 12 Glyphotaelius pellucidus | | 6 | - | 17 | - | 10 |
| 13 Holocentropus stagnalis | | 23 | - | 33 | - | 60 |
| 14 Limnephilus flavicornis | | 23 | - | 50 | 60 | 20 |
| 15 Agraylea multipunctata | | 9 | - | - | 20 | 20 |
| 16 Limnephilus subcentralis | | 14 | 9 | - | - | 40 |
| 17 Triaenodes bicolor | | 34 | 9 | 83 | 40 | 30 |
| 18 Limnephilus politus | | 6 | 9 | 17 | - | - |
| 19 Anabolia nervosa | | 9 | 9 | - | 40 | - |
| 20 Oecetis ochracea | | 6 | 9 | 17 | - | - |
| 21 Limnephilus vittatus | | 3 | - | - | - | - |
| 22 Agrypnia pagetana | | 11 | - | 50 | - | 10 |
| 23 Oecetis furva | | 17 | - | 83 | - | 10 |
| 24 Mystacides longicornis | | 23 | - | 33 | 100 | 10 |
| 25 Cyrnus trimaculatus | | 3 | - | - | 20 | - |
| 26 Molanna angustata | | 6 | - | - | 40 | - |
| 27 Athripsodes aterrimus | | 14 | - | 50 | 40 | - |
| 28 Cyrnus insolutus | | 6 | - | - | 40 | - |
| 29 Lype phaeopa | | 3 | - | - | 20 | - |
| 30 Oxyethira spec. | | 3 | - | - | 20 | - |
| 31 Limnephilus lunatus | | 3 | - | - | - | 10 |
| Number of taxa | 1<br>32133431315631442229308562645489337 | 31<br>4±3 | 14<br>3±2 | 18<br>7±2 | 17<br>7±3 | 13<br>3±1 |

-: 0% < RA ≤ 10% ▬: 10% < RA ≤ 25% ▌: RA > 25%.

The species richness of the sampling sites ranged between 0-12 taxa (Table 2). The number of taxa was correlated most obviously with the pH of the sampling sites (Spearman rank correlation test : $p = 0.0017$). Regression analysis was used to study the relationship between the pH and the number of taxa (N) of the sites. The quadratic model $N = -0.27\ pH^2 + 4.05\ pH - 9.00$ fits the data well ($n = 35$, $R^2 = 0.31$, $p = 0.003$).

Multivariate analysis was applied in order to study the structure of trichopteran communities and the importance of the underlying environmental parameters. The ordination of presence-absence data of caddis larvae by means of species-centered PCA is presented in Figure 1a. In this biplot the sites with few species will be situated near the projection point of the real origin (-0.0832, -0.0948), while sites with many species are likely to be farther away. The first two principal components account for 34% of the total variation. In the biplot the squared length of the site vectors (i.e. approximated number of taxa) showed a positive correlation ($p < 0.05$) with the real number of taxa. Thus biplot approximates $\alpha$-diversity well. The scores of the sites on the first principal component (PC1) showed significant positive Spearman rank correlations with the depth, and the nitrate and sulphate concentrations of the waters. The loadings of the sites on PC2 reflected significant positive correlations with pH, alkalinity, EC, and the calcium, magnesium, chloride, silica, sodium, potassium, dissolved organic carbon (DOC) and dissolved organic phosphorus (DOP) contents. They exhibited significant negative correlations with the area, the acidity, and the aluminium and ammonium content of the systems.

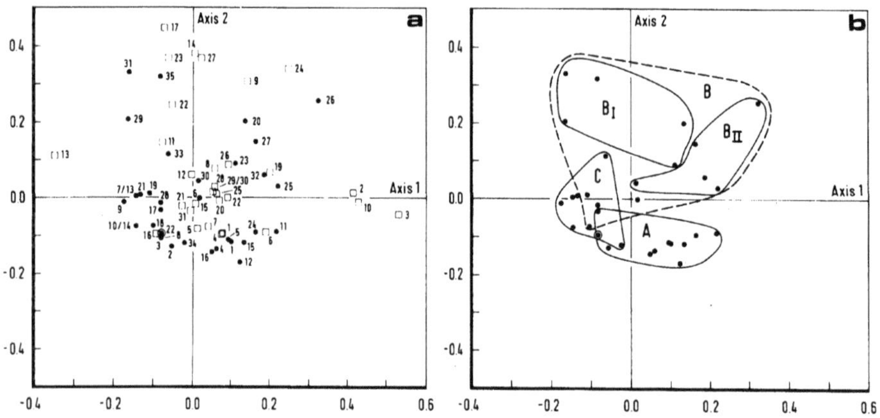


Figure 1. a) Species-centered PCA biplot of sites (●) and trichopteran taxa (□). Site and taxon numbers follow those listed with Table 1 and 2. b) Projections of clusters A, BI, BII, and C in PCA-plot of sites.

With hierarchical classification by means of the average linkage technique four groups of water bodies can be recognized (Fig. 2a). The clusters are projected in the PCA-plot (Fig. 1b). The presence of Trichoptera in these water types is presented in Table 2. Group BI and BII comprise sites with significantly higher number of species if compared with clusters A and C (Wilcoxon test, $p < 0.05$). The trichopteran communities of the latter systems are poor in species. Waters of group A are characterized by *Cyrnus flavidus, Ecnomus tenellus* and *Dasystegia varia. Oecetis furva, O. lacustris* and *Triaenodes bicolor* show a high presence in waters belonging to cluster BI. The waters of group BII always contained *Mystacides longicornis,* whereas *Cyrnus flavidus* and *Limnephilus marmoratus* frequently occurred in sites belonging to this category. In waters of group C only *Holocentropus stagnalis*and *Limnephilus subcentralis* were frequently observed, whereas, most other taxa exhibited low presence percentages. The Kruskal-Wallis test indicated that the depth, alkalinity, pH, EC, and the calcium, magnesium, sodium, potassium, silica, ammonium, sulphate, chloride, phosphate, nitrate, DOP, DON and DOC content differed significantly ($p < 0.05$) among the distinguished types of water. Group A approximately represents strongly acidified waters with low ionic contents and trophic levels, and relatively high ammonium and metal concentrations. Group B consists of circumneutral and alkaline soft waters with relatively high ionic contents and trophic levels. Wilcoxon tests ($p < 0.05$) indicate that the group levels for maximum depth and sulphate concentrations in BII are significantly higher than in A, BI and C. Group C is rather heterogeneous and more or less linked to group A and B. It seems to be composed of transitional systems and acid and/or otherwise disturbed waters. For cluster C the group levels of phosphate, silica and potassium are significantly higher if compared with those of A, whereas the latter show significantly higher values for maximum depth, nitrate and ammonium.

With inverse cluster analysis three distinct species groups can be distinguished (Fig. 2b). All taxa of group 1 (*Cyrnus flavidus, Dasystegia obsoleta, D. varia, Ecnomus tenellus, Limnephilus marmoratus, Oecetis ochracea* and *Phryganea* spec.) are acidobiontic or pH-indifferent. Most of them show high presence percentages. The taxa of group 1 occur in water type A, BI and BII. Species of group 2 (*Agrypnia pagetana, Athripsodes aterrimus, Limnephilus flavicornis, Mystacides longicornis, Oecetis lacustris, O. furva* and *Triaenodes bicolor*) are especially encountered in BI and BII waters. The taxa of group 3 (*Agraylea multipunctata, Anabolia nervosa, Cyrnus insolutus, Lype phaeopa* and *Oxyethira* spec.) are mainly recorded in water type BII.

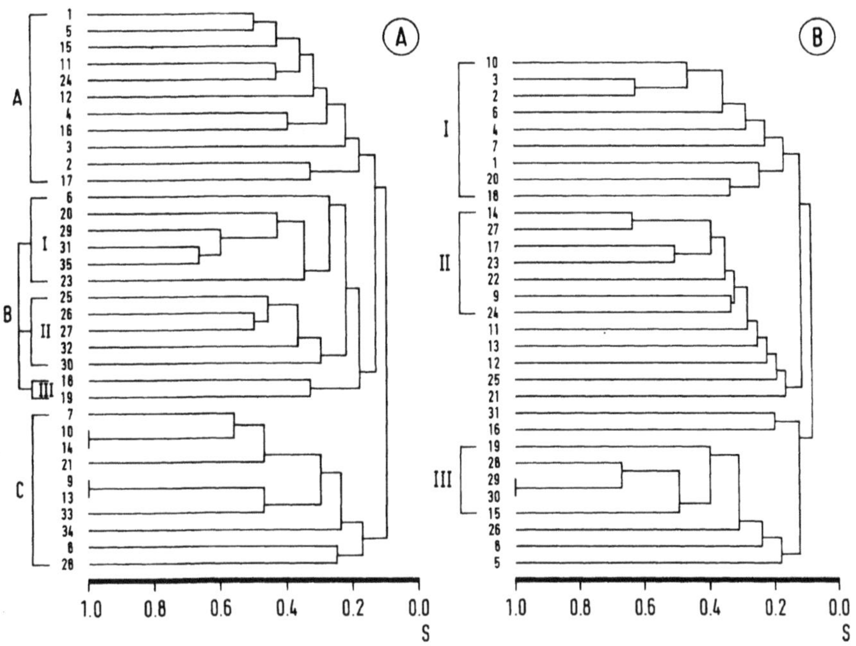


Figure 2. a) Cluster analysis of 35 waters based on their trichopteran assemblages. b) Dendrogram showing species groupings from inverse cluster analysis. Site and species numbers follow those listed with Table 1 and 2. S : Similarity (Jaccard coefficient).

The abundance and biomass of caddis larvae in four selected moorland pools during 1984 are presented in Table 3. The Trichoptera accounted for 1.6-7.3% and 0.4-12.7% of the total number of macroinvertebrates in the water compartment and the sediment, respectively. The yearly average biomass and number of caddis larvae in the sediment were highest in acid moorland pools.

## DISCUSSION

In this study biogeographic data of Trichoptera have been collected in lentic soft waters differing in pH. The environment of many sampling sites (e.g. group A) has become acidified, due to atmospheric deposition of acidifying components (Leuven *et al.* 1986a). It is evident that acidified waters harbour trichopteran communities which are characterized by a low species richness and a high dominance of a few acidibiontic and/or pH-indifferent taxa. Circumneutral and alkaline waters generally reflect a large species richness and show an equal distribution of abundances over the species. Direct gradient analysis has illustrated that the species richness of trichopteran assemblages is strongly related to the pH of the waters. Multivariate analyses have revealed that apart from pH several other limnological factors (e.g. depth, ionic content, trophic level) are correlated with the distribution of caddis larvae in soft waters. However, the importance of most of these factors for the survival of Trichoptera taxa is unknown. The relations between the limnological factors and the structure of trichopteran communities may be (in)direct or even nonsense. Until now relatively few laboratory experiments have been performed to determine effects of selected environmental factors on survival of Trichoptera. Bell (1971) studied pH-tolerance of different stages during the life cycle of aquatic insects and found caddis larvae to be the most tolerant. For *Brachycentrus americanus* and *Hydropsyche betteni* the pH-values at which 50 per cent of the mature larvae died during 30 days were 2.45 and 3.38, respectively. However, emergence of their pupae was more critical to low pH. Havas and Hutchinson (1982) pointed out that the larvae of *Limnephilus pallens* were able to survive extended periods at pH 3.5, though they did not occur in acidic waters. These authors just as well suggest that the distribution of Trichoptera may be restricted by their sensitivity during emergence. Zitschke *et al.* (1983) pointed out that acidification of experimental channels reduced emergence of Trichoptera. Furthermore, structural modifications of the habitats and changes in food sources, availability of casing material and predator-prey relations in acidifying waters may affect distribution of Trichoptera (Havas and Hutchinson 1982 ; Otto and Svensson 1983). More ecophysiological research is required to separate physiological factors from ecological ones.

Table 3. The (relative) abundance and biomass of caddis larvae (± SD) in four moorland pools differing in pH.

| No. | Sampling sites | Number of individuals.m$^{-2}$ | average biomass (mg.m$^{-2}$) | RA in water (%) | RA in sediment(%) |
|---|---|---|---|---|---|
| 1 | Galgeven | 143±52 | 38±48 | 3.8 | 12.7 |
| 15 | Staalbergven | 93±53 | 41±14 | 7.3 | 6.0 |
| 17 | Groot Huisven | 46±57 | 49±14 | 4.5 | 4.0 |
| 36 | Grootmeer | 5± 7 | 2± 2 | 1.6 | 0.4 |

RA: average relative abundance (% of total number of macroinvertebrates)

In spite of a lack of reference data prior to acid stress, the biogeographic data can be used to predict the effects of acidification on trichopteran communities. The presented species richness-pH relations indicate that acidification of lentic soft waters may result in a 50% reduction of the number of taxa. Similar results were obtained in lotic waters (Otto and Svensson, 1983).

A total of 31 Trichoptera was observed in the sampling sites. Geijskes and Higler (1980) pointed out that 177 Trichoptera taxa were encountered in The Netherlands. So, some 18% of Dutch trichopteran fauna was recorded in soft waters. The average species richness of the sampled soft waters agrees rather well with the number of taxa observed in Swedish lakes (Wiederholm and Eriksson 1977 ; Mossberg and Nyberg 1979). Many taxa have been observed in strongly acidified waters (Table 2). *Cyrnus flavidus, Dasystegia varia, D. obsoleta, Holocentropus dubius, Limnephilus* spp. and *Phryganea* spec. appeared also to be acid tolerant in Sweden (Wiederholm and Eriksson 1977 ; Mossberg and Nyberg 1979). Twelve taxa seems to be restricted to acid soft waters. Most of these taxa showed, however, low presence percentages and only *Dasystegia obsoleta* and *Oligotricha striata* can be regarded as acidobiontic. In other types of water, e.g. ditches, the remaining 10 species have also been observed at high pH (Higler, pers. comm.).

Wiederholm and Eriksson (1977) and Mossberg and Nyberg (1979) determined the standing stock of caddis larvae in small acid lakes. Their data show similar abundance and biomass values as found in Dutch soft waters. In an alkaline moorland pool (Grootmeer) with a well developed fish community the standing stock of Trichoptera was remarkably low (Table 3). Trichoptera are known to be susceptible to fish predation. High biomasses of caddis larvae were recorded in acidified waters with no fish present or with sparse fish populations. It is unknown whether these high standing stocks result in a high emergence of Trichoptera. Zitschke *et al.* (1983) pointed out that emergence of caddisflies from acidified experimental channels was reduced in comparison with ambient ones.

ACKNOWLEDGEMENTS

A part of this study has been financed by the Ministry of Housing, Physical Planning and Environment, Directorates Air (project LB 131) and Water (project W 027/011642). The authors are greatly indebted to Prof. Dr. C. den Hartog for critically reading the manuscript, to Dr. L.W.G. Higler for help with identification of Trichoptera, and to the Department of Graphics of the Faculty of Sciences (Catholic University, Nijmegen) for preparing the illustrations.

REFERENCES

Bell, H.L. 1971. Effect of low pH on the survival and emergence of aquatic insects. **Wat. Res.**, 5 : 313-319.

Geijskes, D.G. & Higler, L.W.G. 1980. Handleiding voor het project kokerjuffers of schietmotten (Trichoptera). Instructions European Invertebrate Survey (The Netherlands), 5 : 1-24.

Havas, M. & Hutchinson, T.C. 1982. Aquatic invertebrates from the Smoking Hills, N.W.T. : effect of pH and metals on mortality. **Can. J. Fish. Aquat. Sci.**, 39 : 890-903.

Leuven, R.S.E.W., Kersten, H.L.M., Schuurkes, J.A.A.R., Roelofs, J.G.M. & Arts, G.H.P. 1986a, Evidence for recent acidification of lentic soft waters in The Netherlands. **Water, Air and Soil Pollution**, in press.

Leuven, R.S.E.W., van der Velden, J.A., Vanhemelrijk, J.A.M. & van der Velde, G. 1986b. Impact of acidification on chironomid communities in poorly buffered waters in The Netherlands. **Ent. scand.**, Suppl., in press.

Matthias, U. 1983. Effects of acidification on the composition of mountain brook biocoenoses. **Arch. Hydrobiol.**, Suppl., 65 : 407-483.

Mossberg, P. & Nyberg, P. 1979. Bottom fauna of small acid forest lakes. **Inst. Freshw. Res. Drottingholm Rep.**, 58 : 77-87.

Otto, C. & Svensson, B.S. 1983. Properties of acid brown water streams in South Sweden. **Arch. Hydrobiol.**, 99:15-36

SAS. 1982. User's Guide : Statistics. Cary : SAS Inst. Inc.

Sutcliffe, D.W. & Carrick, T.R. 1973. Studies on mountain streams in the English Lake District. **Freshwat. Biol.**, 3 : 437-462.

Wiederholm, T. & Eriksson, L. 1977. Benthos of an acid lake. **Oikos**, 29 : 261-267.

Zischke, J.A., Arthur, J.W., Nordlie, K.J., Hermanutz, R.O., Standen, D.A. & Henry, T.P. 1983.Acidification effects on macroinvertebrates and fathead minnows (*Pimephales promelas*) in outdoor channels. **Wat. Res.**, 17 : 47-63.

# RECOVERY OF THE TRICHOPTERA FAUNA NEAR MT. ST. HELENS FIVE YEARS AFTER THE 1980 ERUPTION.

N.H. ANDERSON and R.W. WISSEMAN

SUMMARY

The stream fauna of the upper Clearwater Creek drainage was almost eliminated by the lateral blast and associated deposits of tephra and organic debris from the 1980 eruption of Mt. St. Helens. Although 37 species have been recorded in benthic collections, the density of caddisflies in 1985 was only 30 $m^{-2}$.

Light-trap collections of adults for nine nights between July and September 1985 were compared for a site on Clearwater Creek, in the blast zone, and on a tributary where the riparian vegetation remained intact. More than 3,400 adults were collected and 55% of these were from the Clearwater Creek site. Of the 69 species identified, 45 were taken at Clearwater Creek and 56 at the Forested Tributary ; 33 species were common to both sites. Based on both the diversity and abundance of adults, it is apparent that low density of the benthic caddisfly fauna cannot be attributed to a lack of colonizing adults.

## INTRODUCTION

On May 18, 1980, an avalanche and lateral volcanic explosion from Mt. St. Helens devastated 390 $km^2$ of forests in southwestern Washington (Lipman and Mullineaux 1981). The aquatic biota in several major drainage basins was eliminated and the physical structure of stream channels greatly altered.

In cooperation with the U.S. Forest Service, stream biologists from Oregon State University began studies of aquatic systems two to three months after the eruption. Benthic collections in succeeding years indicated that some orders of aquatic insects were recovering quite rapidly in Clearwater Creek. However, after five years, caddisfly larvae were still uncommon, although almost 40 species had been collected.

A black light was used in 1985 to monitor flight activity of caddisflies at two sites : (1) along Clearwater Creek - which had been devastated by the eruption, and (2) along a tributary where the riparian vegetation remained intact and the impact was largely limited to the short-term effects of a heavy deposition of ash. The objective was to compare the number of species and their relative abundance at the two sites and to determine whether benthos density in Clearwater Creek was limited by a lack of colonizing adults.

## METHODS AND SITE DESCRIPTION

This study was conducted in the Clearwater Valley, about 15 km NE of Mt. St. Helens. The main site was at the Upper Clearwater Bridge, at an altitude of 750 m. In this area the forest had been leveled by the blast. The stream channel received a heavy impact from blast deposits ; about 30-70 cm of ash and pumice and large amounts of downed timber. This is a third-order stream, with discharge ranging from 0.2 to >10 $m^3$ $sec^{-1}$. Substrates in the immediate area are largely gravel and sand whereas upstream, where the gradient increases, there is a larger proportion of cobble and much of the mobile material has been removed. After salvage logging in 1983, a debris torrent during the winter of 1983-84 produced a debris jam 50 m long and 5m high in the stream channel.

The Forest Tributary site, located about 3 km downstream, was less affected by the eruption. It was on the edge of the blast zone and the riparian vegetation (primarily >100 yr. old Douglas-fir, *Pseudotsuga menzesii* ) remained intact. Though the site received heavy ashfall (ca. 30 cm), this was largely removed by freshets during the winter of 1980-81. This is a high gradient, second-order tributary with a stairstep profile and a boulder-cobble substrate.

The light trap had a 15-watt fluorescent UV tube operated by a 12-volt motorcycle battery. A Vapona$^R$ strip in the collecting bucket served as a quick knockdown agent. The trap was operated from dusk to dawn, once a week, from mid-July to mid-September. It was set up at the Upper Bridge for one night and then removed to the Tributary site for the next night ; on some occasions, two or three days intervened between collections at the two sites. Air temperature was monitored continuously at the Upper Bridge site.

Benthos collections from Clearwater Creek were taken on 10 occasions between 1980 and 1984, and nine times (May-October) in 1985. Quantitative collections with a Hess-type sampler were usually taken, but most caddis larvae and pupae were obtained from qualitative collections (hand-picking, kick net, or washed from wood). Collections from the Forest Tributary were limited to two qualitative samples in August 1985.

RESULTS

**Comparisons between sites**

Total captures for nine trap nights were 1866 adults at the upper Bridge and 1552 at the Forest Tributary (Table 1). Excluding the unidentified *Limnephilus* females, there were 69 species collected : 56 at the Tributary and 45 at the Upper Bridge ; 33 species were common to both sites. The number of species trapped each week ranged from 8 to 34 at the Tributary and from 10 to 29 at the Upper Bridge (Table 2).

The Tributary fauna was both more diverse and more balanced than that at the Upper Bridge. At the latter site, *Onocosmoecus unicolor* accounted for 54 % of the total catch. The trap captures at the two sites are compared below using indices of diversity and equitability, and coefficient of similarity (Southwood 1978) :

| | Upper Bridge | Forest Tributary |
|---|---|---|
| Shannon-Wiener Diversity (H) | 2.668 | 4.061 |
| Maximum Species Diversity ($H_{max}$) | 5.491 | 5.807 |
| Equitability ($E = \frac{H}{H_{max}}$) | 0.486 | 0.699 |
| Simpson Diversity (D) | 0.676 | 0.888 |
| Coefficient of Similarity (CN) | 0.39 | |

The highest diversity at both sites occured in late July when the temperature at dusk was 22.5°C and decreased to 10°C by morning. In the previous interval (July 18 and 19) with warmer nights, both high numbers and high diversity occurred at the Tributary but there were considerably lower counts at the Upper Bridge (Table 2). This difference may be related to wind conditions as gusts to 15 $kmhr^{-1}$ were noted at the Upper Bridge when the trap was set whereas on the next evening at the Tributary the weather was recorded as calm. The September peak at the Upper Bridge was primarily due to *Onocosmoecus unicolor* (Sept 4, 185 ; Sept10, 701). In contrast only 13 and 20 *O. unicolor* were collected at the Tributary on Sept. 6 and 12.

In addition to variation in weather conditions on successive nights, differences in trap collections at the two sites would be influenced by the marked differences in exposure. The open nature of the Upper Bridge site would increase the distance that the light was visible, while the riparian corridor along the Forest Tributary would decrease wind and temperature fluctuations.

**Flight periods**

The numbers per trap night of the common species (> 40 adults), pooled for both sites are shown in Figure 1. Collections were made only during summer and early fall, so these data do not reflect the entire flight period ; early-season species are either under-represented or absent. The common species represent < 20% of the species pool but 87% of the adults collected.

Most species had a flight period of six or more weeks. *Ecclisomyia* spp. and *Oligophlebodes ruthae* may have been on the wing before trapping was initiated. Except for *Hydropsyche oslari* and *Rhyacophila angelita* , peak occurrence was in July (mid-summer species) or as a pronounced autumnal flight period. Late-season limnephilids (*O. unicolor, Allocosmoecus partitus,* and *Dicosmoecus atripes*) and *Rhyacophila verrula* in the two September collections accounted for 38% of all individuals in the light traps.

Table 1. Comparison of the Trichoptera fauna of Clearwater Creek, Skamania Co., WA., with that of a forested tributary in nine light trap collections, July to September, 1985. Benthos collections indicate presence (X) or absence (-).

| | Clearwater Creek Light Trap | | Clearwater Creek Benthos | | Forest Tributary Light Trap | | Forest Tributary Benthos |
|---|---|---|---|---|---|---|---|
| | Interval | No. | 1980-84 | 1985 | Interval | No. | 1985 |
| RHYACOPHILIDAE | | | | | | | |
| *Himalopsyche phryganea* | 12/8 | 1 | - | - | 19/7-15/8 | 6 | X |
| *Rhyacophila* Alberta Gr. | | | X | X | | | - |
| *R. kincaidi* | 26/7-2/8 | 2 | | | | - | |
| *R. tucula* | 4/9 | 1 | | | 22/8-12/9 | 6 | |
| *R. angelita* | 2/8-4/9 | 6 | X | X | 19/7-12/9 | 64 | X |
| *R. arnaudi* | | - | - | X | | - | - |
| *R.* Betteni Gr. | | | X | X | | | X |
| *R. chilsia* | | - | | | 15/8 | 1 | |
| *R. perda* | | - | | | 27/7-22/8 | 3 | |
| *R. vaccua* | | - | | | 22/8-12/9 | 8 | |
| *R. blarina* | | - | X | - | | - | - |
| *R. grandis* | 21/8 | 1 | - | - | | - | - |
| *R. iranda* | | - | - | - | 19/7 | 3 | - |
| *R. jenniferae* | 12/8-28/8 | 3 | - | X | | - | X |
| *R. narvae* | | - | X | - | 19/7 | 1 | X |
| *R. oreta* | | - | - | - | 19/7-27/7 | 3 | - |
| *R.* Rotunda Gr. | | | X | X | | | X |
| *R. norcuta* | | - | | | 19/7 | 1 | |
| *R. vaefes* | 26/7-10/9 | 7 | - | X | 19/7-29/8 | 12 | - |
| *R. vagrita* | 28/8-10/9 | 5 | X | X | 6/9 | 2 | - |
| *R. valuma* | 18/7 | 1 | X | - | 19/7-27/7 | 11 | X |
| *R. vao* | 11/7-4/9 | 8 | X | X | 12/7-29/8 | 139 | X |
| *R. verrula* | 2/8-10/9 | 190 | X | X | 12/7-12/9 | 65 | - |
| *R. vocala* | 11/7-26/7 | 2 | X | X | 19/7-27/7 | 3 | X |
| GLOSSOSOMATIDAE | | | | | | | |
| *Anagapetus bernea* | | - | - | - | 12/7-19/7 | 13 | X |
| *Glossosoma* | | | X | X | | | X |
| *G. penitum* | | - | | | 19/7 | 1 | |
| *G. pyroxum* | 26/7 | 1 | | | | - | |
| HYDROPTILIDAE | | | | | | | |
| *Agraylea saltesea* | 11/7-26/7 | 7 | - | X | | - | - |
| *Hydroptila* sp. | | - | - | X | | - | - |
| *Ochrotrichia* sp. | | - | X | - | 12/7 | 2 | - |
| *Stactobiella* sp. | | - | X | - | | - | - |
| PHILOPOTAMIDAE | | | | | | | |
| *Dolophilodes* | | - | - | | | | X |
| *D. dorcus* | 26/7 | 1 | | | 12/7-22/8 | 54 | |
| *D. pallidipes* | | - | | | 15/8-29/8 | 6 | |
| POLYCENTROPODIDAE | | | | | | | |
| *Polycentropus* | | | - | X | | | - |
| *P. denningi* | 11/7-12/8 | 26 | | | 19/7-22/8 | 9 | |
| *P. halidus* | 11/7-4/9 | 17 | | | 27/7-29/8 | 8 | |
| HYDROPSYCHIDAE | | | | | | | |
| *Arctopsyche grandis* | 11/7-12/8 | 14 | X | X | 12/7-27/7 | 16 | X |
| *Hydropsyche* | | | - | X | | | X |
| *H. amblis* | | - | | | 19/7-27/7 | 3 | |
| *H. oslari* | 11/7-4/9 | 249* | | | 19/7-29/8 | 430* | |
| *Parapsyche elsis* | | 31 | - | X | 12/7-22/8 | 33 | X |
| LIMNEPHILIDAE | | | | | | | |
| *Allocosmoecus partitus* | 4/9-10/9 | 31 | X | X | 22/8-12/9 | 96 | - |
| *Apatania* sp. | | - | X | - | | - | - |
| *Chyranda centralis* | 18/7-10/9 | 28 | - | - | 27/7-12/9 | 12 | - |
| *Clistoronia magnifica* | 26/7 | 4 | - | - | | - | - |
| *Cryptochia* sp. | | - | X | - | | - | - |
| *Dicosmoecus atripes* | 26/7-10/9 | 71 | - | X | 27/7-12/9 | 15 | X |
| *D. frontalis* | 26/7 | 1 | - | - | 2/8-15/8 | 2 | - |
| *D. gilvipes* | 26/7 | 2 | - | - | | - | - |
| *Ecclisocosmoecus scylla* | 26/7-4/9 | 10 | - | - | 15/8-12/9 | 24 | - |
| *Ecclisomyia* | | | X | X | | | X |
| *E. conspersa* | 11/7-21/8 | 34 | | | 12/7-27/7 | 35 | |
| *E. maculosa* | 11/7-12/8 | 20 | | | 12/7-15/8 | 153 | |

* May include *H. amblis* ♀♀.

Table 1. Cont.

| | Clearwater Creek Light Trap | | Clearwater Creek Benthos | | Forest Tributary Light Trap | | Forest Tributary Benthos |
|---|---|---|---|---|---|---|---|
| | Interval | No. | 1980-84 | 1985 | Interval | No. | 1985 |
| *Hesperophylax* sp. | 12/8 | 1 | - | - | | - | - |
| *Homophylax andax* | 11/7-10/9 | 14 | - | - | 27/7-12/9 | 7 | - |
| *Hydatophylax hesperus* | | - | - | - | 19/7-4/8 | 2 | - |
| *Lenarchus vastus* | 10/9 | 1 | - | - | 27/7 | 4 | - |
| *Limnephilus alconura* | | - | - | - | 19/7-27/7 | 3 | - |
| *L. fagus* | | - | - | - | 19/7-25/8 | 2 | - |
| *L. lunonus* | | - | - | - | 19/7 | 1 | - |
| *L. nogus* | | - | - | - | 19/7-12/9 | 2 | - |
| *L. occidentalis* | 11/7 | 1 | - | - | 19/7-27/7 | 21 | - |
| *L. sitchensis* | 18/7-4/9 | 3 | - | - | 19/7-12/9 | 26 | - |
| *L.* spp. (undet. ♀♀) | 18/7 | 1 | - | - | 19/7-27/7 | 2 | - |
| *Neophylax* | | | X | X | | | X |
| *N. splendens* | 10/9 | 2 | | | | | - |
| *N. n.sp.* | | - | | | 12/7-19/7 | 2 | |
| *Oligophlebodes* | | | - | X | | | - |
| *O. ruthae* | 11/7-26/7 | 10 | | | 12/7-15/8 | 76 | |
| *O. sierra* | 26/7-28/8 | 4 | | | | - | |
| *Onocosmoecus unicolor* | 18/7-10/9 | 1009 | - | X | 27/7-12/9 | 89 | - |
| *Pedomoecus sierra* | 2/8 | 1 | - | - | 27/7 | 1 | - |
| *Psychoglypha bella* | | - | X | X | | - | - |
| *P. browni* | | - | X | X | 27/7-12/9 | 3 | X |
| UENOIDAE | | | | | | | |
| *Neothremma didactyla* | | - | - | X | 19/7-27/7 | 4 | - |
| LEPIDOSTOMATIDAE | | | | | | | |
| *Lepidostoma* | | | - | X | | | X |
| *L. cascadense* | 11/7-2/8 | 17 | | | 12/7-29/8 | 53 | |
| *L. hoodi* | | - | | | 15/8 | 1 | |
| *L. jewetti* | | - | | | 22/8 | 1 | |
| *L. recina* | 26/7 | 1 | | | | - | |
| *L. unicolor* | | - | | | 27/7 | 1 | |
| *L. veroda* | | - | | | 12/7 | 1 | |
| BRACHYCENTRIDAE | | | | | | | |
| *Brachycentrus americanus* | 26/7-21/8 | 22 | X | X | 27/7-29/8 | 8 | - |
| *Micrasema bactro* | 11/7-26/7 | 2 | X | - | 19/7-27/7 | 2 | X |
| LEPTOCERIDAE | | | | | | | |
| *Mystacides alafimbriata* | 11/7 | 1 | - | - | | - | - |
| *Triaenodes tarda* | 26/7 | 2 | - | - | | - | - |
| TOTAL ADULTS | | 1866 | | | | 1552 | |
| TOTAL TAXA | | 46 | 24 | 29 | | 57 | 21 |

Table 2. Number of species and number of adult Trichoptera collected at light traps on each sampling night, Clearwater Valley, 1985.

| Date Bridge, or (Tributary) | Upper Clearwater Bridge No. of Species | Upper Clearwater Bridge No. of Adults | Upper Clearwater Bridge Air Temp. (Range C°) | Forest Tributary No. of Species | Forest Tributary No. of Adults | Forest Tributary Air Temp. (Range C°) | Spp. Common to Both Sites This Date | Spp. Common to Both Sites Accum. |
|---|---|---|---|---|---|---|---|---|
| July 11 (12) | 16 | 44 | 10.5-20.0 | 13 | 56 | 9.0-18.0 | 7 | 7 |
| July 18 (19) | 12 | 86 | 11.5-24.5 | 33 | 421 | 13.0-25.5 | 10 | 15 |
| July 26 (27) | 29 | 241 | 10.0-22.5 | 34 | 332 | 10.0-22.5 | 19 | 23 |
| Aug. 2 ( 4) | 15 | 56 | 9.0-14.0 | 15 | 52 | 8.5-15.0 | 10 | 27 |
| Aug. 12 (15) | 20 | 142 | 8.5-17.5 | 23 | 130 | 9.5-17.0 | 13 | 29 |
| Aug. 21 (22) | 12 | 73 | 7.0-14.0 | 23 | 286 | 7.0-17.0 | 10 | 29 |
| Aug. 28 (29) | 11 | 55 | 6.5-15.5 | 16 | 89 | 9.0-14.5 | 7 | 29 |
| Sept. 4 ( 6) | 15 | 298 | 11.0-12.5 | 8 | 42 | 7.5- 8.5 | 8 | 33 |
| Sept.10 (12) | 10 | 871 | 6.0- 9.0 | 13 | 144 | 7.5- 8.5 | 6 | 33 |

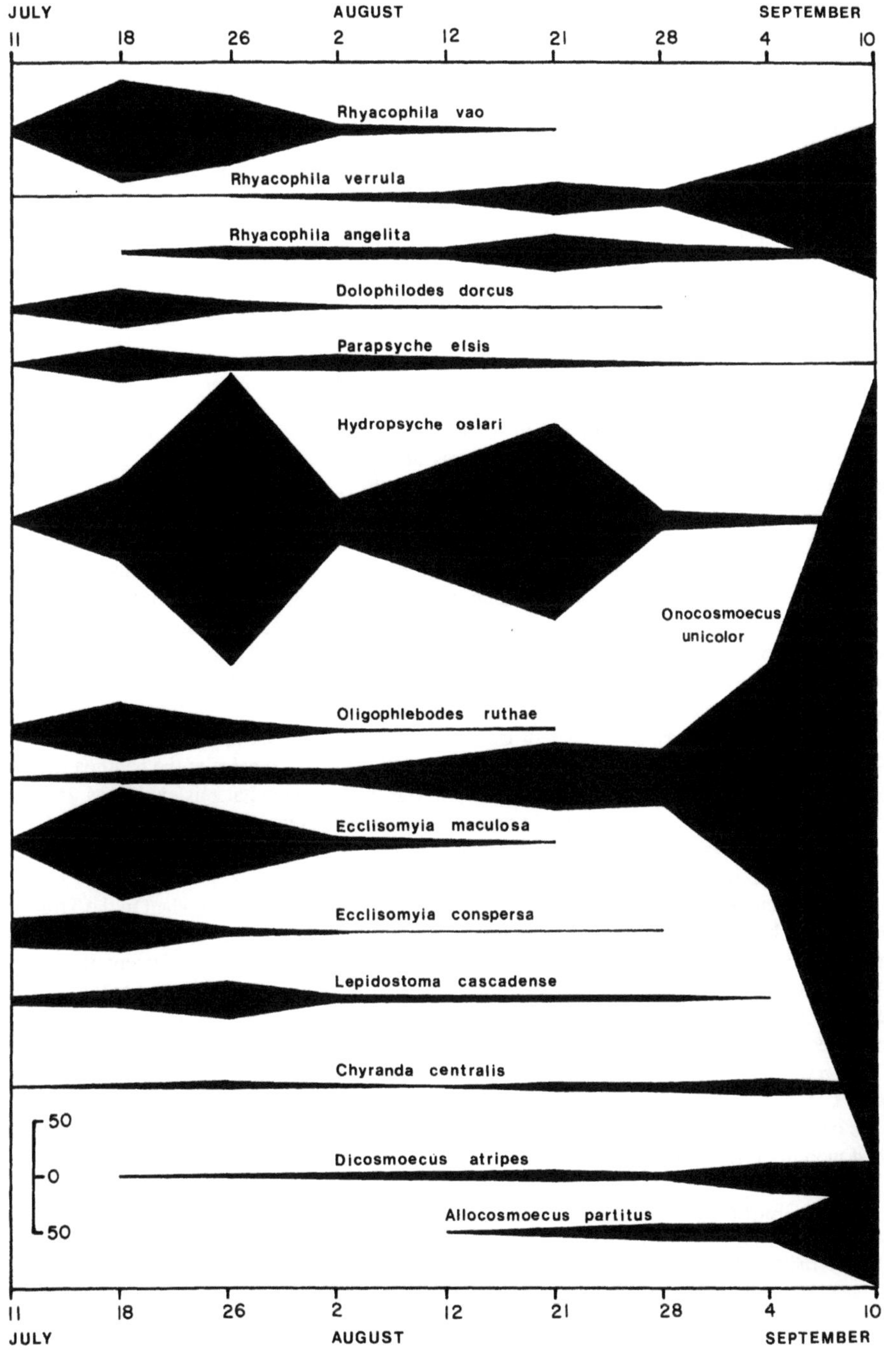


Figure 1. Number of adults of abundant Trichoptera species collected per night for nine intervals, July 11 to Sept. 10, 1985 in the Clearwater Valley, Skamania Co., WA. Data are combined for the Upper Bridge and Forest Tributary sites.

**Comparison of light-trap and benthos taxa**

Taxa collected in benthos samples are listed in Table 1. Species identifications of larvae are based on metamorphotype associations or assignment to a most probable species based on the known occurrence of adults. Some *Rhyacophila* are listed at the species-groupe level because of the presence of two or more species of adults within a group.

As most caddisflies in benthos collections were from qualitative samples, the density of individual species is unknown and data are expressed on a presence-absence basis. From 1980 to 1984, the total caddis in benthos collections from Clearwater Creek was less than 250, which is estimated to be < $10m^{-2}$. In 360 Hess samples collected between May and October, in 1985, the density of caddisflies was about $30m^{-2}$, and $20m^{-2}$ on wood substrates. The increase in number of taxa collected in 1985 is a reflection of increased sampling intensity over the entire season, and perhaps some further recovery of the stream fauna. The majority of caddis in the benthos were early-instar larvae of rhyacophilids and hydropsychids. Despite the intensity of sampling, only a single *Onocosmoecus* larvae was collected from all samples in Clearwater Creek.

About 20 taxa were found in the limited collections from the Forest Tributary. As these samples were from late summer when much of the fauna is either on the wing or in eggs or early instars, we may well have missed many of the species that breed there.

## DISCUSSION

Based on the scarcity of Trichoptera in Clearwater Creek benthos samples from 1980 to 1984, the species richness and abundance at lights in 1985 was unexpected. Surveys of the high-gradient tributary streams in 1985 by C. P. Hawkins have suggested that these are an important source of colonists to Clearwater Creek for many of the insect orders. Although the fauna of the Upper Bridge and Forest Tributary sites differed in terms of species diversity and equitability, the common species were collected at both sites. Of the 25 species with > 10 adults collected at lights (which account for 95% of the individuals), only *Anagapetus bernea* was limited to one site. The broad distribution of adults is indicative of the dispersive abilities of the caddisflies and their potential for recolonization . Amongst the active dispersers are 10-14 species of pond or temporary-habitat breeders that would not be part of the stream fauna. These include the leptocerids and thelLimnephilid genera *Limnephilus, Lenarchus, Clistoronia,* and *Hesperophylax,* and the spring-seep breeders, *Chyranda* and *Homophylax*.

We have recorded at least 37 species of larvae or pupae from Clearwater Creek (Table 1) ; 24 species from 1980-84 and 29 in 1985. The intensive sampling in 1985 added 13 species to the benthos list while eight taxa collected in 1980-84 were not recorded in 1985. The list is somewhat misleading as an index of the recovery of the fauna after the eruption because many species are represented by very few individuals. The mean density for all caddis larvae of 30 $m^{-2}$ in benthos samples and 20 $m^{-2}$ on wood substrates is indicative of the limited recovery after five years.

Limnephilid and lepidostomatid shredders would be expected to be more abundant in the Tributary than in Clearwater Creek because of greater needle and leaf litter inputs to the former. The benthos collections at the Tributary were too limited in space and time to document the occurrence of larvae of many of the species. The limnephilid shredders are mostly strong fliers that will occur at lights away from the breeding habitat. The abundance of *Onocosmoecus unicolor* adults was surprising as we have only one larval record. Obviously we have not collected at the appropriate time, or found the larval habitat in the Clearwater Creek basin.

Light trap data on the species pool and abundance of adults shows that recovery of the fauna in Clearwater Creek is not limited by adult colonists. Species of shredders could be food limited, but this should not be the case for other functional feeding groups. We believe that the shifting substrates and associated streambed mobility is the primary factor limiting caddisfly populations in the stream.

## ACKNOWLEDGEMENTS

We are especially indebted to Jay Bigelow for maintaining the light-trap program. Thanks are due to Richard Meyerhoff and Chuck Hawkins for data on the benthos populations of caddisflies. The initial stream recovery studies at Mount St. Helens were supported by a Cooperative Agreement with the U.S. Forest Service, N° PNW-80-277. The 1985 field work was supported by NSF Grant DEB 85-069. This is Technical Paper N° 7940 of the Oregon Agricultural Experiment Station.

REFERENCES

Lipman, P.W. & Mullineaux D.R. (eds). 1981. The 1980 Eruption of Mount St. Helens, Washington. **US. Geol. Surv.**, Prof. Paper 1250.
Southwood, T.R.E. 1978. Ecological Methods. John Wiley & Sons, New York.

# TRICHOPTERA OF REGULATED ROCKY MOUNTAIN STREAMS

J.V. WARD


## SUMMARY

Regulation by impoundment markedly altered the trichopteran fauna of otherwise pristine Rocky Mountain streams. The number of taxa was typically reduced at regulated sampling stations below both surface-release and deep-release dams situated over a wide range of elevation. The relative contribution of caddisflies to total zoobenthos was diminished at all regulated stations examined, except below a surface-release dam at mid elevation where *Cheumatopsyche* attained high densities. Shredders, scrapers and collector-gatherers decreased, and filter feeders and predators increased in relative abundance below deep-release reservoirs. Filter feeders comprised virtually the entire fauna below a surface-release dam at mid elevation, whereas predators dominated the surface-release station at high elevation. These changes in the trichopteran fauna are largely attributable to the direct and indirect manifestations of altered temperature, flow, and substrate conditions induced by stream regulation.


## INTRODUCTION

Macroinvertebrate communities are markedly influenced by the modified environmental conditions in lotic reaches regulated by upstream impoundments (Ward and Stanford 1979, Ward 1982, Armitage 1984). Shifts in community composition have been attributed to downstream alterations of thermal and flow regima, substrate conditions, trophic dynamics , and biotic interactions.

This paper examines the responses of Trichoptera to stream regulation at lotic sites occurring over a wide range of elevation in the Rocky Mountains of Colorado. Special attention is given to effects of stream regulation on diversity, relative abundance, and functional feeding groups.

## METHODS AND STUDY SITES

The trichopteran fauna was examined at locations ranging in elevation from near treeline down to the foothills (Fig.1). Each study site consisted of a regulated sampling station a short distance below a dam and an unregulated (reference) station either above the reservoir or on a tributary. Additional comparative data are available for trichopteran communities along a river that flows unimpeded from alpine tundra to the plains (Ward 1981, 1986).

Because the streams and sampling methods have been described in detail (Ward 1976, Ward and Short 1978, Zimmerman and Ward 1984, Ward *et al.* 1986), only brief descriptions are included herein. Nine sites were sampled : 1) South St.Vrain Creek, a high elevation site regulated by Brainard Lake, a surface-release reservoir ; 2) Joe Wright Creek, a high elevation site regulated by Chambers Lake, a bottom-release reservoir ; 3) The Blue River, a moderately high elevation site regulated by Dillon Reservoir, a deep-release impoundment ; 4) The upper Colorado River, a mid-elevation site regulated by Lake Granby, a deep-release reservoir ; 5) The Fryingpan River, a mid-elevation site regulated by Ruedi Reservoir, a deep-release impoundment ; 6) Trout Creek, a mid-elevation site regulated by Manitou Lake, a surface-release reservoir ; 7) The Big Thompson River, a mid-elevation site regulated by Lake Estes which releases a combination of surface and bottom water ; 8) North St. Vrain Creek, a foothills site regulated by Buttonrock Reservoir, a deep-release impoundment ; and 9) The South Platte River, a foothills site regulated by Cheesman Lake, a deep-release reservoir. All study sites were located on trout streams of high water quality. Samples were collected with a Surber sampler (929 $cm^2$) from rubble riffles, the predominant benthic habitat at all locations. At regulated locations, only data from sampling stations immediately below the reservoirs are included. Downstream faunal recovery patterns are not addressed in this paper (see Ward and Stanford 1984).

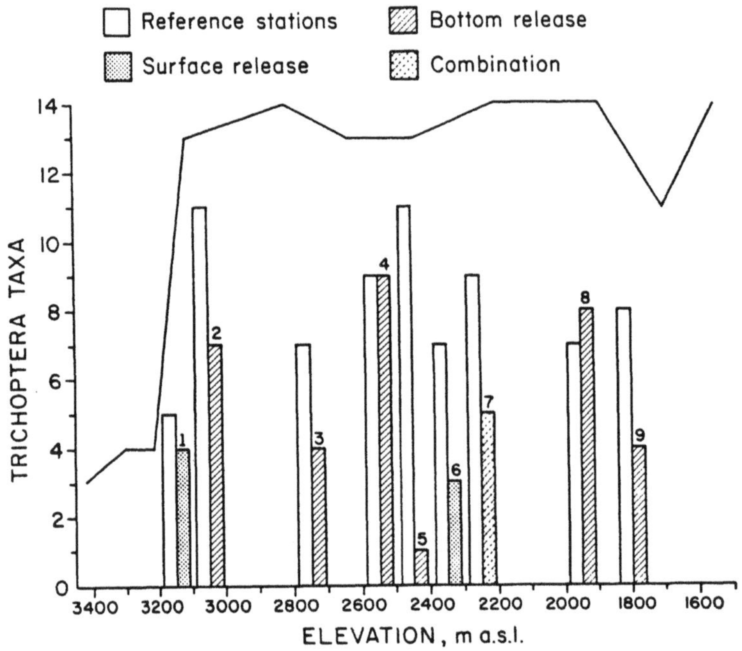


Figure 1. Number of taxa identified from the nine study sites. Sites are arranged according to the approximate elevations of reference stations. Line shows the species richness pattern for a pristine river system (Ward 1986).

## RESULTS

Thirty-four taxa, two of which were found only as adults (*Psychomyia flavida, Micrasema bactro* ), were identified from samples collected at the nine study sites. *Rhyacophila,* with seven species, was the most diverse genus. The number of taxa was typically reduced at regulated stations, a trend that appears to be independent of release depth or elevation (Fig. 1). Genera that were abundant at several reference stations, but that were reduced or absent in regulated segments, include *Brachycentrus, Glossosoma, Arctopsyche,* and *Lepidostoma. Cheumatopsyche* larvae comprised virtually the entire trichopteran fauna at Site 6 below the mid-elevation, surface-release reservoir, but was not collected from the reference station. *Rhyacophila* and *Hydropsyche* were the only genera commonly occurring at both regulated and reference stations.

Mean densities of total zoobenthos ranged from 406 to 7665 organisms $m^{-2}$ at reference stations and from 1259 to 39,979 organisms $m^{-2}$ at regulated stations. Trichopterans comprised from < 1% to 50% of the total fauna (Fig. 2). At all but one site, the relative abundance of trichopterans was lower at regulated than at reference stations. The most dramatic difference occurred at Site 5, a mid elevation stream regulated by a large deep-release reservoir, where caddisflies comprised 42% of the total zoobenthos above the reservoir, but <1% of the fauna below the dam. Only at the mid elevation location regulated by a surface- release reservoir (Site 6) did caddisflies comprise a larger proportion of the total fauna below the dam (50%, versus 18% at the reference station).

Trichopterans should reflect any changes in the availability of food resources because their larvae collectively encompass a broader range of feeding types than the other wholly aquatic orders of insects (Wiggins and Mackay 1978). The trophic structure of the trichopteran fauna, based on functional feeding groups (Merritt and Cummins 1984), is altered by stream regulation. Composite data from all bottom-release sites (Fig. 3) show reductions in the relative abundance of shredders, scrapers, and collector-gatherers, and increases in filter-feeders and predators, at regulated compared to reference stations.

Surface-release impoundments have even greater effects on functional feeding groups, effects that differed markedly between the high and mid elevation sites (Fig. 4). Filter feeders, predators, and scrapers comprised virtually the entire fauna above the high elevation reservoir, whereas predators (*Rhyacophila* spp.) dominate the trichopteran community below the dam. The caddisfly fauna above the mid elevation, surface-release reservoir is balanced between filter-feeders, scrapers, shredders, and predators. Filter feeders (*Cheumatopsyche*) comprised >99% of the caddisflies at the regulated station below the dam.

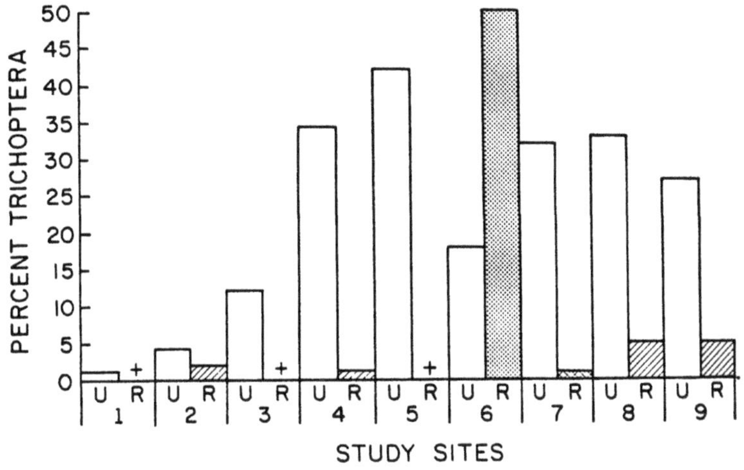


Figure 2. Contribution by Trichoptera to total zoobenthos numbers at the nine study sites. U = unregulated (reference) stations. R = regulated stations. + = < 1%.

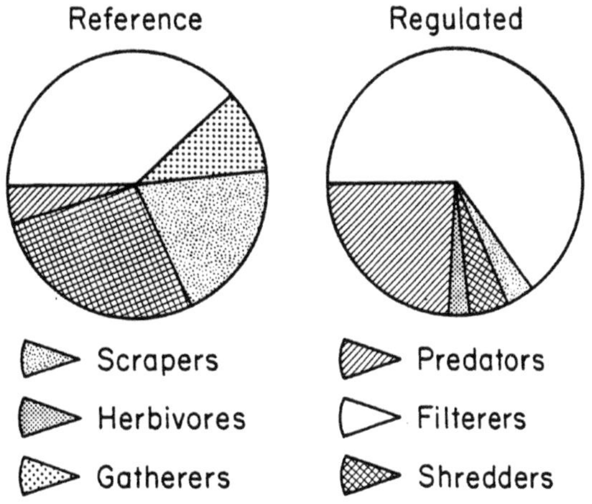


Figure 3. Trophic structure of trichopteran fauna by functional feeding groups (Merritt and Cummins 1984) as a composite of the bottom-release sites. Herbivores refers to macrophyte herbivores. Gatherers refers to collector-gatherers. Values <1 % are not indicated.

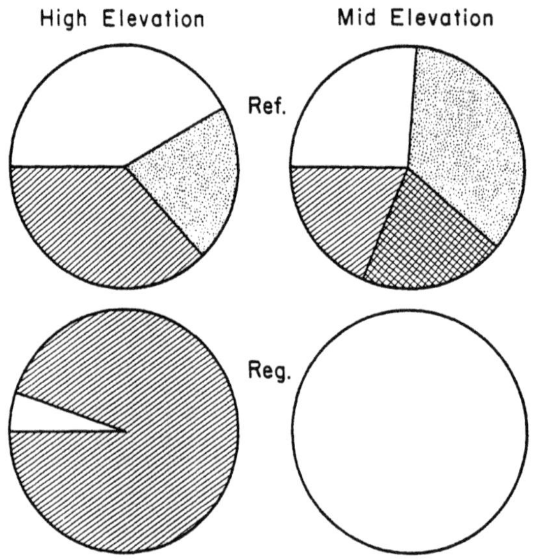


Figure 4. Trophic structure of trichopteran fauna at surface-release sites. Symbols as in Figure 3.

DISCUSSION

In the absence of confounding factors (e.g., pollution, oxygen depletion), downstream faunal changes induced by stream regulation may be largely attributable to the direct and indirect manifestations of altered temperature, flow, and substrate (Ward and Stanford 1979). Most regulation schemes moderate long-term flow fluctuations by reducing the frequency and magnitude of major flood events (Ward 1982). Flow moderation confers bed stability and reduces scouring. Reservoirs alter seston transport (Webster *et al.* 1979, Voshell and Parker 1985) and may export lentic plankton to stream reaches previously lacking this food resource (Ward 1975). Surface-release reservoirs (and natural lakes) elevate summer temperatures and delay vernal warming and autumnal cooling in the receiving stream. Deep-release reservoirs that stratify, decrease natural temperature fluctuations, producing winter-warm and summer-cool conditions in the receiving stream (Ward and Stanford 1979).

Temperature has been ascribed a major role in determining the diversity and abundance of aquatic insects (Vannote and Sweeney 1980, Ward and Stanford 1982). Modified thermal regimes below the dams contributed to reductions in the diversity of trichopterans. Elevated summer temperatures below surface-release reservoirs may exceed the tolerance limits of cool-adapted mountain caddisflies, or place oligothermal species at a competitive disadvantage (Fraley 1979). Reduced thermal heterogeneity below the deep-release reservoirs may eliminate species that rely on temperature changes to cue life cycle events such as induction and termination of diapause, maturation and emergence (Ward and Stanford 1979).

Serial successions of closely-related species of caddisflies along river courses (e.g. Bournaud *et al.* 1982) have been explained as species specific differences in temperature-metabolic rate responses operating in concert with competition (Hildrew and Edington 1979). The result is a spatial partitioning of species along the downstream thermal gradient. Disruption of the natural temperature gradient by impoundment causes thermal discontinuities, resulting in longitudinal shifts of species distributions (Hauer and Stanford 1982, Ward and Stanford 1983).

Modification of the flow regime induced major changes in the trophic structure at regulated stations. Truncation of detrital transport by impoundment reduced downstream food availability for detritivorous shredders. However, note the virtual absence of shredders (amongst Trichoptera) at both reference and regulated stations (Fig. 4) at the highest elevation (Site 1). *Lepidostoma*, the only shredder to normally attain high densities in streams of this region, rarely extends above 3000 m and exhibits maximum densities between 1700 and 2400 m (Ward, 1986).

Moderation of flow extremes appeared to exert a major effect on caddisflies through alterations in epilithon composition and abundance. Mineral scrapers that graze on diatom films (e.g., glossosomatids) were reduced or eliminated, whereas macrophyte herbivores (e.g., hydroptilids) increased as stabilized flows enabled mosses and filamentous algae to colonize rock surfaces (Dufford *et al.* 1986). Samples from the upper surfaces of rubble at regulated stations contained from 2 to 500 times greater epilithon standing crops (ash-free dry weight) than samples from reference stations.

The plankton that develops in impoundments potentially increases the availability and quality of food for filter feeders (Parker and Voshell 1983, Richardson 1984). However, although filter feeders were relatively more abundant below the bottom-release dams of the present study, their absolute numbers (densities) were much lower than at reference stations. Müller (1962) also reported lower densities of filter feeders below deep-release reservoirs. He found that filterers were enhanced if plankton-rich surface water was added to the bottom water released from the dam.

Because surface-release reservoirs draw water from the plankton-rich stratum, it is not surprising that filter feeders dominated the fauna of the receiving streams and attained extremely high densities. Below the mid elevation surface-release reservoir, *Cheumatopsyche* comprised essentially 100% of the trichopteran fauna, with mean density exceeding 6000 larva $m^{-2}$. Filter feeders attained even higher densities at the regulated station below the high elevation surface-release reservoir, but the predominant species were simuliids ($\overline{X}$ density > 36,000 $m^{-2}$), not trichopterans. *Arctopsyche grandis* was abundant at the reference site, but was rare below the dam. The Hydropsychinae that attain such high densities in lake outflows and below surface-release dams cannot tolerate the low temperatures at such high elevations (Ward, 1986). Mountain streams of western North American generally lack the diversity of filter-feeding caddisflies that characterize the eastern deciduous forest (Wiggins and Mackay 1978).

The increase in the relative abundance of predators below bottom-release dams and below the high elevation surface-release reservoir is difficult to interpret. It may be that suitable prey species are more abundant at regulated stations. Chironomids, a group favored by stream regulation, comprised the majority of food items ingested by the predaceous species of *Rhyacophila* examined by Thut (1969). The increased predation pressure at regulated stations may in turn be responsible for reducing or eliminating certain vulnerable prey species.

ACKNOWLEDGMENTS

L.D. Cline, H.L. Copeland, R.G. Dufford, J.H. Harvey, and R.A. Short assisted with laboratory analyses. D.Ruiter identified adults. Mrs Nadine Kuehl typed the manuscript. Support provided in part by a research grant from the Colorado Experiment Station.

REFERENCES

Armitage, P.D. 1984. Environmental changes induced by stream regulation and their effect on lotic macroinvertebrate communities. Regulated Rivers, ed. A. Lillehammer and S.J. Saltveit, Oslo Univ. Press, Olso.

Bournaud, M., Tachet, H. & Perrin, J.F. 1982. Les Hydropsychidae (Trichoptera) du Haut-Rhône entre Genève et Lyon. **Annls. Limnol.**, 18 : 61-80.

Dufford, R.L., Zimmermann, H.J., Cline, L.E. & Ward, J.V. 1986. Responses of epilithic algae to regulation of Rocky Mountain streams. Advances in Regulated Streams Ecology, ed. J.F. Craig and J.B. Kemper, Plenum, N.Y.

Fraley, J.J. 1979. Effects of elevated stream temperatures below a shallow reservoir on a cold water macroinvertebrate fauna. Ecology of Regulated Streams, ed. J.V. Ward and J.A. Stanford, Plenum, N.Y.

Hauer, F.R. & Stanford, J.A. 1982. Ecological responses of hydropsychid caddisflies to stream regulation. **Can. J. Fish. Aquat. Sci.**, 39 : 1235-1242.

Hildrew, A.G. & Edington, J.M. 1979. Factors facilitating coexistence of hydropsychid caddis larvae (Trichoptera) in the same river system. **J. Anim. Ecol.**, 48 : 557-576.

Merritt, R.W. & Cummins, K.W. (ed). 1984. An Introduction to the Aquatic Insects of North America. 2nd Ed. Kendall/Hunt, Dubuque, Iowa.

Müller, K. 1962. Limnologisch-fischereibiologische Untersuchungfen in regulierten Gewässern Schwedisch-

Lapplands. **Oikos,** 13 : 125-154.

Parker, C.P. & Voshell, J.R. 1983. Production of filter-feeding Trichoptera in an impounded and afree-flowing river. **Can. J. Zool.,** 61 : 70-87.

Richardson, J.S. 1984. Effects of seston quality on the growth of a lake-outlet filter feeder. **Oikos,** 43 : 386-390.

Thut, R.N. 1969. Feeding habits of larvae of seven *Rhyacophila* (Trichoptera : Rhyacophilidae) species with notes on other life-history features. **Annls. Entomol. Soc. Amer.,** 62 : 894-898.

Vannote, R.L. & Sweeney, B.W. 1980. Geographical analysis of thermal equilibria : a conceptual model for evaluating the effect of natural and modified thermal regimes on aquatic insect communities. **Amer. Nat.,** 115 : 667-695.

Voshell, J.R. & Parker, C.R. 1985. Quantity and quality of seston in an impounded and free-flowing river in Virginia, U.S.A. **Hydrobiologia,** 122 : 271-280.

Ward, J.V. 1975. Downstream fate of zooplankton from a hypolimnial release mountain reservoir. **Verh. Internat. Verein. Limnol.,** 19 : 1798-1804.

Ward, J.V. 1976. Comparative limnology of differentially regulated sections of a Colorado mountain river. **Arch. Hydrobiol.** 78 : 319-342.

Ward, J.V. 1981. Altitudinal distribution and abundance of Trichoptera in a Rocky Mountain stream. **Proc. 3rd Int. Symp. Trichoptera,** ed. G.P. Morretti, Dr. W. Junk., The Hague.

Ward, J.V. 1982. Ecological aspects of stream regulation : Responses in downstream lotic reaches.**Wat. Poll. Mgmt. Rev.,** 2 : 1-26.

Ward, J.V. 1986. Altitudinal zonation in a Rocky Mountain stream. **Arch. Hydrobiol.** Suppl. 74 : 133-199

Ward, J.V. & Short, R.A. 1978. Macroinvertebrate community structure of four special lotic habitats in Colorado, U.S.A. **Verh. Internatn. Verein. Limnol.,** 20 : 1382-1387.

Ward, J.V. & Stanford, J.A., 1979. Ecological factors controlling stream zoobenthos with emphasis on thermal modification of regulated streams. Ecology of Regulated Streams, ed. J.V. Ward andJ.A. Stanford, Plenum, N.Y.

Ward, J.V. & Stanford, J.A. 1982. Thermal responses in the evolutionary ecology of aquatic insects. **Ann. Rev. Entomol.,** 27 : 97-117.

Ward, J.V. & Stanford, J.A. 1983. The serial discontinuity concept of lotic ecosystems. Dynamics of Lotic Ecosystems, ed. T.D. Fontaine and S.M. Bartell, Ann Arbor Sci. Pub., Ann Arbor, Michigan.

Ward, J.V. & Stanford, J.A. 1984. The regulated stream as a testing ground for ecological theory. Regulated Rivers, ed. A. Lillehammer and S.J. Salveit, Olso Univ. Press, Oslo.

Ward, J.V., Zimmermann, H.J. & Cline L.D. 1986. Lotic zoobenthos of the Colorado system. The Ecology of River Systems, ed. B.R. Davies and K.F. Walker, Dr. W. Junk Pub., Dordrecht.

Webster, J.R., Benfield, E.F. & Cairns, J. Jr. 1979. Model predictions on effects of impoundment on particulate organic matter transport in a river system. Ecology of Regulated Streams, ed. J.V. Ward and J.A. Stanford, Plenum, N.Y.

Wiggins, G.B. & Mackay, R.J. 1978. Some relationships between systematics and trophic ecology in Nearctic aquatic insects, with special reference to Trichoptera. **Ecology,** 59 : 1211-1220.

Zimmermann, H.J. & Ward J.V. 1984. A survey of regulated streams in the Rocky Mountains of Colorado, U.S.A. Regulated Rivers, ed. A. Lillehammer and S.J. Saltveit, Olso Univ. Press, Oslo.

AUTHOR INDEX

This index covers only contributors to this volume. Authors cited are not indexed

# GEOGRAPHICAL INDEX

The names of lotic or lentic waters are in italic.
Abbreviations : L.: lake ; Mt.: mountain ; P.: pond ; R.:river ; S.: stream.
Bold numbers indicate a contribution especially devoted to a given region or to a given habitat.

# SUBJECT INDEX

Numbers correspond to the first page of each contribution

# TAXONOMIC INDEX

Bold numbers indicate a contribution especially devoted to a given species or to a given taxa.

www.ingramcontent.com/pod-product-compliance
Ingram Content Group UK Ltd.
Pitfield, Milton Keynes, MK11 3LW, UK
UKHW051117190926
475688UK00005B/206
* 9 7 8 9 0 6 1 9 3 6 2 0 6 *